W9-ADT-917

TRANSLATIONS OF MATHEMATICAL MONOGRAPHS

VOLUME 63

Selfadjoint Operators in Spaces of Functions of Infinitely Many Variables

by YU. M. BEREZANSKIĬ

American Mathematical Society · Providence · Rhode Island

САМОСОПРЯЖЕННЫЕ ОПЕРАТОРЫ В ПРОСТРАНСТВАХ ФУНКЦИЙ БЕСКОНЕЧНОГО ЧИСЛА ПЕРЕМЕННЫХ

Ю. М. БЕРЕЗАНСКИЙ

КИЕВ
«НАУКОВА ДУМКА»

Translated from the Russian by H. H. McFaden
Translation edited by Ben Silver

1980 *Mathematics Subject Classification* (1985 *Revision*). Primary 47-02, 47A70, 47B15, 47B25; Secondary 26E10, 28C20, 28C15, 35D99, 35J99, 35P05, 35R15, 42N10, 44A35, 45H05, 46A12, 36E35, 46F05, 46F25, 46G10, 46G12, 46M05, 47B10, 47B40, 47C15, 47D05, 47E05, 47F05, 60G99, 81C10, 81C12, 81E99.

ABSTRACT. Questions in the spectral theory of selfadjoint and normal operators acting in spaces of functions of infinitely many variables are studied, in particular, the theory of expansions in generalized eigenfunctions of such operators. Both individual operators and arbitrary commuting families of them are considered. A theory of generalized functions of infinitely many variables is constructed. The circle of questions presented has evolved in recent years, especially in connection with problems in quantum field theory. This book should be useful to mathematicians and physicists interested in the indicated questions, as well as to graduate students and students in advanced university courses.

Bibliography: 295 titles.

Library of Congress Cataloging-in-Publication Data

Berezanskiĭ, I͡U. M. (I͡Uriĭ Makarovich)

Selfadjoint operators in spaces of functions of infinitely many variables.

(Translations of mathematical monographs; v. 63)

Translation of: Samosopri͡azhennye operatory v. prostranstvakh funkt͡siĭ beskonechnogo chisla peremennykh.

Bibliography: p.

Includes index.

1. Distributions, Theory of (Functional analysis) 2. Spectral theory (Mathematics) I. Title. II. Series.

QA324.B4413 1986 515.7′82 85-30841

ISBN 0-8218-4515-2 (alk. paper)

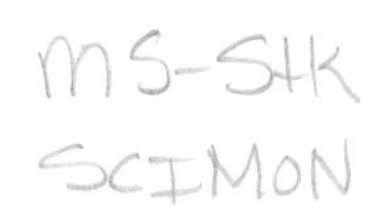

Contents

In memory of my mother
Lidiya Fominchna Berezanskaya
(1895–1984)

Foreword to the English Translation

I am, of course, glad for the translation of this book into English. However, several years have passed since its publication in Russian, and in the course of that time some of the results in the book have been further developed, some constructions and proofs have been transformed and their presentation has become more transparent, and so on. All this made it reasonable to revise the original text for the English translation. I carried out such a revision (though it turned out not to be very easy), with the following sections affected.

§§1, 2, and 4 of Chapter 2 and §3 of Chapter 3 were written anew; §§5 and 6 of Chapter 2 were essentially supplemented, and a number of minor additions and corrections were made in other parts of the book. The Comments on the Literature and the Bibliography were supplemented somewhat. (Additions to the latter are sufficiently local and do not encompass most of the papers completed after 1978; after all, this is a translation of a 1978 book and not a newly written monograph.)

In 1986 I hope to submit to press a book written jointly with Yu. G. Kondrat′ev and entitled *Spectral methods in infinite-dimensional analysis*, where applications to problems in harmonic analysis and quantum field theory will be given for the techniques developed below. (A similar book was planned as long ago as 1977, but was not realized for a number of reasons.) This second book will naturally be a continuation of the present monograph.

I want to express my sincere appreciation to the American Mathematical Society both for the translation itself and for the possibility of making the indicated changes. I am grateful to the young Kiev mathematicians T. V. Tsikalenko, A. V. Knyazyuk, N. V. Belichkova, and V. L. Ostrovskiĭ for help with the manuscript.

September 1984 *Yu. M. Berezanskiĭ*

Foreword

Interest in the analysis of functions of infinitely many variables has increased considerably in recent years in connection with the problems and the advances (conditional, perhaps, from the point of view of physicists) in quantum field theory on the one hand, and with the purely mathematical desire to comprehend the situation in the analysis of point functions on infinite-dimensional spaces on the other hand. There is another area of mathematics—probability theory (more precisely, the theory of random processes)—where questions connected, to a varying extent, with similar problems have been studied for a long time. This abundance of directions influencing the formulation of concepts and results, the ever greater number of diverse problems arising here, and also the differences in the languages have led to a lack of coordination in the understanding of the main concepts, even of what a function of infinitely many variables is. At the same time, this lack of coordination is in the essence of the matter: for a whole realm of problems in which infinitely many variables are "entangled" it is not possible to pick a single language that is adequate to the requirements.

In this book we consider a certain mathematical apparatus which is useful in studying selfadjoint operators acting in spaces of functions of infinitely many variables. More precisely, the operators considered act in infinite tensor products of spaces, isomorphic images of such spaces, or completions of them in certain quasi-inner (i.e., degenerate inner) products. Examples of such operators are given in the last chapter. Another book (which the author hopes to complete shortly) deals with other classes of such operators: operators associated with positive-definite functions of infinitely many variables, with the infinite-dimensional moment problem, or with Hamiltonians and the functions of Wightman and Schwinger in quantum field theory.

Accordingly, a function of infinitely many variables is understood here to be a vector in an infinite tensor product of one-dimensional spaces. Such an understanding is convenient in considering many questions in the spectral theory of operators and in field theory. The exposition is constructed from this conception.

The first chapter is devoted to a presentation of the theory of spaces of functions (ordinary and generalized) of infinitely many variables. It is convenient and useful to construct this theory abstractly, on the level of tensor products of Hilbert spaces. An abstract theory of generalized functions of finite and infinite order is constructed in §1. More precisely, there we study the theory of spaces with positive norm, their projective limits, and corresponding riggings of the original spaces. The next section deals with the theory of finite and infinite tensor products and riggings of them, with the main attention centered on separable subspaces of a complete von Neumann infinite product.

In §3 we consider specific spaces of functions of finitely many variables. The concept of a Sobolev space is basic for these constructions. Finally, in §4 we present examples of spaces of functions of infinitely many variables obtained by "multiplying" Sobolev spaces with the help of the procedures in §2 based on infinite tensor products. Here it should be emphasized that the theory of generalized functions of infinitely many variables is constructed by means of suitable riggings of such products. The classical point of view in constructing generalized functions now causes difficulties, since it is impossible to introduce a standard Lebesgue-type measure dx on the infinite direct product $\mathbb{R}^\infty = \mathbb{R}^1 \times \mathbb{R}^1 \times \cdots$ of real lines $\mathbb{R}^1$. Therefore, it is not possible to understand an ordinary function $f(x)$ $(x \in \mathbb{R}^\infty)$ as a generalized function equal to the functional $l(\varphi) = \int_{\mathbb{R}^\infty} f(x)\varphi(x)\,dx$ on the test functions $\varphi(x)$.

The second chapter deals with a study of arbitrary families of commuting selfadjoint (or normal) operators. Ordinarily, the case of infinitely many variables arises in field theory problems as a secondary phenomenon, the primary phenomenon (at least on an intuitive level) being the case of infinitely many operators. Therefore, this chapter is in a certain sense the main one in the book. We consider families of commuting operators from the point of view of the possibility of expansion in their joint generalized eigenfunctions, i.e., the possibility of simultaneously diagonalizing them. Of course, (in any case if the operators are bounded) it is possible to span a commutative normed algebra by the family and to diagonalize it, i.e., to pass to the Gel'fand isomorphism. However, in computations it is often convenient not to span such an algebra but to proceed as for expansions in generalized eigenvectors of finitely many commuting operators A_x $(x = 1, \ldots, p)$. What is more, in some sense this treatment of families of unbounded operators replaces the rather complex theory of commutative topological algebras in an important situation.

Difficulties arise in such a treatment of arbitrary families $(A_x)_{x \in X}$ of commuting operators when X is an index set of arbitrary cardinality replacing $\{1, \ldots, p\}$, and they are especially perceptible when X is more than countable. At the same time, it is frequently essential to consider an X that is more than countable; for example, in the same field theory the field operators $A(\varphi)$ are indexed by the functions φ $(A_x = A(\varphi), x = \varphi)$. The second chapter is concerned with just such a general family $(A_x)_{x\in X}$ of commuting selfadjoint or normal operators A_x.

The joint resolution E of the identity for this family $(A_x)_{x\in X}$ is constructed in §1 of Chapter 2. Its construction reduces to a generalization of Kolmogorov's theorem on the existence of an infinite product of probability measures. The resolution is a resolution of the identity on the Tychonoff product $\mathbb{R}^X$ of the lines $\mathbb{R}^1$ (or the product $\mathbb{C}^X$ of the planes $\mathbb{C}^1$), i.e., on the set of all real- or complex-valued functions $X \ni x \mapsto \lambda(x)$ on the index set X, topologized in a certain way. The concept and properties of the support of an operator-valued measure E, i.e., the minimal closed (in the Tychonoff topology) set of full outer measure, is essential for what follows.

In §2 we present the theory of expansions in joint generalized eigenvectors φ of the family $(A_x)_{x\in X}$. This is one of the central sections in the book. Here the proof of an equality of the type $A_x\varphi = \lambda(x)\varphi$ (where x runs through the whole of X and $\lambda(\cdot)$ varies over a set of full outer E-measure) involves difficulties when X has arbitrary cardinality. The joint generalized vectors φ in this equality must, of course, form a complete system.

In §3 the Gel′fand theory of commutative normed algebras is used to present a connection between the families $(A_x)_{x\in X}$ under consideration and the theory of random processes indexed by points $x \in X$. In one direction it is trivial: every random variable a_x, i.e., every function $M \ni \omega \mapsto a_x(\omega) \in \mathbb{C}^1$ on a set M of elementary events ω can be interpreted as the operator of multiplication by a_x in L_2 with respect to a probability measure. In the other direction it is sufficiently clear.

Representation of algebraic structures by commuting operators is considered in §4. The problem arising here can be explained as follows. In the case of arbitrary commuting operators A_x the joint resolution E of the identity can be concentrated in essence on all the functions $\lambda(\cdot)$ in $\mathbb{R}^X$ or $\mathbb{C}^X$. However, if there are algebraic relations between the operators A_x for the different x, then the set of possible $\lambda(\cdot)$ becomes smaller due to the equality $A_x\varphi = \lambda(x)\varphi$ $(x \in X)$. For example, if X is a linear space and $A_{\alpha x+\beta y} = \alpha A_x + \beta A_y$ $(x, y \in X;\ \alpha, \beta \in \mathbb{C}^1)$, then also $\lambda(\alpha x + \beta y) = \alpha\lambda(x) + \beta\lambda(y)$, i.e., $\lambda(\cdot)$ is a linear functional. In this way it is thus possible to establish a Stone-type theorem on spectral representation of a group U_x of unitary operators, where x runs through an abelian group X which is not locally compact in general.

Some additional facts on the theory of expansions are given in §5: for example, the theory of the often-encountered Carleman operators, and so on. The last section, §6, stands somewhat by itself. There we present a method (which often turns out to be convenient) for proving that operators are selfadjoint, and also that they commute, on the basis of an investigation of uniqueness for solutions of the Cauchy problem for evolution equations.

The third chapter is devoted to some applications of the apparatus developed. It can be regarded as illustrative, though all the situations described in it are important. (The book mentioned above, now in preparation, is essentially a continuation of this chapter.)

In §1 we present a theory of commuting selfadjoint operators acting with respect to different variables, and functions of such operators. The scheme of this construction is analogous to that of the construction of a differential expression $L(-i(\partial/\partial x_1), \ldots, -i(\partial/\partial x_p))$ with constant coefficients from the elementary differentiations $-i(\partial/\partial x_1), \ldots, -i(\partial/\partial x_p)$ and a polynomial $L(\lambda_1, \ldots, \lambda_p)$. The presentation is abstract on the level of infinite tensor products (p is, of course, equal to ∞), and $-i(\partial/\partial x_n)$ is replaced by a general selfadjoint operator A_n acting in a space of functions of a single variable. In particular, the infinite-dimensional Fourier-Wiener transformation is considered from this point of view. In §2 we study the special case of the functions in §1 when the role of $L(\lambda_1, \lambda_2, \ldots)$ is played by the function $\lambda_1 + \lambda_2 + \cdots$. This leads to the so-called operators admitting separation of infinitely many variables. Such operators can occur as free Hamiltonians. (In this case the A_n are second-order differential operators.)

In §3 evolution equation techniques are used to investigate selfadjointness of perturbations of a differential operator with infinitely many separating variables by a potential depending on infinitely many of these variables. Unlike for ordinary differential operators in an unbounded domain, when a "singularity" arises in connection with the behavior of the coefficients at infinity, another "singularity" appears here in connection with the character of growth of the potential "in the direction of increase of the number of variables."

We note that the space of L_2-functions of infinitely many variables with respect to a Gaussian measure is isomorphic to the boson Fock space important in quantum field theory (the Segal isomorphism). This isomorphism allows us to study corresponding operators in such an L_2-space rather than operators in Fock space. And the former space is an instance of the spaces to which the theory developed here is applicable. With its help a number of our constructions (for example, the theory of generalized functions) can be carried over to Fock space. However, these results will not be presented here.

The exposition is preceded by an introduction in which the notation is given and some concepts to be used are made precise. To read the book it suffices to be familiar with a standard university course in functional analysis, including the theory of unbounded operators and the theory of generalized functions.

For the most part, references to the literature are located in "Comments on the Literature". The enumeration of formulas, theorems, and lemmas is independent in each chapter. The end of a proof is indicated by the symbol ■.

Individual parts of this book have appeared in special courses which the author has given in the Mechanics-Mathematics Department of Kiev State

University. The impetus to write the book stems from thinking through lectures for the School on Spectral Theory organized by the C.I.M.E. in Varenna (August–September 1973).

In conclusion I want to express my thanks to those participants in seminars at the Institute of Mathematics of the Academy of Sciences of the Ukrainian SSR who took part in discussions of many results in the book, and especially to my students Yu. G. Kondrat′ev, Yu. S. Samoĭlenko, and G. F. Us for a number of essential remarks.

March 1977 *Yu. M. Berezanskiĭ*

Introduction

We dwell on certain concepts in whose definitions there has been a lack of coordination, and we introduce the notation needed.

1. The usual set theory notation will be used: $\in, \notin, \subseteq, \subset, \not\subseteq, =, \neq, \cup, \cap, \setminus$, $\varnothing, \times$. We write $A \subseteq B$ if every $x \in A$ is in B, and $A \subset B$ if $A \subseteq B$ and it is known that $A \neq B$. The set of all points $x \in R$ having a definite property $P(x)$ will be denoted by $\{x \in R | P(x)\}$. Indices are sometimes omitted in unions, summations, etc.; for example, $\sum_{i=1}^{N} 1 = N$.

If $R \ni x \mapsto f(x) \in Q$ is a mapping $f = f(\cdot)$ of an abstract space (i.e., set) R into an abstract space Q, then $f(A)$ denotes the image of a subset $A \subseteq R$, $f^{-1}(B)$ denotes the full preimage of a subset $B \subseteq Q$, i.e., $f^{-1}(B) = \{x \in R | f(x) \in B\} = f^{-1}(B \cap f(B))$, and $f \restriction A$ denotes the restriction of f to a subset $A \subseteq R$.

2. A topological space R is always understood to be a Hausdorff topological space. Thus, R has a family Σ of subsets $U, V, \ldots$—basic neighborhoods whose union covers the whole of R; U is called a basic neighborhood of any point $x \in U$. The family Σ satisfies the following two axioms:

1) for any $x, y \in R$ with $x \neq y$ there exist disjoint neighborhoods U and V of x and y;

2) for any $x \in R$ and any two neighborhoods U and V of it there exists a third neighborhood W of this point such that $W \subseteq U \cap V$.

The rest of the topological concepts are the usual ones. The closure of a set A is denoted by $\tilde{A}$. Any open set O is called a neighborhood of each point $x \in O$; a neighborhood of a point x will also be denoted by $U(x), O(x), \ldots$. If $A \subseteq R$, then A can be endowed with the relative topology, which is given by basic neighborhoods of the form $U \cap A$, with $U \in \Sigma$. This topology is also said to be induced from the topology of the space R, and A itself, as a topologized space, is called a subspace of R. The concept of compactness is equivalent to that of bicompactness. If a locally compact space can be expressed in the form $R = \bigcup_1^\infty R_n$, where the R_n are compact and $R_1 \subseteq R_2 \subseteq \cdots$, then R can be extended in the standard way to a compact space $\mathbf{R}$—a compactification of R: let $\mathbf{R} = R \cup \{\infty\}$; here ∞ is an adjoined point whose neighborhoods are understood to be the sets of the form $R\backslash Q$, where $Q \subseteq R$ is compact. A metric space is understood to be a complete metric space, and a ball in such a space is understood to be an open ball unless stated otherwise.

3. The real (complex) N-dimensional space is denoted by $\mathbb{R}^N$ (by $\mathbb{C}^N$), $N = 1, 2, \ldots$. In this space a basic neighborhood is understood to be any open set, and $\langle x, y\rangle$ denotes the inner product of vectors x and y in these spaces. A domain $G \subseteq \mathbb{R}^N$ is understood to be an arbitrary open set in $\mathbb{R}^N$. Its boundary $\partial G = \tilde{G}\backslash G$ is said to be l times piecewise continuously differentiable (or of class C^l) if it can be broken up into at most countably many disjoint sets $D_1, D_2, \ldots$ with each D_n given by an equation of the form $x_j = f(x_1, \ldots, x_{j-1}, x_{j+1}, \ldots, x_N)$, where j takes one of the values $1, \ldots, N$ depending on n, and f is an l times continuously differentiable function in some ball in $\mathbb{R}^{N-1}$ ($l = 0, 1, \ldots$). There can be no more than countably many of the sets D_n in each ball in $\mathbb{R}^N$.

4. If R is a topological space, then $C(R)$ denotes the collection of all continuous complex-valued functions $R \ni x \mapsto f(x) \in \mathbb{C}^1$. Let R be a locally compact space. A function $R \ni x \mapsto f(x) \in \mathbb{C}^1$ is said to be compactly supported if it is zero outside some compact set (depending on f), and locally bounded if it is bounded on each compact set by a constant (depending on the set). The collection of all complex-valued continuous compactly supported functions on a locally compact space R is denoted by $C_0(R)$.

The role of R can certainly be played by $\mathbb{R}^N$. Or R can be taken to be a domain $G \subseteq \mathbb{R}^N$, topologized by the relative topology induced by $\mathbb{R}^N$. The class $C_0(G)$ now consists of all the continuous functions on G which vanish "on strips near ∂G" and for $|x| \geq r$, where $r > 0$ is sufficiently large (if G is unbounded), i.e., $f \in C_0(G)$ is compactly supported near ∂G and at ∞.

Let $G \subseteq \mathbb{R}^N$ be a domain, and $\tilde{G}$ its closure. Then

$$C^l(\tilde{G}) = C^l(G \cup \partial G) \qquad (l = 0, 1, \ldots, \infty;\ C^0 \equiv C)$$

denotes the collection of all functions $\tilde{G} \ni x \mapsto f(x) \in \mathbb{C}^1$, which are restrictions to $\tilde{G}$ of l times continuously differentiable functions on $\mathbb{R}^N$, and $C_0^l(G)$ denotes the class of all compactly supported functions in $C^l(\tilde{G})$, i.e., $C_0^l(G) = C^l(\tilde{G}) \cap C_0(G)$. The collections of all real-valued functions in these and other classes encountered will be indicated by the subscript Re. The notation usually used for derivatives and powers is

$$D^\mu = D_1^{\mu_1} \cdots D_N^{\mu_N}, \quad D_n = \partial/\partial x_n \qquad (n = 1, \ldots, N),\ \mu = (\mu_1, \ldots, \mu_N),$$
$$\mu_1, \ldots, \mu_N = 0, 1, \ldots, \qquad |\mu| = \mu_1 + \cdots + \mu_N,$$
$$s^\mu = s_1^{\mu_1} \cdots s_N^{\mu_N} \qquad (s = (s_1, \ldots, s_N)).$$

5. For a family $(R_\alpha)_{\alpha \in \mathrm{A}}$ of abstract spaces R_α the product $\times_{\alpha \in \mathrm{A}} R_\alpha$ is understood to be the collection of all functions x of the form $A \ni \alpha \mapsto x(\alpha) = x_\alpha \in R_\alpha$ on the index set A. If the R_α are topological spaces, then the Tychonoff topology is usually introduced in $\times_{\alpha\in A} R_\alpha$ (see §1 in Chapter 2). A complex-valued function $\times_{\alpha\in A} R_\alpha \ni x \mapsto f(x) \in \mathbb{C}^1$ is said to be cylindrical if it depends on finitely many variables. This means that there exist a finite set $\{\alpha_1, \ldots, \alpha_p\} \subseteq A$ (depending on f) and a function f_c, $\times_{n=1}^{p} R_{\alpha_n} \ni (x_{\alpha_1}, \ldots, x_{\alpha_p}) \mapsto f_c(x_{\alpha_1}, \ldots, x_{\alpha_p}) \in \mathbb{C}^1$, such that $f(x) = f_c(x_{\alpha_1}, \ldots, x_{\alpha_p})$ $(x \in \times_{\alpha\in A} R_\alpha)$. One of the simplest situations is that in which $A = \{1, 2, \ldots\}$ and $R_\alpha = \mathbb{R}^1$ $(\alpha = 1, 2, \ldots)$, so that $\times_{\alpha\in A} R_\alpha = \mathbb{R}^1 \times \mathbb{R}^1 \times \cdots$; the last product is denoted by $\mathbb{R}^\infty$. A cylindrical function on $\mathbb{R}^\infty$ has the form $\mathbb{R}^\infty \ni x = (x_1, x_2, \ldots) \mapsto f(x) = f_c(x_1, \ldots, x_p)$ with some p and f_c. Analogous notation can be introduced with $\mathbb{R}^1$ replaced by $\mathbb{C}^1$.

6. Let R be an abstract space of points $x, \mathfrak{R}$ a σ-algebra of subsets of it, and μ a (nonnegative) finite measure on $\mathfrak{R}$. The corresponding space of p-integrable $(p \geq 1)$ complex-valued functions on R is denoted by $L_p(R, d\mu(x))$. It is sometimes convenient to write $R \ni x \mapsto f(x) \in \mathbb{C}^1$ for functions f in such spaces and, more generally, for functions defined almost everywhere on R and almost everywhere finite, though this expression is of course conditional. Let $L_{p,\mathrm{loc}}(R, d\mu(x))$ denote the collection of all locally p-integrable functions on R, i.e., almost-everywhere defined and almost-everywhere finite functions that are $\mathfrak{R}$-measurable and such that $\int_B |f(x)|^p \, d\mu(x) < \infty$ for every $B \in \mathfrak{R}$ of finite measure.

Consider a sequence (finite or infinite) of spaces $R_1, R_2, \ldots$, corresponding σ-algebras $\mathfrak{R}_1, \mathfrak{R}_2, \ldots$, and measures $\mu_1, \mu_2, \ldots$. A measure μ on $\times_1^q R_n$ $(q < \infty)$ called the direct or tensor product of the original measures μ_n can be constructed from these objects by a well-known procedure. (The measures must be probability measures in the case $q = \infty$, i.e., $\mu_n(R_n) = 1$, $n = 1, 2, \ldots$.) This measure is denoted by $\mu = \times_1^n \mu_n$; it is also called a product measure.

If R is a topological space, then $\mathcal{B}(R)$ denotes the σ-algebra of Borel subsets of R, i.e., $\mathcal{B}(R)$ is the smallest σ-algebra containing all open (or closed) subsets of R.

Lebesgue measure on $\mathbb{R}^N$ is denoted by m, and integration with respect to it by $dm(x)$ or dx. If $G \subseteq \mathbb{R}^N$ is a domain, then the space $L_p(G, dm(x))$ is denoted by $L_p(G)$. In the case $R = G$ the notation $L_{p,\mathrm{loc}}(G)$ has a sense somewhat different from the previous sense: it stands for the collection of all complex-valued functions on G such that $\int_B |f(x)|^p \, dx < \infty$ for any m-measurable bounded set B at a positive distance from ∂G. (This is of course the previous class when $G = \mathbb{R}^N$.)

One more term involving measures: if R is a space, $\mathfrak{R}$ is a σ-algebra on it, and $\mathfrak{R} \ni \alpha \mapsto \mu(\alpha) \geq 0$ is a σ-finite measure, then a set $\alpha \in \mathfrak{R}$ is called a set of full measure if $\mu(R \backslash \alpha) = 0$; it is also said that μ is concentrated on α. If only R and $\mathfrak{R}$ are given, then a measurable function on R is understood to be such a function taking only finite values; if also a measure is given on $\mathfrak{R}$, then a measurable function can also take infinite values on a set of measure zero. The characteristic function of a set $A \subseteq R$ is denoted by κ_A.

Suppose that $R, \mathfrak{R}$, and μ are given and, moreover, let $p \in L_{1,\mathrm{loc}}(R, d\mu(x))$ be a fixed almost-everywhere nonnegative function. Such a function is called a weight. Integration with respect to the measure $\mathfrak{R} \ni B \mapsto \int_B p(x)\,d\mu(x)$ is denoted by $p(x)\,d\mu(x)$; $L_p(R, p(x)\,d\mu(x))$ and $L_{p,\mathrm{loc}}(R, p(x)\,d\mu(x))$ are the corresponding classes.

7. We dwell on mappings on measures. Suppose that a σ-finite measure $\mathfrak{R} \ni \alpha \mapsto \mu(\alpha) \geq 0$ is defined on a σ-algebra $\mathfrak{R}$ in a space R. Assume that R' is another abstract space and $R \ni x \mapsto \varphi(x) \in R'$ is a fixed mapping of R onto the whole of R'. Consider the collection $\mathfrak{R}_\varphi = \{\alpha' \subseteq R' | \varphi^{-1}(\alpha') \in \mathfrak{R}\}$ of sets; it is easy to see that $\mathfrak{R}_\varphi$ is a σ-algebra. Define a measure μ_φ on $\mathfrak{R}_\varphi$ by setting $\mu_\varphi(\alpha') = \mu(\varphi^{-1}(\alpha'))$, $\alpha' \in \mathfrak{R}_\varphi$. It is obvious that all the properties of a measure are satisfied; the measure μ_φ is called the φ-image of the measure μ, or it is said that we transport μ to μ_φ by means of the mapping φ. If a function $R' \ni x' \mapsto F(x') \in \mathbb{C}^1$ is measurable with respect to the σ-algebra $\mathfrak{R}_\varphi$, then the function $R \ni x \mapsto F(\varphi(x)) \in \mathbb{C}^1$ is measurable with respect to the σ-algebra $\mathfrak{R}$, and if one of these functions is integrable with respect to the corresponding measure, then so is the other, and the integrals are equal:

$$\int_R F(\varphi(x))\,d\mu(x) = \int_{R'} F(x')\,d\mu_\varphi(x').$$

We make some additional remarks. It can be assumed that φ is a mapping into R'. Of course, everything is preserved if $\varphi(R)$ is regarded as the new space R'. But it is possible, aside from this, to introduce a σ-algebra of subsets of R' and a corresponding measure μ_φ by setting

$$\mathfrak{R}_\varphi = \{\alpha' \subseteq R' | \varphi^{-1}(\alpha') = \varphi^{-1}(\alpha' \cap \varphi(R)) \in \mathfrak{R}\}$$

and

$$\mu_\varphi(\alpha') = \mu(\varphi^{-1}(\alpha' \cap \varphi(R))) \qquad (\alpha' \in \mathfrak{R}_\varphi).$$

In other words, we adjoin to the σ-algebra constructed by the former procedure on $\varphi(R)$ all the subsets of $R' \setminus \varphi(R)$ and consider that they have μ_φ-measure zero.

Further, we can assume that $\mathfrak{R}'$ is a previously given σ-algebra of subsets of R' with the property that $\varphi^{-1}(\alpha') \in \mathfrak{R}$ for any $\alpha' \in \mathfrak{R}'$, and let $\mu_\varphi(\alpha') = \mu(\varphi^{-1}(\alpha'))$, $\alpha' \in \mathfrak{R}'$. (For example, if $R \ni x \mapsto \varphi(x) \in R' = \mathbb{C}^1$ is a measurable function, then it is possible to set $\mathfrak{R}' = \mathcal{B}(R')$.)

For these modifications we preserve the previous name of the φ-image of the measure; it is also clear that what was claimed about integration of functions with respect to a measure and its φ-image remains valid.

The same definitions will be used also for operator-valued measures in what follows.

8. As a rule, complex linear topological spaces X are considered below; they will often be separable. Real spaces are distinguished by the subscript

Re if necessary. The linear and the closed linear span of a set $M \subseteq X$ are denoted by l.s.(M) and c.l.s.(M), respectively.

A set M is said to be total if c.l.s.$(M) = X$. A subspace of X is always understood to be a closed linear subset of X. If X_1 and X_2 are two topological spaces, then we say that a topological imbedding $X_1 \subseteq X_2$ is valid (or $X_1 \to X_2$) if the indicated set inclusion is valid, X_1 is dense in X_2, and the topology in X_1 is stronger than that in X_2 (roughly speaking, convergence in X_1 implies convergence in X_2). Suppose that the set-theoretic inclusion $X_1 \subseteq X_2$ holds. The operator $X_1 \ni x \mapsto Ox = x \in X_2$ is called the imbedding operator. Its continuity and the denseness of X_1 in X_2 are equivalent to the validity of the topological imbedding $X_1 \subseteq X_2$.

The collection of continuous antilinear functionals l on X is denoted by X', and the collection of continous linear functionals by X^*. The topologies in dual spaces will be specified each time (if necessary). The collection of all continuous linear operators acting from the whole of a linear topological space X_1 into an analogous space X_2 is denoted by $\mathcal{L}(X_1 \to X_2)$. The condition that A acts from X_1 into X_2 is written in the form $A\colon X_1 \to X_2$. Here A need not be defined on the whole space X_1.

A function of p variables in X is called a multilinear (p-linear) form if it is linear in each variable (in particular, a 2-linear form if $p = 2$). Functions of two variables that are linear in the first and antilinear in the second, i.e., sesquilinear forms, are called bilinear forms.

9. A normed linear (Banach) space E is always understood to be a complete space. (The same applies to a Hilbert space H.) Everything said about linear topological spaces, of course, applies to normed linear spaces and Hilbert spaces. In the notation for the norm of a vector and for an inner product, the space is indicated by an index.

Let Φ be a linear space with two norms $\|\varphi\|_{E_1}$ and $\|\varphi\|_{E_2}$ such that $\|\varphi\|_{E_1} \leq c\|\varphi\|_{E_2}$ ($\varphi \in \Phi$) for some $c > 0$. Denote by E_1 (E_2) the completion of Φ in the norm $\|\cdot\|_{E_1}$ ($\|\cdot\|_{E_2}$). In general it is impossible to regard E_2 as a part of E_1. The inclusion $E_2 \subseteq E_1$ is valid if and only if the following consistency condition holds for the norms: if the sequence $(\varphi_n)_1^\infty$ ($\varphi_n \in \Phi$) is Cauchy in the norm $\|\cdot\|_{E_2}$ and converges to zero in the norm $\|\cdot\|_{E_1}$, then it must also converge to zero in the norm $\|\cdot\|_{E_2}$. In some cases in what follows we write $E_2 \subseteq E_1$ without verifying the consistency condition. (The reader can easily verify it: the norms under discussion will be Sobolev norms.)

In the notation for the norm of an operator $A \in \mathcal{L}(E_1 \to E_2)$ we do not ordinarily write an index indicating the spaces, but sometimes we use the subscript $E_1 \to E_2$. We shall usually consider operators $A\colon E_1 \to E_2$ with domains dense in E_1. (The domain and range of an operator A are denoted by $\mathfrak{D}(A)$ and $\mathfrak{R}(A)$, respectively.) The restriction of an operator A to $F \subseteq \mathfrak{D}(A)$ is denoted by $A \upharpoonright F$. The kernel of such an operator A is defined to be the linear set $\operatorname{Ker} A = \{x \in E_1 | Ax = 0\}$. The algebraic inverse operator exists on

$\mathfrak{R}(A)$ if and only if $\operatorname{Ker} A = 0$. Unless stated otherwise, an inverse operator A^{-1} is understood to be an operator defined on the whole of E_2 which is in $\mathcal{L}(E_2 \to E_1)$ and such that $A^{-1}Ax = x$ $(x \in \mathfrak{D}(A))$; here it is necessary that $\mathfrak{R}(A) = E_2$.

Let H_1 and H_2 be Hilbert spaces. If $A \in \mathcal{L}(H_1 \to H_2)$ is such that $(Ax, Ay)_{H_2} = (x, y)_{H_1}$ for $x, y \in H_1$, then it is called an isometric operator. An isometry is defined to be an isometric operator A such that $\mathfrak{R}(A) = H_2$. An isometry obviously establishes an isomorphism between H_1 and H_2.

An operator A densely defined in a Hilbert space H is said to be Hermitian if $(Ax, y)_H = (x, Ay)_H$ for $x, y \in \mathfrak{D}(A)$, selfadjoint if the adjoint operator A^* is equal to A, and essentially selfadjoint if the closure $\tilde{A}$ of A is selfadjoint. Let A be a closed operator acting in H. Every linear set $F \subseteq \mathfrak{D}(A)$ such that $(A \upharpoonright F)^\sim = A$ is called a core of A.

The sum of two operators $A_1, A_2\colon E_1 \to E_2$ is understood to be the operator $(A_1 + A_2)x = A_1x + A_2x$, where $x \in \mathfrak{D}(A_1 + A_2) = \mathfrak{D}(A_1) \cap \mathfrak{D}(A_2)$. If $A_1\colon E_1 \to E_2$ and $A_2\colon E_2 \to E_3$, then the product of these operators is defined to be $A_2A_1x = A_2(A_1x)$, where $(Ax, y)_H = (x, Ay)_H$ $(x, y \in \mathfrak{D}(A))$.

A densely defined closed operator A acting in a Hilbert space H is said to be normal if $A^*A = AA^*$.

We add some remarks concerning Hilbert spaces. Suppose that a quasi-inner product is introduced in a complex linear space K, i.e., a function $K \ni x, y \mapsto (x, y) \in \mathbb{C}^1$, satisfying all the requirements of an inner product except that it may be degenerate: $N = \{x \in K | (x, x) = 0\} \supseteq \{0\}$. A Hilbert space can be constructed by a standard procedure from a quasi-inner product: first take the quotient of K by N, and then take the completion.

An operator A acting in H is said to be nonnegative if $(Ax, x)_H \geq 0$ for $x \in \mathfrak{D}(A)$, and positive if $(Ax, x)_H > 0$ for $x \in \mathfrak{D}(A)$ with $x \neq 0$. We shall deal with multilinear forms, i.e., forms linear in each variable. A bilinear form is understood to be linear in the first variable and antilinear in the second—such a form is frequently called sesquilinear; linearity in each of two variables will be referred to as two-linearity.

10. The concepts of a Hilbert-Schmidt operator A, its Hilbert norm $|A|$, and the trace $\operatorname{Tr}(A)$ of an operator are used in what follows. Here the definitions are the usual ones: if H_1 and H_2 are two Hilbert spaces the first of which is separable, then an operator $A \in \mathcal{L}(H_1 \to H_2)$ is called a Hilbert-Schmidt operator or a quasinuclear operator if $\sum_1^\infty \|Ae_j\|_{H_2}^2 = |A|^2 < \infty$ for some orthonormal basis $(e_j)_1^\infty$ in H_1. A nonnegative operator $A \in \mathcal{L}(H_1 \to H_1)$ is called a finite trace operator or a nuclear operator if $\sum_1^\infty (Ae_j, e_j)_{H_1} = \operatorname{Tr}(A) < \infty$. These definitions are easily seen not to depend on the choice of the orthonormal basis $(e_j)_1^\infty$. We do not give the simple but important relations between the usual norm, the Hilbert norm, and the trace of an operator.

An imbedding $H_1 \subseteq H_2$ is said to be quasinuclear if the imbedding operator $O\colon H_1 \to H_2$ is quasinuclear.

11. Three more isolated remarks. An operator A acting in the space $L_2(R, d\mu(x))$ is called an integral operator if it has the form

$$(Af)(x) = \int_R K(x,y)f(y)\,d\mu(y),$$

where $K(x,y)$ is a $\mu \times \mu$-measurable almost-everywhere finite function of $(x,y) \in R \times R$, and the formula itself is valid on some linear set of $f \in \mathfrak{D}(A)$ dense in $L_2(R, d\mu(x))$. The function $K(x,y)$ is called the kernel of A.

Convergence of an integral of a vector-valued function with respect to an ordinary measure is understood to be with respect to the norm of the space where the values of the function lie. We shall also consider strong (i.e., with respect to the norm of the space) derivatives of a vector-valued function.

CHAPTER 1

Spaces of Generalized Functions

In this chapter we study the theory of generalized functions of finitely and infinitely many variables. The presentation is based on the theory of spaces with negative norm, which first appeared in special situations in papers of Leray and Lax and then later found a solid place in various areas of analysis. The transition from a single variable to infinitely many variables in this construction of the theory of generalized functions is based on the concept of tensor product: a chain $H_- \supseteq H_0 \supseteq H_+$ is taken, with H_0 consisting of ordinary functions of one variable, H_+ of test functions, and H_- of the corresponding generalized functions, and then the tensor product of it by itself is taken infinitely many times. This leads to the required chain $\bigotimes_{n=1}^{\infty} H_- \supseteq \bigotimes_{n=1}^{\infty} H_0 \supseteq \bigotimes_{n=1}^{\infty} H_+$ of functions of what is now an infinite number of variables. Of course, this is not the only possible way of constructing a theory of generalized functions. We chose this path because of its convenience for those spectral theory problems to which the essence of the book is devoted.

The general theory of spaces with positive and negative norms (i.e., the theory of rigging Hilbert spaces by Hilbert spaces) is constructed in §1. The theory of tensor products of Hilbert spaces and riggings of them is presented in §2. We note certain features of the constructions. The tensor product of finitely many spaces is presented in the standard way, while in passing to the product of infinitely many spaces we first construct what is, in essence, a separable subspace of the complete von Neumann tensor product. The complete von Neumann product itself is constructed in §2.10 as the orthogonal sum of the separable subspaces already introduced (incidently, the complete von Neumann product itself is not used subsequently in the book, and its construction is given only to make the picture complete).

The separable subspaces mentioned are constructed with the help of stabilizing sequences of unit vectors along which the spaces being multiplied are "strung". This turns out to be convenient from the point of view of the applications we need. The construction can actually be generalized somewhat to a construction of so-called weighted infinite tensor products, in which weight factors can be used to regulate the nature of decrease of a vector "along the direction of increase of the number of variables". Although the weighting procedure can be avoided by renorming the spaces, it is technically convenient. We mention also that §§2.8 and 2.9 contain a variant of the Schwartz kernel theorem that, in our view, is simplest and most appropriate to the essence of the matter. (A very simple variant of it for the case of the space L_2 is contained in §3.11.)

In §3 we present the theory of riggings of L_2 by Sobolev spaces on a finite or infinite domain in $\mathbb{R}^N$. Special attention is given to conditions ensuring that the imbedding of one space in another is quasinuclear, i.e., ensuring that the imbedding operator is a Hilbert-Schmidt operator. Quasinuclearity of the imbedding is an important requirement that appears in various areas of analysis (the theory of expansions in generalized eigenfunctions, the theory of Gaussian measure in infinite-dimensional spaces, the Schwartz kernel theorem, and so on). Here it is shown how the classical Sobolev-Schwartz theory of generalized functions enters into our scheme.

In §4 we construct a theory of generalized functions of infinitely many variables based on the idea of infinite tensor multiplication of chains of spaces of test functions and generalized functions of a single variable. Some of the results here are illustrative. However, some of the spaces of test functions and generalized functions constructed in this way play an essential role both in spectral theory and in applications to quantum field theory. This applies first and foremost to the space $\mathcal{A}(\mathbb{R}^\infty)$ presented in §3.5.

§1. The concept of a space with negative norm

The following chain of spaces plays a key role in the Sobolev-Schwartz theory of generalized functions:

$$\mathcal{D}'(\mathbb{R}^N) \supseteq L_2(\mathbb{R}^N) \supseteq \mathcal{D}(\mathbb{R}^N), \tag{1.1}$$

where $L_2(\mathbb{R}^N)$ is constructed from Lebesgue measure, $\mathcal{D}(\mathbb{R}^N)$ is the space of test functions and consists of the compactly supported infinitely differentiable functions, and $\mathcal{D}'(\mathbb{R}^N)$ is the space of generalized functions, namely, the continuous antilinear functionals on $\mathcal{D}(\mathbb{R}^N)$. The role of the space $L_2(\mathbb{R}^N)$ reduces to the fact that the inner product $(f,g)_{L_2(\mathbb{R}^N)}$ in it can be extended by continuity to a bilinear form giving the action of a generalized function $\alpha \in \mathcal{D}'(\mathbb{R}^N)$ on a test function $u \in \mathcal{D}(\mathbb{R}^N)$; this bilinear form can be denoted by $(\alpha, u)_{L_2(\mathbb{R}^N)}$. We remark that $\mathcal{D}'(\mathbb{R}^N)$ is the closure of $L_2(\mathbb{R}^N)$ in a certain topology.

Chains of the type (1.1) with general Hilbert spaces as the spaces will be considered below.

1.1. ***Definition of the negative space.*** Let H_0 be a complex Hilbert space with inner product $(\cdot,\cdot)_{H_0}$ and norm $\|\cdot\|_{H_0}$, and let f and g be elements of it. Assume that H_0 contains a dense linear subset H_+ which is itself a Hilbert space with respect to another inner product $(\cdot,\cdot)_{H_+}$ and that the norm $\|\cdot\|_{H_+}$ in H_+ is such that

$$\|u\|_{H_0} \leq \|u\|_{H_+} \qquad (u \in H_+). \tag{1.2}$$

(The more general case when $\|\cdot\|_{H_0} \leq c\|\cdot\|_{H_+}$ for some $c < \infty$ can be reduced to (1.2) by renorming H_+.) The elements of H_+, which play the role of the test functions, are denoted by $u, v, \ldots$.

Each element $f \in H_0$ generates a continuous antilinear functional l_f on H_+ according to the formula

$$l_f(u) = (f,u)_{H_0} \qquad (u \in H_+); \tag{1.3}$$

its continuity follows from the estimate

$$|l_f(u)| = |(f,u)_{H_0}| \leq \|f\|_{H_0}\|u\|_{H_0} \leq \|f\|_{H_0}\,\|u\|_{H_+}.$$

We introduce a new norm $\|\cdot\|_{H_-}$ in H_0 by taking the norm of f to be the norm of the functional l_f corresponding to it:

$$\|f\|_{H_-} = \|l_f\| = \sup_{u\in H_+} \frac{|(f,u)_{H_0}|}{\|u\|_{H_+}}. \tag{1.4}$$

Of the norm properties it is necessary to check only that $f = 0$ when $\|f\|_{H_-} = 0$. But $(f,u)_{H_0} = 0$ for all $u \in H_+$ if $\|f\|_{H_-} = 0$, and then $f = 0$, because H_+ is dense in H_0.

Completing H_0 in the norm (1.4), we get a normed linear space H_- called the space with negative norm. Accordingly, we have constructed a chain

$$H_- \supseteq H_0 \supseteq H_+ \tag{1.5}$$

of spaces with negative, zero, and positive norms. Their elements will be denoted by $\alpha, \beta, \ldots \in H_-$, $f, g, \ldots \in H_0$ and $u, v, \ldots \in H_+$, and will sometimes be called generalized functions, ordinary functions, and smooth functions, respectively. (The sense of this terminology becomes clear in what follows.) We also say that (1.5) is a rigging of the space H_0 by the spaces H_+ and H_-.

Since the mapping $H_0 \ni f \mapsto l_f \in (H_+)'$ is linear and one-to-one, it is easy to see that H_- can be regarded as situated in the dual space of antilinear functionals on H_+: $H_- \subseteq (H_+)'$. Therefore, it makes sense to denote the expression $\alpha(u)$, which is similar to the form $(\alpha, u)_{L_2(\mathbf{R}^N)}$ in the Sobolev-Schwartz theory of generalized functions, by $(\alpha,u)_{H_0} = (\overline{u,\alpha})_{H_0}$ ($\alpha \in H_-$, $u \in H_+$). The bilinear form $(\alpha,u)_{H_0}$ is the extension by continuity of the form $H_+ \times H_+ \ni (v,u) \mapsto (v,u)_{H_0} \in \mathbf{C}^1$ to $H_- \times H_+$. This form is also said to determine a pairing of the spaces H_+ and H_- (by means of H_0). The generalization

$$|(\alpha,u)_{H_0}| \leq \|\alpha\|_{H_-}\|u\|_{H_+} \qquad (\alpha \in H_-,\ u \in H_+) \tag{1.6}$$

of the Cauchy-Schwarz-Bunyakovskiĭ inequality is obviously valid. (For $\alpha = f$ inequality (1.6) means that $|l_f(u)| \leq \|l_f\|\,\|u\|_{H_+}$, and then it can be extended by continuity from H_0 to H_-.)

EXAMPLE. Let $H_0 = L_2(\mathbb{R}^N)$ and $H_+ = L_2(\mathbb{R}^N, p(x)\,dx)$, where $1 \leq p(x) \in C(\mathbb{R}^N)$ $(x \in \mathbb{R}^N)$. Then $H_- = L_2(\mathbb{R}^N, p^{-1}(x)\,dx)$.

1.2. *The Hilbert property of the negative space.*

THEOREM 1.1. *The negative space H_- is a Hilbert space.*

PROOF. Let us construct an inner product in H_-. (This construction will also be used in what follows.) Consider the bilinear form

$$H_0 \times H_+ \ni (f, u) \mapsto B(f, u) = (f, u)_{H_0} \in \mathbb{C}^1. \tag{1.7}$$

It is continuous, $|B(f,u)| \leq \|f\|_{H_0} \|u\|_{H_0} \leq \|f\|_{H_0} \|u\|_{H_+}$, and thus admits a representation

$$B(f, u) = (f, Au)_{H_0} = (A^* f, u)_{H_+},$$

where $A\colon H_+ \to H_0$ and $A^*\colon H_0 \to H_+$ are mutually adjoint continuous operators. According to (1.7), A is equal to the imbedding operator $O\colon H_+ \to H_0$. Let $I = O^*$. Then

$$(f, u)_{H_0} = (f, Ou)_{H_0} = (If, u)_{H_+} \qquad (f \in H_0,\ u \in H_+,\ I\colon H_0 \to H_+.) \tag{1.8}$$

We introduce in H_0 the quasi-inner product

$$(f, g)_{H_-} = (If, Ig)_{H_+} = (f, Ig)_{H_0} = (If, g)_{H_0} \qquad (f, g \in H_0). \tag{1.9}$$

According to (1.4), (1.8), and (1.9),

$$\begin{aligned}\|f\|_{H_-} &= \sup_{u \in H_+} \frac{|(f,u)_{H_0}|}{\|u\|_{H_+}} = \sup_{u \in H_+} \frac{|(If,u)_{H_+}|}{\|u\|_{H_+}} \\ &= \|If\|_{H_+} = \sqrt{(f,f)_{H_-}} \qquad (f \in H_0).\end{aligned}$$

Since $\|\cdot\|_{H_-}$ is a norm in H_0, (1.9) actually defines not just a quasi-inner but an inner product. This inner product is carried over from H_0 to H_- as a result of completion, and the latter space becomes a Hilbert space. ■

Accordingly, H_- has the inner product

$$(\alpha, \beta)_{H_-}, \|\alpha\|_{H_-} = \sqrt{(\alpha, \alpha)_{H_-}} \qquad (\alpha, \beta \in H_-).$$

Since $\|I\| = \|O\| \leq 1$, it follows that $\|f\|_{H_-} = \|If\|_{H_+} \leq \|f\|_{H_0}$ $(f \in H_0)$.

1.3. *Coincidence of the negative space with the dual of the positive space.* The first equality in (1.9) shows that I is a densely defined isometric operator from H_- to H_+. Closing it by continuity, we get an isometric operator $\mathbf{I}\colon H_- \to H_+$ acting from the whole of H_- to H_+, and $I = \mathbf{I} \restriction H_0$.

It is easy to see that $\mathbf{I}$ *is an isometry between the whole of H_- and the whole of H_+* (i.e., the range $\mathfrak{R}(\mathbf{I})$ of $\mathbf{I}$ is H_+). Indeed, $\mathfrak{R}(\mathbf{I})$ is dense in H_+: if $u \in H_+$ is such that $u \perp \mathfrak{R}(\mathbf{I})$ in H_+, then for any $f \in H_0$ we have by (1.8) that $O = (\mathbf{I}f, u)_{H_+} = (If, u)_{H_+} = (f, u)_{H_0}$, and so $u = 0$. On the other hand, $\mathfrak{R}(\mathbf{I})$ is closed in H_+. Therefore, $\mathfrak{R}(\mathbf{I}) = H_+$. ■

Further,

$$(\alpha, u)_{H_0} = (\mathbf{I}\alpha, u)_{H_+} \qquad (\alpha \in H_-,\ u \in H_+). \tag{1.10}$$

Indeed, suppose that $H_0 \ni f_n \to \alpha$ as $n \to \infty$ in H_-. Then, by (1.6), (1.8), and the continuity of $\mathbf{I}$,

$$(\alpha, u)_{H_0} = \lim_{n\to\infty} (f_n, u)_{H_0} = \lim_{n\to\infty} (\mathbf{I} f_n, u)_{H_+} = (\mathbf{I}\alpha, u)_{H_+}. \quad \blacksquare$$

THEOREM 1.2.

$$H_- = (H_+)'.$$

PROOF. It is only necessary to establish that any functional $l \in (H_+)'$ has the form $l(u) = (\alpha, u)_{H_0}$ $(u \in H_+)$ for some $\alpha \in H_-$. By the Riesz theorem, there exists an $\alpha \in H_+$ such that $l(u) = (a, u)_{H_+}$ $(u \in H_+)$. Since $\mathfrak{R}(\mathbf{I}) = H_+$, (1.10) gives us for $\alpha = \mathbf{I}^{-1}a \in H_-$ that

$$l(u) = (a, u)_{H_+} = (\mathbf{I}\mathbf{I}^{-1}a, u)_{H_+} = (\mathbf{I}\alpha, u)_{H_+} = (\alpha, u)_{H_0} \qquad (u \in H_+). \quad \blacksquare$$

Thus, the pairing between H_+ and its dual space $(H_+)' = H_-$ is determined by means of H_0.

Note that all the given results extend in an obvious way to real spaces H_0, H_+, and H_-.

1.4. *Construction of isometries between the spaces of the chain.*

THEOREM 1.3. *The isometry* $\mathbf{I}\colon H_- \to H_+$ *admits a decomposition into a product of two isometries*:

$$\mathbf{I} = J\mathbf{J} \quad (\mathbf{J}\colon H_- \to H_0, J\colon H_0 \to H_+); \qquad OJ = \mathbf{J} \restriction H_0. \tag{1.11}$$

PROOF. Let O and I be the operators introduced in §1.2, and let $A = OI\colon H_0 \to H_0$. This operator is clearly bounded, and it is nonnegative: by (1.8),

$$(Af, f)_{H_0} = (OIf, f)_{H_0} = (If, f)_{H_0} = (If, If)_{H_+} \geq 0 \qquad (f \in H_0).$$

Let $B = \sqrt{A}\colon H_0 \to H_0$, and regard this operator as acting from a dense subset of H_- into H_0. Then it is isometric;

$$\begin{aligned}(Bf, Bg)_{H_0} &= (B^2 f, g)_{H_0} = (OIf, g)_{H_0} \\ &= (If, Ig)_{H_+} = (f, g)_{H_-} \qquad (f, g \in H_0).\end{aligned}$$

Closing B by continuity, we get an isometric operator $\mathbf{J}\colon H_- \to H_0$.

Let us prove the equality $\mathfrak{R}(\mathbf{J}) = H_0$. It suffices to see that if $f \in H_0$ and $f \perp \mathfrak{R}(\mathbf{J})$ in H_0, then $f = 0$. For any $g \in H_0$ we have that

$$0 = (\mathbf{J}g, f)_{H_0} = (Bg, f)_{H_0} = (g, Bf)_{H_0},$$

whence $Bf = 0$. But then $OIf = B^2 f = 0$, i.e., $If = 0$, so $f = 0$. Thus, $\mathfrak{R}(\mathbf{J}) = H_0$.

We show that $\mathfrak{R}(B) \subseteq H_+$. To do this it suffices to see that

$$B\mathbf{J} = O\mathbf{I}. \tag{1.12}$$

For any $f \in H_0$ we have that $B\mathbf{J}f = B^2 f = OIf = O\mathbf{I}f$. Since H_0 is dense in H_- and the operators $\mathbf{J}\colon H_- \to H_0, B\colon H_0 \to H_0$ and $O\mathbf{I}\colon H_- \to H_0$ are continuous, this gives us (1.12).

Denote by J the operator B, regarded as an operator from H_0 to H_+. Then (1.12) shows that (1.11) is valid. It remains to prove that $J\colon H_0 \to H_+$ is an isometric operator with H_+ as its range. But this follows at once from (1.11), because $J = \mathbf{I}\mathbf{J}^{-1}$, and $\mathbf{J}^{-1}\colon H_0 \to H_-$ and $\mathbf{I}\colon H_- \to H_+$ are isometries. ∎

We remark that the equality $OJ = \sqrt{OI}$ follows from the proof of the theorem. (Here $OI\colon H_0 \to H_0$ and is nonnegative, and the square root is taken in the usual sense in H_0.)

1.5. ***Adjointness with respect to the zero space.*** For operators acting continuously between the spaces of the chain (1.5), it is natural to introduce the concept of the adjoint with respect to H_0. For example, if $A\colon H_+ \to H_0$, then the adjoint operator $A^+\colon H_0 \to H_-$ is defined by

$$(Au, f)_{H_0} = (u, A^+ f)_{H_0} \qquad (u \in H_+,\ f \in H_0). \tag{1.13}$$

If $A\colon H_+ \to H_-$, then $A^+\colon H_+ \to H_-$ is defined by the analogous equality

$$(Au, v)_{H_0} = (u, A^+ v)_{H_0} \qquad (u, v \in H_+). \tag{1.14}$$

The operator A^+ is defined similarly for A acting between the other spaces of the chain.

The existence of A^+ follows easily from the existence of the usual adjoint operator A^*. For instance, in the case of (1.13) we have from (1.10) that

$$(Au, f)_{H_0} = (u, A^* f)_{H_+} = (u, \mathbf{I}^{-1} A^* f)_{H_0},$$

i.e.,

$$A^+ = \mathbf{I}^{-1} A^* \qquad (A^*\colon H_0 \to H_+). \tag{1.15}$$

In the case of (1.14) we use the relation $(\alpha, u)_{H_0} = (\alpha, \mathbf{I}^{-1} u)_{H_-}$ $(\alpha \in H_-,\ u \in H_+)$, which is a consequence of (1.10). This gives us

$$(Au, v)_{H_0} = (Au, \mathbf{I}^{-1} v)_{H_-} = (u, A^* \mathbf{I}^{-1} v)_{H_+} = (u, \mathbf{I}^{-1} A^* \mathbf{I}^{-1} v)_{H_0},$$

i.e.,

$$A^+ = \mathbf{I}^{-1} A^* \mathbf{I}^{-1} \qquad (A^*\colon H_- \to H_+). \tag{1.16}$$

For operators $A\colon H_+ \to H_-$ it is possible to generalize the concept of selfadjointness: A will be said to be selfadjoint if $A^+ = A$, i.e., $(Au, v)_{H_0} = (u, Av)_{H_0}$ $(u, v \in H_+)$. An ordinary bounded selfadjoint operator A defined in H_0 will of course be selfadjoint also in this sense when it is understood as an operator from H_+ to H_-. An operator $A\colon H_+ \to H_-$ is selfadjoint in this generalized sense if and only if the operator $\mathbf{I}A\colon H_+ \to H_+$ is selfadjoint in H_+ (or when the operator $A\mathbf{I}^{-1}\colon H_- \to H_-$ is selfadjoint in H_-). This follows from the equality (see (1.10))

$$(\mathbf{I}Au, v)_{H_+} = (Au, v)_{H_0} = (u, Av)_{H_0} = (u, \mathbf{I}Av)_{H_+} \qquad (u, v \in H_+).$$

The concept of nonnegativity of an operator defined in H_0 can be generalized similarly: an operator $A\colon H_+ \to H_-$ is said to be nonnegative ($A \geq 0$) if $(Au, u)_{H_0} \geq 0$ ($u \in H_+$). A nonnegative operator $A\colon H_+ \to H_-$ is called an operator with finite trace $\operatorname{Tr}(A)$ if $\operatorname{Tr}(A) = \sum_1^\infty (Ae_j, e_j)_{H_0} < \infty$ for some orthonormal basis $(e_j)_1^\infty$ in H_+. It is easy to see that $\operatorname{Tr}(A)$ does not vary if the basis $(e_j)_1^\infty$ is changed.

The inequality $|A| \leq \operatorname{Tr}(A)$ *holds for a nonnegative* $A\colon H_+ \to H_-$. Indeed, if $(e_j)_1^\infty$ is an orthonormal basis in H_+, then $(\mathbf{I}^{-1}e_j)_1^\infty$ is an orthonormal basis in H_-; therefore,

$$|A|^2 = \sum_{j=1}^{\infty} \|Ae_j\|_{H_-}^2 = \sum_{j,k=1}^{\infty} |(Ae_j, \mathbf{I}^{-1}e_k)_{H_-}|^2$$
$$= \sum_{j,k=1}^{\infty} |(Ae_j, e_k)_{H_0}|^2 \leq \sum_{j,k=1}^{\infty} (Ae_j, e_j)_{H_0}(Ae_k, e_k)_{H_0} = (\operatorname{Tr}(A))^2.$$

(We have used the isometry $\mathbf{I}$, (1.10), and the inequality

$$|(Au, v)_{H_0}|^2 \leq (Au, u)_{H_0}(Av, v)_{H_0} \qquad (u, v \in H_+),$$

i.e., the Cauchy-Schwarz-Bunyakovskiĭ inequality for the quasi-inner product $(u, v) = (Au, v)_{H_0}$.) ■

It is interesting to note that the operator $\mathbf{I}^{-1}\colon H_+ \to H_-$ is selfadjoint and nonnegative in the sense introduced—this follows from the relation (see (1.10))

$$(\mathbf{I}^{-1}u, v)_{H_0} = (u, v)_{H_+} = (u, \mathbf{I}^{-1}v)_{H_0} \qquad (u, v \in H_+).$$

The operators $J^{-1}\colon H_+ \to H_0$ and $\mathbf{J}^{-1}\colon H_0 \to H_-$ are adjoint with respect to H_0:

$$\mathbf{J}^{-1} = (J^{-1})^+, \tag{1.17}$$

as follows from the relation (see Theorem 1.3 and (1.10))

$$(J^{-1}u, f)_{H_0} = (u, Jf)_{H_+} = (u, \mathbf{I}^{-1}Jf)_{H_0}$$
$$= (u, \mathbf{J}^{-1}J^{-1}Jf)_{H_0} = (u, \mathbf{J}^{-1}f)_{H_0}.$$

Of course,

$$\mathbf{I} = \mathbf{I}^+, \quad \mathbf{J} = J^+, \quad \mathbf{I} \geq 0. \tag{1.18}$$

The scheme of action of the operators and the main equalities are as follows:

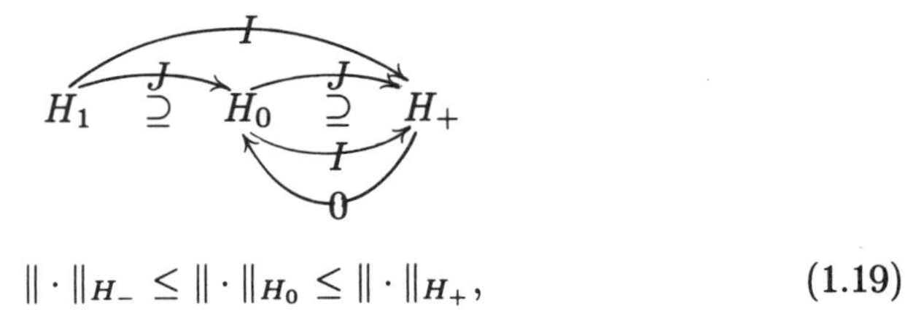

$$\|\cdot\|_{H_-} \leq \|\cdot\|_{H_0} \leq \|\cdot\|_{H_+}, \tag{1.19}$$

$$(\mathbf{I}\alpha,\beta)_{H_0}=(\alpha,\mathbf{I}\beta)_{H_0},\quad (\alpha,u)_{H_0}=(\mathbf{I}\alpha,u)_{H_+},\quad \mathbf{I}=J\mathbf{J},$$
$$(Jf,\alpha)_{H_0}=(f,\mathbf{J}\alpha)_{H_0},\qquad (\alpha,\beta\in H_-, f\in H_0, u\in H_+).\tag{1.20}$$

1.6. *Construction of a chain from a subspace of the positive space.* Suppose that the chain (1.19) is given, and G_+ is a fixed subspace of H_+ that is dense in H_0. Taking G_+ and H_0 as the positive and zero spaces, respectively, we can construct a negative space and get the chain

$$G_-\supseteq H_0\supseteq G_+.\tag{1.21}$$

It may be asked how G_- is connected with H_-. Consider the orthogonal complement $N_-\subseteq H_-$ of G_+ with respect to $(\cdot,\cdot)_{H_0}$, i.e., $N_-=\{\alpha\in H_-|(\alpha,v)_{H_0}=0,\ v\in G_+\}$. Then G_- *is isomorphic to the orthogonal complement of* N_- *in* H_- (i.e., it can be assumed that $G_-=H_-\ominus N_-$). Indeed, with each $\alpha\in H_-$ we associate a $\beta_\alpha\in G_-$ by setting $(\beta_\alpha,v)_{H_0}=\beta_\alpha(v)=(\alpha,v)_{H_0}$ $(v\in G_+)$. The mapping $H_-\ni\alpha\mapsto\beta_\alpha\in G_-$ is linear and has kernel N_-. Its range fills the whole of G_-: if $\beta\in G_-$ and $\mathbf{I}_1$ and $\mathbf{I}_2$ are the isometries $\mathbf{I}$ connected with the respective chains (1.19) and (1.21), then $\alpha=\mathbf{I}_1^{-1}\mathbf{I}_2\beta\in H_-$ (as follows from (1.10)) will have the property that $\beta_\alpha=\beta$. On the other hand, let P be the projection of H_- onto $H_-\ominus N_-$. The map $H_-\ni\alpha\mapsto P\alpha$ also has N_- as its kernel; therefore, there is a natural one-to-one linear mapping $H_-\ominus N_-\ni P\alpha\mapsto\alpha\mapsto\beta_\alpha\in G_-$ between the Hilbert spaces $H_-\ominus N_-$ and G_-. This mapping preserves the norm. Indeed, it follows from (1.10) that $\mathbf{I}_1^{-1}G_+=H_-\ominus N_-$, and so, by (1.10),

$$\|\beta_\alpha\|_{G_-}=\sup_{v\in G_+}\frac{|(\alpha,v)_{H_0}|}{\|v\|_{H_+}}=\sup_{v\in G_+}\frac{|(\alpha,\mathbf{I}_1^{-1}v)_{H_-}|}{\|\mathbf{I}_1^{-1}v\|_{H_-}}=\|P\alpha\|_{H_-}.$$

Accordingly, the mapping is an isomorphism between $H_-\ominus N_-$ and G_-. ∎

1.7. *Construction of a chain from the negative space.* Above we constructed the chain (1.19) from a given pair of spaces H_0 and H_+. Let us show that it is possible to construct it also from the pair H_0 and H_-. The proof of the following theorem gives a construction of the corresponding H_+.

THEOREM 1.4. *Suppose that H_- is a given Hilbert space and H_0 is a dense linear subset of it that is itself a Hilbert space with respect to another inner product, with $\|f\|_{H_-}\le\|f\|_{H_0}$ $(f\in H_0)$. Then it is possible to construct a positive space $H_+\subseteq H_0$ such that the corresponding negative space is H_-.*

PROOF. Consider the bilinear form

$$H_0\times H_0\ni(f,g)\mapsto B(f,g)=(f,g)_{H_-}\in\mathbb{C}^1.$$

Due to the inequality $\|f\|_{H_-}\le\|f\|_{H_0}$ $(f\in H_0)$, it is continuous and hence can be represented in the form

$$(f,g)_{H_-}=(Kf,g)_{H_0}\qquad(f,g\in H_0),$$

where $K\colon H_0 \to H_0$ is a continuous operator; obviously, $\|K\| \leq 1$ and $K \geq 0$. Its range $\mathfrak{R}(K)$ is dense in H_0: if $0 = (Kf, h)_{H_0} = (f, h)_{H_-}$ $(f \in H_0)$, then $h = 0$ in view of the denseness of H_0 in H_-. The inverse operator K^{-1} exists on $\mathfrak{R}(K)$: if $Kf = 0$, then $(f, g)_{H_-} = (Kf, g)_{H_0} = 0$ $(g \in H_0)$, i.e., $f = 0$.

Let

$$(u, v)_{H_+} = (K^{-1}u, v)_{H_0} \qquad (u, v \in \mathfrak{R}(K)). \tag{1.22}$$

Since $\|K\| \leq 1$, it follows from the spectral decomposition for K that $\|u\|^2_{H_+} = (K^{-1}u, u)_{H_0} \geq \|u\|^2_{H_0}$ $(u \in \mathfrak{R}(K))$. This implies that the space H_+ obtained as the completion of $\mathfrak{R}(K)$ with respect to $(\cdot, \cdot)_{H_-}$ satisfies all the requirements imposed on the positive space with respect to H_0. We construct the negative space G_- from H_0 and this H_+. It is required to prove that $G_- = H_-$.

For a proof note that $K = OI$, where the last two operators are constructed from the chain $G_- \supseteq H_0 \supseteq H_+$. This equality follows from the relation $(If, v)_{H_+} = (f, v)_{H_0}$ $(f \in H_0,\ v \in H_+ \supseteq \mathfrak{R}(K))$ (which follows from (1.22)), and the denseness of $\mathfrak{R}(K)$ in H_+. Therefore,

$$(f, g)_{H_-} = (Kf, g)_{H_0} = (If, g)_{H_0} = (f, g)_{G_-}$$

for $f, g \in H_0$, which gives us the equality $H_- = G_-$ in view of the denseness of H_0 in H_- and in G_-. ■

The situation frequently arises when a quasi-inner product $(f, g)_{H_-}$ such that $(f, f)_{H_-} \leq \|f\|^2_{H_0}$ is introduced on the vectors $f, g, \ldots$ in some Hilbert space H_0. If $(\cdot, \cdot)_{H_-}$ is an inner product, then by completing H_0 in $(\cdot, \cdot)_{H_-}$ we obtain a Hilbert space $H_- \supseteq H_0$. We can apply Theorem 1.4 to the pair H_0, H_- and interpret H_- as a negative space with the zero space H_0. In the case of a quasi-inner product we can consider the linear set $G = \{f \in H_0 \mid (f, f)_{H_-} = 0\}$. By the estimate $(f, f)_{H_-} \leq \|f\|^2_{H_0}$, it is closed in H_0, i.e., it is a subspace. Let $G_0 = H_0 \ominus G$ (the orthogonal difference in H_0). Clearly, the form $(\cdot, \cdot)_{H_-}$ now defines an inner product on G_0, and we come to the case considered. The completion G_- of G_0 with respect to $(\cdot, \cdot)_{H_-}$ can thus also be interpreted as a negative space with zero space G_0.

EXAMPLE. Let $H_0 = L_2(R, d\mu(x))$, where R is an abstract space with a finite measure μ, and let $R \times R \ni (x, y) \mapsto K(x, y) \in \mathbb{C}^1$ be a bounded $\mu \times \mu$-measurable kernel. Define

$$(f, g)_{H_K} = \int_R \int_R K(x, y) f(y) \overline{g(x)}\, d\mu(x)\, d\mu(y) \qquad (f, g \in L_2(R, d\mu(x))), \tag{1.23}$$

where K is positive-definite, i.e., $(f, f)_{H_K} \geq 0$ $(f \in L_2(R, d\mu(x)))$. Then (1.23) defines a quasi-inner product on H_0. According to the foregoing, the Hilbert space H_K corresponding to this quasi-inner product can be understood as the negative space with respect to a zero space coinciding with H_0 or with some subspace G_0 of it and with respect to some positive space H_+. (We set $(\cdot, \cdot)_{H_-} = \varepsilon(\cdot, \cdot)_{H_K}$ with a sufficiently small constant $\varepsilon > 0$.) This is an important interpretation of H_K as a space of generalized functions (see also §3.12).

1.8. ***Construction of a chain from an operator.*** It can be concluded from (1.19) that

$$(u,v)_{H_+} = (J^{-1}u, J^{-1}v)_{H_0} \qquad (u,v \in H_+),$$

$$(\alpha,\beta)_{H_-} = (\mathbf{J}\alpha, \mathbf{J}\beta)_{H_0} \qquad (\alpha,\beta \in H_-). \tag{1.24}$$

According to (1.24), it is possible to construct the chain (1.19) by specifying initially along with H_0 not H_+ or H_- but some operator defined in H_0 and analogous to J or, what is more convenient, J^{-1}. This approach to the construction of the chain (1.19) often turns out to be useful.

Thus, suppose that a closed operator D with dense domain $\mathfrak{D}(D)$ is defined in some Hilbert space H_0 such that

$$\|Du\|_{H_0} \geq \|u\|_{H_0} \qquad (u \in \mathfrak{D}(D)). \tag{1.25}$$

Obviously, $\mathfrak{D}(D)$ is a complete Hilbert space with respect to the inner product

$$(u,v)_{H_+} = (Du, Dv)_{H_0}. \tag{1.26}$$

We can regard it as the positive space H_+ and then construct the corresponding H_-.

The operator J is constructed from the chain $H_- \supseteq H_0 \supseteq H_+$. Comparing the first of the equalities (1.24) with (1.26), we conclude that $J_0^{-1} = \sqrt{D^*D}$, where J_0^{-1} is understood to be the operator J^{-1} in H_0 with the domain H_+. (We have used the positivity of J_0^{-1} and the theorem on metrically equal operators.) If, moreover, D is a positive selfadjoint operator, then $J_0^{-1} = D$. In this case $OJ = D^{-1}$, and so, according to the second of the equalities in (1.24), $(f,g)_{H_-} = (D^{-1}f, D^{-1}g)_{H_0}$ $(f,g \in H_0)$; the negative space H_- is obtained as the completion of H_0 with respect to the last inner product.

1.9. ***Weighted orthogonal sums of chains.*** We give a simple but technically convenient construction, namely, we construct weighted orthogonal sums of the chains (1.5). First we introduce the concept of a weighted orthogonal sum of Hilbert spaces. Let $(H_n)_1^\infty$ be a sequence of Hilbert spaces H_n, and $\delta = (\delta_n)_1^\infty$ $(\delta_n > 0)$ a weight sequence. The weighted orthogonal sum $\mathcal{H}_\delta = \bigoplus_{n=1;\delta}^\infty H_n$ is understood to be the Hilbert space of sequences $f = (f_n)_1^\infty$ such that $\sum_1^\infty \|f_n\|_{H_n}^2 \delta_n < \infty$, with the natural linear operations and the inner product $(f,g)_{\mathcal{H}_\delta} = \sum_{n=1}^\infty (f_n, g_n)_{H_n} \delta_n$. In the case $\delta = 1$, i.e., $\delta_n = 1$ $(n = 1, 2, \ldots)$, $\mathcal{H}_1$ becomes the usual orthogonal sum.

Let

$$H_{-,n} \supseteq H_{0,n} \supseteq H_{+,n} \qquad (n = 1, 2, \ldots) \tag{1.27}$$

be a sequence of chains of the form (1.5), *and let* $\delta = (\delta_n)_1^\infty$ $(\delta_n \geq 1)$ *be a weight. Then*

$$\mathcal{H}_{-,\delta^{-1}} = \bigoplus_{n=1;\delta^{-1}}^{\infty} H_{-,n} \supseteq \bigoplus_{n=1;1}^{\infty} H_{0,n} \supseteq \bigoplus_{n=1;\delta}^{\infty} H_{+,n} = \mathcal{H}_{+,\delta}$$

$$\bigoplus_{n=1;1}^{\infty} H_{0,n} = \mathcal{H}_{0,1}$$

$$(\delta^{-1} = (\delta_n^{-1})_{n=1}^{\infty}) \quad (1.28)$$

is also a chain. Indeed, $\mathcal{H}_{+,\delta}$ is dense in $\mathcal{H}_{0,1}$, and $\|u\|_{\mathcal{H}_{0,1}} \leq \|u\|_{\mathcal{H}_{+,\delta}}$ $(u \in \mathcal{H}_{+,\delta})$, in view of the condition $\delta_n \geq 1$. Therefore, we can construct a chain $G_- \supseteq \mathcal{H}_{0,1} \supseteq \mathcal{H}_{+,\delta}$. We prove that $G_- = \mathcal{H}_{-,\delta^{-1}}$. Let I be connected with the last chain, and let I_n be the analogous operators connected with (1.27). Then

$$\begin{aligned}(If, u)_{\mathcal{H}_{+,\delta}} &= (f, u)_{\mathcal{H}_{0,1}} = \sum_{n=1}^{\infty}(f_n, u_n)_{H_{0,n}} = \sum_{n=1}^{\infty}(I_n f_n, u_n)_{H_{+,n}} \\ &= \sum_{n=1}^{\infty}(\delta_n^{-1} I_n f_n, u_n) H_{+,n}\delta_n \qquad (f \in \mathcal{H}_{0,1},\ u \in \mathcal{H}_{+,\delta}).\end{aligned}$$

Hence, $(If)_n = \delta_n^{-1} I_n f_n$ $(n = 1, 2, \ldots)$; therefore, for $f, g \in \mathcal{H}_{0,1}$

$$\begin{aligned}(f, g)_{G_-} &= (If, g)_{\mathcal{H}_{0,1}} = \sum_{n=1}^{\infty}(I_n f_n, g_n)_{H_{0,n}}\delta_n^{-1} \\ &= \sum_{n=1}^{\infty}(f_n, g_n)_{H_{-,n}}\delta_n^{-1} = (f, g)_{\mathcal{H}_{-,\delta^{-1}}}.\end{aligned}$$

Since the space $\mathcal{H}_{0,1}$ is a dense subset of G_- and $\mathcal{H}_{-,\delta^{-1}}$, the last equality gives us that $G_- = \mathcal{H}_{-,\delta^{-1}}$. ■

Of course, finite (and not infinite) sequences of Hilbert spaces can appear in the foregoing constructions.

1.10. *Projective limits of Banach spaces.* Let us consider a collection $(B_\tau)_{\tau \in \mathrm{T}}$ of Banach spaces B_τ(τ varies over an arbitrary index set T). We assume that $\Phi = \bigcap_{\tau \in \mathrm{T}} B_\tau$ is dense in each B_τ and provide this linear space with the following topology. A basic neighborhood of zero in Φ is understood to be a set

$$U(0; \tau_1, \ldots, \tau_m, \varepsilon_1, \ldots, \varepsilon_m) = \{\varphi \in \Phi |\ \|\varphi\|_{B_{\tau_1}} < \varepsilon_1, \ldots, \|\varphi\|_{B_{\tau_m}} < \varepsilon_m\};$$

a base of neighborhoods of zero is obtained by varying $\tau_1, \ldots, \tau_m$ arbitrarily over T and varying $\varepsilon_1 > 0, \ldots, \varepsilon_m > 0$ $(m = 1, 2, \ldots)$. A basic neighborhood of any vector $\psi \in \Phi$ is taken to be a translated basic neighborhood of zero:

$$U(\psi; \tau_1, \ldots, \tau_m, \varepsilon_1, \ldots, \varepsilon_m) = \{\varphi \in \Phi | \varphi - \psi \in U(0; \tau_1, \ldots, \tau_m, \varepsilon_1, \ldots, \varepsilon_m)\}.$$

It is easy to prove that the collection of all sets

$$U(\psi;\tau_1,\dots,\tau_m,\varepsilon_1,\dots,\varepsilon_m)$$

satisfies the axiomatics of a base of neighborhoods in a Hausdorff topological space, and the linear operations are continuous in this topology. The linear topological space constructed in this way is called the projective limit of the spaces B_τ, denoted by $\Phi = \operatorname{pr}\lim_{\tau\in \mathrm{T}} B_\tau$. If T is countable, then Φ is called a countably normed space, and the notation $\Phi = \operatorname{pr}\lim_{\tau\to\infty} B_\tau$ is also used in this case.

The norms are always assumed below to be directed in the following way: for any $\tau',\tau''\in \mathrm{T}$ there exists a $\tau'''\in \mathrm{T}$ such that, topologically, $B_{\tau'''}\subseteq B_{\tau'}$ and $B_{\tau'''}\subseteq B_{\tau''}$. This implies that each neighborhood

$$U(0;\tau_1,\dots,\tau_m,\varepsilon_1,\dots,\varepsilon_m)$$

contains a neighborhood $U(0;\tau,\delta)$ for some $\tau\in\mathrm{T}$ and $\delta>0$. Therefore, a base of neighborhoods for Φ is formed by the collection of all neighborhoods of the form $U(\psi;\tau,\varepsilon)$ ($\psi\in\Phi$, $\tau\in\mathrm{T}$, $\varepsilon>0$). In the special case when T is countable, for example, $\mathrm{T}=\{0,1,\dots\}$ and $\|\varphi\|_{B_0}\le\|\varphi\|_{B_1}\le\cdots$ ($\varphi\in\Phi$), we say that the system of norms giving the topology on Φ is monotone. Of course, now

$$B_0\supseteq B_1\supseteq\cdots\supseteq \operatorname*{pr\,lim}_{\tau\in\infty} B_\tau.$$

Recall also that a sequence $(\varphi_n)_1^\infty$ of vectors $\varphi_n\in\Phi$ is said to converge to $\varphi\in\Phi$ if for any neighborhood $U(\varphi)$ there is an index $N=N(U(\varphi))$ such that $\varphi_n\in U(\varphi)$ for $n>N$. Specification of all the convergent sequences in the linear topological space Φ does not in general uniquely determine the topology, i.e., it is possible to give the linear set Φ nonequivalent topologies which turn it into two different topological spaces Φ' and Φ'' with the same collection of convergent sequences.

Projective limits of Banach spaces often arise in the following situation. We have a linear set E with a system of norms $\|u\|_{B_\tau}$ ($u\in E$), where $\tau\in\mathrm{T}$. Denote by B_τ the completion of E with respect to $\|\cdot\|_B$. Then $\Phi=\bigcap_{\tau\in\mathrm{T}} B_\tau$ contains E and is thus dense in each B_τ. To have the norms directed now means that for any $\tau',\tau''\in\mathrm{T}$ there exist $\tau'''\in\mathrm{T}$ and constants $c',c''\in(0,\infty)$ such that

$$\|u\|_{B_{\tau'}}\le c'\|u\|_{B_{\tau'''}}\quad\text{and}\quad \|u\|_{B_{\tau''}}\le c''\|u\|_{B_{\tau'''}}\qquad (u\in E);$$

moreover, the norms $\|\cdot\|_{B_{\tau'}},\|\cdot\|_{B_{\tau'''}}$ and $\|\cdot\|_{B_{\tau''}},\|\cdot\|_{B_{\tau'''}}$ are consistent. Consequently, $B_{\tau'''}\subseteq B_{\tau'}$ and $B_{\tau'''}\subseteq B_{\tau''}$, and it is possible to define $\operatorname{pr}\lim_{\tau\in\mathrm{T}} B_\tau$. Monotonicity of the norms is of course equivalent to the inequalities $\|u\|_{B_0}\le\|u\|_{B_1}\le\cdots$ ($u\in E$).

In the case when the $B_\tau=H_\tau$ ($\tau\in\mathrm{T}$) are Hilbert spaces we speak of a projective limit Φ of Hilbert spaces or, if T is countable, of a countably Hilbert space Φ. For a given Φ, understood as a projective limit of Banach spaces, it is often possible to construct Hilbert spaces with Φ serving as their projective limit (§§3.9 and 3.10). We recall an important definition: a projective limit $\Phi = \operatorname{pr\,lim}_{\tau \in \mathrm{T}} H_\tau$ of Hilbert spaces is called a nuclear space if for each $\tau \in \mathrm{T}$ there is a $\tau' \in \mathrm{T}$ such that $H_{\tau'} \subseteq H_\tau$, and this imbedding is quasinuclear.

We shall frequently deal with the following situation. A collection $(H_\tau)_{\tau \in \mathrm{T}}$ of Hilbert spaces is given, with each a dense subset of some fixed Hilbert space H_0 $(0 \in \mathrm{T})$, and $\|u\|_{H_0} \leq \|u\|_{H_\tau}$ $(u \in H_\tau)$. (More generally, the imbedding $H_\tau \to H_0$ is continuous, in which case the last inequality can be attained by renorming H_τ.) In this case it is possible to construct the chain

$$H_{-\tau} \supseteq H_0 \supset H_\tau, \tag{1.29}$$

where $H_{-\tau}$ is the negative space constructed from the zero space H_0 and the positive space H_τ. On the other hand, if $\Phi = \bigcap_{\tau \in \mathrm{T}} H_\tau$ is dense in each H_τ, then it is possible to construct $\operatorname{pr\,lim}_{\tau \in \mathrm{T}} H_\tau$.

We have the following simple but important theorem.

THEOREM 1.5. *If $\Phi = \operatorname{pr\,lim}_{\tau \in \mathrm{T}} B_\tau$ is a projective limit of Banach spaces, then the dual space of antilinear functionals is given by $\Phi' = \bigcup_{\tau \in \mathrm{T}} B'_\tau$. In particular, if Φ is a projective limit of Hilbert spaces H_τ $(\tau \in \mathrm{T})$ for which the chains* (1.29) *have been constructed, then $\Phi' = \bigcup_{\tau \in \mathrm{T}} H_{-\tau}$.*

PROOF. If $l \in B'_{\tau'}$, then $l \restriction \Phi$ is an antilinear functional on Φ. The continuity of l at 0 in the topology of B_τ implies that for any $\varepsilon > 0$ there exists a $U(0; \tau, \delta)$ such that $|l(\varphi)| < \varepsilon$ for $\varphi \in U(0; \tau, \delta)$; but this means that l is continuous at 0 in the topology of Φ. Thus, $l \in \Phi'$.

Conversely, let $l \in \Phi'$. It follows from the continuity of l at zero that for $\varepsilon = 1$ there exists a basic neighborhood of 0 in the topology of Φ, i.e., a neighborhood $U(0; \tau, \delta)$ $(\tau \in \mathrm{T},\ \delta > 0)$ such that $|l(\varphi)| < 1$ for $\varphi \in U(0; \tau, \delta)$. Consider the space B_τ. The antilinear functional l is defined on the dense linear subset Φ of B_τ and is bounded by 1 on the intersection of Φ with the ball $\|\cdot\|_{B_\tau} < \delta$ in B_τ. Thus, it is continuous on Φ, equipped with the norm from the space B_τ. Extending it by continuity to the whole of B_τ, we get that $l \in B'_\tau$. ∎

If we have a sequence $(H_\tau)_0^\infty$ of Hilbert spaces with a monotone consistent system of norms and $\Phi = \bigcap_0^\infty H_\tau$ is dense in each H_τ, then it is easy to prove the relations

$$\begin{gathered}\bigcup_{\tau=0}^{\infty} H_{-\tau} = (\operatorname{pr\,lim}_{\tau \to \infty} H_\tau)' \\ \supseteq \cdots \supseteq H_{-2} \supseteq H_{-1} \supseteq H_0 \supseteq H_1 \supseteq H_2 \supseteq \cdots \supseteq \operatorname{pr\,lim}_{\tau \to \infty} H_\tau = \bigcap_{\tau=0}^{\infty} H_\tau, \\ \cdots \leq \|\varphi\|_{H_{-2}} \leq \|\varphi\|_{H_{-1}} \leq \|\varphi\|_{H_0} \leq \|\varphi\|_{H_1} \leq \|\varphi\|_{H_2} \leq \cdots \\ (\varphi \in \Phi).\end{gathered} \tag{1.30}$$

Here H_{-1} is constructed according to (1.29); each space in the inclusions of (1.30) is dense on the space to the left of it.

We make some additional simple general observations. Up to this point we have started out from a rigging of H_0 by Hilbert spaces H_+ and H_-, i.e., from a chain of the form (1.19). However, a rigging of H_0 by linear topological spaces is often needed. Namely, suppose that a linear topological space Φ with elements $\varphi, \psi, \dots$ is topologically (i.e., densely and continuously) imbedded in a Hilbert space H_0. As before, each element $f \in H_0$ gives rise to a continuous antilinear functional l_f on Φ according to the formula $l_f(\varphi) = (f, \varphi)_{H_0}$ ($\varphi \in \Phi$). Identifying f with l_f, we get an imbedding of H_0 in the space Φ' of continuous antilinear functionals on Φ. (The identification is unambiguous: if $l_f = 0$, then $f = 0$.) If Φ' is endowed with the weak topology, then the imbedding $H_0 \to \Phi'$ is obviously continuous. Accordingly, we have constructed a chain

$$\Phi' \supseteq H_0 \supseteq \Phi. \tag{1.31}$$

One also says that (1.31) is a rigging of H_0 by the spaces Φ and Φ'.

Mainly, Φ is a projective limit of Hilbert spaces in the situations encountered below. (The chain (1.30) is an example of such a rigging.) We mention the following in this connection.

Assume that $\Phi = \operatorname{pr\,lim}_{\tau \in \mathrm{T}} H_\tau$ in the rigging (1.31). Without loss of generality it can be assumed that each H_τ is topologically imbedded in H_0, with $\|\varphi\|_{H_0} \leq \|\varphi\|_{H_\tau}$ ($\varphi \in H_\tau; \tau \in \mathrm{T}$), and $0 \in \mathrm{T}$. (Instead of the spaces H_τ consider the Hilbert spaces $F_\tau = H_0 \cap H_\tau$ with the inner product $(\varphi, \psi)_{F_\tau} = (\varphi, \psi)_{H_0} + (\varphi, \psi)_{H_\tau}$ ($\varphi, \psi \in F_\tau$), and $F_0 = H_0$. Since the mapping $\Phi \ni \varphi \mapsto \varphi \in H_0$ is continuous at 0, there exist a $\tau \in \mathrm{T}$ and a $c_\tau \in [0, \infty)$ such that $\|\varphi\|_{H_0} \leq c_\tau \|\varphi\|_{H_\tau}$ for $\varphi \in \Phi$, and hence also for $\varphi \in F_\tau$. This implies that the norms in H_τ ($\tau \in \mathrm{T}$) and in F_τ ($\tau \in \mathrm{T} \cup \{0\}$) determine the same topology in Φ, and the F_τ have the required properties.) Construct the chain (1.29) for each $\tau \in \mathrm{T}$. Then, by Theorem 1.5,

$$\Phi' = \bigcup_{\tau \in \mathrm{T}} H_{-\tau}. \tag{1.32}$$

The relation (1.32) enables us to equip Φ' with the topology of the inductive limit of the spaces $H_{-\tau}$: a system of basic neighborhoods is formed by all possible intersections of finitely many balls

$$\begin{aligned} &V(\alpha_1, \dots, \alpha_p; \tau_1, \dots, \tau_p; \varepsilon_1, \dots, \varepsilon_p) \\ &\quad = \{\alpha \in \Phi' | \alpha \in \bigcap_{n=1}^{p} H_{-\tau_n}, \ \|\alpha - \alpha_1\|_{H_{-\tau_1}} < \varepsilon_1, \dots, \|\alpha - \alpha_p\|_{H_{-\tau_p}} < \varepsilon_p\} \\ &\qquad (\alpha_n \in H_{-\tau_n}, \ \tau_n \in \mathrm{T}, \ \varepsilon_n \in (0, \infty); \ n = 1, \dots, p; p = 1, 2, \dots). \end{aligned}$$

(All the required axioms are again satisfied; it must be kept in mind that for any $\tau', \tau'' \in \mathrm{T}$ there exists a $\tau''' \in \mathrm{T}$ such that the inclusions $H_{-\tau'''} \supseteq H_{-\tau'}$ and $H_{-\tau'''} \supseteq H_{-\tau''}$ are topological.) Accordingly, $\Phi' = \operatorname{ind\,lim}_{\tau \in \mathrm{T}} H_{-\tau}$.

As in the case of the chain (1.19), the inner product in H_0 determines the pairing between Φ and Φ' in (1.31). Namely, if $\alpha \in \Phi'$, then $\alpha(\varphi) = (\alpha, \varphi)_{H_0}$ $(\varphi \in \Phi)$, where $\Phi' \times \Phi \ni (\alpha, \varphi) \mapsto (\alpha, \varphi)_{H_0} \in \mathbb{C}^1$ is the extension of $(f, \varphi)_{H_0}$ by continuity as $H_0 \ni f \to \alpha$. (This follows immediately from (1.32) and the pairing between H_τ and $H_{-\tau}$.)

If in (1.19) the imbedding $H_+ \to H_0$ is quasinuclear (i.e., the imbedding operator is Hilbert-Schmidt), then the corresponding rigging or chain is said to be quasinuclear. If Φ is nuclear, then the rigging or chain (1.31) is said to be nuclear.

§2. Finite and infinite tensor products of Hilbert spaces

The well-known concept of the tensor product of a finite number of Hilbert spaces is presented below, along with a generalization of it to the case of an infinite number of spaces: the concept of a weighted tensor product, which encompasses the usual infinite tensor product with fixed stabilization. This generalization is needed to construct expansions in generalized eigenfunctions for operators acting in a space of functions of infinitely many variables; it makes possible the construction of spaces of such functions imbedded one in another with Hilbert-Schmidt imbedding operators. It is the weight that enables us to suitably damp the behavior of functions in the "direction of growth of the number of variables". Instead of using a weighting it is possible to achieve the same effect by suitably renorming the spaces being multiplied and then taking their ordinary infinite tensor product with a definite stabilization. (These two paths are equivalent.) However, the use of weightings is technically more convenient. Tensor products of chains $H_- \supseteq H_0 \supseteq H_+$ are also presented. We confine the exposition to separable infinite-dimensional spaces.

2.1. ***Tensor products of finitely many Hilbert spaces.*** Let $(H_n)_1^p$ be a finite sequence of separable Hilbert spaces H_n, and $(e_j^{(n)})_{j=0}^\infty$ an orthonormal basis in H_n $(n = 1, \ldots, p)$. We form the formal product

$$e_\alpha = e_{(\alpha_1, \ldots, \alpha_p)} = e_{\alpha_1}^{(1)} \otimes \cdots \otimes e_{\alpha_p}^{(p)} \qquad (\alpha = (\alpha_n)_{n=1}^p;\ \alpha_1, \ldots, \alpha_p = 0, 1, \ldots). \tag{2.1}$$

(i.e., we consider the ordered sequence $(e_{\alpha_1}^{(1)}, \ldots, e_{\alpha_p}^{(p)})$ and span a Hilbert space by the formal vectors (2.1) by regarding them as forming an orthonormal basis of it. The separable Hilbert space obtained is called the tensor product of the spaces $H_1, \ldots, H_p$ and denoted by $H_1 \otimes \cdots \otimes H_p = \bigotimes_1^p H_n$. Thus, its vectors have the form

$$f = \sum_{\alpha \in N^p} f_\alpha e_\alpha \quad (f_\alpha \in \mathbb{C}^1), \qquad \sum_{\alpha \in N^p} |f_\alpha|^2 = \|f\|^2_{\otimes_1^p H_n} < \infty, \tag{2.2}$$

where N^p denotes the countable set of all indices α in (2.1). Of course,

$$(f,g)_{\otimes_1^p H_n} = \sum_{\alpha \in N^p} f_\alpha \bar{g}_\alpha,$$

where $g = \sum_{\alpha \in N^p} g_\alpha e_\alpha$ is a second vector of the form (2.2).

Let $f^{(n)} = \sum_{j=1}^\infty f_j^{(n)} e_j^{(n)} \in H_n$ $(n = 1, \ldots, p)$ be some vectors. Define

$$f = f^{(1)} \otimes \cdots \otimes f^{(p)} = \sum_{\alpha_1, \ldots, \alpha_p = 1}^\infty f_{\alpha_1}^{(1)} \cdots f_{\alpha_p}^{(p)} e_{\alpha_1}^{(1)} \otimes \cdots \otimes e_{\alpha_p}^{(p)}. \tag{2.3}$$

The coefficients $f_\alpha = f_{\alpha_1}^{(1)} \cdots f_{\alpha_p}^{(p)}$ in the expansion (2.3) satisfy condition (2.2); therefore, the vector (2.3) belongs to $\bigotimes_1^p H_n$, and

$$\|f\|_{\otimes_1^p H_n} = \prod_{n=1}^p \|f^{(n)}\|_{H_n}. \tag{2.4}$$

It is clear that the function

$$\prod_{n=1}^p H_n \ni (f^{(1)}, \ldots, f^{(p)}) \mapsto f^{(1)} \otimes \cdots \otimes f^{(p)} \in \bigotimes_{n=1}^p H_n$$

is linear in each variable, and the linear span of the vectors (2.3) is dense in $\bigotimes_1^p H_n$. If M_n is a linear subset of H_n $(n = 1, \ldots, p)$, then we set

$$\text{a.} \bigotimes_{n=1}^p M_n = \text{l.s.}\{f^{(1)} \otimes \cdots \otimes f^{(p)} | f^{(1)} \in M_1, \ldots, f^{(p)} \in M_p\},$$

$$\bigotimes_{n=1}^p M_n = \text{c.l.s.}\{f^{(1)} \otimes \cdots \otimes f^{(p)} | f^{(1)} \in M_1, \ldots, f^{(p)} \in M_p\}.$$

This definition of a tensor product depends, of course, on the choice of the orthonormal bases $(e_j^{(n)})_{j=0}^\infty$ in the spaces H_n $(n = 1, \ldots, p)$, but it is easy to see that a change in the bases leads to a tensor product isomorphic to the original product, with preservation of its structure.

EXAMPLE. Let $H_n = L_2(R_n, d\mu_n(x_n))$, where R_n is a space with a measure μ_n $(\mu_n(R_n) \leq +\infty)$ $(n = 1, \ldots, p)$. Then

$$\bigotimes_{n=1}^p H_n = L_2\left(\bigtimes_{n=1}^p R_n, d(\mu_1 \times \cdots \times \mu_p)(x)\right).$$

To prove this equality we must place a vector $e_\alpha = (e_{\alpha_1}^{(1)}(x_1)) \otimes \cdots \otimes (e_{\alpha_p}^{(p)}(x_p)) \in \bigotimes_1^p H_n$ of the form (2.1) in correspondence with the function

$$e_\alpha(x) = e_{\alpha_1}^{(1)}(x_1) \cdots e_{\alpha_p}^{(p)}(x_p) \in L_2\left(\bigtimes_{n=1}^p R_n, d(\mu_1 \times \cdots \times \mu_p)(x)\right).$$

Since such functions form an orthonormal basis in the space

$$L_2\left(\bigtimes_{n=1}^{p} R_n, d(\mu_1 \times \cdots \times \mu_p)(x)\right),$$

this correspondence generates the required isomorphism between $\bigotimes_1^p H_n$ and $L_2(\bigtimes_1^p R_n, d(\mu_1 \times \cdots \times \mu_p)(x))$.

2.2. *Tensor products of finitely many operators.* We proceed to the definition of the tensor product of a finite number of bounded operators—an operator constructed from the operators being multiplied in the same way that a mixed derivative is constructed from the derivatives in each variable.

THEOREM 2.1. *Let $(H_n)_1^p$ and $(G_n)_1^p$ be two sequences of Hilbert spaces, and $(A_n)_1^p$ a sequence of bounded operators $A_n\colon H_n \to G_n$. The tensor product $A_1 \otimes \cdots \otimes A_p = \bigotimes_1^p A_n$ is defined by*

$$\begin{aligned}\left(\bigotimes_{n=1}^{p} A_n\right) f &= \left(\bigotimes_{n=1}^{p} A_n\right)\left(\sum_{\alpha \in N^p} f_\alpha e_{\alpha_1}^{(1)} \otimes \cdots \otimes e_{\alpha_p}^{(p)}\right) \\ &= \sum_{\alpha \in N^p} f_\alpha (A_1 e_{\alpha_1}^{(1)}) \otimes \cdots \otimes (A_p e_{\alpha_p}^{(p)}) \qquad \left(f \in \bigotimes_{n=1}^{p} H_n\right). \end{aligned} \tag{2.5}$$

The series on the right-hand side of (2.5) converges weakly in $\bigotimes_1^p G_n$ and defines a bounded operator $\bigotimes_1^p A_n\colon \bigotimes_1^p H_n \to \bigotimes_1^p G_n$, with

$$\left\| \bigotimes_{n=1}^{p} A_n \right\| = \prod_{n=1}^{p} \|A_n\|. \tag{2.6}$$

PROOF. It suffices to carry out a proof for $p = 2$, since proofs for $p = 3, p = 4$, etc. follow easily from this because the tensor product is associative (i.e., $H_1 \otimes \cdots \otimes H_p = (H_1 \otimes \cdots \otimes H_{p-1}) \otimes H_p$).

Accordingly, let $p = 2$. Denote by $(l_j^{(n)})_{j=0}^\infty$ an orthonormal basis in G_n $(n = 1, 2)$ and let

$$g = \sum_{\beta \in N^2} g_{(\beta_1, \beta_2)} l_{\beta_1}^{(1)} \otimes l_{\beta_2}^{(2)} \in G_1 \otimes G_2.$$

To start with, we take f to be a vector with only finitely many nonzero coordinates $f_\alpha = f_{(\alpha_1, \alpha_2)}$. Fix α_2 and $\beta_1 = 0, 1, \dots,$ and denote by $f(\alpha_2)$ the vector in H_1 with coordinates $(f_{(\alpha_1,\alpha_2)})_{\alpha_1=0}^\infty$ in the basis $(e_{\alpha_1}^{(1)})_{\alpha_1=0}^\infty$ and by $g(\beta_1)$ the vector in G_2 with coordinates $(g_{(\beta_1,\beta_2)})_{\beta_2=0}^\infty$ in the basis $(l_{\beta_2}^{(2)})_{\beta_2=0}^\infty$.

We get that

$$\begin{aligned}
&\left|\left(\sum_{\alpha\in N^2} f_\alpha(A_1e^{(1)}_{\alpha_1})\otimes(A_2e^{(2)}_{\alpha_2}), g\right)_{G_1\otimes G_2}\right|^2 \\
&\quad = \left|\sum_{\alpha_1,\alpha_2=0}^{\infty}\sum_{\beta_1,\beta_2=0}^{\infty} f_{(\alpha_1,\alpha_2)}\bar{g}_{(\beta_1,\beta_2)}(A_1e^{(1)}_{\alpha_1}, l^{(1)}_{\beta_1})_{G_1}(A_2e^{(2)}_{\alpha_2}, l^{(2)}_{\beta_2})_{G_2}\right|^2 \\
&\quad = \left|\sum_{\alpha_1,\alpha_2=0}^{\infty}\sum_{\beta_1,\beta_2=0}^{\infty} (A_1e^{(1)}_{\alpha_1}, l^{(1)}_{\beta_1})_{G_1} f_{(\alpha_1,\alpha_2)}\overline{(A_2^*l^{(2)}_{\beta_2}, e^{(2)}_{\alpha_2})_{H_2}g_{(\beta_1,\beta_2)}}\right|^2 \\
&\quad = \left|\sum_{\alpha_2=0}^{\infty}\sum_{\beta_1=0}^{\infty}(A_1f(\alpha_2), l^{(1)}_{\beta_1})_{G_1}\overline{(A_2^*g(\beta_1), e^{(2)}_{\alpha_2})_{H_2}}\right|^2 \\
&\quad \le \sum_{\alpha_2=0}^{\infty}\sum_{\beta_1=0}^{\infty}|(A_1f(\alpha_2), l^{(1)}_{\beta_1})_{G_1}|^2\sum_{\alpha_2=0}^{\infty}\sum_{\beta_1=0}^{\infty}|(A_2^*g(\beta_1), e^{(2)}_{\alpha_2})_{H_2}|^2 \\
&\quad = \sum_{\alpha_2=0}^{\infty}\|A_1\,f(\alpha_2)\|^2_{G_1}\sum_{\beta_1=0}^{\infty}\|A_2^*g(\beta_1)\|^2_{H_2} \\
&\quad \le \|A_1\|^2\,\|A_2^*\|^2\sum_{\alpha_2=0}^{\infty}\|f(\alpha_2)\|^2_{H_1}\sum_{\beta_1=0}^{\infty}\|g(\beta_1)\|^2_{G_2} \\
&\quad = \|A_1\|^2\,\|A_2\|^2\sum_{\alpha_1,\alpha_2=0}^{\infty}|f_{(\alpha_1,\alpha_2)}|^2\sum_{\beta_1,\beta_2=0}^{\infty}|g_{(\beta_1,\beta_2)}|^2.
\end{aligned}$$

This inequality obviously implies the weak convergence of the series

$$\sum_{\alpha_1,\alpha_2=0}^{\infty} f_{(\alpha_1,\alpha_2)}(A_1e^{(1)}_{\alpha_1})\otimes(A_2e^{(2)}_{\alpha_2})$$

in $G_1\otimes G_2$ now for arbitrary $f\in H_1\otimes H_2$, along with an upper estimate of its norm (in $G_1\otimes G_2$) in terms of $\|A_1\|\,\|A_2\|\times\|f\|_{H_1\otimes H_2}$. Thus, the operator $A_1\otimes A_2\colon H_1\otimes H_2\to G_1\otimes G_2$ is unambiguously defined by (2.5), it is bounded, and its norm is at most $\|A_1\|\,\|A_2\|$.

On the other hand, (2.5) and (2.4) give us that

$$\begin{aligned}
\|(A_1\otimes A_2)(f^{(1)}\otimes f^{(2)})\|_{G_1\otimes G_2} &= \|(A_1f^{(1)})\otimes(A_2f^{(2)})\|_{G_1\otimes G_2} \\
&= \|A_1f^{(1)}\|_{G_1}\|A_2f^{(2)}\|_{G_2} \\
&\qquad (f^{(1)}\in H_1,\ f^{(2)}\in H_2).
\end{aligned}$$

Choosing unit vectors $f^{(1)}$ and $f^{(2)}$ properly, we get that the last product differs little from $\|A_1\|\,\|A_2\|$. Therefore, the inequality $\|A_1\otimes A_2\|<\|A_1\|\,\|A_2\|$ is impossible, i.e., the relation (2.6) is proved for $p=2$. ■

We underscore that (2.5) leads easily to the equality

$$\left(\bigotimes_{n=1}^{p} A_n\right)(f^{(1)}\otimes\cdots\otimes f^{(p)}) = (A_1 f^{(1)})\otimes\cdots\otimes(A_p f^{(p)})$$

$$(f^{(1)}\in H_1,\ldots,f^{(p)}\in H_p). \quad (2.7)$$

This equality uniquely determines the operator $\bigotimes_1^p A_n$. It is also clear that the mapping

$$\mathop{\times}_{n=1}^{p}(\mathcal{L}(H_n\to G_n)) \ni (A_1,\ldots,A_p)\mapsto \bigotimes_{n=1}^{p} A_n \in \mathcal{L}\left(\bigotimes_{n=1}^{p} H_n \mapsto \bigotimes_{n=1}^{p} G_n\right)$$

is linear in each variable. Note that with the help of (2.7) it is easy to get that for $A_n\in\mathcal{L}(H_n\to G_n)$ and $B_n\in\mathcal{L}(G_n\to F_n)$ $(n=1,\ldots,p)$

$$\left(\bigotimes_{n=1}^{p} B_n\right)\left(\bigotimes_{n=1}^{p} A_n\right) = \bigotimes_{n=1}^{p}(B_nA_n) \quad\text{and}\quad \left(\bigotimes_{n=1}^{p} A_n\right)^* = \bigotimes_{n=1}^{p} A_n^*. \quad (2.8)$$

Suppose that each of the operators A_n in Theorem 2.1 is a Hilbert-Schmidt operator. Then $\bigotimes_1^p A_n$ is also a Hilbert-Schmidt operator, and

$$\left|\bigotimes_{n=1}^{p} A_n\right| = \prod_{n=1}^{p} |A_n|. \quad (2.9)$$

Indeed, according to (2.5) and (2.4),

$$\left|\bigotimes_{n=1}^{p} A_n\right|^2 = \sum_{\alpha\in N^p}\left\|\left(\bigotimes_{n=1}^{p} A_n\right)e_\alpha\right\|^2_{\otimes_1^p G_n}$$

$$= \sum_{\alpha\in N^p}\|(A_1e^{(1)}_{\alpha_1})\otimes\cdots\otimes(A_pe^{(p)}_{\alpha_p})\|^2_{\otimes_1^p G_n} = \sum_{\alpha\in N^p}\prod_{n=1}^{p}\|A_ne^{(n)}_{\alpha_n}\|^2_{G_n}$$

$$= \prod_{n=1}^{p}\left(\sum_{\alpha_n=0}^{\infty}\|A_ne^{(n)}_{\alpha_n}\|^2_{G_n}\right) = \prod_{n=1}^{p}|A_n|^2 < \infty. \quad \blacksquare$$

In what follows we shall also have to consider tensor products $\bigotimes_1^p A_n$ of unbounded operators $A_n\colon H_n\supseteq\mathfrak{D}(A_n)\to G_n$ $(n=1,\ldots,p)$. Here we note only that they are unambiguously defined by (2.5) on the domain $\mathfrak{D}(\bigotimes_1^p A_n) = \text{a.}\ \bigotimes_1^p(\mathfrak{D}(A_n))$.

2.3. ***Tensor products of infinitely many Hilbert spaces.*** We proceed to the definition in the simplest case of the tensor product of the Hilbert spaces $(H_n)_1^\infty$, i.e., a separable subspace of the complete von Neumann tensor product.

Let $(H_n)_1^\infty$ be a sequence of Hilbert spaces, and $e=(e^{(n)})_1^\infty$ $(e^{(n)}\in H_n)$ a fixed sequence of unit vectors in the H_n. In each H_n consider an orthonormal basis $(e_j^{(n)})_{j=0}^\infty$ such that $e_0^{(n)} = e^{(n)}$, and form the formal product

$$e_\alpha = e_{(\alpha_1,\alpha_2,\dots)} = e_{\alpha_1}^{(1)} \otimes e_{\alpha_2}^{(2)} \otimes \cdots \qquad (\alpha = (\alpha_n)_{n=1}^\infty), \tag{2.10}$$

where $\alpha_1, \alpha_2, \dots = 0, 1, \dots$, and $\alpha_{n+1} = \alpha_{n+2} = \cdots = 0$ beginning with some index n depending on α. Denote by A the countable set of all multi-indices α of the described form. The infinite tensor product $\mathcal{H}_e = \bigotimes_{n=1;e}^\infty H_n$ of the Hilbert spaces H_n with stabilizing sequence e is defined to be the Hilbert space spanned by the basis $(e_\alpha)_{\alpha\in \mathrm{A}}$, which is regarded as orthonormal by definition. Thus, the vectors in $\mathcal{H}_e$ have the form

$$f = \sum_{\alpha\in\mathrm{A}} f_\alpha e_\alpha \qquad (f_\alpha \in \mathbb{C}^1), \tag{2.11}$$

where

$$\sum_{\alpha\in\mathrm{A}} |f_\alpha|^2 = \|f\|_{\mathcal{H}_e}^2 < \infty, \quad (f, g)_{\mathcal{H}_e} = \sum_{\alpha\in\mathrm{A}} f_\alpha \bar{g}_\alpha. \tag{2.12}$$

Of course, we get the same tensor product when we vary the bases $(e_j^{(n)})_{j=1}^\infty$ (such that $e_0^{(n)} = e^{(n)}$).

It will often be convenient for us to represent the set A as a union of disjoint sets, each consisting of "finite" sequences. Namely, for an $\alpha \in \mathrm{A}$ let $\nu(\alpha)$ denote the minimal $m = 1, 2, \dots$ such that $\alpha_{m+1} = \alpha_{m+2} = \cdots = 0$. Let $\mathrm{A}_n = \{\alpha \in \mathrm{A} | \nu(\alpha) = n\}$ $(n = 1, 2, \dots)$. Obviously, $\mathrm{A}_n \cap \mathrm{A}_m = \varnothing$ $(n \neq m)$ and $\mathrm{A} = \bigcup_1^\infty \mathrm{A}_n$. This is the required partition. Correspondingly, summations in formulas of the type (2.11) and (2.12) are often represented in the form

$$\begin{aligned}\sum_{\alpha\in\mathrm{A}} c_\alpha &= \sum_{n=1}^\infty \sum_{\alpha\in\mathrm{A}_n} c_\alpha \\ &= c_{(0,0,\dots)} + \sum_{n=1}^\infty \left\{ \sum_{\alpha_1,\dots,\alpha_{n-1}=0}^\infty \sum_{\alpha_n=1}^\infty c_{(\alpha_1,\alpha_2,\dots,\alpha_n,0,0,\dots)} \right\}.\end{aligned}$$

On the other hand, $\bigcup_1^m \mathrm{A}_n$ consists of all the multi-indices

$$\alpha = (\alpha_1, \dots, \alpha_m, 0, 0, \dots),$$

where $\alpha_1, \dots, \alpha_m = 0, 1, \dots$. Therefore, a summation over A can also be written in the form

$$\sum_{\alpha\in\mathrm{A}} c_\alpha = \lim_{m\to\infty} \sum_{n=1}^m \sum_{\alpha\in\mathrm{A}_n} c_\alpha = \lim_{m\to\infty} \sum_{\alpha_1,\dots,\alpha_m=0}^\infty c_{(\alpha_1,\dots,\alpha_m,0,0,\dots)}.$$

The space $\mathcal{H}_e$ thus constructed is called a separable subspace of the complete von Neumann tensor product of the spaces H_n.

Consider the product $\mathcal{H}_e = \bigotimes_{n=1;e}^{\infty} H_n$. Let $f^{(n)} = \sum_{j=0}^{\infty} f_j^{(n)} e_j^{(n)} \in H_n$ $(n = 1, \ldots, p)$, and, as in (2.30), define

$$\begin{aligned} f &= f^{(1)} \otimes \cdots \otimes f^{(p)} \otimes e^{(p+1)} \otimes e^{(p+2)} \otimes \cdots \\ &= \sum_{\alpha_1, \ldots, \alpha_p = 0}^{\infty} f_{\alpha_1}^{(1)} \cdots f_{\alpha_p}^{(p)} e_{\alpha_1}^{(1)} \otimes \cdots \otimes e_{\alpha_p}^{(p)} \otimes e^{(p+1)} \otimes e^{(p+2)} \otimes \cdots . \end{aligned} \tag{2.13}$$

It is easy to see that the coefficients f_α in (2.13) satisfy condition (2.12); therefore, as in the case of a finite tensor product, $f \in \mathcal{H}_e$ and its norm can be computed by the same formula (2.4).

As in the case of finite tensor products, we also use the following natural notation. Let M_n be a linear subset of H_n and $g^{(n)} \in H_n$ $(n = 1, 2, \ldots)$. We set

$$\begin{aligned} &\left(\text{a.} \bigotimes_{n=1}^{p} M_n\right) \otimes g^{(p+1)} \otimes \cdots \otimes g^{(q)} \otimes e^{(q+1)} \otimes e^{(q+2)} \otimes \cdots \\ &\quad = \text{l.s.}\{f^{(1)} \otimes \cdots \otimes f^{(p)} \otimes g^{(p+1)} \otimes \cdots \otimes g^{(q)} \\ &\qquad\qquad \otimes e^{(q+1)} \otimes e^{(q+2)} \otimes \cdots \mid f^{(1)} \in M_1, \ldots, f^{(p)} \in M_p\}; \end{aligned}$$

$$\begin{aligned} \text{a.} \bigotimes_{n=1;e}^{\infty} M_n = \text{l.s.}\{f^{(1)} \otimes \cdots \otimes f^{(p)} \otimes e^{(p+1)} \otimes e^{(p+2)} \otimes \cdots \mid f^{(1)} \in M_1, \\ \ldots, f^{(p)} \in M_p;\ p = 1, 2, \ldots\}. \end{aligned}$$

If instead of the l.s. we take the c.l.s. in $\mathcal{H}_e$ in these formulas, then we get subspaces of $\mathcal{H}_e$. Denote them by $(\bigotimes_1^p M_n) \otimes g^{(p+1)} \otimes \cdots \otimes g^{(q)} \otimes e^{(q+1)} \otimes e^{(q+2)} \otimes \cdots$ and $\bigotimes_{n=1;e}^{\infty} M_n$, respectively. Observe also that by associating the vectors (2.3) and (2.13) with each other and then extending this correspondence by linearity and continuity, we get a natural isomorphism between the Hilbert spaces $\bigotimes_1^p H_n$ and $(\bigotimes_1^p H_n) \otimes e^{(p+1)} \otimes e^{(p+2)} \otimes \cdots$. The image of a vector $f_p \in \bigotimes_1^p H_n$ in $(\bigotimes_1^p H_n) \otimes e^{(p+1)} \otimes e^{(p+2)} \otimes \cdots$ is denoted by $f_p \otimes e^{(p+1)} \otimes e^{(p+2)} \otimes \cdots$.

EXAMPLES. 1. Let $H_n = L_2(R_n, d\mu_n(x_n))$, where R_n is a space with a probability measure μ_n, i.e., $\mu_n(R_n) = 1$ $(n = 1, 2, \ldots)$. Then

$$\bigotimes_{n=1;e}^{\infty} H_n = L_2\left(\bigtimes_{n=1}^{\infty} R_n, d(\mu_1 \times \mu_2 \times \cdots)(x_1, x_2, \ldots)\right)$$

for $e^{(n)}(x_n) = 1$ $(x_n \in R_n,\ n = 1, 2, \ldots)$. Indeed, associate the cylindrical function $e_{\alpha_1}^{(1)}(x_1) \cdots e_{\alpha_p}^{(p)}(x_p)$ with a vector

$$e_\alpha = e_{\alpha_1}^{(1)}(x_1) \otimes \cdots \otimes e_{\alpha_p}^{(p)}(x_p) \otimes e^{(p+1)} \otimes e^{(p+2)} \otimes \cdots \in \bigotimes_{n=1;e}^{\infty} H_n$$

of the form (2.10). Considering the example in §2.1, we conclude that this correspondence generates an isomorphism between the space $(\bigotimes_1^p H_n) \otimes e^{(p+1)} \otimes e^{(p+2)} \otimes \cdots$ and a subspace of the space

$$L_2\left(\bigtimes_{n=1}^{\infty} R_n, d(\mu_1 \times \mu_2 \times \cdots)(x_1, x_2, \ldots)\right)$$

consisting of the cylindrical functions, namely, the space

$$L_2\left(\bigtimes_{n=1}^{\infty} R_n, d(\mu_1 \times \cdots \times \mu_p)(x_1, \ldots, x_p)\right).$$

It is easy to arrive at the required equality by considering that the cylindrical functions are dense in $L_2(\bigtimes_1^\infty R_n, d(\mu_1 \times \mu_2 \times \cdots)(x_1, x_2, \ldots))$ and that $(\bigotimes_1^p H_n) \otimes e^{(p+1)} \otimes e^{(p+2)} \otimes \cdots$ is dense in $\bigotimes_{n=1;e}^\infty H_n$ (when $p = 1, 2, \ldots$ varies).

2. Let $H_n = l_2$ $(n = 1, 2, \ldots)$, where l_2 is the space of square-summable sequences $f = (f_j)_0^\infty$. The stabilizing sequence is chosen as follows: $e^{(n)} = (1, 0, 0, \ldots)$ for each n. Then

$$\bigotimes_{n=1;e}^{\infty} H_n = \left\{ f = (f_\alpha)_{\alpha \in \mathrm{A}} \,\middle|\, \sum_{\alpha \in \mathrm{A}} |f_\alpha|^2 < \infty \right\}.$$

Here A, as before, is the set of all finite multi-indices.

2.4. ***Imbeddings of tensor products.*** We show that if we have two collections of Hilbert spaces, and each space in the first collection is imbedded in a corresponding space in the second collection, then the imbedding property is preserved for the tensor products of the spaces. First consider the case when there are finitely many spaces.

Let $(H_n)_1^p$ and $(G_n)_1^p$ be two sequences of Hilbert spaces, where $H_n \subseteq G_n$ and the imbedding operator $O_n \colon H_n \to G_n$ is continuous $(n = 1, \ldots, p)$. Then $\bigotimes_1^p H_n \subseteq \bigotimes_1^p G_n$, and the corresponding imbedding operator is $O = \bigotimes_1^p O_n$. This assertion is derivable immediately from Theorem 1.2 by considering the mapping

$$\bigotimes_{n=1}^{p} H_n \ni \sum_{\alpha \in N^p} f_\alpha e_{\alpha_1}^{(1)} \otimes \cdots \otimes e_{\alpha_p}^{p}$$
$$\mapsto \sum_{\alpha \in N^p} f_\alpha (O_1 e_{\alpha_1}^{(1)}) \otimes \cdots \otimes (O_p e_{\alpha_p}^{(p)}) \in \bigotimes_{n=1}^{p} G_n.$$

If each imbedding $H_n \to G_n$ is quasinuclear, i.e., O_n is a Hilbert-Schmidt operator, then the same is true for the imbedding $\bigotimes_1^p H_n \to \bigotimes_1^p G_n$, as follows from (2.9).

We pass to the corresponding assertions for infinitely many Hilbert spaces.

THEOREM 2.2. *Let $(H_n)_1^\infty$ and $(G_n)_1^\infty$ be two sequences of Hilbert spaces, where $H_n \subseteq G_n$, the imbedding operator $O_n\colon H_n \to G_n$ is continuous ($n = 1, 2, \ldots$), and $\prod_1^\infty \|O_n\| = c < \infty$. If the sequence $e = (e^{(n)})_1^\infty$ ($e^{(n)} \in H_n$) of vectors is such that $\|e^{(n)}\|_{H_n} = \|e^{(n)}\|_{G_n} = 1$, then $\bigotimes_{n=1;e}^\infty H_n \subseteq \bigotimes_{n=1;e}^\infty G_n$, and this imbedding is continuous and has norm at most c.*

PROOF. Let $(e_j^{(n)})_{j=0}^\infty$ ($e_0^{(n)} = e^{(n)}$) be an orthonormal basis in H_n ($n = 1, 2, \ldots$). We form from it the basis vectors e_α according to (2.10) and consider a vector $f \in \bigotimes_{n=1;e}^\infty H_n$; it has the representation (2.11). From f we construct the vector $f_p = \sum_{\nu(\alpha)\le p} f_\alpha e_\alpha$; it has the form $h_p \otimes e^{(p+1)} \otimes e^{(p+2)} \otimes \cdots$, where $h_p \in \bigotimes_1^p H_n$. But $\bigotimes_1^p H_n \subseteq \bigotimes_1^p G_n$ by the foregoing, so $h_p \in \bigotimes_1^p G_n$ and, consequently, $f_p \in \bigotimes_1^p G_n$ and it has the form $\sum_{\nu(\alpha)\le p} g_\alpha l_\alpha$, with the l_α constructed for the spaces G_n in a way similar to that for the e_α, and with $g_\alpha \in \mathbb{C}^1$ ($p = 1, 2, \ldots$). Obviously,

$$\|f_p\|^2_{\bigotimes_{n=1;e}^\infty G_n} = \sum_{\nu(\alpha)\le p} |g_\alpha|^2 = \|h_p\|^2_{\bigotimes_1^p G_n}.$$

Since the norm of a tensor product of finitely many operators is equal to the product of their norms (see 2.6)) and since $\|O_n\| \ge 1$ (as follows from the relation $\|e^{(n)}\|_{H_n} = \|e^{(n)}\|_{G_n}$), we have that

$$\begin{aligned}\|f_p\|_{\bigotimes_{n=1;e}^\infty G_n} = \|h_p\|^2_{\bigotimes_1^p G_n} &\le \left(\prod_{n=1}^p \|O_n\|^2\right) \|h_p\|^2_{\bigotimes_1^p H_n} \\ &\le \left(\prod_{n=1}^\infty \|O_n\|\right)^2 \|f_p\|^2_{\bigotimes_{n=1;e}^\infty H_n}.\end{aligned} \tag{2.14}$$

It remains to show that $(f_p)_1^\infty$ is a Cauchy sequence in $\bigotimes_{n=1;e}^\infty G_n$. This can be seen at once by writing an inequality of the form (2.14) for $f_p - f_q$ and using the fact that $\sum_{\alpha\in \mathrm{A}} |f_\alpha|^2 < \infty$. The theorem is proved. ■

It can be shown that the norm of the imbedding operator is equal to c.

THEOREM 2.3. *Assume that the conditions of Theorem 2.2 hold, and that the imbeddings O_n are quasinuclear. (It is not assumed beforehand that the product $\prod_1^\infty \|O_n\|$ converges.) Then the imbedding of*

$$\bigotimes_{n=1;e}^\infty H_n \quad in \quad \bigotimes_{n=1;e}^\infty G_n$$

exists and is quasinuclear if and only if

$$\sum_{n=1}^\infty (|O_n|^2 - 1) < \infty, \tag{2.15}$$

i.e., the infinite product $\prod_1^\infty |O_n|$ *converges. The number to which this product converges is equal to the Hilbert norm of the imbedding operator*

$$\mathcal{O}\colon \bigotimes_{n=1;e}^{\infty} H_n \to \bigotimes_{n=1;e}^{\infty} G_n.$$

We note that $|O_n| \geq \|O_n\| \geq 1$, and so the series (2.15) has nonnegative terms, and convergence of $\prod_1^\infty |O_n|$ implies convergence of $\prod_1^\infty \|O_n\|$.

PROOF. Let e_α be vectors of the form (2.10). Then

$$\begin{aligned}
|\mathcal{O}|^2 &= \sum_{\alpha\in\mathrm{A}} \|e_\alpha\|^2_{\otimes_{n=1;e}^\infty G_n} = \lim_{m\to\infty} \sum_{n=1}^{m} \sum_{\alpha\in\mathrm{A}_n} \|e_\alpha\|^2_{\otimes_{n-1;e}^\infty G_n} \\
&= \lim_{m\to\infty} \sum_{\alpha_1,\dots,\alpha_m=0}^{\infty} \|e_{(\alpha_1,\dots,\alpha_m,0,0,\dots)}\|^2_{\otimes_{n=1;e}^\infty G_n} \\
&= \lim_{m\to\infty} \sum_{\alpha_1,\dots,\alpha_m=0}^{\infty} \left(\prod_{k=1}^{\infty} \|e_{\alpha_k}\|^2_{G_k}\right) \\
&= \lim_{m\to\infty} \prod_{k=1}^{m} |O_k|^2 = \prod_{k=1}^{\infty} |O_k|^2.
\end{aligned}$$

(We have used the fact that $\|e^{(m+1)}\|_{G_{m+1}} = \|e^{(m+2)}\|_{G_{m+2}} = \cdots = 1$.) The theorem follows from the fact that convergence of the latter infinite product is equivalent to convergence of the series (2.15). ■

Note that the Hilbert norm of the operator $\mathcal{O}$ also satisfies the formula

$$\begin{aligned}
|\mathcal{O}|^2 &= \sum_{\alpha\in\mathrm{A}} \|e_\alpha\|^2_{\otimes_{n=1;e}^\infty G_n} = \sum_{n=1}^{\infty} \sum_{\alpha\in\mathrm{A}_n} \|e_\alpha\|^2_{\otimes_{n=1;e}^\infty G_n} \\
&= \|e_{(0,0,\dots)}\|^2_{\otimes_{n=1;e}^\infty G_n} \\
&\quad + \sum_{n=1}^{\infty} \left\{ \sum_{\alpha_1,\dots,\alpha_{n-1}=0}^{\infty} \sum_{\alpha_n=1}^{\infty} \|e^{(1)}_{\alpha_1} \otimes \cdots \otimes e^{(n)}_{\alpha_n} \otimes e^{(n+1)} \right. \\
&\qquad\qquad \left. \otimes\, e^{(n+2)} \otimes \cdots \|^2_{\otimes_{n=1;e}^\infty G_n} \right\} \\
&= 1 + \sum_{n=1}^{\infty} \left\{ \sum_{\alpha_1,\dots,\alpha_{n-1}=0}^{\infty} \sum_{\alpha_n=1}^{\infty} \left(\prod_{k=1}^{n} \|e^{(k)}_{\alpha_k}\|^2_{G_k} \right) \right\} \\
&= 1 + \sum_{n=1}^{\infty} \left(\prod_{k=1}^{n-1} |O_k|^2 \right) (|O_n|^2 - 1).
\end{aligned}$$

It is also easy to prove directly that the last sum is equal to the infinite product: by passing to partial sums it is simple to establish the more general relation

$$1+\sum_{n=1}^{\infty}\left(\prod_{k=1}^{n-1}c_k\right)(c_n-1)=\prod_{k=1}^{\infty}c_k \qquad (c_k\geq 1).$$

It is essential to underscore that satisfaction of the condition (2.15) and thus quasinuclearity of the imbedding $\bigotimes_{n=1;e}^{\infty}H_n\to\bigotimes_{n=1;e}^{\infty}G_n$ can always be achieved by properly renorming the spaces H_n $(n=1,2,\ldots)$; namely, we have the following simple general lemma.

LEMMA 2.1. *Let H and G be two Hilbert spaces such that $H\subseteq G$, with the imbedding $O\colon H\to G$ quasinuclear, and suppose that there exists a vector $e\in H$ such that $\|e\|_H=\|e\|_G=1$. Then for any $\varepsilon>0$ there is an inner product on H equivalent to the original one and such that, as before, $\|e\|_H=1$, while $|O|^2$ becomes less than $1+\varepsilon$.*

PROOF. Let $(e_j)_0^{\infty}$ be an orthonormal basis in H such that $e_0=e$. Then

$$|O|^2=\sum_{j=0}^{\infty}\|e_j\|_G^2=1+\sum_{j=1}^{\infty}\|e_j\|_G^2.$$

We introduce a new inner product $(\cdot,\cdot)_{H\#}$ by setting

$$(f,g)_{H\#}=(f,e)_H\overline{(g,e)}_H+c(Pf,Pg)_H \qquad (f,g\in H),$$

where P is the projection on the orthogonal complement of e in H, and $c>0$ is a constant. It is equivalent to the original inner product $(\cdot,\cdot)_H$, and the vectors $(e_j^{\#})_0^{\infty}$ with $e_0^{\#}=e_0$ and $e_j^{\#}=c^{-1/2}e_j$ $(j=1,2,\ldots)$, form an orthonormal basis in it. By choosing a sufficiently large c it is possible to satisfy the inequality $\sum_1^{\infty}\|e_j^{\#}\|_G^2<\varepsilon$. Now $|O|^2<1+\varepsilon$. ■

2.5. *Tensor products of chains.* Consider $p=2,3,\ldots$ chains of the form (1.5):

$$H_{-,n}\supseteq H_{0,n}\supseteq H_{+,n} \qquad (n=1,\ldots,p). \tag{2.16}$$

According to §2.4,

$$\bigotimes_{n=1}^{p}H_{-,n}\supseteq\bigotimes_{n=1}^{p}H_{0,n}\supseteq\bigotimes_{n=1}^{p}H_{+,n}. \tag{2.17}$$

It is not hard to see that $\bigotimes_1^p H_{-,n}$ *can be understood as the negative space with respect to the zero space* $\bigotimes_1^p H_{0,n}$ *and the positive space* $\bigotimes_1^p H_{+,n}$, *i.e.*, (2.17) *is a chain.* Indeed, let G_- be the negative space with respect to the zero space $\bigotimes_1^p H_{0,n}$ and the positive space $\bigotimes_1^p H_{+,n}$. (The fact that these spaces can be taken as a zero space and a positive space follows from §2.1 and (2.6).) Let O and I be the corresponding operators connected with the chain $G_-\supseteq\bigotimes_1^p H_{0,n}\supseteq\bigotimes_1^p H_{+,n}$, and let O_n and I_n be the same operators connected with (2.16). Then $O=\bigotimes_1^p O_n$, and the second of the formulas

(2.8) gives us that $I = O^* = \bigotimes_1^p O_n^* = \bigotimes_1^p I_n$. But G_- is the completion of $\bigotimes_1^p H_{0,n}$ with respect to the inner product

$$(f,g)_{G_-} = (If,g)_{\otimes_1^p H_{0,n}},$$

and $\bigotimes_1^p H_{-,n}$ obviously coincides with the completion of the same space with respect to

$$(f,g) = \left(\left(\bigotimes_{n=1}^{p} I_n\right) f, g\right)_{\otimes_1^p H_{0,n}}.$$

Since $I = \bigotimes_1^p I_n$, it follows that $G_- = \bigotimes_1^p H_{-,n}$. ■

This proof gives us for the corresponding operators that

$$I = \bigotimes_{n=1}^{p} I_n, \quad \mathbf{I} = \bigotimes_{n=1}^{p} \mathbf{I}_n, \quad J = \bigotimes_{n=1}^{p} J_n, \quad \mathbf{J} = \bigotimes_{n=1}^{p} \mathbf{J}_n. \tag{2.18}$$

Let us proceed to the assertion for an infinite number of chains.

LEMMA 2.2. *Consider the chain* (1.5). *Assume that there exists a vector* $e \in H_+$ *such that* $\|e\|_{H_+} = \|e\|_{H_0} = 1$. *Then* $Ie = e$ *and* $\|e\|_{H_-} = 1$.

PROOF. Denote by P the projection of H_+ onto the orthogonal complement of the unit vector e in this space. For any $u \in H_+$ we have that $\|u\|_{H_+}^2 = |(u,e)_{H_+}|^2 + \|Pu\|_{H_+}^2$. In particular,

$$\|Ie\|_{H_+}^2 |(Ie,e)_{H_+}|^2 + \|PIe\|_{H_+}^2 = |(e,e)_{H_0}|^2 + \|PIe\|_{H_+}^2 = 1 + \|PIe\|_{H_+}^2.$$

On the other hand, since $\|I\| = 1$, it follows that $\|Ie\|_{H_+}^2 \le \|e\|_{H_0}^2 = 1$. Therefore, $PIe = 0$, i.e., $Ie = (Ie,e)_{H_+} e = e$. Further,

$$\|e\|_{H_-}^2 = (Ie,e)_{H_0} = (e,e)_{H_0} = 1. \quad ■$$

Suppose that an infinite sequence of chains (2.16) is given ($n = 1, 2, \ldots$). We assume that there exists a sequence $e = (e^{(n)})_1^\infty$ of vectors $e^{(n)} \in H_{+,n}$, such that $\|e^{(n)}\|_{H_{+,n}} = \|e^{(n)}\|_{H_{0,n}} = 1$. Then also $\|e^{(n)}\|_{H_{-,n}} = 1$, by Lemma 2.1. Thus, the sequence $e = (e^{(n)})_1^\infty$ can be taken as a stabilizing sequence for three infinite products: the positive, zero, and negative products in (2.16). We have the following theorem, which is analogous to the assertion in §1.9.

THEOREM 2.4. *The sequence of spaces*

$$\mathcal{H}_{-,e} = \bigotimes_{n=1;e}^{\infty} H_{-,n} \supseteq \underset{\underset{\displaystyle\mathcal{H}_{0,e}}{\|}}{\bigoplus_{n=1;e}^{\infty} H_{0,n}} \supseteq \bigotimes_{n=1;e}^{\infty} H_{+,n} = \mathcal{H}_{+,e} \tag{2.19}$$

is a chain.

PROOF. According to Theorem 2.2, $\mathcal{H}_{-,e} \supseteq \mathcal{H}_{0,e} \supseteq \mathcal{H}_{+,e}$, $\mathcal{H}_{+,e}$ is dense in the other two spaces, and $\|u\|_{\mathcal{H}_{0,e}} \leq \|u\|_{\mathcal{H}_{+,e}}$ $(u \in \mathcal{H}_{+,e})$. Denote by G_- the negative space with respect to the zero space $\mathcal{H}_{0,e}$ and the positive space $\mathcal{H}_{+,e}$. Let I be the operator connected with the chain $G_- \supseteq \mathcal{H}_{0,e} \supseteq \mathcal{H}_{+,e}$. It suffices to prove that $(f,g)_{G_-} = (If, Ig)_{\mathcal{H}_{+,e}}$ is equal to $(f,g)_{\mathcal{H}_{-,e}}$, for $f, g \in \mathcal{H}_{0,e}$ or even for f, g of the form (2.13).

Let $(l_j^{(n)})_{j=0}^{\infty}$ be an orthogonal basis in the space $\mathcal{H}_{+,n}$ such that $l_0^{(n)} = e^{(n)}$ $(n = 1, 2, \ldots)$, and construct the corresponding vectors l_α $(\alpha \in \mathrm{A})$. For $u \in \mathcal{H}_{+,e}$ we have that $u_\alpha = (u, l_\alpha)_{\mathcal{H}_{+,e}}$ $(\alpha \in \mathrm{A})$. In particular,

$$(If)_\alpha = (If, l_\alpha)_{\mathcal{H}_{+,e}} = (f, l_\alpha)_{\mathcal{H}_{0,e}} \qquad (f \in \mathcal{H}_{0,e};\ \alpha \in \mathrm{A}). \tag{2.20}$$

Take $f = f^{(1)} \otimes \cdots \otimes f^{(p)} \otimes e^{(p+1)} \otimes e^{(p+2)} \otimes \cdots$ $(p = 1, 2, \ldots)$ in (2.20), where $f^{(k)} \in H_{0,k}$. With the notation $\nu(\alpha) = n$ we have for $n \leq p$ that

$$\begin{aligned}(If)_\alpha &= \prod_{j=1}^{n} (f^{(j)}, l_{\alpha_j}^{(j)})_{H_{0,j}} \prod_{j=n+1}^{p} (f^{(j)}, e^{(j)})_{H_{0,j}} \\ &= \prod_{j=1}^{n} (f^{(j)}, \mathbf{I}_j^{-1} l_{\alpha_j}^{(j)})_{H_{-,j}} \prod_{j=n+1}^{p} (f^{(j)}, \mathbf{I}_j^{-1} e^{(j)})_{H_{-,j}} \\ &= \prod_{j=1}^{n} (f^{(j)}, \mathbf{I}_j^{-1} l_{\alpha_j}^{(j)})_{H_{-,j}} \prod_{j=n+1}^{p} (f^{(j)}, e^{(j)})_{H_{-,j}}. \end{aligned} \tag{2.21}$$

Here $\mathbf{I}_j$ denotes the isometry $\mathbf{I}$ connected with the chain (2.16). In (2.21) we have used the equality

$$\mathbf{I}_k^{-1} l_0^{(k)} = \mathbf{I}_k^{-1} e^{(k)} = e^{(k)} \qquad (k = 1, 2, \ldots), \tag{2.22}$$

which follows from the relation $Ie^{(k)} = e^{(k)}$ for the corresponding operators I_k; this relation is a consequence of Lemma 2.2.

For $n > p$ the expression for $(If)_\alpha$ contains the factor

$$(e^{(n)}, l_{\alpha_n}^{(n)})_{H_{0,n}} = (e^{(n)}, \mathbf{I}_n^{-1} l_{\alpha_n}^{(n)})_{H_{-,n}},$$

where $\alpha_n = 1, 2, \ldots$. But $\mathbf{I}_n$ implements an isometry between $H_{-,n}$ and $H_{+,n}$; therefore, $(\mathbf{I}_n^{-1} l_j^{(n)})_{j=0}^{\infty}$ forms an orthonormal basis in $H_{-,n}$ and, by (2.22), this factor is equal to zero. Thus, $(If)_\alpha = 0$ for $\nu(\alpha) > p$.

Let $g = g^{(1)} \otimes \cdots \otimes g^{(p)} \otimes e^{(p+1)} \otimes e^{(p+2)} \otimes \cdots$ $(p = 1, 2, \ldots)$, $g^{(k)} \in H_{0,k}$. Considering the foregoing and (2.21), we find that

$$\begin{aligned}
&(If, Ig)_{\mathcal{H}_{+,e}} \\
&= \sum_{\alpha \in \mathrm{A}} (If)_\alpha \overline{(Ig)_\alpha} = \sum_{n=1}^{p} \sum_{\alpha \in \mathrm{A}_n} (If)_\alpha \overline{(Ig)_\alpha} \\
&= \sum_{\alpha_1=0}^{\infty} (f^{(1)}, \mathbf{I}_1^{-1} l_{\alpha_1}^{(1)})_{H_{-,1}} \overline{(g^{(1)}, \mathbf{I}_1^{-1} l_{\alpha_1}^{(1)})_{H_{-,1}}} \\
&\quad \times \prod_{j=2}^{p} (f^{(j)}, e^{(j)})_{H_{-,j}} \overline{(g^{(j)}, e^{(j)})_{H_{-,j}}} \\
&\quad + \sum_{n=2}^{p} \left(\sum_{\alpha_1, \ldots, \alpha_{n-1}=0}^{\infty} \sum_{\alpha_n=1}^{\infty} \prod_{j=1}^{n} (f^{(j)}, \mathbf{I}_j^{-1} l_{\alpha_j}^{(j)})_{H_{-,j}} \overline{(g^{(j)}, \mathbf{I}_1^{-1} l_{\alpha_j}^{(j)})_{H_{-,j}}} \right. \\
&\qquad\qquad \left. \times \prod_{j=n+1}^{p} (f^{(j)}, e^{(j)})_{H_{-,j}} \overline{(g^{(j)}, e^{(j)})_{H_{-,j}}} \right) \\
&= (f^{(1)}, g^{(1)})_{H_{-,1}} f_0^{(2)} \cdots f_0^{(p)} \overline{g_0^{(2)}} \cdots \overline{g_0^{(p)}} \\
&\quad + \sum_{n=2}^{k} (f^{(1)}, g^{(1)})_{H_{-,1}} \cdots (f^{(n-1)}, g^{(n-1)})_{H_{-,n-1}} \\
&\qquad \times [(f^{(n)}, g^{(n)})_{H_{-,n}} - f_0^{(n)} \overline{g_0^{(n)}}] f_0^{(n+1)} \cdots f_0^{(p)} \overline{g_0^{(n+1)}} \cdots \overline{g_0^{(p)}} \\
&= (f, g)_{\mathcal{H}_{-,e}}.
\end{aligned}$$

(Here $f_0^{(k)} = (f^{(k)}, e^{(k)})_{H_{-,k}}$ and $g_0^{(k)} = (g^{(k)}, e^{(k)})_{H_{-,k}}$.) ■

2.6. ***Weighted tensor products of Hilbert spaces.*** Let $(H_n)_1^\infty$ be a sequence of Hilbert spaces, $e = (e^{(n)})_1^\infty$ $(e^{(n)} \in H_n)$ a fixed sequence of unit vectors, $\delta = (\delta_n)_1^\infty$ $(\delta_n > 0)$ a fixed sequence of numbers, and $(e_j^{(n)})_{j=0}^\infty$ an orthonormal basis in H_n such that $e_0^{(n)} = e^{(n)}$. We form the vectors e_α of type (2.10) and define the weighted infinite tensor product $\mathcal{H}_{e,\delta} = \bigotimes_{n=1;e,\delta}^\infty H_n$ of the Hilbert spaces H_n with stabilizing sequence e and weight δ to be the Hilbert space spanned by the basis $(\delta_{\nu(\alpha)}^{-1/2} e_\alpha)_{\alpha \in \mathrm{A}}$, which is regarded as orthonormal by definition. (Here $\nu(\alpha)$ and A are the same as in §2.3.) Consequently, the vectors in $\mathcal{H}_{e,\delta}$ have the form

$$f = \sum_{\alpha \in \mathrm{A}} f_\alpha e_\alpha \qquad (f_\alpha \in \mathbb{C}^1),$$

$$\sum_{\alpha \in \mathrm{A}} |f_\alpha|^2 \delta_{\nu(\alpha)} = \|f\|^2_{\mathcal{H}_{e,\delta}} < \infty, \qquad (f, g)_{\mathcal{H}_{e,\delta}} = \sum_{\alpha \in \mathrm{A}} f_\alpha \bar{g}_\alpha \delta_{\nu(\alpha)}. \tag{2.23}$$

Introducing A_n in the same way as in §2.3, we get

$$f=\sum_{n=1}^{\infty}\sum_{\alpha\in\mathrm{A}_n} f_\alpha e_\alpha, \qquad \sum_{n=1}^{\infty}\delta_n\left(\sum_{\alpha\in\mathrm{A}_n}|f_\alpha|^2\right)=\|f\|^2_{\mathcal{H}_{e,\delta}}<\infty,$$

$$(f,g)_{\mathcal{H}_{e,\delta}}=\sum_{n=1}^{\infty}\delta_n\left(\sum_{\alpha\in\mathrm{A}_n} f_\alpha\bar{g}_\alpha\right). \tag{2.24}$$

If $\delta=1$, i.e., $\delta_n=1$ ($n=1,2,\dots$), then $\mathcal{H}_{e,1}=\mathcal{H}_e=\bigotimes_{n=1;e}^{\infty}H_n=\bigotimes_{n=1;e,1}^{\infty}H_n$.

As in the case $\delta=1$, vectors $f=f^{(1)}\otimes\cdots\otimes f^{(p)}\otimes e^{(p+1)}\otimes e^{(p+2)}\otimes\cdots$ can be defined by (2.13); it is easy to see that they belong to $\mathcal{H}_{e,\delta}$ for any weight δ.

We note that the construction of the weighted tensor product can be somewhat broadened in the following natural direction. Let $(H_n)_1^\infty$ be a sequence of Hilbert spaces, $e=(e^{(n)})_1^\infty$ ($e^{(n)}\in H_n$) a stabilizing sequence, and $d=(d_\alpha)_{\alpha\in\mathrm{A}}$ a sequence of positive numbers. In the usual way we construct the vectors e_α of the form (2.10) from bases in the spaces H_n. The weighted infinite tensor product $\mathcal{H}_{e,d}=\bigotimes_{n=1;e,d}^{\infty}H_n$ of the Hilbert spaces H_n with stabilizing sequence e and weight d is defined as the Hilbert space spanned by the basis $(d_\alpha^{-1/2}e_\alpha)_{\alpha\in\mathrm{A}}$, which is regarded as orthonormal by definition.

We now show that a renorming of the spaces H_n (see Lemma 2.1) can be reduced to a properly chosen weighting of the tensor product.

Accordingly, suppose that in each H_n ($n=1,2,\dots$) we introduce the new inner product

$$(f^{(n)},g^{(n)})_{H_n^\#}=(f^{(n)},e^{(n)})_{H_n}\overline{(g^{(n)},e^{(n)})_{H_n}}+c_n(P_nf^{(n)},P_ng^{(n)})_{H_n}$$
$$(f^{(n)},g^{(n)}\in H_n).$$

Here $e^{(n)}\in H_n$ is the vector in the stabilizing sequence, P_n is the projection onto the orthogonal complement of $e^{(n)}$ in H_n, and $c_n>0$ is an arbitrary constant. The tensor product of the renormed spaces is denoted by $\mathcal{H}_e^\#$. An orthonormal basis in $\mathcal{H}_e^\#$ is formed by the vectors

$$e_\alpha^\#=e_{\alpha_1}^{\#(1)}\otimes\cdots\otimes e_{\alpha_p}^{\#(p)}\otimes e^{(p+1)}\otimes e^{(p+2)}\otimes\cdots,$$

where $(e_j^{\#(n)})_{j=0}^\infty$ is an orthonormal basis in H_n in the new scalar product. We have that

$$\|e_\alpha^\#\|_{\mathcal{H}_e}=\prod_{j=1}^{p}c_j^{-1/2}, \quad \text{i.e.,} \quad \left\|e_\alpha^\#\prod_{j=1}^{p}c_j^{1/2}\right\|_{\mathcal{H}_e}=1;$$

thus $\mathcal{H}_e^\#=\mathcal{H}_{e,\delta}$, where $\delta=(\delta_n)_1^\infty$, $\delta_n=(c_1\cdots c_n)^{-1}$.

The converse is obvious: if some weighted infinite tensor product $\mathcal{H}_{e,\delta}$ with weight $\delta = (\delta_n)_1^\infty$ has been constructed, then it can be understood as the tensor product of the renormed spaces without a weighting, where the renorming constants are chosen as follows: $c_1 = 1/\delta_1$ and $c_n = \delta_{n-1}/\delta_n$ $(n = 2, 3, \ldots)$. (However, if a sequence $d = (d_\alpha)_{\alpha \in \mathrm{A}}$ appears as the weight sequence, then in general $\mathcal{H}_{e,d}$ cannot now be understood as an unweighted tensor product of the spaces, renormed by means of Lemma 2.1.)

Let us dwell on imbeddings of weighted tensor products. A simple modification of the computations after the proof of Theorem 2.3 gives us the following assertion. *Suppose that* $(H_n)_1^\infty$ *and* $(G_n)_1^\infty$ *are two sequences of Hilbert spaces with* $H_n \subseteq G_n$*, the imbedding operators* $O_n \colon H_n \to G_n$ *are quasinuclear* $(n = 1, 2, \ldots)$*, and* $\delta = (\delta_n)_1^\infty$ *is a weight. Then the imbedding of* $\bigotimes_{n=1;e,\delta}^\infty H_n$ *in* $\bigotimes_{n=1;e,1}^\infty G_n$*, exists and is quasinuclear if and only if*

$$1 + \sum_{n=1}^\infty \frac{1}{\delta_n} \left(\prod_{k=1}^{n-1} |O_k|^2 \right) (|O_n|^2 - 1) < \infty. \tag{2.25}$$

(The left-hand side of this inequality is equal to the square of the Hilbert norm of the imbedding operator $\mathcal{O} \colon \bigotimes_{n=1;e,\delta}^\infty H_n \to \bigotimes_{n=1;e,1}^\infty G_n$.)

Indeed, let e_α be vectors of the form (2.10). We have

$$\begin{aligned}
|\mathcal{O}|^2 &= \sum_{\alpha \in \mathrm{A}} \|\delta_{\nu(\alpha)}^{-1/2} e_\alpha\|^2_{\otimes_{n=1;e,1}^\infty G_n} \\
&= \|e^{(1)} \otimes e^{(2)} \otimes \cdots\|^2_{\otimes_{n=1;e,1}^\infty G_n} \\
&\quad + \sum_{n=1}^\infty \frac{1}{\delta_n} \Bigg\{ \sum_{\alpha_1, \ldots, \alpha_{n-1}=0}^\infty \sum_{\alpha_n=1}^\infty \|e_{\alpha_1}^{(1)} \otimes \cdots \otimes e_{\alpha_n}^{(n)} \otimes e^{(n+1)} \\
&\qquad\qquad \otimes e^{(n+2)} \otimes \cdots \|_{\otimes_{n=1;e,1}^\infty G_n} \Bigg\} \\
&= 1 + \sum_{n=1}^\infty \frac{1}{\delta_n} \left\{ \sum_{\alpha_1, \ldots, \alpha_{n-1}=0}^\infty \sum_{\alpha_n=1}^\infty \prod_{k=1}^n \|e_{\alpha_k}^{(k)}\|^2_{G_k} \right\} \\
&= 1 + \sum_{n=1}^\infty \frac{1}{\delta_n} \left(\prod_{k=1}^{n-1} |O_k|^2 \right) (|O_n|^2 - 1),
\end{aligned}$$

which yields the assertion. ■

In the case $\delta = 1$ we get a condition equivalent to (2.15).

REMARK 1. It follows immediately from (2.25) that the imbedding

$$\bigotimes_{n=1;e,\delta}^\infty H_n \to \bigotimes_{n=1;e,1}^\infty G_n$$

can always be made quasinuclear by proper choice of the weight δ (under the condition that all the imbeddings $H_n \to G_n$ are quasinuclear). What is more, if the imbeddings $H_n \to G_n$ $(n = 1, 2, \ldots)$ are quasinuclear and η is a weight, then it is always possible to find a weight δ such that the imbedding $\bigotimes_{n=1;e,\delta}^{\infty} H_n \to \bigotimes_{n=1;e,\eta}^{\infty} G_n$ is quasinuclear. To prove this, interpret $\bigotimes_{n=1;e,\eta}^{\infty} G_n$ as the unweighted tensor product $\bigotimes_{n=1;e,1}^{\infty} G_n^{\#}$, where the $G_n^{\#}$ are the spaces G_n, suitably renormed, and then use the assertion already established.

Suppose now that an infinite sequence of chains (2.16) is given, where $n = 1, 2, \ldots$, and there exists a common stabilizing sequence $e = (e^{(n)})_1^{\infty}$ $(e^{(n)} \in H_{+,n})$ with $\|e^{(n)}\|_{H_{+,n}} = \|e^{(n)}\|_{H_{0,n}} = 1$. Then $\|e^{(n)}\|_{H_{-,n}} = 1$ by Lemma 2.2. Thus, the sequence $e = (e^{(n)})_1^{\infty}$ can be taken as a stabilizing sequence for the three infinite products of the positive, zero, and negative spaces in (2.16). We have the following result, which generalizes Theorem 2.4 to weighted tensor products.

THEOREM 2.5. *If $\delta = (\delta_n)_1^{\infty}$ $(\delta_n \geq 1)$ is a weight, then*

$$\mathcal{H}_{-,e,\delta^{-1}} = \bigotimes_{n=1;e,\delta^{-1}}^{\infty} H_{-,n} \supseteq \bigotimes_{n=1;e,1}^{\infty} H_{0,n} \supseteq \bigotimes_{n=1;e,\delta}^{\infty} H_{+,n} = \mathcal{H}_{+,e,\delta} \quad (2.26)$$

$$\bigotimes_{n=1;e,1}^{\infty} H_{0,n} = \mathcal{H}_{0,e,1}$$

$(\delta^{-1} = (\delta_n^{-1}))$ is a chain.

LEMMA 2.3. *Theorem 2.2 is preserved if the tensor product $\bigotimes_{n=1;e}^{\infty} H_n$ is replaced by the weighted tensor product $\bigotimes_{n=1;e,\delta}^{\infty} H_n$, where $\delta = (\delta_n)_1^{\infty}$ $(\delta_n \geq 1)$.*

PROOF. This goes just as the proof of Theorem 2.2, except that it is necessary to rewrite (2.14) as follows, introducing the factors $\delta_n \geq 1$:

$$\begin{aligned}
\|f_p\|^2_{\bigotimes_{n=1;e}^{\infty} G_n} &\leq \left(\prod_{n=1}^{p} \|O_n\|\right)^2 \|h_p\|^2_{\bigotimes_{n=1}^{\infty} H_n} = \left(\prod_{n=1}^{p} \|O_n\|\right)^2 \sum_{\nu(\alpha) \leq p} |f_\alpha|^2 \\
&\leq \left(\prod_{n=1}^{p} \|O_n\|\right)^2 \sum_{\nu(\alpha) \leq p} |f_\alpha|^2 \, \delta_{\nu(\alpha)} \\
&\leq \left(\prod_{n=1}^{\infty} \|O_n\|\right)^2 \|f_p\|^2_{\bigotimes_{n=1;e,\delta}^{\infty} H_n}. \quad \blacksquare
\end{aligned}$$

The proof of Theorem 2.5 is analogous to that of Theorem 2.4, except that in the beginning we use not Theorem 2.2 but the given result (the space $\bigotimes_{n=1;e}^{\infty} H_{+,n}$ being replaced by $\bigotimes_{n=1;e,\delta}^{\infty} H_{+,n}$). In our case $u \in \mathcal{H}_{+,e,\delta}$; thus

$u_\alpha = \delta_{\nu(\alpha)}^{-1}(u, l_\alpha)_{\mathcal{H}_{+,e,\delta}}$, and consequently, instead of (2.20), we get

$$(If)_\alpha = \delta_{\nu(\alpha)}^{-1}(If, l_\alpha)_{\mathcal{H}_{+,e,\delta}} = \delta_{\nu(\alpha)}^{-1}(f, l_\alpha)_{\mathcal{H}_{0,e,1}} \qquad (f \in \mathcal{H}_{0,e,1};\ \alpha \in \mathrm{A}).$$

Carrying out the computation (2.21) with its help, we obtain the previous answer, multiplied by δ_n^{-1}. This implies that

$$(If, Ig)_{\mathcal{H}_{+,e,\delta}} = (f, g)_{\mathcal{H}_{-,e,\delta^{-1}}}. \quad \blacksquare$$

2.7. *Tensor products of infinitely many operators.* It is possible to develop a theory of infinite tensor products of operators that is similar to the construction of finite tensor products presented in §2.2. However, we shall not do this in the general case, but confine ourselves to certain special constructions needed in what follows.

Suppose that $(H_n)_1^\infty$ and $(G_n)_1^\infty$ are two sequences of Hilbert spaces, $e = (e^{(n)})_1^\infty$ $(e^{(n)} \in H_n,\ \|e^{(n)}\|_{H_n} = 1)$ and $l = (l^{(n)})_1^\infty$ $(l^{(n)} \in G_n,\ \|l^{(n)}\|_{G_n} = 1)$ are sequences of unit vectors, and $\delta = (\delta_n)_1^\infty$ $(\delta_n > 0)$ and $\eta = (\eta_n)_1^\infty$ $(\eta_n > 0)$ are sequences of weights; we can form the weighted tensor products

$$\bigotimes_{n=1;e,\delta}^{\infty} H_n \quad \text{and} \quad \bigotimes_{n=1;l,\eta}^{\infty} G_n.$$

Further, let $(A_n)_1^\infty$ be a sequence of continuous operators $A_n\colon H_n \to G_n$. It will be assumed that $A_n e^{(n)} = l^{(n)}$ for sufficiently large n, i.e., the operators A_n carry the stabilizing sequence e (except possibly for a finite number of its terms) into the stabilizing sequence l.

In this case we define the product $A_1 \otimes A_2 \otimes \cdots = \bigotimes_1^\infty A_n$ to be the operator carrying each vector

$$f = f^{(1)} \otimes \cdots \otimes f^{(p)} \otimes e^{(p+1)} \otimes e^{(p+2)} \otimes \cdots$$

into the vector

$$\left(\bigotimes_{n=1}^{\infty} A_n\right) f = (A_1 f^{(1)}) \otimes \cdots \otimes (A_p f^{(p)}) \otimes (A_{p+1} e^{(p+1)}) \otimes (A_{p+2} e^{(p+2)}) \otimes \cdots$$

$$(f^{(n)} \in H_n,\ p = 1, 2, \ldots).$$

The last vector has the form

$$g^{(1)} \otimes \cdots \otimes g^{(q)} \otimes l^{(q+1)} \otimes l^{(q+2)} \otimes \cdots$$

and is thus in $\bigotimes_{n=1;l,\eta}^\infty G_n$. This mapping extends by linearity to a mapping a. $\bigotimes_{n=1;e}^\infty H_n \to$ a. $\bigotimes_{n=1;l}^\infty G_n$. The product $\bigotimes_1^\infty A_n$ is said to converge if the operators A_n are such that the mapping obtained can be extended by continuity to a continuous operator from $\bigotimes_{n=1;e,\delta}^\infty H_n$ to $\bigotimes_{n=1;l,\eta}^\infty G_n$. Of course, it is possible to formally write a relation of type (2.5):

$$\left(\bigotimes_{n=1}^{\infty} A_n\right)\left(\sum_{\alpha \in \mathrm{A}} f_\alpha e_{\alpha_1}^{(1)} \otimes e_{\alpha_2}^{(2)} \otimes \cdots\right) = \sum_{\alpha \in \mathrm{A}} f_\alpha (A_1 e_{\alpha_1}^{(1)}) \otimes (A_2 e_{\alpha_2}^{(2)}) \otimes \cdots.$$

We shall not explain conditions for the existence of $\bigotimes_1^\infty A_n$, noting only that under the conditions of Theorem 2.2 and Lemma 2.3 the imbedding operator

$$\mathcal{O}\colon \bigotimes_{n=1;e,\delta}^{\infty} H_n \to \bigotimes_{n=1;e,\eta}^{\infty} G_n$$

of the spaces occurring there can obviously be represented as the infinite tensor product of the imbedding operators, i.e., $\mathcal{O} = \bigotimes_1^\infty O_n$, and this product converges. This is valid also under the conditions (2.15) or (2.25), and the imbedding is then quasinuclear.

In what follows we need a more complicated construction of the product $\bigotimes_1^\infty A_n$ for operators A_n that do not carry all but finitely many terms of e into l. We give a special result in this direction.

Consider the sequence of chains (2.16) ($n = 1, 2, \ldots$) along with their product, the chain (2.26). Assume that the imbeddings $O_n\colon H_{+,n} \to H_{0,n}$ are quasinuclear, while the weight $\delta = (\delta_n)_1^\infty$ is such that

$$\delta_n \geq 1, \quad \sum_{n=1}^{\infty} \frac{1}{\delta_n} \left(\prod_{k=1}^{n} |O_k|^2 \right) < \infty. \tag{2.27}$$

Condition (2.27) implies (2.25), and hence the imbedding operator $\mathcal{O}\colon \mathcal{H}_{+,e,\delta} \to \mathcal{H}_{0,e,1}$ has the representation $\mathcal{O} = \bigotimes_1^\infty O_n$. By taking the adjoints of these imbedding operators in the sense of the zero spaces (see §1.5), it is easy to see that O_n^+ and $\mathcal{O}^+$ are the imbedding operators of $H_{0,n}$ in $H_{-,n}$ and $\mathcal{H}_{0,e,1}$ in $\mathcal{H}_{-,e,\delta^{-1}}$, respectively. Consequently, $\mathbf{O}_n = O_n^+ O_n$ and $\mathbf{O} = \mathcal{O}^+\mathcal{O}$ are the imbedding operators of $H_{+,n}$ in $H_{-,n}$ and $\mathcal{H}_{+,e,\delta}$ in $\mathcal{H}_{-,e,\delta^{-1}}$, respectively. The convergence of the product $\mathcal{O} = \bigotimes_1^\infty O_n$ easily yields the convergence of the products

$$\mathcal{O}^+ = \bigotimes_{n=1}^{\infty} O_n^+\colon \mathcal{H}_{0,e,1} \to \mathcal{H}_{-,e,\delta^{-1}} \quad \text{and} \quad \mathbf{O} = \bigotimes_{n=1}^{\infty} \mathbf{O}_n\colon \mathcal{H}_{+,e,\delta} \to \mathcal{H}_{-,e,\delta^{-1}},$$

and hence also of the product $A_1 \otimes \cdots \otimes A_p \otimes \mathbf{O}_{p+1} \otimes \mathbf{O}_{p+2} \otimes \cdots\colon H_{+,e,\delta} \to \mathcal{H}_{-,e,\delta^{-1}}$, where p is fixed and $A_n \in \mathcal{L}(H_{+,n} \to H_{-,n})$.

THEOREM 2.6. *Suppose that the imbeddings $O_n\colon H_{+,n} \to H_{0,n}$ are quasinuclear and the conditions (2.27) hold. Consider an arbitrary sequence of nonnegative operators $(A_n)_1^\infty$, $A_n\colon H_{+,n} \to H_{-,n}$, such that $\operatorname{Tr}(A_n) \leq 1$. Then the weak limit of the operators*

$$A_1 \otimes \cdots \otimes A_p \otimes \mathbf{O}_{p+1} \otimes \mathbf{O}_{p+2} \otimes \cdots\colon \mathcal{H}_{+,e,\delta} \to \mathcal{H}_{-,e,\delta^{-1}}, \tag{2.28}$$

as $p \to \infty$ exists and is a nonnegative operator with finite trace not exceeding the sum in (2.27). This operator is denoted by $\bigotimes_1^\infty A_n$.

PROOF. We denote by $A^{(p)}$ the operator (2.28) and prove that the limit $\lim_{p\to\infty}(A^{(p)}e_\alpha, e_\beta)_{\mathcal{H}_{0,e,1}}$ ($\alpha, \beta \in \mathrm{A}$) exists, where the e_α are constructed from orthonormal bases $(e_j^{(n)})_{j=0}^\infty$ $(e_0^{(n)} = e^{(n)})$ in the spaces $H_{+,n}$. If $m = \max\{\nu(\alpha), \nu(\beta)\}$, then for $p > m$

$$(A^{(p)}e_\alpha, e_\beta)_{\mathcal{H}_{0,e,1}} = \left(\prod_{n=1}^m (A_n e_{\alpha_n}^{(n)}, e_{\beta_n}^{(n)})_{H_{0,n}}\right)\left(\prod_{n=m+1}^p (A_n e^{(n)}, e^{(n)})_{H_{0,n}}\right). \tag{2.29}$$

(Here we have used the equality $(\mathbf{O}_n e^{(n)}, e^{(n)})_{H_{0,n}} = \|e^{(n)}\|^2_{H_{0,n}} = 1$.) Since $0 \le (A_n e^{(n)}, e^{(n)})_{H_{0,n}} \le 1$, the second product in (2.29) has a limit as $p \to \infty$, possibly equal to zero. This proves the existence of the desired limit.

Thus, $\lim_{p\to\infty}(A^{(p)}u, v)_{\mathcal{H}_{0,e,1}}$ exists for u, v in a dense subset of $\mathcal{H}_{+,e,\delta}$. To prove the theorem it suffices to see that $A^{(p)} \ge 0$ and $\operatorname{Tr}(A^{(p)}) \le c$ $(p = 1, 2, \dots)$, where c denotes the sum in (2.27). (Recall that $\|A^{(p)}\| \le |A^{(p)}| \le \operatorname{Tr}(A^{(p)})$; see §1.5.)

Since $A^{(p)}$ is continuous, we need only prove it is nonnegative on the vectors $u = u_q \otimes e^{(q+1)} \otimes e^{(q+2)} \otimes \cdots \in \mathcal{H}_{+,e,\delta}$, where $u_q \in \bigotimes_1^q H_{+,n}$ $(q = 1, 2, \dots)$. In (2.28) let $\mathbf{O}_{p+1}, \mathbf{O}_{p+2}, \dots$ be denoted by $A_{p+1}, A_{p+2}, \dots$; then

$$(A^{(p)}u, u)_{\mathcal{H}_{0,e,1}} = \left(\left(\bigotimes_{n=1}^q A_n\right) u_q, u_q\right)_{\otimes_1^q H_{0,n}} \left(\prod_{n=q+1}^\infty (A_n e^{(n)}, e^{(n)})_{H_{0,n}}\right) \ge 0.$$

Here we have used the fact that the operators $\mathbf{O}_n\colon H_{+,n} \to H_{-,n}$ are nonnegative and that if $A_n\colon H_{+,n} \to H_{-,n}$ are nonnegative $(n = 1, \dots, q)$, then so is $\bigotimes_1^q A_n\colon \bigotimes_1^q H_{+,n} \to \bigotimes_1^q H_{-,n}$.

Let us establish the estimate $\operatorname{Tr}(A^{(p)}) \le c$. Using the above notation, we get that

$$\begin{aligned}
\operatorname{Tr}(A^{(p)}) &= \sum_{\alpha\in \mathrm{A}} (A^{(p)}(\delta_{\nu(\alpha)}^{-1/2} e_\alpha), \delta_{\nu(\alpha)}^{-1/2} e_\alpha)_{\mathcal{H}_{0,e,1}} \\
&= \sum_{n=1}^\infty \delta_n^{-1} \left(\sum_{\alpha\in \mathrm{A}_n} (A^{(p)} e_\alpha, e_\alpha)_{\mathcal{H}_{0,e,1}}\right) \\
&= \delta_1^{-1} \sum_{\alpha_1=0}^\infty (A_1 e_{\alpha_1}^{(1)}, e_{\alpha_1}^{(1)})_{H_{0,1}} \left(\prod_{k=2}^\infty (A_k e^{(k)}, e^{(k)})_{H_{0,k}}\right) \\
&\quad + \sum_{n=2}^\infty \delta_n^{-1} \left[\sum_{\alpha_1,\dots,\alpha_{n-1}=0}^\infty \sum_{\alpha_n=1}^\infty \left(\prod_{k=1}^n (A_k e_{\alpha_k}^{(k)}, e_{\alpha_k}^{(k)})_{H_{0,k}}\right)\right. \\
&\qquad \left.\times \left(\prod_{k=n+1}^\infty (A_k e^{(k)}, e^{(k)})_{H_{0,k}}\right)\right]
\end{aligned}$$

$$= \delta_1^{-1} \operatorname{Tr}(A_1) \left(\prod_{k=2}^{\infty} (A_k e^{(k)}, e^{(k)})_{H_{0,k}} \right)$$
$$+ \sum_{n=2}^{\infty} \delta_n^{-1} \left(\prod_{k=1}^{n-1} \operatorname{Tr}(A_k) \right) (\operatorname{Tr}(A_n) - (A_n e^{(n)}, e^{(n)})_{H_{0,n}})$$
$$\times \left(\prod_{k=n+1}^{\infty} (A_k e^{(k)}, e^{(k)})_{H_{0,k}} \right)$$
$$\leq \sum_{n=1}^{\infty} \delta_n^{-1} \left(\prod_{k=1}^{n} \operatorname{Tr}(A_k) \right) \left(\prod_{k=n+1}^{\infty} (A_k e^{(k)}, e^{(k)})_{H_{0,k}} \right). \tag{2.30}$$

It was used above that A_k is nonnegative, $(A_k e^{(k)}, e^{(k)})_{H_{0,k}} \leq \operatorname{Tr}(A_k) \leq 1$ for $k \leq p$, and $(A_k e^{(k)}, e^{(k)})_{H_{0,k}} = (\mathbf{O}_k e^{(k)}, e^{(k)})_{H_{0,k}} = 1$ for $k > p$. Further, $\operatorname{Tr}(A_k) \leq 1 \leq |O_k|^2$ for $k \leq p$, while for $k > p$

$$\operatorname{Tr}(A_k) = \operatorname{Tr}(\mathbf{O}_k) = \sum_{j=0}^{\infty} (\mathbf{O}_k e_j^{(k)}, e_j^{(k)})_{H_{0,k}} = \sum_{j=0}^{\infty} \|O_k e_j^{(k)}\|^2_{H_{0,k}} = |O_k|^2.$$

Therefore, the estimate (2.30) can be continued to

$$\operatorname{Tr}(A^{(p)} \leq \sum_{n=1}^{\infty} \delta_n^{-1} \left(\prod_{k=1}^{n} |O_k|^2 \right) = c. \quad \blacksquare$$

REMARK 1. Condition (2.27) can be weakened by carrying out the estimate (2.30) more accurately. We mention only that in the case $\delta = 1$ it can be replaced by inequality (2.15), and $\operatorname{Tr}(\bigotimes_1^\infty A_n) \leq \prod_1^\infty |O_k|^2$. To prove this it should be noted that the computation (2.30) now gives

$$\operatorname{Tr}(A^{(p)}) = \left(\prod_{k=1}^{p} \operatorname{Tr}(A_k) \right) \left(\prod_{k=p+1}^{\infty} |O_k|^2 \right),$$

which yields the assertion. At the end of subsection 10 below we give conditions ensuring that the operators (2.28) converge strongly.

2.8. ***The kernel theorem.*** We introduce the concept of a generalized kernel. Consider $p < \infty$ chains

$$H_{-,n} \supseteq H_{0,n} \supseteq H_{+,n} \qquad (n = 1, \ldots, p) \tag{2.31}$$

and form their tensor product, i.e., form the chain

$$\bigotimes_{n=1}^{p} H_{-,n} \supseteq \bigotimes_{n=1}^{p} H_{0,n} \supseteq \bigotimes_{n=1}^{p} H_{+,n}. \tag{2.32}$$

By analogy with kernels $F(x_1, \ldots, x_p)$, the elements $F, G, \ldots$ of the space $\bigotimes_1^p H_{0,n}$ can also be called (ordinary) kernels; elements $U, V, \ldots \in \bigotimes_1^p H_{+,n}$

and $A, B, \ldots \in \bigotimes_1^p H_{-,n}$ are called smooth and generalized kernels, respectively. (Here it is convenient to use capital letters.) We prove a kernel theorem which shows that every continuous multilinear form can be written in a natural way by means of a generalized kernel.

Accordingly, consider a continuous multilinear (more precisely, p-linear) form $F(f^{(1)}, \ldots, f^{(p)})$: a continuous function

$$\bigoplus_{n=1}^{p} H_{0,n} \ni (f^{(1)}, \ldots, f^{(p)}) \mapsto F(f^{(1)}, \ldots, f^{(p)}) \in \mathbb{C}^1,$$

which is linear in each variable $f^{(n)}$ when the others are fixed. *It is continuous if and only if for some $c < \infty$*

$$|F(f^{(1)}, \ldots, f^{(p)})| \le c \prod_{n=1}^{p} \|f^{(n)}\|_{H_{0,n}} \qquad (f^{(n)} \in H_{0,n}). \tag{2.33}$$

Indeed, it follows from the multilinearity of F that (2.33) implies its continuity at any point of $\bigoplus_1^p H_{0,n}$. Conversely, suppose that F is multilinear and continuous at 0. Assume that (2.33) does not hold. Then for any $m = 1, 2, \ldots$ there is a point $(f^{(1),m}, \ldots, f^{(p),m}) \in \bigoplus_1^p H_{0,n}$ such that

$$|F(f^{(1),m}, \ldots, f^{(p),m})| > m \prod_{n=1}^{p} \|f^{(n),m}\|_{H_{0,n}}.$$

This and the multilinearity give us that $|F(g^{(1),m}, \ldots, g^{(p),m})| > 1$, where

$$\begin{aligned}(g&^{(1),m}, \ldots, g^{(p),m}) \\ &= (m^{-1/p}\|f^{(1),m}\|_{H_{0,1}}^{-1} f^{(1),m}, \ldots, m^{-1/p}\|f^{(p),m}\|_{H_{0,p}}^{-1} f^{(p),m}) \to 0\end{aligned}$$

as $m \to \infty$. Passing to the limit as $m \to \infty$ in the last inequality and using the continuity of F at 0 and the fact that $F(0, \ldots 0) = 0$, we arrive at a contradiction. ■

Fix an orthonormal basis $(e_j^{(n)})_{j=0}^{\infty}$ in each space $H_{0,n}$. Let

$$f^{(n)} = \sum_{\alpha_n=0}^{\infty} f_{\alpha_n}^{(n)} e_{\alpha_n}^{(n)}$$

be the expansion of a vector $f^{(n)} \in H_{0,n}$ in this basis $(n = 1, \ldots, p)$, and let $\alpha = (\alpha_1, \ldots, \alpha_p) \in N^p$. The multilinearity and continuity give us that F can be represented as a convergent series in terms of its coordinates F_α and the coordinates of the vectors $f^{(n)}$:

$$F(f^{(1)}, \ldots, f^{(p)}) = \sum_{\alpha \in N^p} F_\alpha f_{\alpha_1}^{(1)} \cdots f_{\alpha_p}^{(p)},$$

$$F_\alpha = F(e_{\alpha_1}^{(1)}, \ldots, e_{\alpha_p}^{(p)}) \in \mathbb{C}^1. \tag{2.34}$$

The proof of the kernel theorem is based on the following two lemmas.

LEMMA 2.4. *Let* $\bigoplus_1^p H_{0,n} \ni (f^{(1)}, \dots, f^{(p)}) \mapsto F(f^{(1)}, \dots, f^{(p)})$ *be a continuous multilinear form, and let* $A_n\colon H_{0,n} \to H_{0,n}$ $(n = 2, \dots, p)$ *be Hilbert-Schmidt operators. Consider the continuous multilinear form*

$$\bigoplus_{n=1}^{p} H_{0,n} \ni (f^{(1)}, \dots, f^{(p)}) \mapsto G(f^{(1)}, f^{(2)}, \dots, f^{(p)}) = F(f^{(1)}, A_2 f^{(2)}, \dots, A_p f^{(p)}).$$

Then the coordinates G_α *of* G *satisfy*

$$\sum_{\alpha \in N^p} |G_\alpha|^2 < \infty. \tag{2.35}$$

Conversely, if $0 \neq A_n\colon H_{0,n} \to H_{0,n}$ $(n = 2, \dots, p)$ *are continuous and such that for any form* F *the coordinates of the form* G *satisfy the condition* (2.35), *then all the operators* A_n *are Hilbert-Schmidt operators.*

PROOF. Fix $f^{(2)}, \dots, f^{(p)}$; then $H_{0,1} \ni f^{(1)} \mapsto l(f^{(1)}) = F(f^{(1)}, \dots, f^{(p)})$ is a continuous linear functional on $H_{0,1}$ with norm at most $c \prod_2^p \|f^{(n)}\|_{H_{0,n}}$, by (2.33). Since $l(f^{(1)}) = (f^{(1)}, a)_{H_{0,1}}$, where the coordinates a_{α_1} of the vector $a \in H_{0,1}$ in the basis $(e_{\alpha_1}^{(1)})_{\alpha_1=0}^\infty$ have the form $\overline{F(e_{\alpha_1}^{(1)}, f^{(2)}, \dots, f^{(p)})}$, it follows that

$$\sum_{\alpha_1=0}^{\infty} |F(e_{\alpha_1}^{(1)}, f^{(2)}, \dots, f^{(p)})|^2 = \|l\|^2 \leq c^2 \prod_{n=2}^{p} \|f^{(n)}\|_{H_{0,n}}^2 \qquad (f^{(2)} \in H_{0,2}, \dots,\ f^{(p)} \in H_{0,p}).$$

By this estimate,

$$\begin{aligned}
\sum_{\alpha \in N^p} |G_\alpha|^2 &= \sum_{\alpha_1, \dots, \alpha_p = 0}^{\infty} |G(e_{\alpha_1}^{(1)}, \dots, e_{\alpha_p}^{(p)})|^2 \\
&= \sum_{\alpha_1, \alpha_2, \dots, \alpha_p = 0}^{\infty} |F(e_{\alpha_1}^{(1)}, A_2 e_{\alpha_2}^{(2)}, \dots, A_p e_{\alpha_p}^{(p)})|^2 \\
&\leq c^2 \sum_{\alpha_2, \dots, \alpha_p = 0}^{\infty} \left(\prod_{n=2}^{p} \|A_n e_{\alpha_n}^{(n)}\|_{H_{0,n}}^2 \right) = c^2 \prod_{n=2}^{p} |A_n|^2 < \infty.
\end{aligned}$$

Conversely, suppose that condition (2.35) holds for any continuous multilinear form F. We show that $|A_n| < \infty$ $(n = 2, \dots, p)$. To do this, note that if we introduce the matrix $(A_{n,\alpha_n\beta_n})_{\alpha_n,\beta_n=0}^\infty$ of A_n in the basis $(e_j^{(n)})_{j=0}^\infty$ (i.e., $A_{n,\alpha_n\beta_n} = (A_n e_{\beta_n}^{(n)}, e_{\alpha_n}^{(n)})_{H_{0,n}}$), then

$$G_\alpha = \sum_{\beta_2, \dots, \beta_p = 0}^{\infty} F_{(\alpha_1, \beta_2, \dots, \beta_p)} \left(\prod_{n=2}^{p} A_{n,\beta_n\alpha_n} \right). \tag{2.36}$$

Let the form F be constructed by taking the numbers

$$F_\alpha = \delta_{\alpha_2\gamma_2} \cdots \delta_{\alpha_{n-1}\gamma_{n-1}} \delta_{\alpha_n\alpha_1} \delta_{\alpha_{n+1}\gamma_{n+1}} \cdots \delta_{\alpha_p\gamma_p}$$

as its coordinates, where δ_{jk} is the Kronecker symbol, and $n = 2, \dots, p$ and $\gamma_2, \dots, \gamma_p = 0, 1, \dots$ are fixed. Substituting these values of F_α in (2.36), we find that

$$G_\alpha = A_{2,\gamma_2\alpha_2} \cdots A_{n-1,\gamma_{n-1}\alpha_{n-1}} A_{n,\alpha_1\alpha_n} A_{n+1,\gamma_{n+1}\alpha_{n+1}} \cdots A_{p,\gamma_p\alpha_p}.$$

It now follows from (2.35) that

$$\begin{aligned}
\sum_{\alpha\in N^p} |G_\alpha|^2 &= \left(\sum_{\alpha_2=0}^{\infty} |A_{2,\gamma_2\alpha_2}|^2\right) \cdots \left(\sum_{\alpha_{n-1}=0}^{\infty} |A_{n-1,\gamma_{n-1}\alpha_{n-1}}|^2\right) \\
&\quad \times \left(\sum_{\alpha_1,\alpha_n=0}^{\infty} |A_{n,\alpha_1\alpha_n}|^2\right) \left(\sum_{\alpha_{n+1}=0}^{\infty} |A_{n+1,\gamma_{n+1}\alpha_{n+1}}|^2\right) \\
&\quad \cdots \left(\sum_{\alpha_p=0}^{\infty} |A_{p,\gamma_p\alpha_p}|^2\right) \\
&= \|A_2^* e^{(2)}_{\gamma_2}\|^2_{H_{0,2}} \cdots \|A^*_{n-1} e^{(n-1)}_{\gamma_{n-1}}\|^2_{H_{0,n-1}} |A_n|^2 \\
&\quad \times \|A^*_{n+1} e^{(n+1)}_{\gamma_{n+1}}\|^2_{H_{0,n+1}} \cdots \|A^*_p e^{(p)}_{\gamma_p}\|^2_{H_{0,p}} \\
&< \infty.
\end{aligned} \tag{2.37}$$

Fix $m = 2, \dots, n-1, n+1, \dots, p$. There is always a $\gamma_m = 0, 1, \dots$ such that $A^*_m e^{(m)}_{\gamma_m} \neq 0$, because otherwise $A^*_m = 0$, which would contradict the condition that $A_m \neq 0$. Taking these values as γ_m in (2.37), we get that the factor in front of A_n^2 in (2.37) is nonzero; hence $|A_n| < \infty$. ∎

LEMMA 2.5. *Let*

$$\bigoplus_{n=1}^{p} H_{0,n} \ni (f^{(1)}, \dots, f^{(p)}) \mapsto G(f^{(1)}, \dots, f^{(p)})$$

be a continuous multilinear form. It can be represented as

$$G(f^{(1)}, \dots, f^{(p)}) = (f^{(1)} \otimes \cdots \otimes f^{(p)}, K)_{\otimes_1^p H_{0,n}},$$

where $K = \bigotimes_1^p H_{0,n}$, *if and only if condition* (2.35) *holds for its coordinates* G_α.

PROOF. Suppose that (2.35) holds. We introduce the vectors (2.1) and let $K = \sum_{\alpha\in N^p} \overline{G}_\alpha e_\alpha \in \bigotimes_1^p H_{0,n}$. Then clearly

$$(f^{(1)} \otimes \cdots \otimes f^{(p)}, K)_{\otimes_1^p H_{0,n}} = \sum_{\alpha\in N^p} f^{(1)}_{\alpha_1} \cdots f^{(p)}_{\alpha_p} G_\alpha = G(f^{(1)}, \dots, f^{(p)})$$

$$(f^{(n)} \in H_{0,n}).$$

Conversely, if the required representation exists, then

$$G_\alpha = G(e^{(1)}_{\alpha_1}, \dots, e^{(p)}_{\alpha_p}) = (e^{(1)}_{\alpha_1} \otimes \cdots \otimes e^{(p)}_{\alpha_p}, K)_{\otimes_1^p H_{0,n}} = \overline{K}_\alpha,$$

and (2.34) is satisfied in view of the inclusion $K \in \bigotimes_1^p H_{0,n}$. ■

THEOREM 2.7. *Let the chains* (2.31) *be constructed in such a way that the imbeddings* $H_{+,n} \to H_{0,n}$ $(n = 2, \dots, p)$ *are quasinuclear. Then for each continuous multilinear form*

$$\bigoplus_{n=1}^{p} H_{0,n} \ni (f^{(1)}, \dots, f^{(p)}) \mapsto F(f^{(1)}, \dots, f^{(p)})$$

it is possible to construct a generalized kernel $\Phi \in H_{0,1} \otimes H_{-,2} \otimes \cdots \otimes H_{-,p}$, *such that*

$$F(f^{(1)}, u^{(2)}, \dots, u^{(p)}) = (f^{(1)} \otimes u^{(2)} \otimes \cdots \otimes u^{(p)}, \Phi)_{\otimes_1^p H_{0,n}}$$
$$(f^{(1)} \in H_{0,1},\ u^{(2)} \in H_{+,2} \dots, u^{(p)} \in H_{+,p}). \quad (2.38)$$

This kernel is uniquely determined by F.

Conversely, if the representation (2.38) *is valid with* $\Phi \in H_{0,1} \otimes H_{-,2} \otimes \cdots \otimes H_{-p}$ *for each form* F *of the indicated type, then the imbeddings* $H_{+,n} \to H_{0,n}$ $(n = 2, \dots, p)$ *are quasinuclear.*

PROOF. Consider the operators O_n and J_n connected with the chain (2.31); for $n = 1$ assume that $H_{+,1} = H_{0,1} = H_{-,1}$. For $f^{(1)} \in H_{0,1}$ and $u^{(2)} \in H_{+,2}, \dots, u^{(p)} \in H_{+,p}$ we have

$$F(f^{(1)}, u^{(2)}, \dots, u^{(p)}) = F(f^{(1)}, O_2 J_2 J_2^{-1} u^{(2)}, \dots, O_p J_p J_p^{-1} u^{(p)})$$
$$= G(f^{(1)}, J_2^{-1} u^{(2)}, \dots, J_p^{-1} u^{(p)}), \quad (2.39)$$

where the form G is introduced by the equality

$$G(f^{(1)}, f^{(2)}, \dots, f^{(p)}) = F(f^{(1)}, O_2 J_2 f^{(2)}, \dots, O_p J_p f^{(p)})$$
$$(f^{(n)} \in H_{0,n};\ n = 1, \dots, p). \quad (2.40)$$

The operator $O_n \colon H_{+,n} \to H_{0,n}$ is Hilbert-Schmidt for $n = 2, \dots, p$, and so is $A_n = O_n J_n \colon H_{0,n} \to H_{0,n}$. Therefore, by Lemma 2.4, the coordinates G_α of G satisfy condition (2.35), and consequently, G has a representation

$$G(f^{(1)}, \dots, f^{(p)}) = (f^{(1)} \otimes \cdots \otimes f^{(p)}, K)_{\otimes_1^p H_{0,n}}$$

with $K \in \bigotimes_1^p H_{0,n}$, by Lemma 2.5.

It is now possible to extend (2.39) as follows (the operation $+$ below is taken with respect to the chain (2.32)):

$$F(f^{(1)}, u^{(2)}, \dots, u^{(p)}) = (f^{(1)} \otimes (J_2^{-1} u^{(2)}) \otimes \cdots \otimes (J_p^{-1} u^{(p)}), K)_{\otimes_1^p H_{0,n}}$$
$$= ((1 \otimes J_2^{-1} \otimes \cdots \otimes J_p^{-1})(f^{(1)} \otimes u^{(2)} \otimes \cdots \otimes u^{(p)}), K)_{\otimes_1^p H_{0,n}}$$
$$= (f^{(1)} \otimes u^{(2)} \otimes \cdots \otimes u^{(p)}, (1 \otimes J_2^{-1} \otimes \cdots \otimes J_p^{-1})^+ K)_{\otimes_1^p H_{0,n}}$$
$$= (f^{(1)} \otimes u^{(2)} \otimes \cdots \otimes u^{(p)}, (1 \otimes \mathbf{J}_2^{-1} \otimes \cdots \otimes \mathbf{J}_p^{-1}) K)_{\otimes_1^p H_{0,n}}. \quad (2.41)$$

Here we have used the relation $(\bigotimes_1^p J_n^{-1})^+ = \bigotimes_1^p \mathbf{J}_n^{-1}$, which follows from (1.15), (2.18), (2.8), and (1.18). To prove (2.38) it remains to set

$$\Phi = (1 \otimes \mathbf{J}_2^{-1} \otimes \cdots \otimes \mathbf{J}_p^{-1})K.$$

The uniqueness of the kernel with respect to the form F follows from the fact that the l.s. of the vectors $f^{(1)} \otimes u^{(2)} \otimes \cdots \otimes u^{(p)}$ ($f^{(1)} \in H_{0,1}$, $u^{(n)} \in H_{+,n}$) is dense in $H_{0,1} \otimes H_{+,2} \otimes \cdots \otimes H_{+,p}$.

We prove the last assertion of the theorem. If the representation (2.38) holds for F, then there is a representation similar to the preceding one for the form G introduced by (2.40):

$$\begin{aligned}
G(f^{(1)}, f^{(2)}, \ldots, f^{(p)}) &= F(f^{(1)}, O_2 J_2 f^{(2)}, \ldots, O_p J_p f^{(p)}) \\
&= F(f^{(1)}, J_2 f^{(2)}, \ldots, J_p f^{(p)}) \\
&= (f^{(1)} \otimes J_2 f^{(2)} \otimes \cdots \otimes J_p f^{(p)}, \Phi)_{\otimes_1^p H_{0,n}} \\
&= ((1 \otimes J_2 \otimes \cdots \otimes J_p)(f^{(1)} \otimes f^{(2)} \otimes \cdots \otimes f^{(p)}), \Phi)_{\otimes_1^p H_{0,n}} \\
&= (f^{(1)} \otimes \cdots \otimes f^{(p)}, (1 \otimes J_2 \otimes \cdots \otimes J_p)^+ \Phi)_{\otimes_1^p H_{0,n}} \\
&= (f^{(1)} \otimes \cdots \otimes f^{(p)}, (1 \otimes \mathbf{J}_2 \otimes \cdots \otimes \mathbf{J}_p) \Phi)_{\otimes_1^p H_{0,n}} \\
&\qquad (f^{(n)} \in H_{0,n};\ n = 1, \ldots, p).
\end{aligned}$$

But $(1 \otimes \mathbf{J}_2 \otimes \cdots \otimes \mathbf{J}_p)\Phi = K \in \bigotimes_1^p H_{0,n}$, and so Lemma 2.5 allows us to conclude that the coordinates G_α of G satisfy (2.35). Since F is arbitrary, while $A_n = O_n J_n \neq 0$, the second part of Lemma 2.4 leads us to the conclusion that $|A_n| < \infty$, i.e., $|O_n| < \infty$ ($n = 2, \ldots, p$). ∎

Of course, Theorem 2.7 is also valid when instead of the first chain in the sequence (2.31) we single out any other one. The kernel Φ is now different in general. (All these kernels coincide as functionals on $\bigotimes_1^p H_{+,n}$.) In somewhat rougher language, the result can be formulated in a more symmetric way: if each imbedding $H_{+,n} \to H_{0,n}$ ($n = 1, \ldots, p$) in (2.31) is quasinuclear, then for any continuous multilinear form $\bigoplus_1^p H_{0,n} \ni (f^{(1)}, \ldots, f^{(p)}) \mapsto F(f^{(1)}, \ldots, f^{(p)})$ it is possible to construct a unique kernel $\Phi \in \bigotimes_1^p H_{-,n}$ such that

$$F(u^{(1)}, \ldots, u^{(p)}) = (u^{(1)} \otimes \cdots \otimes u^{(p)}, \Phi)_{\otimes_1^p H_{0,n}} \qquad (u^{(n)} \in H_{+,n}). \tag{2.42}$$

2.9. *The case of bilinear forms. Positive-definite kernels.* We use Theorem 2.7 to establish an analogous assertion for bilinear forms. In a special case it can be interpreted as a proof that each bounded operator on L_2 can be written as an integral operator with a generalized kernel.

Consider the chain

$$H_- \supseteq H_0 \supseteq H_+. \tag{2.43}$$

Assume that there is an involution in H_+ which is also an involution in H_0. More precisely, an antilinear mapping $H_+ \ni u \mapsto u^* \in H_+$ is defined in H_+ such that $(u^*)^* = u$, $(u^*, v^*)_{H_+} = \overline{(u,v)_{H_+}}$ and $(u^*, v^*)_{H_0} = \overline{(u,v)_{H_0}}$ $(u, v \in H_+)$. It is then not hard to see that also

$$(u^*, v^*)_{H_-} = \overline{(u,v)_{H_-}} \qquad (u, v \in H_+). \tag{2.44}$$

Indeed, $(Iu^*, v)_{H_+} = (u^*, v)_{H_0} = \overline{(u, v^*)_{H_0}} = \overline{(Iu, v^*)_{H_+}} = ((Iu)^*, v)_{H_+}$ for $u, v \in H_+$, and so $Iu^* = (Iu)^*$. Therefore,

$$\begin{aligned}(u^*, v^*)_{H_-} &= (Iu^*, v^*)_{H_0} = ((Iu)^*, v^*)_{H_0} \\ &= \overline{(Iu, v)_{H_0}} = \overline{(u,v)_{H_-}}.\end{aligned}$$

The equality (2.44) shows that the involution * can be extended by continuity to an involution (for which the same notation is kept) defined in H_-: $H_- \ni \alpha \mapsto \alpha^* \in H_-$. It is also clear that the mapping $H_- \supseteq H_0 \ni f \mapsto f^* \in H_0$ is the involution in H_0. Obviously,

$$\mathbf{I}\alpha^* = (\mathbf{I}\alpha)^* \qquad (\alpha \in H_-). \tag{2.45}$$

If the involution described above has been introduced in the spaces of the chain (2.43), then we say that (2.43) is a chain with an involution *.

Let $B(f,g)$ be a bilinear form on H_0, i.e., a continuous function $H_0 \oplus H_0 \ni (f,g) \mapsto B(f,g) \in \mathbb{C}^1$, linear in the first variable and antilinear in the second. We have the following theorem.

THEOREM 2.8. *Suppose that the imbedding $H_+ \to H_0$ in the chain* (2.43) *is quasinuclear. Then for each continuous bilinear form $H_0 \oplus H_0 \ni (f,g) \mapsto B(f,g)$ it is possible to construct a generalized kernel* $\mathrm{B}_1 \in H_0 \otimes H_-$, *such that*

$$B(u,g) = (\mathrm{B}_1, g \otimes u^*)_{H_0 \otimes H_0} \qquad (u \in H_+,\ g \in H_0). \tag{2.46}$$

This kernel is uniquely determined by B.

Conversely, if each form B of the indicated type has a representation (2.46) *with* $\mathrm{B}_1 \in H_0 \otimes H_-$, *then the imbedding $H_+ \to H_0$ is quasinuclear.*

PROOF. We construct a continuous two-linear form F by setting $F(f,g) = \overline{B(g^*, f)}$ $(f, g \in H_0)$. According to Theorem 2.7, there exists a generalized kernel $\Phi \in H_0 \otimes H_-$ such that $F(f,u) = (f \otimes u, \Phi)_{H_0 \otimes H_0}$ $(f \in H_0,\ u \in H_+)$. Therefore, for $u \in H_+$ and $g \in H_0$,

$$B(u,g) = \overline{F(g,u^*)} = \overline{(g \otimes u^*, \Phi)_{H_0 \otimes H_0}} = (\Phi, g \otimes u^*)_{H_0 \otimes H_0},$$

i.e., the representation (2.46) is established with $B_1 = \Phi$. The denseness of the l.s. of the vectors $g \otimes u^*$ ($g \in H_0$, $u \in H_+$) in $H_0 \otimes H_+$ implies that B_1 is uniquely determined by B.

Conversely, suppose that the representation (2.46) is valid for each continuous bilinear form B. Consider an arbitrary continuous two-linear form $H_0 \oplus H_0 \ni (f, g) \mapsto F(f, g)$ and let $B(f, g) = \overline{F(g^*, f)}$. The form B is continuous and bilinear; therefore, $B(u, g) = (B_1, g \otimes u^*)_{H_0 \otimes H_0}$ ($u \in H_+$, $g \in H_0$) by assumption. But then

$$F(g, u) = \overline{B(u^*, g)} = \overline{(B_1, g \otimes u)}_{H_0 \otimes H_0} = (g \otimes u, B_1)_{H_0 \otimes H_0},$$

i.e., the representation (2.38) is valid with $\Phi = B_1$. By the second part of Theorem 2.7, the imbedding $H_+ \to H_0$ is quasinuclear. ■

This theorem could be proved independently of Theorem 2.7, and even somewhat more simply (in this connection see the proof of Theorem 3.3). Note that, as in the case of Theorem 2.7, B has the representation

$$B(f, v) = (B_2, v \otimes f^*)_{H_0 \otimes H_0} \qquad (f \in H_0, \ v \in H_+), \tag{2.47}$$

along with the representation (2.46), where $B_2 \in H_- \otimes H_0$ is uniquely determined by B.

It follows from (2.46) and (2.47) that

$$(B_1, u \otimes v)_{H_0 \otimes H_0} = B(v^*, u) = (B_2, u \otimes v)_{H_0 \otimes H_0}$$

for $u, v \in H_+$, i.e., the restrictions of the functionals B_1 and B_2 to $H_+ \otimes H_+$ coincide. Denoting this common restriction by B, we get that *if the imbedding $H_+ \to H_0$ is quasinuclear, then for each continuous bilinear form $H_0 \oplus H_0 \ni (f, g) \mapsto B(f, g)$ it is possible to construct a unique kernel* $B \in H_- \otimes H_-$ *such that*

$$B(u, v) = (B, v \otimes u^*)_{H_0 \otimes H_0} \qquad (u, v \in H_+). \tag{2.48}$$

Let $A\colon H_0 \to H_0$ be a continuous operator. We construct from it the continuous bilinear form $B(f, g) = (Af, g)_{H_0}$ ($f, g \in H_0$) and from that a kernel $B \in H_- \otimes H_-$ according to (2.48). This kernel is called the kernel of the operator A. Thus,

$$(Au, v)_{H_0} = (B, v \otimes u^*)_{H_0 \otimes H_0} \qquad (u, v \in H_+). \tag{2.49}$$

If $H_0 = L_2(R, d\mu(x))$, where R is a space with measure μ and A is a Hilbert-Schmidt operator on it, i.e., an operator of the form

$$(Af)(x) = \int_R K(x, y) f(y)\, d\mu(y),$$
$$K \in L_2(R \times R, d\mu(x) \otimes d\mu(y)) = H_0 \otimes H_0, \tag{2.50}$$

then

$$(Af, g)_{H_0} = \int_R \int_R K(x, y) f(y) \overline{g(x)}\, d\mu(x)\, d\mu(y)$$
$$= (K, g \otimes f^*)_{H_0 \otimes H_0} \qquad (f, g \in H_0), \tag{2.51}$$

where $f^*(x) = \overline{f(x)}$. Comparing (2.49) and (2.51), we conclude that $\mathrm{B} = K$. (It was the desire to obtain this equality that determined the arrangement of the variables in (2.46)–(2.48).) Formula (2.49) shows that an arbitrary bounded operator A on $L_2(R, d\mu(x))$ always has the representation (2.50) in a certain sense with $K = \mathrm{B}$, i.e., A can always be regarded as an "integral operator with generalized kernel".

It is now appropriate to introduce some new notation. Suppose that (2.43) is a chain with involution *. A generalized kernel $\mathrm{K} \in H_- \otimes H_-$ is said to be Hermitian or positive-definite, respectively, if

$$(\mathrm{K}, v \otimes u^*)_{H_0 \otimes H_0} = \overline{(\mathrm{K}, u \otimes v^*)}_{H_0 \otimes H_0} \qquad (u, v \in H_+)$$

or

$$(\mathrm{K}, u \otimes u^*)_{H_0 \otimes H_0} \geq 0 \qquad (u \in H_+);$$

in other words, the continuous bilinear form

$$H_+ \oplus H_+ \ni (u, v) \mapsto B(u, v) = (\mathrm{K}, v \otimes u^*)$$

should be Hermitian or positive-definite. It is clear that a positive-definite kernel is always Hermitian. If $\mathrm{K} \in H_0 \otimes H_0$ or $\mathrm{K} \in H_+ \otimes H_+$, then u and v can of course be replaced by $f, g \in H_0$ in the relations written above, or even by $\alpha, \beta \in H_-$. If B is the kernel of a continuous operator $A\colon H_0 \to H_0$ (see (2.49)), then it is Hermitian or positive-definite if and only if A is selfadjoint or nonnegative.

2.10. ***Complete von Neumann tensor products of infinitely many Hilbert spaces.*** Roughly speaking, this product is defined as the orthogonal sum of the spaces $\mathcal{H}_{e,1} = \bigotimes_{n=1;e,1}^{\infty} H_n$ over all possible stabilizing sequences e. For a precise definition we need some general considerations.

Fix the space $\mathcal{H}_{e,\delta} = \bigotimes_{n=1;e,\delta}^{\infty} H_n$, where e is a stabilizing sequence and δ is a weight. We define the vector $f^{(1)} \otimes f^{(2)} \otimes \cdots$ $(f^{(n)} \in H_n)$ as the weak limit (if it exists) in $\mathcal{H}_{e,\delta}$ of the vectors

$$f[p] = f^{(1)} \otimes \cdots \otimes f^{(p)} \otimes e^{(p+1)} \otimes e^{(p+2)} \otimes \cdots \tag{2.52}$$

as $p \to \infty$. Since l.s. $(e_\alpha)_{\alpha \in \mathrm{A}}$ is dense in $\mathcal{H}_{e,\delta}$, the weak limit of the vectors $f[p]$ exists if and only if 1) the norms $\|f[p]\|_{\mathcal{H}_{e,\delta}}$ are uniformly bounded with respect to $p = 1, 2, \ldots$, and 2) $\lim_{p\to\infty}(f[p], e_\alpha)_{\mathcal{H}_{e,\delta}}$ exists for each $\alpha \in \mathrm{A}$. These conditions can be given a more analytic form; however, we shall not do this in the general case, but confine ourselves to the case of the unweighted product $\mathcal{H}_{e,1}$.

It will be assumed in what follows that $f^{(n)} \neq 0$ $(n = 1, 2, \ldots)$, since otherwise the situation is trivial and $f^{(1)} \otimes f^{(2)} \otimes \cdots = 0$. The conditions 1) and 2) for $\mathcal{H}_{e,1}$ are as follows: the products $\prod_1^n \|f^{(n)}\|_{H_n}$ are uniformly bounded with respect to $p = 1, 2, \ldots$, and for each $\alpha_1, \ldots, \alpha_m = 0, 1, \ldots$ the product

$$\left(\prod_{n=1}^{m}(f^{(n)}, e_{\alpha_n}^{(n)})_{H_n}\right)\left(\prod_{n=m+1}^{\infty}(f^{(n)}, e^{(n)})_{H_n}\right)$$

converges to a finite number (possibly zero) for each $m = 1, 2, \ldots$. Since convergence of the last product is not affected by its first m factors, it can be said that *the weak limit of the vectors* (2.52) *exists in* $\mathcal{H}_{e,1}$ *as* $p \to \infty$, *i.e., the vector* $f^{(1)} \otimes f^{(2)} \otimes \cdots$ *exists, if and only if the products* $\prod_1^p \|f^{(n)}\|_{H_n}$ *are uniformly bounded with respect to* $p = 1, 2, \ldots$, *and all the products* $\prod_{n=q}^{\infty}(f^{(n)}, e^{(n)})_{H_n}$ $(q = 1, 2, \ldots)$ *converge to finite numbers.*

LEMMA 2.6. *The strong limit of the vectors* (2.52) *exists in* $\mathcal{H}_{e,1}$ *as* $p \to \infty$ *if and only if the products* $\prod_{n=1}^{\infty} \|f^{(n)}\|_{H_n}$ *and* $\prod_{n=q}^{\infty}(f^{(n)}, e^{(n)})_{H_n}$ $(q = 1, 2, \ldots)$ *converge to finite numbers, and* $\prod_1^{\infty} \|f^{(n)}\|_{H_n} = 0$ *when*

$$\prod_{n=q}^{\infty}(f^{(n)}, e^{(n)})_{H_n} = 0 \quad \textit{for each } q.$$

PROOF. Suppose that the conditions of the lemma are satisfied. Then the weak limit $\lim f[p] = g$ exists in view of the foregoing. We compute the norm $\|g\|_{\mathcal{H}_{e,1}}$. For $\alpha = (\alpha_1, \ldots, \alpha_m, 0, 0, \ldots)$ we have

$$\begin{aligned} g_\alpha = (g, e_\alpha)_{\mathcal{H}_{e,1}} &= \lim_{p\to\infty}(f[p], e_\alpha)_{\mathcal{H}_{e,1}} \\ &= \left(\prod_{n=1}^{m}(f^{(n)}, e_{\alpha_n}^{(n)})_{H_n}\right)\left(\prod_{n=m+1}^{\infty}(f^{(n)}, e^{(n)})_{H_n}\right). \end{aligned} \tag{2.53}$$

Therefore,

$$\begin{aligned} \|g\|^2_{\mathcal{H}_{e,1}} &= \sum_{\alpha\in A}|g_\alpha|^2 = \lim_{m\to\infty}\sum_{\alpha_1,\ldots,\alpha_m=0}^{\infty}|g_{(\alpha_1,\ldots,\alpha_m,0,0,\ldots)}|^2 \\ &= \lim_{m\to\infty}\left(\prod_{n=1}^{m}\|f^{(n)}\|^2_{H_n}\right)\left(\prod_{n=m+1}^{\infty}|(f^{(n)}, e^{(n)})_{H_n}|^2\right). \end{aligned} \tag{2.54}$$

If $\prod_{n=q}^{\infty}(f^{(n)}, e^{(n)})_{H_n} \neq 0$ for some $q = 1, 2, \ldots$, then

$$\prod_{n=m+1}^{\infty}(f^{(n)}, e^{(n)})_{H_n} \underset{m\to\infty}{\longrightarrow} 1,$$

and it follows from (2.54) that

$$\|g\|^2_{\mathcal{H}_{e,1}} = \lim_{m\to\infty}\prod_{n=1}^{m}\|f^{(n)}\|^2_{H_n} = \lim_{p\to\infty}\|f[p]\|^2_{\mathcal{H}_{e,1}}.$$

If $\prod_{n=q}^{\infty}(f^{(n)}, e^{(n)})_{H_n} = 0$ for each q, then it follows from (2.54) that $g = 0$. But in this case, by the condition in the lemma,

$$\lim_{p\to\infty} \|f[p]\|_{\mathcal{H}_{e,1}} = \prod_{n=1}^{\infty} \|f^{(n)}\|_{H_n} = 0.$$

Therefore we always have

$$\|g\|_{\mathcal{H}_{e,1}} = \lim_{p\to\infty} \|f[p]\|_{\mathcal{H}_{e,1}},$$

and, consequently, $f[p] \to g$ strongly as $p \to \infty$.

Conversely, suppose that $f[p] \to g$ strongly as $p \to \infty$. Then

$$\|g\|_{\mathcal{H}_{e,1}} = \lim_{p\to\infty} \|f[p]\|_{\mathcal{H}_{e,1}} = \lim_{p\to\infty} \prod_{n=1}^{p} \|f^{(n)}\|_{H_n},$$

which implies that the product $\prod_1^{\infty} \|f^{(n)}\|_{H_n}$ converges to the finite limit $\|g\|_{\mathcal{H}_{e,1}}$. Further,

$$\lim_{p\to\infty} (f[p], f[q-1])_{\mathcal{H}_{e,1}} = \lim_{p\to\infty} \left(\prod_{n=1}^{q-1} \|f^{(n)}\|^2\right)_{H_n} \left(\prod_{n=q}^{p} (f^{(n)}, e^{(n)})_{H_n}\right)$$

for each $q = 1, 2, \dots$; therefore, the product $\prod_{n=q}^{\infty}(f^{(n)}, e^{(n)})_{H_n}$ also converges to a finite limit due to the condition that $f^{(n)} \neq 0$, $n = 1, \dots, q-1$. If $\prod_{n=q}^{\infty}(f^{(n)}, e^{(n)})_{H_n} = 0$ for each q, then $g_\alpha = 0$ for each $\alpha \in \mathrm{A}$ by (2.53), and hence $g = 0$. But then

$$\|g\|_{\mathcal{H}_{e,1}} = \prod_{n=1}^{\infty} \|f^{(n)}\|_{H_n} = 0. \quad \blacksquare$$

Note that if the conditions in Lemma 2.6 hold for the two sequences $(f^{(n)})_1^{\infty}$ and $(g^{(n)})_1^{\infty}$ $(f^{(n)}, g^{(n)} \in H_n)$, then

$$(f^{(1)} \otimes f^{(2)} \otimes \cdots, g^{(1)} \otimes g^{(2)} \otimes \cdots)_{\mathcal{H}_{e,1}} = \prod_{n=1}^{\infty} (f^{(n)}, g^{(n)})_{H_n}. \tag{2.55}$$

This follows from the fact that the inner product on the left-hand side of (2.55) is equal to

$$\lim_{p\to\infty} (f[p], g[p])_{\mathcal{H}_{e,1}} = \lim_{p\to\infty} \prod_{n=1}^{p} (f^{(n)}, g^{(n)})_{H_n}.$$

COROLLARY. *If the $f^{(n)}$ in (2.52) are taken to be unit vectors, then the strong limit $\lim_{p\to\infty} f[p]$ exists if and only if for some $q = 1, 2, \dots$ the product $\prod_{n=q}^{\infty}(f^{(n)}, e^{(n)})_{H_n}$ converges to a finite nonzero number.*

LEMMA 2.7. *Suppose that $l = (l^{(n)})_1^{\infty}$ is a sequence of unit vectors $l^{(n)} \in H_n$ such that the following strong limit exists in the space $\mathcal{H}_{e,1}$:*

$$l^{(1)} \otimes l^{(2)} \otimes \cdots = \lim_{p\to\infty} l^{(1)} \otimes \cdots \otimes l^{(p)} \otimes e^{(p+1)} \otimes e^{(p+2)} \otimes \cdots . \tag{2.56}$$

Then the strong limits

$$\begin{aligned}\tilde{l}_\alpha &= l^{(1)}_{\alpha_1} \otimes l^{(2)}_{\alpha_2} \otimes \cdots \\ &= \lim_{p\to\infty} l^{(1)}_{\alpha_1} \otimes \cdots \otimes l^{(p)}_{\alpha_p} \otimes e^{(p+1)} \otimes e^{(p+2)} \otimes \cdots \qquad (\alpha \in \mathrm{A}) \quad (2.57)\end{aligned}$$

exist in the same space, where $(l^{(n)}_j)_{j=0}^\infty$ *is an orthonormal basis in* H_n *such that* $l^{(n)}_0 = l^{(n)}$ $(n = 1, 2, \ldots)$. *These limits form an orthonormal basis* $(\tilde{l}_\alpha)_{\alpha\in\mathrm{A}}$ *in* $\mathcal{H}_{e,1}$.

PROOF. By the corollary to Lemma 2.6, the limits (2.56) and (2.57) exist if the respective products $\prod_{n=q_1}^\infty (l^{(n)}, e^{(n)})_{H_n}$ and $\prod_{n=q_2}^\infty (l^{(n)}_{\alpha_n}, e^{(n)})_{H_n}$ converge to finite nonzero numbers for some $q_1, q_2 = 1, 2, \ldots$. For $\alpha \in \mathrm{A}$ we have $\alpha_n = 0$ and $l^{(n)}_{\alpha_n} = l^{(n)}$ for sufficiently large n, so these two products coincide when $q_1 = q_2$ is large enough; hence, the limit (2.57) exists if (2.56) exists, and conversely.

Further, it follows from (2.55) that the vectors $\tilde{l}_\alpha$ $(\alpha \in \mathrm{A})$ form an orthogonal system in $\mathcal{H}_{e,1}$. It is a basis.

To prove this, we fix a vector $e_\beta = e^{(1)}_{\beta_1} \otimes e^{(2)}_{\beta_2} \otimes \cdots$ $(\beta \in \mathrm{A})$ and compute the squared norm $\|Pe_\beta\|^2_{\mathcal{H}_{e,1}}$ of the projection of e_β on c.l.s. $((\tilde{l}_\alpha)_{\alpha\in\mathrm{A}})$. By (2.55),

$$\begin{aligned}\|Pe_\beta\|^2_{\mathcal{H}_{e,1}} &= \sum_{\alpha\in\mathrm{A}} |(e_\beta, \tilde{l}_\alpha)_{\mathcal{H}_{e,1}}|^2 = \sum_{\alpha\in\mathrm{A}} \left(\prod_{n=1}^\infty |(e^{(n)}_{\beta_n}, l^{(n)}_{\alpha_n})_{H_n}|^2 \right) \\ &= \lim_{m\to\infty} \sum_{\alpha_1,\ldots,\alpha_m=0}^\infty \left(\prod_{n=1}^m |(e^{(n)}_{\beta_n}, l^{(n)}_{\alpha_n})_{H_n}|^2 \prod_{n=m+1}^\infty |(e^{(n)}_{\beta_n}, l^{(n)})_{H_n}|^2 \right) \\ &= \lim_{m\to\infty} \left[\prod_{n=m+1}^\infty |(e^{(n)}_{\beta_n}, l^{(n)})_{H_n}|^2 \sum_{\alpha_1,\ldots,\alpha_m=0}^\infty \left(\prod_{n=1}^m |(e^{(n)}_{\beta_n}, l^{(n)}_{\alpha_n})_{H_n}|^2 \right) \right] \\ &= \lim_{m\to\infty} \left[\prod_{n=m+1}^\infty |(e^{(n)}_{\beta_n}, l^{(n)})_{H_n}|^2 \right].\end{aligned}$$

We have $\beta_n = 0$ and $e^{(n)}_{\beta_n} = e^{(n)}$ for n sufficiently large; therefore, the expression in the last square brackets is equal to $\prod_{n=m+1}^\infty |(e^{(n)}, l^{(n)})_{H_n}|^2$ for m sufficiently large, and consequently, the last limit is equal to 1. Accordingly, $\|Pe_\beta\|_{\mathcal{H}_{e,1}} = 1 = \|e_\beta\|_{\mathcal{H}_{e,1}}$, i.e., $e_\beta \in$ c.l.s.$((\tilde{l}_\alpha)_{\alpha\in\mathrm{A}})$ $(\beta \in \mathrm{A})$. Thus, c.l.s.$((\tilde{l}_\alpha)_{\alpha\in\mathrm{A}}) = \mathcal{H}_{e,1}$. ■

The construction of the complete von Neumann product of Hilbert spaces $(H_n)_1^\infty$ is presented as the following theorem.

THEOREM 2.9. *Let* E *be the collection of all stabilizing sequences* $e = (e^{(n)})_1^\infty$ $(e^{(n)} \in H_n,\ \|e^{(n)}\|_{H_n} = 1)$. *An* $l \in E$ *is said to be equivalent to an*

$e \in E$, written $l \sim e$, if each strong limit

$$l^{(1)'} \otimes l^{(2)'} \otimes \cdots = \lim_{p \to \infty} l^{(1)'} \otimes \cdots \otimes l^{(p)'} \otimes e^{(p+1)} \otimes e^{(p+2)} \otimes \cdots$$

exists in $\mathcal{H}_{e,1}$, where $(l^{(n)'})_1^\infty$ is the sequence $(l^{(n)})_1^\infty$, "diluted" by the vectors $e^{(n)}$, i.e., each $l^{(n)'}$ is equal either to $l^{(n)}$ or to $e^{(n)}$.

The relation $\sim$ is an equivalence relation, and the whole of E breaks up into the union of the disjoint classes E_ω of mutually equivalent stabilizing sequences: $E = \bigcup_\omega E_\omega$. If a representative $e(\omega)$ is chosen in each class E_ω, then the complete von Neumann product of the spaces $(H_n)_1^\infty$ is understood to be the orthogonal sum

$$\bigoplus_\omega \mathcal{H}_{e(\omega),1} = \bigoplus_\omega \left(\bigotimes_{n=1;e(\omega),1}^{\infty} H_n \right).$$

This construction is invariant with respect to the choice of $e(\omega) \in E_\omega$ in the sense that if $l(\omega) \in E_\omega$, then between the spaces $\mathcal{H}_{l(\omega),1}$ and $\mathcal{H}_{e(\omega),1}$ it is possible to establish an isometry which assigns to each formal unit basis vector

$$\mathcal{H}_{l(\omega),1} \ni l_\alpha(\omega) = (l_{\alpha_1}^{(1)}(\omega)) \otimes (l_{\alpha_2}^{(2)}(\omega)) \otimes \cdots$$

the unit basis vector

$$\begin{aligned} \mathcal{H}_{e(\omega),1} \ni \tilde{l}_\alpha(\omega) &= (l_{\alpha_1}^{(1)}(\omega)) \otimes (l_{\alpha_2}^{(2)}(\omega)) \otimes \cdots \\ &= \lim_{p \to \infty} (l_{\alpha_1}^{(1)}(\omega)) \otimes \cdots \otimes (l_{\alpha_p}^{(p)}(\omega)) \otimes (e^{(p+1)}(\omega)) \\ &\qquad \otimes (e^{(p+2)}(\omega)) \otimes \cdots, \end{aligned}$$

where the limit is understood as a strong limit in $\mathcal{H}_{e(\omega),1}$ ($\alpha \in \mathrm{A}$).

PROOF. According to the corollary to Lemma 2.6, the strong limit $l^{(1)'} \otimes l^{(2)'} \otimes \cdots$ exists for each choice of $l^{(n)'} = l^{(n)}$ and $e^{(n)}$ if and only if for some common $q = 1, 2, \ldots$ the products $\prod_{n=q}^\infty (l^{(n)'}, e^{(n)})_{H_n}$ all converge to finite nonzero numbers. This convergence is equivalent to the convergence of the series

$$\sum_{n=q}^{\infty} \ln((l^{(n)'}, e^{(n)})_{H_n}). \tag{2.58}$$

If $l^{(n)'} = e^{(n)}$, then $(l^{(n)'}, e^{(n)})_{H_n} = 1$, and the corresponding term in (2.58) is annihilated. Therefore, convergence of all possible series (2.58) with $l^{(n)'}$ equal to $l^{(n)}$ or $e^{(n)}$ is equivalent to convergence of the series $\sum_{n=q}^\infty \ln((l^{(n)}, e^{(n)})_{H_n})$ in the sense that any series formed from this series by replacing the terms $\ln(l^{(n_k)}, e^{(n_k)})_{H_{n_k}}$ ($k = 1, 2, \ldots$) by zeros, where $(n_k)_1^\infty$ is some subsequence of the sequence $q, q+1, \ldots$, is convergent. But this is known to be equivalent to the absolute convergence of the series $\sum_{n=q}^\infty \ln((l^{(n)}, e^{(n)})_{H_n})$. On the other hand, by the inequality $\frac{1}{2}|z| < |\ln(1+z)| < \frac{3}{2}|z|$, which is valid for complex z

such that $|z| < \frac{1}{2}$, absolute convergence of the last series is equivalent to the condition

$$\sum_{n=1}^{\infty} |(l^{(n)}, e^{(n)})_{H_n} - 1| < \infty. \tag{2.59}$$

Accordingly, $l \sim e$ if and only if (2.59) holds.

Further,

$$\|l^{(n)} - e^{(n)}\|_{H_n}^2 = 2(1 - \operatorname{Re}(l^{(n)}, e^{(n)})_{H_n}),$$

and so (2.59) implies the inequality

$$\sum_{n=1}^{\infty} \|l^{(n)} - e^{(n)}\|_{H_n}^2 < \infty. \tag{2.60}$$

It is now easy to see that $\sim$ is an equivalence relation. Indeed, $e \sim e$, and it follows from (2.59) that $e \sim l$ if $l \sim e$. Let us prove transitivity: if $l \sim e$ and $e \sim g$, then $l \sim g$. We have that

$$\begin{aligned} |(l^{(n)}, g^{(n)})_{H_n} - 1| &\le |(l^{(n)}, e^{(n)})_{H_n} - 1| + |(l^{(n)}, g^{(n)} - e^{(n)})_{H_n}| \\ &\le |(l^{(n)}, e^{(n)})_{H_n} - 1| + |(l^{(n)} - e^{(n)}, g^{(n)} - e^{(n)})_{H_p}| \\ &\quad + |(e^{(n)}, g^{(n)} - e^{(n)})_{H_n}| \\ &\le |(l^{(n)}, e^{(n)})_{H_n} - 1| + |(e^{(n)}, g^{(n)})_{H_n} - 1| \\ &\quad + \|l^{(n)} - e^{(n)}\|_{H_n} \|g^{(n)} - e^{(n)}\|_{H_n} \qquad (n = 1, 2, \ldots). \end{aligned}$$

This estimate, (2.59), and the convergence of (2.60) for l, e and e, g give us (2.59) for l, g. Hence, $l \sim g$.

Thus, it is possible to break E up into the union of the classes of mutually equivalent stabilizing sequences and to construct the complete von Neumann product. The last assertion of the theorem follows immediately from Lemma 2.7. ■

We remark that in the case of weighted Hilbert spaces H_n it is certainly possible to repeat the construction of the infinite tensor product $\mathcal{H}_{e,1} = \bigotimes_{n=1;e,1}^{\infty} H_n$ with stabilizing sequence $e = (e^{(n)})_1^{\infty}$ and then form the complete von Neumann product. The concept of equivalence $l \sim e$ can now be given the simple form: *$l \sim e$ if and only if the strong limit*

$$l^{(1)} \otimes l^{(2)} \otimes \cdots = \lim_{p \to \infty} l^{(1)} \otimes \cdots \otimes l^{(p)} \otimes e^{(p+1)} \otimes e^{(p+2)} \otimes \cdots$$

exists in $\mathcal{H}_{e,1}$. Indeed, this limit exists if and only if the product

$$\prod_{n=q}^{\infty} (l^{(n)}, e^{(n)})_{H_n}$$

converges to a finite nonzero number for some $q = 1, 2, \ldots$. But the numbers $(l^{(n)}, e^{(n)})_{H_n}$ are real with $|(l^{(n)}, e^{(n)})_{H_n}| \le 1$; therefore, $(l^{(n)}, e^{(n)})_{H_n} = 1 + a_n$, where $a_n \le 0$, and the last product converges if and only if $\sum_{n=q}^{\infty} a_n$

converges, i.e., the series (2.59) converges. Thus, the equivalence concept $l \sim e$ given here is the same as that indicated in the condition of Theorem 2.8. ■

In conclusion we mention conditions ensuring strong convergence of the operators (2.28). Namely, suppose that the imbeddings $O_n \colon H_{+,n} \to H_{0,n}$ ($n = 1, 2, \ldots$) are quasinuclear and that condition (2.15) holds. Consider a sequence $(A_n)_1^\infty$, $A_n \colon H_{+,n} \to H_{-,n}$, of nonnegative operators such that the product $\prod_{n=q}^\infty (A_n e^{(n)}, e^{(n)})_{H_{0,n}}$ converges for some $q = 1, 2, \ldots$ along with $\prod_1^\infty \operatorname{Tr}(A_n)$. Then the strong limit $\bigotimes_1^\infty A_n$ of the operators (2.28) exists as $p \to \infty$, and its trace is finite and equal to $\prod_1^\infty \operatorname{Tr}(A_n)$.

The proof of this fact can be obtained on the basis of the following remarks. Suppose that $A_n \in \mathcal{L}(H_n \to H_n)$ ($n = 1, 2, \ldots$) are continuous and nonnegative, and that the product $\prod_{n=q}^\infty (A_n e^{(n)}, e^{(n)})_{H_n}$ ($e^{(n)}$ is a unit vector in H_n) converges for some $q = 1, 2, \ldots$ along with $\prod_1^\infty \operatorname{Tr}(A_n)$. Then the following strong limit exists as $p \to \infty$:

$$\begin{aligned}
&\lim_{p\to\infty} A_1 f^{(1)} \otimes \cdots \otimes A_q f^{(q)} \otimes A_{q+1} e^{(q+1)} \otimes \cdots \otimes A_p e^{(p)} \otimes e^{(p+1)} \otimes \cdots \\
&\quad = A_1 f^{(1)} \otimes \cdots \otimes A_q f^{(q)} \otimes A_{q+1} e^{(q+1)} \otimes \cdots \\
&\quad = \left(\bigotimes_{n=1}^\infty A_n \right) (f^{(1)} \otimes \cdots \otimes f^{(q)} \otimes e^{(q+1)} \otimes \cdots) \\
&\qquad\qquad (f^{(n)} \in H_n;\ n = 1, \ldots, q).
\end{aligned}$$

(Verify the conditions in Lemma 2.6 with the help of the estimate

$$(A_n e^{(n)}, e^{(n)})_{H_n} \le \|A_n\| \le \operatorname{Tr}(A_n).)$$

The transition to operators acting from the positive space to the negative space is realized as before. The formula for $\operatorname{Tr}(\bigotimes_1^\infty A_n)$ can be obtained by passing to the limit as $p \to \infty$ in the expression for $\operatorname{Tr}(A^{(p)})$ (see the Remark after Theorem 2.6).

2.11. ***Triplets of spaces.*** The following generalization of the concept of a chain $H_- \supseteq H_0 \supseteq H_+$ of spaces is sometimes useful. Given three Hilbert spaces H, H_0, and H', we say that they form a triplet of spaces if the following conditions hold:

a) $\Phi = H \cap H_0 \cap H'$ is dense in each of these spaces;

b) the bilinear form $B(\varphi, \psi) = (\varphi, \psi)_{H_0}$ ($\varphi, \psi \in \Phi$) admits the estimate

$$|B(\varphi, \psi)| \le \|\varphi\|_{H'} \, \|\psi\|_H$$

and can thus be extended by continuity to a continuous bilinear form $H' \oplus H \ni (\alpha, v) \mapsto B(\alpha, v) \in \mathbb{C}^1$ (denote it again by $(\alpha, v)_{H_0}$);

c) for each $u \in H$ there is a unique vector $\alpha_u \in H'$ such that $(u, v)_H = (\alpha_u, v)_{H_0}$ ($v \in H$).

Of course, if we have a chain, then H_+, H_0, H_- form a triplet of spaces. At the same time, it is easy to give an example of a triplet that is not a chain:

let

$$H = L_2(\mathbb{R}^N, p(x)\,dx), \quad \text{where } 0 < p(x) \in C(\mathbb{R}^N),$$
$$H_0 = L_2(\mathbb{R}^N) \quad \text{and} \quad H' = L_2(\mathbb{R}^N, p^{-1}(x)\,dx).$$

(A chain is obtained when $p(x) \geq 1$ $(x \in \mathbb{R}^N)$; see §1.1.)

Let us rephrase condition c) in the definition of a triplet. We introduce a continuous operator $A\colon H' \to H$ such that $(\alpha, v)_{H_0} = (A\alpha, v)_H$ $(\alpha \in H',\ v \in H)$; clearly, $\|A\| \leq 1$. Then condition c) means that for each $u \in H$ the equation $A\alpha = u$ has a unique solution $\alpha = \alpha_u \in H'$. In other words, A maps the whole of H' onto the whole of H, and $\operatorname{Ker} A = 0$. By Banach's theorem, this is equivalent to the existence of the continuous operator A^{-1} mapping the whole of H onto the whole of H'. This is the required rephrasing of condition c).

It is also clear that condition c) means that H' coincides with the collection of all continuous antilinear functionals on H, written by means of the form $(\cdot,\cdot)_{H_0}$. Therefore, we also say that H' is the dual of H with respect to H_0.

THEOREM 2.10. *If H, H_0, H' is a triplet, then so is H', H_0, H. If $H_1, H_{0,1}, H_1'$ and $H_2, H_{0,2}, H_2'$ are triplets, then so is $H_1 \otimes H_2, H_{0,1} \otimes H_{0,2}, H_1' \otimes H_2'$.*

PROOF. Let H, H_0, H' be a triplet. For H', H_0, H condition a) holds trivially. Further, since $B(\varphi, \psi) = \overline{B(\psi, \varphi)}$, it follows from condition b) for H, H_0, H' that

$$|B(\varphi, \psi)| = |B(\psi, \varphi)| \leq \|\varphi\|_H\, \|\psi\|_{H'} \qquad (\varphi, \psi \in \Phi),$$

i.e., condition b) holds for H', H_0, H; therefore, the form $B(\varphi, \psi) = (\varphi, \psi)_{H_0}$ extends by continuity to a form $H \oplus H' \ni (u, \beta) \mapsto B_1(u, \beta) \in \mathbb{C}^1$. We prove that condition c) also holds for H', H_0, H. Let $A_1\colon H \to H'$ be a continuous operator such that $B_1(u, \beta) = (A_1 u, \beta)_H$ $(u \in H, \beta \in H')$. It is required to prove that the inverse operator A_1^{-1} exists and is continuous.

For $\psi, \varphi \in \Phi$ we have that

$$(\varphi, \psi)_{H_0} = \overline{(\psi, \varphi)}_{H_0} = \overline{B_1(\psi, \varphi)} = (\varphi, A_1\psi)_{H'}.$$

Therefore, $(A\varphi, \psi)_H = (\varphi, \psi)_{H_0} = (\varphi, A_1\psi)_{H'}$ $(\varphi, \psi \in \Phi)$. The operators $A\colon H' \to H$ and $A_1\colon H \to H'$ in the equality $(A\varphi, \psi)_H = (\varphi, A_1\psi)_{H'}$ $(\varphi, \psi \in \Phi)$ are continuous, so we can pass to the limit in this equality as $\varphi \to \alpha$ (in H') and $\psi \to v$ (in H). The result is that $(A\alpha, v)_H = (\alpha, A_1 v)_{H'}$ $(\alpha \in H',\ v \in H)$, i.e., $A_1 = A^*$. But if A^{-1} exists, then so does $(A^*)^{-1} = A_1^{-1}$. The first assertion is proved.

We prove the second assertion. Let $H = H_1 \otimes H_2$, $H_0 = H_{0,1} \otimes H_{0,2}$, $H' = H_1' \otimes H_2'$, and $\Phi_j = H_j \cap H_{0,j} \cap H_j'$ $(j = 1, 2)$. Since $\Phi_1 \otimes \Phi_2 \subseteq H \cap H_0 \cap H' = \Phi$ and is dense in each of the spaces H, H_0, and H', Φ is also dense in each of these spaces, i.e., condition a) holds for H, H_0, H'.

Consider the form $B_j(\varphi^{(j)}, \psi^{(j)}) = (\varphi^{(j)}, \psi^{(j)})_{H_{0,j}}$ $(\varphi^{(j)}, \psi^{(j)} \in \Phi_j)$, its extension to a bilinear form B_j of the variables $(\alpha^{(j)}, v^{(j)}) \in H'_j \oplus H_j$, and a corresponding operator $A_j \colon H'_j \to H_j$ such that $(\alpha^{(j)}, v^{(j)})_{H_{0,j}} = (A_j\alpha^{(j)}, v^{(j)})_{H_j}$ $(\alpha^{(j)} \in H'_j,\ v^{(j)} \in H_j,\ j = 1,2)$. The product $A_1 \otimes A_2$ acts continuously from the whole of $H'_1 \otimes H'_2$ to $H_1 \otimes H_2$ and has inverse $(A_1 \otimes A_2)^{-1} = A_1^{-1} \otimes A_2^{-1}$. Properties b) and c) will be proved if we establish that the bilinear form $B(\varphi, \psi) = (\varphi, \psi)_{H_0}$ $(\varphi, \psi \in \Phi)$ admits the estimate $|B(\varphi, \psi)| \leq \|\varphi\|_{H'} \|\psi\|_H$ $(\varphi, \psi \in \Phi)$ and the corresponding operator A coincides with $A_1 \otimes A_2$.

If $\varphi = \sum_1^p \varphi_n^{(1)} \otimes \varphi_n^{(2)}$ and $\psi = \sum_1^p \psi_m^{(1)} \otimes \psi_m^{(2)}$ $(\varphi_n^{(j)},\ \psi_m^{(j)} \in \Phi_j;\ j = 1,2)$, then

$$\begin{aligned} B(\varphi, \psi) = (\varphi, \psi)_{H_0} &= \sum_{n,m=1}^{p} (\varphi_n^{(1)}, \psi_m^{(1)})_{H_{0,1}} (\varphi_n^{(2)}, \psi_m^{(2)})_{H_{0,2}} \\ &= \sum_{n,m=1}^{p} (A_1\varphi_n^{(1)}, \psi_m^{(1)})_{H_1} (A_2\varphi_n^{(2)}, \psi_m^{(2)})_{H_2} \\ &= \sum_{n,m=1}^{p} ((A_1\varphi_n^{(1)}) \otimes (A_2\varphi_n^{(2)}), \psi_m^{(1)} \otimes \psi_m^{(2)})_H = ((A_1 \otimes A_2)\varphi, \psi)_H. \end{aligned} \tag{2.61}$$

From this we conclude that

$$\begin{aligned} |B(\varphi, \psi)| &= |((A_1 \otimes A_2)\varphi, \psi)_H| \\ &\leq \|A_1 \otimes A_2\| \, \|\varphi\|_{H'} \, \|\psi\|_H \leq \|\varphi\|_{H'} \, \|\psi\|_H. \end{aligned}$$

This inequality extends by continuity to the inequality

$$|B(\varphi, \psi)| \leq \|\varphi\|_{H'} \, \|\psi\|_H \qquad (\varphi, \psi \in \Phi).$$

It can now be asserted that there exists an operator $A \colon H' \to H$ such that, in particular, $B(\varphi, \psi) = (A\varphi, \psi)_H$ $(\varphi, \psi \in \Phi)$. Substituting this relation in (2.61), we conclude that $A = A_1 \otimes A_2$. ■

We mention also that in condition c) it is possible to require not uniqueness of the vector $\alpha_u \in H'$ but instead the analogous condition both for H' and H. Clearly, conditions a)–c) yield the requirement just formulated, because H', H_0, H is also a triplet. Conversely, the last condition yields condition c). Indeed, assuming that α_u is not uniquely determined, we find a nonzero $\gamma \in H'$ such that $(\gamma, v)_{H_0} = 0$ $(v \in H)$. According to our requirement, for any $\alpha \in H'$ there exists a $u_\alpha \in H$ such that $(\beta, \alpha)_{H'} = (\beta, u_\alpha)_{H_0}$ $(\beta \in H')$.

Setting $\alpha = \beta = \gamma$ here, we get that $(\gamma, \gamma)_{H'} = (\gamma, u_\gamma)_{H_0} = 0$, i.e., $\gamma = 0$, which is absurd.

§3. Spaces of functions of finitely many variables

Here we consider examples (essential for what follows) of chains

$$H_- \supseteq H_0 \supseteq H_+, \tag{3.1}$$

consisting of spaces of functions of finitely many variables in bounded and unbounded domains.

3.1. ***Spaces of square-integrable functions with a weight.*** We consider a simple example of a chain (3.1) generalizing the example in §1.1. Despite the triviality of the construction, such chains are frequently encountered.

Let $H_0 = L_2(R, d\mu(x))$, where R is a space with measure μ defined on some σ-algebra of sets in R, and $\mu(R) \leq \infty$. We take a measurable weight $p(x) \geq 1$ $(x \in R)$ that is finite almost everywhere, and construct the space $L_2(R, p(x)\, d\mu(x))$. As is easy to see, it can be taken to be the positive space H_+. We claim that *the corresponding negative space H_- is equal to $L_2(R, p^{-1}(x)\, d\mu(x))$*, i.e.,

$$L_2(R, p^{-1}(x)\, d\mu(x)) \supseteq L_2(R, d\mu(x)) \supseteq L_2(R, p(x)\, d\mu(x)) \tag{3.2}$$

is a chain.

Indeed, let us find the operator $I\colon H_0 \to H_+$. We have that

$$\int_R f(x)\overline{u(x)}\, d\mu(x) = (f, u)_{H_0} = (If, u)_{H_+} = \int_R (If)(x)\overline{u(x)}p(x)\, d\mu(x)$$

$$(f \in L_2(R, d\mu(x)),\ u \in L_2(R, p(x)\, d\mu(x))).$$

Due to the arbitrariness of u in the last equality, it follows that

$$(If)(x) = p^{-1}(x) f(x) \qquad (f \in L_2(R, d\mu(x))). \tag{3.3}$$

This implies that $(f, g)_{H_-} = (If, g)_{H_0} = \int_R f(x)\overline{g(x)}p^{-1}(x)\, d\mu(x)$ for $f, g \in L_2(R, d\mu(x))$. Taking the completion, we get that

$$H_- = L_2(R, p^{-1}(x)\, d\mu(x)). \quad \blacksquare$$

Thus, the functions in H_+ are "rapidly decreasing", while those in H_- are "rapidly increasing" (if it is assumed, for example, that $R = \mathbb{R}^N$, $d\mu(x) = dx$, and $p(x) \to \infty$ as $|x| \to \infty$). To say that an $\alpha \in H_-$ is a generalized function now means only that it is not $L_2(R, d\mu(x))$.

3.2. ***Sobolev spaces on a bounded domain.*** Let G be a bounded domain in $\mathbb{R}^N$ $(N = 1, 2, \ldots)$ with boundary ∂G that is once piecewise continuously differentiable.(1)The Sobolev space $W_2^l(G)$ $(l = 0, 1, \ldots)$ is defined to be the

(1)The smoothness restriction could, of course, be weakened, but we do not do so for simplicity of the formulations.

completion of $C^\infty(\tilde{G})$ ($\tilde{G} = G \cup \partial G$) in the inner product

$$(u,v)_{W_2^l(G)} = \sum_{|\mu|\le l} (D^\mu u, D^\mu v)_{L_2(G)} \qquad (u,v \in C^\infty(\tilde{G})). \tag{3.4}$$

(Note that if only the terms with $\mu = 0$ or $|\mu| = l$ are kept in the sum (3.4), then the inner product obtained is equivalent to (3.4).) Obviously, $\|u\|_{L_2(G)} \le \|u\|_{W_2^l(G)}$ ($u \in W_2^l(G)$), and $W_2^l(G)$ is dense in $L_2(G)$. Therefore, it is possible to construct a chain (3.1) by setting $H_0 = L_2(G)$ and $H_+ = W_2^l(G)$. Denoting the corresponding negative space H_- by $W_2^{-l}(G)$, we get

$$W_2^{-l}(G) \supseteq L_2(G) \supseteq W_2^l(G) \qquad (l = 0, 1, \dots). \tag{3.5}$$

Obviously, $W_2^{l''}(G) \subseteq W_2^{l'}(G)$ for $l'' \ge l'$; therefore, $W_2^{-l''}(G) \supseteq W_2^{-l'}(G)$. The smaller space is dense in the larger in both these topological imbeddings.

The space $W_2^{-l}(G)$ of generalized functions is called a negative Sobolev space. We explain the nature of this space, confining ourselves to the case when $l = 1$ and the boundary is smooth. Let $\partial G \in C^2$. Consider the operator

$$\left\{ u \in C^\infty(\tilde{G}) \,\middle|\, \frac{\partial u}{\partial \nu} \upharpoonright \partial G = 0 \right\} \ni u \mapsto (-\Delta + 1)u$$

defined in $L_2(G)$, where $\partial/\partial\nu$ is differentiation with respect to the outward normal to ∂G. It is well known that such an operator is essentially selfadjoint (see, for example, Berezanskiĭ [4], Chapter 6, Theorem 1.3). Denote by $A \ge 1$ its closure; it is clear that A^{-1} exists.

The space $W_2^{-1}(G)$ coincides with the completion of $L_2(G)$ in the inner product

$$(f, g)_{W_2^{-1}(G)} = (A^{-1} f, g)_{L_2(G)} \qquad (f, g \in L_2(G)).$$

Indeed, it must be shown that $OI = A^{-1}$, where O and I are the operators connected with the chain (3.5). Let $u \in C^\infty(\tilde{G})$ be such that $\partial u/\partial\nu \upharpoonright \partial G = 0$. Integrating by parts, we get that

$$\begin{aligned}
\int_G f(x)\overline{u(x)}\, dx &= (f, u)_{H_0} = (If, u)_{W_2^1(G)} \\
&= \int_G (If)(x)\overline{u(x)}\, dx + \sum_{j=1}^N \int_G (D_j If)(x)\overline{(D_j u)(x)}\, dx \\
&= \int_G (OIf)(x)\overline{((-\Delta + 1)u)(x)}\, dx \\
&= \int_G (OIf)(x)\overline{(Au)(x)}\, dx \qquad (f \in L_2(G)).
\end{aligned}$$

This implies that $OIf \in \mathfrak{D}(A^*) = \mathfrak{D}(A)$ and $A(OIf) = f$. ■

The operator A^{-1} is an integral operator: there exists a positive-definite kernel $K(x,y)$ (the Green's function) such that

$$(A^{-1}f)(x) = \int_G K(x,y)f(y)\,dy \qquad (f \in C_0(G));$$
$$K(x,y) - e(x,y) \in C(G \times G),$$

where $e(x,y)$ is the fundamental solution of the expression $-\Delta + 1$ (see, for example, Berezanskiĭ [4], Chapter 3, §5). Therefore, $W_2^{-1}(G)$ coincides also with the completion of $C_0(G)$ in the inner product generated by the positive-definite kernel:

$$(f,g)_{W_2^{-1}(G)} = \int_G \int_G K(x,y)f(y)\overline{g(x)}\,dx\,dy \qquad (f,g \in C_0(G)).$$

The nature of the space $W_2^{-l}(G)$ is similar in the general case, except that the Green's function of a more complicated problem for an elliptic equation of order $2l$ appears instead of the Green's function for the problem $-\Delta u + u = f$ with the Neumann boundary condition $\partial u/\partial\nu \restriction \partial G = 0$. However, it is difficult to study the space $W_2^{-l}(G)$ along this path, and we proceed differently.

3.3. ***The delta-function as an element of a negative Sobolev space.*** We give a general definition of the delta-function. Let R be an abstract space, and Φ a linear topological space consisting of certain functions $R \ni x \mapsto \varphi(x) \in \mathbb{C}^1$ and having the property that the convergence $\Phi \ni \varphi_n \to \varphi \in \Phi$ as $n \to \infty$ implies that $\varphi_n(x) \to \varphi(x)$ as $n \to \infty$ for each $x \in R$. Fix $\xi \in R$ and consider the antilinear functional l defined on Φ by $l(\varphi) = \overline{\varphi(\xi)}$ $(\varphi \in \Phi)$. According to our assumption, it is continuous, i.e., $l \in \Phi'$. This functional is called the δ-function concentrated at the point ξ, and is denoted by δ_ξ. In particular, this situation holds when R is compact and $\Phi \subseteq C(R)$ topologically. Note that $\|\delta_\xi\|_{C'(R)} = 1$ $(\xi \in R)$.

THEOREM 3.1. *If $l > N/2$, then the δ-function δ_ξ concentrated at $\xi \in \tilde{G}$ is defined in the space $W_2^{-l}(G)$, and the vector-valued function $\tilde{G} \ni \xi \mapsto \delta_\xi \in W_2^{-l}(G)$ is strongly continuous.*

PROOF. According to the imbedding theorems, $W_2^l(G) \subset C(\tilde{G})$ topologically when $l > N/2$; therefore, δ_ξ $(\xi \in \tilde{G})$ is defined as an element of $W_2^{-l}(G)$.

Let us prove that δ_ξ depends continuously on ξ. First of all observe that, obviously,

$$W_2^{-l}(G) \supset C'(\tilde{G}) \supset L_2(G) \supset C(\tilde{G}) \supset W_2^l(G), \tag{3.6}$$

with all inclusions topological and each space dense in the one to its left. Let S be the imbedding operator $W_2^l(G) \to C(\tilde{G})$; by the imbedding theorems, it is compact. The operator $S^*\colon C'(\tilde{G}) \to W_2^{-l}(G)$ adjoint to it in the usual sense is also an imbedding and is compact.

Assume the opposite: suppose that the vector-valued function $\tilde{G} \ni \xi \mapsto \delta_\xi \in W_2^{-l}(G)$ is not strongly continuous at some point $\xi_0 \in \tilde{G}$. Then there exists a sequence of points $\xi_n \in \tilde{G}$ converging to ξ_0 and a number $\varepsilon_0 > 0$ such that $\|\delta_{\xi_n} - \delta_{\xi_0}\|_{W_2^{-l}(G)} \geq \varepsilon_0$ $(n = 1, 2, \ldots)$. Since $\|\delta_{\xi_n}\|_{C'(\tilde{G})} = 1$ $(n = 1, 2, \ldots)$ and the imbedding $C'(\tilde{G}) \to W_2^{-l}(G)$ is compact, the sequence $(\delta_{\xi_n})_{n=1}^{\infty}$ is precompact in $W_2^{-l}(G)$, so it has a subsequence $(\delta_{\xi_{n_k}})_{k=1}^{\infty}$ such that $\delta_{\xi_{n_k}}$ converges strongly in $W_2^{-l}(G)$ to some element $\alpha \in W_2^{-l}(G)$ as $k \to \infty$. It is easy to see that $\alpha = \delta_{\xi_0}$:

$$(\alpha, u)_{L_2(G)} = \lim_{k\to\infty} (\delta_{\xi_{n_k}}, u)_{L_2(G)} = \lim_{k\to\infty} \overline{u(\xi_{n_k})} = \overline{u(\xi_0)} = (\delta_{\xi_0}, u)_{L_2(G)}$$

for $u \in W_2^l(G)$. Accordingly, $\delta_{\xi_{n_k}} \to \delta_{\xi_0}$ strongly as $k \to \infty$, which contradicts the choice of the points ξ_n. ■

3.4. *Quasinuclearity of imbeddings of Sobolev spaces.* We have the following essential theorem.

THEOREM 3.2. *Let $W_2^{l'}(G)$ and $W_2^{l''}(G)$ be two Sobolev spaces such that $l'' - l' > N/2$ $(l', l'' = 0, 1, \ldots)$. Then the imbedding $W_2^{l''}(G) \to W_2^{l'}(G)$ is quasinuclear.*

PROOF. Consider first the main case, when $l' = 0$. We must prove that the imbedding $O\colon W_2^l(G) \to L_2(G)$ in the chain (3.5) is quasinuclear when $l > N/2$. Let $J\colon L_2(G) \to W_2^l(G)$ be the isometry connected with the chain (3.5). The quasinuclearity of O is equivalent to that of $OJ\colon L_2(G) \to L_2(G)$ Let us establish the latter. For $f \in L_2(G)$ we use (1.19) and (1.20) to obtain

$$\begin{aligned}(OJf)(x) &= (Jf)(x) = (Jf, \delta_x)_{L_2(G)} = (f, J^+\delta_x)_{L_2(G)} \\ &= (f, \mathbf{J}\delta_x)_{L_2(G)} = \int_G f(y)\overline{(\mathbf{J}\delta_x)(y)}\, dy. \end{aligned} \tag{3.7}$$

(Note that $\mathbf{J}\delta_x \in L_2(G)$.) If $K(x, y) = \overline{(\mathbf{J}\delta_x)(y)}$, then

$$\begin{aligned}\int_G\int_G |K(x,y)|^2\, dx\, dy &= \int_G \|\mathbf{J}\delta_x\|^2_{L_2(G)}\, dx = \int_G \|\delta_x\|^2_{W_2^{-l}(G)}\, dx \\ &\leq \max_{x\in\tilde{G}} \|\delta_x\|^2_{W_2^{-l}(G)} m(G) < \infty. \end{aligned} \tag{3.8}$$

Here the continuity (by Theorem 3.1) of the scalar function

$$\tilde{G} \ni x \mapsto \|\delta_x\|_{W_2^{-l}(G)} \in \mathbf{R}^1$$

has been used. The relations (3.7) and (3.8) show that OJ is a Hilbert-Schmidt operator.

The proof of the theorem in the general case is based on the part already proved and the following two general lemmas.

LEMMA 3.1. *Let E be a linear set on which a linear operator $E \ni f \mapsto Tf \in E$ and two inner products $(\cdot,\cdot)_{H_1}$ and $(\cdot,\cdot)_{G_1}$ are defined. Denote by H_1 and G_1 the completions of E in $(\cdot,\cdot)_{H_1}$ and $(\cdot,\cdot)_{G_1}$, and assume that $H_1 \subseteq G_1$ topologically. If the imbedding $O_1\colon H_1 \to G_1$ is quasinuclear, then* *so is the imbedding* $O_2\colon H_2 \to G_2$, *where* H_2 *and* G_2 *are the completions of* E *in the inner products* $(f,g)_{H_2} = (f,g)_{H_1} + (Tf,Tg)_{H_1}$ *and* $(f,g)_{G_2} = (f,g)_{G_1} + (Tf,Tg)_{G_1}$ $(f,g \in E)$.

PROOF. Since $\|f\|_{H_1} \le \|f\|_{H_2}$ $(f \in E)$, H_2 is continuously imbedded in H_1. In turn, the imbedding $H_1 \to G_1$ is quasinuclear by assumption. Therefore, the imbedding $H_2 \to G_1$ is quasinuclear, and consequently, if $(e_j)_1^\infty$ is an orthonormal basis in H_2 (which can consist of vectors $e_j \in E$), then

$$\sum_{j=1}^{\infty} \|e_j\|_{G_1}^2 < \infty. \tag{3.9}$$

We introduce the (generally speaking) quasi-inner products $(f,g)_{H_3} = (Tf,Tg)_{H_1}$ and $(f,g)_{G_3} = (Tf,Tg)_{G_1}$ $(f,g \in E)$ in E. The identification of points of E according to each of these quasi-inner products gives one and the same linear set $\hat{E}$ of full preimages $\hat{f} = T^{-1}f' = \{f \in E | Tf = f'\}$ of the vectors $f' \in E$. Subsequent completion leads to Hilbert spaces H_3 and G_3, and $H_3 \subseteq G_3$ topologically. The imbedding $H_3 \to G_3$ is quasinuclear. Indeed, let $(\hat{l}_j)_1^\infty$ be an orthonormal basis in H_3 constructed from vectors $\hat{l}_j \in \hat{E}$, and let $l_j \in E$ be a representative of the class $\hat{l}_j$. Then

$$(Tl_j, Tl_k)_{H_1} = (l_j, l_k)_{H_3} = (\hat{l}_j, \hat{l}_k)_{H_3} = \delta_{jk} \qquad (j,k = 1,2,\dots),$$

i.e., $(Tl_j)_1^\infty$ is an orthonormal system in H_1, and $\sum_1^\infty \|Tl_j\|_{G_1}^2 < \infty$, because the imbedding $H_1 \to G_1$ is quasinuclear. But $\|Tl_j\|_{G_1} = \|\hat{l}_j\|_{G_3}$, and the last condition means that the imbedding $A\colon H_3 \to G_3$ is quasinuclear.

Since $\|\hat{f}\|_{H_3} = \|Tf\|_{H_1} \le \|f\|_{H_2}$, the mapping $E \ni f \mapsto Bf = \hat{f} \in \hat{E}$ extends by continuity to a continuous mapping $B\colon H_2 \to H_3$. The mapping $AB\colon H_2 \to G_3$ is quasinuclear. Therefore, if $(e_j)_1^\infty$ is an orthonormal basis in H_2 consisting of vectors in E, then

$$\infty > \sum_{j=1}^{\infty} \|ABe_j\|_{G_3}^2 = \sum_{j=1}^{\infty} \|\hat{e}_j\|_{G_3}^2 = \sum_{j=1}^{\infty} \|Te_j\|_{G_1}^2.$$

From this and (3.9), we get

$$\sum_{j=1}^{\infty} \|e_j\|_{G_2}^2 = \sum_{j=1}^{\infty} (\|e_j\|_{G_1}^2 + \|Te_j\|_{G_1}^2) < \infty. \quad \blacksquare$$

REMARK 1. It is easy to see that $|O_2| \le \sqrt{2}\,|O_1|$.

REMARK 2. Suppose that the operator T in the formulation of Lemma 3.1 is invertible, i.e., $\operatorname{Ker} T = 0$. Then the statement of the lemma remains valid if the inner products $(\cdot,\cdot)_{H_2}$ and $(\cdot,\cdot)_{G_2}$ are replaced by $(f,g)_{H_2} = (Tf,Tg)_{H_1}$ and $(f,g)_{G_2} = (Tf,Tg)_{G_1}$ $(f,g \in E)$. The proof of this fact follows from the arguments given above.

LEMMA 3.2. *Let E be a linear set in which inner products $(\cdot,\cdot)_{H_k}$ and $(\cdot,\cdot)_{G_k}$ $(k = 1,\dots,n)$ are defined, and let $(\cdot,\cdot)_H = \sum_{k=1}^n (\cdot,\cdot)_{H_k}$ and $(\cdot,\cdot)_G = \sum_{k=1}^n (\cdot,\cdot)_{G_k}$. Let H_k, G_k, H, and G be the corresponding completions of E. If $H_k \subseteq G_k$ and the imbeddings $H_k \to G_k$ are quasinuclear $(k = 1,\dots,n)$, then so is the imbedding $H \to G$.*

PROOF. Fix $k = 1,\dots,n$. Since $\|f\|_{H_k} \le \|f\|_H$ $(f \in E)$, it follows that $H \subseteq H_k$, and this imbedding is continuous. But then the imbedding $H \to G_k$ is quasinuclear, and thus if $(e_j)_1^\infty$ is an orthonormal basis in H, then $\sum_1^\infty \|e_j\|_{G_k}^2 < \infty$. Summing these inequalities over all k, we get that $\sum_1^\infty \|e_j\|_G^2 < \infty$. ■

We finish the proof of the theorem. It is required to establish that if $l > N/2$ is an integer, then the imbedding $W_2^{m+l}(G) \to W_2^m(G)$ is quasinuclear for any $m = 0, 1, \dots$. With the use of Lemma 3.1 in mind we set $E = C^\infty(\tilde{G}), (Tf)(x) = (D^\nu f)(x)$, where D^ν is a fixed derivative of order $|\nu| \le m$, and

$$(f,g)_{H_1} = (f,g)_{W_2^l(G)} \quad \text{and} \quad (f,g)_{G_1} = (f,g)_{L_2(G)} \qquad (f,g \in E).$$

According to the proven part of the theorem, the imbedding $H_1 = W_2^l(G) \to L_2(G) = G_1$ is quasinuclear; therefore, by Lemma 3.1, so is the imbedding $H_2 \to G_2$, where H_2 and G_2 are the completions of E in the inner products

$$\begin{aligned}(f,g)_{H_2} &= (f,g)_{W_2^l(G)} + (D^\nu f, D^\nu g)_{W_2^l(G)},\\ (f,g)_{G_2} &= (f,g)_{L_2(G)} + (D^\nu f, D^\nu g)_{L_2(G)} \qquad (f,g \in E). \qquad (3.10)\end{aligned}$$

We denote the inner products $(\cdot,\cdot)_{H_2}$ and $(\cdot,\cdot)_{G_2}$ in (3.10) by $(\cdot,\cdot)_{H_\nu}$ and $(\cdot,\cdot)_{G_\nu}$, respectively, and apply Lemma 3.2, with $E = C^\infty(\tilde{G})$; n is now the number of vector indices $\nu = (\nu_1,\dots,\nu_N)$ such that $|\nu| \le m$. By what was proved, $H_\nu \to G_\nu$ is quasinuclear for each such ν; therefore, the imbedding $H \to G$ is also quasinuclear, where H and G are the completions of E in the inner products

$$(f,g)_H = n(f,g)_{W_2^l(G)} + \sum_{|\nu|\le m} (D^\nu f, D^\nu g)_{W_2^l(G)},$$

$$(f,g)_G = n(f,g)_{L_2(G)} + \sum_{|\nu|\le m} (D^\nu f, D^\nu g)_{L_2(G)} \qquad (f,g \in E).$$

The first of these is equivalent to $(\cdot,\cdot)_{W_2^{m+l}(G)}$, and the second is equivalent to $(\cdot,\cdot)_{W_2^m(G)}$. Therefore $H = W_2^{m+l}(G)$, $G = W_2^m(G)$, and the theorem is completely proved. ■

We note that for $l > N/2$ the operator OI for the chain (3.5) is an integral operator with kernel $K(x,y) \in C(\tilde{G} \times \tilde{G})$ (and, of course, with finite trace). The proof of this is just like the main case of Theorem 3.2, except that the operator J must be replaced by I. Then $K(x,y) = \overline{(\mathbf{I}\delta_x)(y)}$, and since $\mathbf{I}$ is an isometry between $W_2^{-l}(G)$ and $W_2^l(G) \subset C(\tilde{G})$, while $\tilde{G} \ni x \mapsto \delta_x \in W_2^{-l}(G)$ is a continuous vector-valued function, the required continuity of K follows from this.

3.5. *The kernel theorem in spaces of square-integrable functions on bounded domains.* Let $H_{0,n} = L_2(G^{(n)})$, where $G^{(n)} \subset \mathbb{R}^{N^{(n)}}$ ($N^{(n)} = 1, 2, \ldots$) is bounded ($n = 1, \ldots, p$). The kernel Theorem 2.7 is applicable to a multilinear form

$$\bigoplus_{n=1}^{p} H_{0,n} \ni (f^{(1)}, \ldots, f^{(p)}) \mapsto F(f^{(1)}, \ldots, f^{(p)}),$$

with the chains

$$W_2^{-l^{(n)}}(G^{(n)}) \supset L_2(G^{(n)}) \supset W_2^{l^{(n)}}(G^{(n)}), \qquad l^{(n)} > N^{(n)}/2,$$

taken (according to Theorem 3.2) as the chains (2.31). Then the kernel Φ in (2.38) lies in the space

$$(L_2(G^{(1)})) \otimes (W_2^{-l^{(2)}}(G^{(2)})) \otimes \cdots \otimes (W_2^{-l^{(p)}}(G^{(p)})).$$

An analogous situation holds also in the case of bilinear forms (see Theorem 2.8).

The next lemma gives, in particular, some information about the nature of the space in which the corresponding generalized kernel is situated.

LEMMA 3.3. *Suppose that $W_2^{l'}(G)$ and $W_2^{l''}(G'')$ ($l', l'' = 0, 1, \ldots$) are two Sobolev spaces on bounded domains $G' \subset \mathbb{R}^{N'}$ and $G'' \subset \mathbb{R}^{N''}$ with once piecewise continuously differentiable boundaries. Then the following topological imbeddings are valid:*

$$\begin{gathered} W_2^{\min(l',l'')}(G' \times G'') \supseteq (W_2^{l'}(G')) \otimes (W_2^{l''}(G'')) \supseteq W_2^{l'+l''}(G' \times G''), \\ W_2^{-\min(l',l'')}(G' \times G'') \subseteq (W_2^{-l'}(G')) \otimes (W_2^{-l''}(G'')) \subseteq W_2^{-(l'+l'')}(G' \times G''), \end{gathered} \tag{3.11}$$

with each space dense in the one containing it.

PROOF. Denote the respective norms of the three spaces in the first chain in (3.11) by $\|\cdot\|_1$, $\|\cdot\|_2$, and $\|\cdot\|_3$, and the respective derivatives with respect to x' and x'' by $D^{\mu'}$ and $D^{\mu''}$. Obviously,

$$\begin{aligned} \|u\|_1^2 \le & \sum_{|\mu'| \le l', |\mu''| \le l''} \|D^{\mu'} D^{\mu''} u\|_{L_2(G' \times G'')}^2 \\ & \le \|u\|_3^2 \qquad (u \in C^\infty((G' \times G'')^\sim)). \end{aligned} \tag{3.12}$$

On the other hand, for a function

$$u((x', x'')) = \sum_{n=1}^{q} u_n'(x') u_n''(x''),$$

with $u'_n \in C^\infty(\tilde{G}')$ and $u''_n \in C^\infty(\tilde{G}'')$ we have that

$$\sum_{|\mu'|\le l', |\mu''|\le l''} \|D^{\mu'}D^{\mu''}u\|^2_{L_2(G'\times G'')}$$

$$= \sum_{|\mu'|\le l', |\mu''|\le l''} \sum_{n,m=1}^{q} (D^{\mu'}u'_n \cdot D^{\mu''}u'_n, D^{\mu'}u'_m \cdot D^{\mu''}u''_m)_{L_2(G'\times G'')}$$

$$= \sum_{n,m=1}^{q} \left(\sum_{|\mu'|\le l'} (D^{\mu'}u'_n, D^{\mu'}u'_m)_{L_2(G')} \right) \times \left(\sum_{|\mu''|\le l''} (D^{\mu''}u''_n, D^{\mu''}u''_m)_{L_2(G'')} \right)$$

$$= \sum_{n,m=1}^{q} (u'_n, u'_m)_{W_2^{l'}(G')}(u''_n, u''_m)_{W_2^{l''}(G'')} = \|u\|_2^2.$$

Thus, (3.12) gives us that $\|u\|_1 \le \|u\|_2 \le \|u\|_3$ for such u. Considering the denseness of these functions u in the corresponding spaces, we get the statement of the lemma regarding the first chain in (3.11). The statement regarding the second chain follows from this by duality. ■

In the case of bilinear (and also two-linear) forms the type of the corresponding kernel can be made more precise. Accordingly, let $H_0 = L_2(G)$ on a bounded domain $G \subset \mathbf{R}^N$ $(N = 1, 2, \ldots)$. Consider a continuous bilinear form $H_0 \oplus H_0 \ni (f, g) \mapsto B(f, g)$. We construct the chain

$$W_2^{-l}(G) \supseteq L_2(G) \supseteq W_2^l(G) \qquad (l > N/2) \tag{3.13}$$

and in it we introduce the natural involution by setting $u^*(x) = \overline{u(x)}$ $(x \in G;\ u \in W_2^l(G))$. By (2.48), there exists a $\mathrm{B} \in (W_2^{-l}(G)) \otimes (W_2^{-l}(G))$ such that

$$B(u, v) = (\mathrm{B}, v \otimes u^*)_{L_2(G\times G)} \qquad (u, v \in W_2^l(G)). \tag{3.14}$$

Let J and $\mathbf{J}$ be the operators connected with the chain (3.13). It is clear that they commute with * (see (2.45)). Therefore, by (3.14),

$$\begin{aligned} B(u, v) &= (\mathrm{B}, (JJ^{-1}v) \otimes (JJ^{-1}u)^*)_{L_2(G\times G)} \\ &= ((J \otimes J)^+\mathrm{B}, (J^{-1}v) \otimes (J^{-1}u)^*)_{L_2(G\times G)} \\ &= ((\mathbf{J} \otimes \mathbf{J})\mathrm{B}, (J^{-1}v) \otimes (J^{-1}u)^*)_{L_2(G\times G)}. \end{aligned}$$

Since $\mathbf{J}\colon W_2^{-l}(G) \to L_2(G)$, it follows that $K = (\mathbf{J} \otimes \mathbf{J})\mathrm{B} \in L_2(G \times G)$. The representation (3.14) can now be written as

$$B(u, v) = \int_G \int_G K(x, y)(J^{-1}v)(y)\overline{(J^{-1}u)(x)}\, dx\, dy \qquad (u, v \in W_2^l(G)). \tag{3.15}$$

THEOREM 3.3. *It is actually true that $K \in C((G \times G)^{\sim})$ in the representation* (3.15) *for the continuous bilinear form*

$$L_2(G) \oplus L_2(G) \ni (f, g) \mapsto B(f, g).$$

PROOF. We need a direct derivation of (2.48). In essence, the derivation coincides with the proof of the corresponding part of Theorem 2.7 but is simpler. Consider a general chain (2.43) with quasinuclear imbedding $H_+ \to H_0$, the operators connected with it, and a continuous bilinear form $H_0 \oplus H_0 \ni (f, g) \mapsto B(f, g)$.

Let $A\colon H_0 \to H_0$ be the operator corresponding to the form B, i.e., $(Af, g)_{H_0} = B(f, g)$ $(f, g \in H_0)$. Then for $u, v \in H_+$

$$\begin{aligned} B(u, v) = (Au, v)_{H_0} &= (AOJJ^{-1}u, OJJ^{-1}v)_{H_0} \\ &= ((OJ)^* A(OJ)J^{-1}u, J^{-1}v)_{H_0}. \end{aligned} \tag{3.16}$$

The operator $C = (OJ)^* A(OJ)\colon H_0 \to H_0$ is quasinuclear; therefore, there exists an $F \in H_0 \otimes H_0$ such that $(Cf, g)_{H_0} = (F, g \otimes f^*)_{H_0 \otimes H_0}$ $(f, g \in H_0)$ (see Lemma 3.4). Using this equality and (3.16), we get that

$$B(u, v) = (F, (J^{-1}v) \otimes (J^{-1}u)^*)_{H_0 \otimes H_0} \qquad (u, v \in H_+). \tag{3.17}$$

(Note that by transferring both the operators J^{-1} here to F we get the representation (2.48), where $\mathrm{B} = (\mathbf{J}^{-1} \otimes \mathbf{J}^{-1})F$.)

Consider now the form B in the theorem. Comparing (3.15) and (3.16), we conclude that $F = K$. Let us compute the kernel $K \in L_2(G \times G)$ of C directly.

According to the remark in §1.4, OJ is a selfadjoint operator in H_0; therefore, $(OJ)^* = OJ$, and $Cf = (OJ)A(OJ)f = JA(OJ)f \in W_2^l(G)$ for $f \in H_0 = L_2(G)$. Thus, in a way similar to that for (3.7),

$$\begin{aligned} (Cf)(x) = (JA(OJ)f)(x) &= (JA(OJ)f, \delta_x)_{L_2(G)} \\ &= (f, (OJ)^* A^* \mathbf{J}\delta_x)_{L_2(G)} = (f, (OJ)A^* \mathbf{J}\delta_x)_{L_2(G)} \\ &= \int_G f(y)\overline{(JA^*\mathbf{J}\delta_x)(y)}\, dy. \end{aligned} \tag{3.18}$$

This implies that $K(x, y) = \overline{(JA^*\mathbf{J}\delta_x)(y)}$. According to Theorem 3.1, the vector-valued function $\tilde{G} \ni x \mapsto \delta_x \in W_2^{-l}(G)$ is continuous. But

$$JA^*\mathbf{J}\colon W_2^{-l}(G) \to W_2^l(G)$$

is continuous; therefore, the vector-valued function $\tilde{G} \ni x \mapsto JA^*\mathbf{J}\delta_x \in W_2^l(G)$ is also continuous, and this gives us that $K \in C((G \times G)^{\sim})$. ∎

We establish the lemma used in proving the theorem; it almost coincides with Lemma 2.5.

LEMMA 3.4. *Let H be a Hilbert space with involution* *, *and* $C: H \to H$ *a bounded operator. The form* $(Cf, g)_H$ *can be written in the form*

$$(F, g \otimes f^*)_{H\otimes H} \qquad (f, g \in H),$$

where $F \in H \otimes H$, *if and only if* C *is quasinuclear.*

PROOF. If $(e_j)_1^\infty$ is an orthonormal basis in H, then

$$(e_k^*)_1^\infty \quad \text{and} \quad (e_j \otimes e_k^*)_{j,k=1}^\infty$$

are also such bases in H and $H \otimes H$, respectively. Therefore, if $(Cf, g)_H = (K, g \otimes f^*)_{H\otimes H}$ $(f, g \in H)$, then

$$\sum_{j,k=1}^\infty |(Ce_k, e_j)_H|^2 = \sum_{j,k=1}^\infty |(K, e_j \otimes e_k^*)_{H\otimes H}|^2 = \|K\|_{H\otimes H}^2 < \infty,$$

and C is a quasinuclear operator. Conversely, let C be quasinuclear, i.e.,

$$\sum_{j,k=1}^\infty |(Ce_k, e_j)_H|^2 < \infty.$$

Then K can be taken to be the element of $H \otimes H$ whose coordinates in the basis $(e_j \otimes e_k^*)_{j,k=1}^\infty$ are the numbers $(Ce_k, e_j)_H$. Obviously,

$$(Cf, g)_H = (K, g \otimes f^*)_{H\otimes H} \qquad (f, g \in H). \quad ■$$

Note that the occurrence of the element K in $C((G\times G)^\sim)$, and not just in $L_2(G \times G)$, has to do with the fact that, with respect to one of the variables, B is not just in $W_2^{-l}(G)$ but in $L_2(G)$ (see (2.46) and (2.47)). We return to the representation (3.14) again in §3.11. A simple procedure for finding the kernel of B will be given there.

3.6. *The spaces* $\overset{\circ}{W}{}_2^l(G)$. Consider the Sobolev space $W_2^l(G)$ $(l = 1, 2, \ldots)$ and denote by $\overset{\circ}{W}{}_2^l(G)$ the closure in $W_2^l(G)$ of the collection $C_0^\infty(G)$ of infinitely differentiable functions with compact support in G. It is known (see, for example, the Introduction to Berezanskiĭ [4]) that $\overset{\circ}{W}{}_2^l(G)$ is a proper subspace of $W_2^l(G)$ consisting of all the functions $u \in W_2^l(G)$ such that $(D^\mu u) \restriction \partial G = 0$, where $|\mu| \le l - 1$. Clearly, $\overset{\circ}{W}{}_2^l(G) = H_+$ is dense in $L_2(G) = H_0$. Constructing the corresponding negative space, we get the chain

$$\overset{\circ}{W}{}_2^{-l}(G) \supseteq L_2(G) \supseteq \overset{\circ}{W}{}_2^l(G) \qquad (l = 1, 2, \ldots). \tag{3.19}$$

Application of the arguments in §1.6 to the chains (3.5) and (3.19) leads to a number of properties of the chain (3.19). We shall not state them, but note only that the imbedding $\overset{\circ}{W}{}_2^l(G) \to L_2(G)$ is quasinuclear when $l > N/2$. This is a consequence of the following simple remark: if the imbedding $H_+ \to H_0$ is quasinuclear and G_+ is a subspace of H_+, then the imbedding $G_+ \to H_0$

is also quasinuclear, being the composition of a continuous imbedding and a quasinuclear imbedding.

3.7. *Sobolev spaces on an unbounded domain.* Let $G \subseteq \mathbb{R}^N$ be an unbounded (in general) domain in $\mathbb{R}^N$ ($N = 1, 2, \ldots$) with once piecewise continuously differentiable boundary. For functions $u, v \in (C_0^\infty(\mathbb{R}^N)) \upharpoonright \tilde{G}$ we introduce the inner product $(u, v)_{W_2^l(G)}$ with the help of (3.4). ($C_0^\infty(\mathbb{R}^N)$ is the collection of infinitely differentiable compactly supported functions on $\mathbb{R}^N$.) As in the case of bounded G, the corresponding completion of $(G_0^\infty(\mathbb{R}^N)) \upharpoonright \tilde{G}$ is called the Sobolev space $W_2^l(G)$ ($l = 0, 1, \ldots$).

Clearly, we can now take $W_2^l(G)$ and $L^2(G)$ as the positive and zero spaces, and construct a chain. The notation (3.5) is retained for this chain. Of course, the remark in §3.2 about imbeddings of Sobolev spaces with large index l remains true in the case of unbounded G.

If G' is a bounded subdomain of G, then

$$\|u \upharpoonright G'\|_{W_2^l(G')} \le \|u\|_{W_2^l(G)} \qquad (u \in (C_0^\infty(\mathbb{R}^N)) \upharpoonright G),$$

and thus $(u \upharpoonright G') \in W_2^l(G')$ for $u \in W_2^l(G)$ as a result of completion. Therefore, the functions in $W_2^l(G)$ have the same local properties as the functions in this space on a bounded domain. However, the global properties of $W_2^l(G)$ can be different; for example, Theorem 3.2 on quasinuclearity of imbeddings is no longer preserved in general.

When introducing the space $W_2^l(\mathbb{R}^N)$, it is useful to pass to the Fourier transformation

$$C_0^\infty(\mathbb{R}^N) \ni u(x) \mapsto \tilde{u}(s) = \frac{1}{(2\pi)^{N/2}} \int_{\mathbb{R}^N} u(x) e^{-i\langle s, x\rangle}\, dx$$
$$(s \in \mathbb{R}^N, \langle s, x\rangle = s_1 x_1 + \cdots + s_N x_N). \quad (3.20)$$

For $u \in C_0^\infty(\mathbb{R}^N)$ it is obvious that $(D^\mu u)^\sim(s) = i^{|\mu|} s^\mu \tilde{u}(s)$ ($s \in \mathbb{R}^N$), where $s^\mu = s_1^{\mu_1} \cdots s_N^{\mu_N}$ and $\mu = (\mu_1, \ldots, \mu_N)$. Therefore, by Parseval's equality,

$$(u, v)_{W_2^l(\mathbb{R}^N)} = \int_{\mathbb{R}^N} \tilde{u}(s)\overline{\tilde{v}(s)} p_l(s)\, dx, p_l(s) = \sum_{|\mu| \le l} s^{2\mu} \ge 1$$
$$(l = 0, 1, \ldots;\ u, v \in C_0^\infty(\mathbb{R}^N)). \quad (3.21)$$

Since $(C_0^\infty(\mathbb{R}^N))^\sim$ is dense in $L_2(\mathbb{R}^N)$, (3.21) shows that $W_2^l(\mathbb{R}^N)$ is isometric to the space $L_2(\mathbb{R}^N, p_l(s)\, ds)$. This gives us by continuity that, after closure, (3.20) establishes an isometry between the spaces of the chain $W_2^{-l}(\mathbb{R}^N) \supseteq L_2(\mathbb{R}^N) \supseteq W_2^l(\mathbb{R}^N)$ and the corresponding spaces of the chain of type (3.2)

$$L_2(\mathbb{R}^N, p_l^{-1}(s)\, ds) \supseteq L_2(\mathbb{R}^N) \supseteq L_2(\mathbb{R}^N, p_l(s)\, ds) \qquad (l = 0, 1, \ldots). \quad (3.22)$$

Note that in (3.22) the imbedding of the positive space in the zero space cannot be quasinuclear for any choice of the weight; this corroborates what was said about Theorem 3.2 not being preserved. We show how to modify the concept of a Sobolev space in order to make the imbedding quasinuclear.

Fix a $q(x) \in C^l(\tilde{G})$ with $q(x) > 0$ for $x \in \tilde{G}$. On the functions $u, v \in (C_0^\infty(\mathbf{R}^N)) \upharpoonright \tilde{G}$ we introduce the inner product

$$(u, v)_{W_2^{(l,q)}(G)} = (u(x)q(x), v(x)q(x))_{W_2^l(G)}; \tag{3.23}$$

the completion of $(C_0^\infty(\mathbf{R}^N)) \upharpoonright \tilde{G}$ in this inner product is denoted by the space $W_2^{(l,q)}(G)$. Clearly, $u(x) \in W_2^{(l,q)}(G)$ if and only if $u(x)q(x) \in W_2^l(G)$. This implies that the local properties of the functions in $W_2^{(l,q)}(G)$ are the same as those of the functions in $W_2^l(G)$, i.e., the corresponding functions on a bounded domain.

Assume in addition that $q(x) \geq 1$ for $x \in \tilde{G}$. Then

$$\|u\|_{W_2^{(l,q)}(G)} = \|uq\|_{W_2^l(G)} \geq \|u\|_{L_2(G)} \qquad (u \in W_2^{(l,q)}(G)).$$

Therefore, $W_2^{(l,q)}(G)$ and $L_2(G)$ can be taken as the positive space and the zero space. Constructing the corresponding negative space, we obtain the chain

$$W_2^{-(l,q)}(G) \supseteq L_2(G) \supseteq W_2^{(l,q)}(G). \tag{3.24}$$

THEOREM 3.4. *If $l > N/2$ is an integer, then the δ-function δ_ξ concentrated at a point $\xi \in \mathbf{R}^N$ is in the space $W_2^{-(l,q)}(\mathbf{R}^N)$, and the vector-valued function $\mathbf{R}^N \ni \xi \mapsto \delta_\xi \in W_2^{-(l,q)}(\mathbf{R}^N)$ is continuous and satisfies*

$$\|\delta_\xi\|_{W_2^{-(l,q)}(\mathbf{R}^N)} \leq cq^{-1}(\xi) \qquad (\xi \in \mathbf{R}^N;\ c < \infty). \tag{3.25}$$

PROOF. For $\xi \in \mathbf{R}^N$ let K be an open ball of radius 1 such that $\xi \in K$. According to the imbedding theorems, $W_2^l(K) \subset C(\tilde{K})$ and

$$|v(x)| \leq c\|v\|_{W_2^l(K)} \qquad (x \in \tilde{K},\ v \in W_2^l(K)). \tag{3.26}$$

Therefore, $W_2^{(l,q)}(\mathbf{R}^N) \subset C(\mathbf{R}^N)$, and (3.26) gives us that for $v = uq$ and $x = \xi$

$$\begin{aligned} |u(\xi)| &= q^{-1}(\xi)|u(\xi)q(\xi)| \leq cq^{-1}(\xi)\|(uq) \upharpoonright K\|_{W_2^l(K)} \\ &\leq cq^{-1}(\xi)\|u\|_{W_2^{(l,q)}(\mathbf{R}^N)} \qquad (\xi \in \mathbf{R}^N;\ u \in W_2^{(l,q)}(\mathbf{R}^N)). \end{aligned} \tag{3.27}$$

This inequality shows that δ_ξ is defined as an element of $W_2^{-(l,q)}(\mathbf{R}^N)$, and (3.25) is valid.

To prove that δ_ξ is continuous, observe that for any open ball K with radius equal to 1

$$\|\delta_\xi - \delta_\eta\|_{W_2^{-(l,q)}(\mathbf{R}^N)} \leq \|q^{-1}(\xi)\delta_\xi - q^{-1}(\eta)\delta_\eta\|_{W_2^{-l}(K)} \qquad (\xi, \eta \in K). \tag{3.28}$$

Indeed, we see in a way similar to that for (3.27) that if $u \in W_2^{(l,q)}(\mathbb{R}^N)$, then

$$\begin{aligned}|(\delta_\xi - \delta_\eta, u)_{L_2(\mathbb{R}^N)}| &= |u(\xi) - u(\eta)| \\ &= |(q^{-1}(\xi)\delta_\xi - q^{-1}(\eta)\delta_\eta, (uq) \upharpoonright K)_{L_2(K)}| \\ &\le \|q^{-1}(\xi)\delta_\xi - q^{-1}(\eta)\delta_\eta\|_{W_2^{-l}(K)} \|(uq) \upharpoonright K\|_{W_2^l(K)} \\ &\le \|q^{-1}(\xi)\delta_\xi - q^{-1}(\eta)\delta_\eta\|_{W_2^{-l}(K)} \|uq\|_{W_2^l(\mathbb{R}^N)} \\ &= \|q^{-1}(\xi)\delta_\xi - q^{-1}(\eta)\delta_\eta\|_{W_2^{-l}(K)} \|u\|_{W_2^{(l,q)}(\mathbb{R}^N)},\end{aligned}$$

which implies (3.28).

The required continuity of δ_ξ follows from (3.28) and the continuity of $q^{-1}(\xi)$ and of the vector-valued function $\tilde{K} \ni \xi \mapsto \delta_\xi \in W_2^{-l}(K)$ (see Theorem 3.1). ■

The assertion in Theorem 3.4 is valid not only for $G = \mathbb{R}^N$ but also for a broader class of domains. A domain $G \subseteq \mathbb{R}^N$ with once piecewise continuously differentiable boundary is said to be regular if there exist a bounded domain $K \subset \mathbb{R}^N$ with boundary of the same class and an $R > 0$ such that for any point $\xi \in \tilde{G}$ with $|\xi| \ge R$ there is a domain K_ξ obtained from K by an orthogonal rotation and a translation such that $\xi \in \tilde{K}_\xi$ and $\tilde{K}_\xi \subseteq \tilde{G}$. *If $G \subseteq \mathbb{R}^N$ is a regular domain, then the assertion of Theorem* 3.4 *remains true for the space $W_2^{-(l,q)}(G)$, with $\xi \in \tilde{G}$ in the formulation of this assertion.* Indeed, inequality (3.26) is now preserved with a constant c not depending on $\xi \in \tilde{G}$ with $|\xi| \ge R$, and hence also an inequality $|u(\xi)| \le cq^{-1}(\xi)\|u\|_{W_2^{(l,q)}(G)}$ $(\xi \in \tilde{G}, |\xi| \ge R)$ of the type (3.27). This implies an estimate of the type (3.25). The proof that δ_ξ is continuous does not differ from that given above. ■

We shall not generalize this theorem to the case of nonregular domains, for example, domains with "cusps" going to infinity, when the c in (3.25) depends on $\xi \in \tilde{G}$ in general.

THEOREM 3.5. *Let $G = \mathbb{R}^N$ or, more generally, let G be a regular domain, and let $q(x) \ge 1$ for $x \in \tilde{G}$. The imbedding $W_2^{(l,q)}(G) \to L_2(G)$ is quasinuclear if*

$$l > N/2, \qquad \int_G q^{-2}(x)\,dx < \infty. \tag{3.29}$$

PROOF. The proof differs little from that of Theorem 3.2 in the main case. Indeed, consider the chain (3.24) and the operators connected with it. It is necessary to prove that the operator $OJ\colon L_2(G) \to L_2(G)$ is quasinuclear. The representation (3.7) is obviously valid for it. Setting $K(x,y) = \overline{(\mathbf{J}\delta_x)(y)}$, we get with the help of (3.25) and (3.29) that for G

$$\begin{aligned}\int_G\int_G |K(x,y)|^2\,dx\,dy &= \int_G \|\mathbf{J}\delta_x\|^2_{L_2(G)}\,dx = \int_G \|\delta_x\|^2_{W_2^{-(l,q)}(G)}\,dx \\ &\le c^2 \int q^{-2}(x)\,dx < \infty. \quad ■\end{aligned}$$

As in the case of bounded G, the operator OI is an integral operator with kernel in $C((G\times G)^{\sim})$ when $l > N/2$ if the domain is regular. What is more, only the continuity of the vector-valued function $\tilde{G} \ni \xi \mapsto \delta_\xi \in W_2^{-(l,q)}(G)$ is used to prove this fact, and the estimate (3.25) is not needed. Therefore, as is clear from the proof of Theorem 3.4, it is true also for nonregular domains.

3.8. ***Sobolev spaces with a weight.*** The spaces $W_2^{(l,q)}(G)$ will play a mostly auxiliary role in what follows. The Sobolev spaces $W_2^l(G, p(x)\,dx)$ with a weight will be encountered much more frequently; they are defined as follows.

Let $G \subseteq \mathbf{R}^N$ be an unbounded (in general) domain in $\mathbf{R}^N$ ($N = 1, 2, \ldots$) with once piecewise continuously differentiable boundary, and let $p(x) \in C(\tilde{G})$, $p(x) > 0$ ($x \in \tilde{G}$), be a fixed weight. On the functions $u, v \in (C_0^\infty(\mathbf{R}^N)) \restriction \tilde{G}$ we introduce the inner product

$$(u, v)_{W_2^l(G,p(x)\,dx)} = \sum_{|\mu|\le l} \int_G (D^\mu u)(x)\overline{(D^\mu v)(x)}p(x)\,dx$$

$$(l = 0, 1, \ldots,); \qquad (3.30)$$

$W_2^l(G, p(x)\,dx)$ is defined as the completion of $(C_0^\infty(\mathbf{R}^N)) \restriction \tilde{G}$ with respect to (3.30). It is clear that the local properties of the functions in $W_2^l(G, p(x)\,dx)$ are the same as for the functions in $W_2^l(G')$ with bounded G'. Comparing (3.23) and (3.30), we obviously get the estimate

$$\|u\|_{W_2^{(l,q)}(G)} \le c_l \|u\|_{W_2^l(G, q_{(l)}^2(x)\,dx)} \qquad (c_l < \infty)$$

for functions $u \in (C_0^\infty(\mathbf{R}^N)) \restriction \tilde{G}$, where we use the notation

$$q_{(l)}(x) = \max_{|\mu|\le l} |(D^\mu q)(x)| \qquad (x \in \tilde{G}). \qquad (3.31)$$

This implies the topological imbedding

$$W_2^l(G, q_{(l)}^2(x)\,dx) \subseteq W_2^{(l,q)}(G) \qquad (l = 0, 1, \ldots), \qquad (3.32)$$

where the first space is dense in the second.

Assume that $p(x) \ge 1$ for $x \in \tilde{G}$; then we can take $W_2^l(G, p(x)\,dx)$ and $L_2(G)$ as the positive and zero spaces, and construct the corresponding negative space $W_2^{-l}(G, p(x)\,dx)$. As a result we get a chain that will be used frequently:

$$W_2^{-l}(G, p(x)\,dx) \supseteq L_2(G) \supseteq W_2^l(G, p(x)\,dx) \qquad (l = 0, 1, \ldots). \qquad (3.33)$$

The role of Theorem 3.2 will be played by the next theorem.

THEOREM 3.6. *Let $G = \mathbf{R}^N$ or, more generally, let G be a regular domain, let $m = 0, 1, \ldots,$ and let $l > N/2$ be an integer. If the weights $q_1, q_2 \in C^l(\tilde{G})$ are such that $0 < q_1(x) \le q_2(x)$ ($x \in \tilde{G}$) and*

$$\int_G \frac{q_1^2(x)}{q_2^2(x)}\,dx < \infty, \qquad (3.34)$$

then the imbedding $W_2^{m+l}(G, q_{2,(l)}^2(x)\,dx) \to W_2^m(G, q_1^2(x)\,dx)$ *is quasinuclear.*

PROOF. Comparing the conditions in this theorem with those in Theorem 3.5, we conclude that the imbedding $H_1 = W_2^{(l,q_2/q_1)}(G) \to L_2(G) = G_1$ is quasinuclear. Now apply to these spaces the remark after Lemma 3.1 with $E = (C_0^\infty(\mathbb{R}^N)) \restriction \tilde{G}$ and $(Tf)(x) = q_1(x)f(x)$ $(x \in \tilde{G},\ f \in E)$; $\operatorname{Ker} T = 0$. As a result, the imbedding $H_2 = W_2^{(l,q_2)}(G) \to L_2(G, q_1^2(x)\,dx) = G_2$ is quasinuclear.

According to (3.32), the imbedding $W_2^l(G, q_{2,(l)}^2(x)\,dx) \to W_2^{(l,q_2)}(G)$ is continuous; therefore, the imbedding $W_2^l(G, q_{2,(l)}^2(x)\,dx) \to L_2(G, q_1^2(x)\,dx)$ is quasinuclear.

The rest of the argument is completely analogous to the proof of Theorem 3.2. Namely, apply Lemma 3.1 with $E = (C_0^\infty(\mathbb{R}^N)) \restriction \tilde{G}, (Tf)(x) = (D^\nu f)(x)$, where D^ν is a fixed derivative of order $|\nu| \le m$, and

$$(f,g)_{H_1} = (f,g)_{W_2^l(G,q_{2,(l)}^2(x)\,dx)} \quad \text{and} \quad (f,g)_{G_1} = (f,g)_{L_2(G,q_1^2(x)\,dx)}$$

$(f, g \in E)$. Considering the proven quasinuclearity of the imbedding $H_1 \to G_1$, we get that the imbedding $H_2 \to G_2$ is also quasinuclear, where H_2 and G_2 are the completions of E in the inner products

$$(f,g)_{H_2} = (f,g)_{W_2^l(G,q_{2,(l)}^2(x)\,dx)} + (D^\nu f, D^\nu g)_{W_2^l(G,q_{2,(l)}^2(x)\,dx)},$$

$$(f,g)_{G_2} = (f,g)_{L_2(G,q_1^2(x)\,dx)} + (D^\nu f, D^\nu g)_{L_2(G,q_1^2(x)\,dx)} \qquad (f,g \in E).$$

Now applying Lemma 3.2, we conclude that the imbedding $H \to G$ is quasinuclear, where H and G are the completions of E in the inner products

$$(f,g)_H = n(f,g)_{W_2^l(G,q_{2,(l)}^2(x)\,dx)} + \sum_{|\nu|\le m} (D^\nu f, D^\nu g)_{W_2^l(G,q_{2,(l)}^2(x)\,dx)},$$

$$(f,g)_G = n(f,g)_{L_2(G,q_1^2(x)\,dx)} + \sum_{|\nu|\le m} (D^\nu f, D^\nu g)_{L_2(G,q_1^2(x)\,dx)} \qquad (f,g \in E).$$

Obviously, $H = W_2^{m+l}(G, q_{2,(m)}^2(x)\,dx)$ and $G = W_2^m(G, q_1^2(x)\,dx)$, which gives us the theorem. ■

In considering the Schwartz spaces $\mathcal{S}(\mathbb{R}^N)$ of test functions we shall need Sobolev spaces with a special weight. Let

$$S_l(\mathbb{R}^N) = W_2^l(\mathbb{R}^N, (1+|x|^2)^l\,dx).$$

Then

$$(u,v)_{S_l(\mathbb{R}^N)} = \sum_{|\mu|\le l} \int_{\mathbb{R}^N} (D^\mu u)(x)\overline{(D^\mu v)(x)}(1+|x|^2)^l\,dx$$

$$(l = 0, 1, \dots;\ u, v \in S_l(\mathbb{R}^N)), \qquad (3.35)$$

and the sequence of norms $\|\cdot\|_{S_l(\mathbb{R}^N)}$ is monotone: $\|\cdot\|_{S_0(\mathbb{R}^N)} \le \|\cdot\|_{S_1(\mathbb{R}^N)} \le \dots$.

THEOREM 3.7. *If* $l'' - l' > N/2$ $(l', l'' = 0, 1, \dots)$, *then the imbedding* $S_{l''}(\mathsf{R}^N) \to S_{l'}(\mathsf{R}^N)$ *is quasinuclear.*

PROOF. Let $m = 0, 1, \dots$, and let $l > N/2$ be an integer. It is required to prove that the imbedding $S_{m+l}(\mathsf{R}^N) \to S_m(\mathsf{R}^N)$ is quasinuclear. We use Theorem 3.6 with $q_1(x) = (1+|x|^2)^{m/2}$ and $q_2(x) = (1+|x|^2)^{(m+l)/2}$; condition (3.34) is obviously satisfied. Consequently, the imbedding

$$W_2^{m+l}(\mathsf{R}^N, q_{2,(l)}^2(x)\,dx) \to S_m(\mathsf{R}^N)$$

is quasinuclear.

The weight $q_2(x)$ satisfies the estimate $q_{2,(l)}(x) \le c_{m,l} q_2(x)$ $(x \in \mathsf{R}^N;$ $c_{m,l} < \infty)$; therefore, $S_{m,l}(\mathsf{R}^N) \subseteq W_2^{m+l}(\mathsf{R}^N, q_{2,(l)}^2(x)\,dx)$ topologically. Taking the composition of the last two imbeddings, we conclude that $S_{m+l}(\mathsf{R}^N) \to S_m(\mathsf{R}^N)$ is quasinuclear. ■

3.9. *The Schwartz space $S(\mathsf{R}^N)$ as a projective limit of Sobolev spaces.* The space $S(\mathsf{R}^N)$ is usually defined as follows. Define a monotone sequence of norms on $C_0^\infty(\mathsf{R}^N)$ $(N = 1, 2, \dots)$ by

$$\|u\|_{S_\tau(\mathsf{R}^N)} = \max_{x\in\mathsf{R}^N} \Big((1+|x|^2)^{\tau/2} \sum_{|\mu|\le\tau} |(D^\mu u)(x)|\Big) \qquad (\tau = 0, 1, \dots). \quad (3.36)$$

Let $S_\tau(\mathsf{R}^N)$ be the completion of $C_0^\infty(\mathsf{R}^N)$ with respect to (3.36). Then

$$S(\mathsf{R}^N) = \operatorname*{pr\,lim}_{\tau\in\mathrm{T}} S_\tau(\mathsf{R}^N),$$

where $\mathrm{T} = \{0, 1, \dots\}$, i.e., $S(\mathsf{R}^N) = \bigcap_{\tau=0}^\infty S_\tau(\mathsf{R}^N)$ and the sets

$$U(0; \tau, \varepsilon) = \{\varphi \in S(\mathsf{R}^N) \mid \|\varphi\|_{S_\tau(\mathsf{R}^N)} < \varepsilon\}$$

with arbitrary $\tau = 0, 1, \dots$ and $\varepsilon > 0$ form a base of neighborhoods of zero in $S(\mathsf{R}^N)$. Thus, $S(\mathsf{R}^N)$ consists of the infinitely differentiable functions on R^N which decrease, along with all their derivatives, more rapidly than any power $|x|^{-l}$ as $|x| \to \infty$; convergence in $S(\mathsf{R}^N)$ is described by the indicated neighborhood base. We show that it can be constructed in a similar way from the Sobolev spaces $S_l(\mathsf{R}^N)$.

THEOREM 3.8. *The space $S(\mathsf{R}^N)$ coincides with the projective limit of the spaces $S_\tau(\mathsf{R}^N)$:*

$$S(\mathsf{R}^N) = \operatorname*{pr\,lim}_{\tau\in\mathrm{T}} S_\tau(\mathsf{R}^N).$$

This space is nuclear.

PROOF. To derive the relation $S(\mathsf{R}^N) = \operatorname{pr\,lim}_{\tau\in\mathrm{T}} S_\tau(\mathsf{R}^N)$ it is necessary to prove two inequalities: for each $\tau \in \mathrm{T}$ there is a $\tau' \in \mathrm{T}$ such that $\|u\|_{S_\tau(\mathsf{R}^N)} \le c_{\tau\tau'} \|u\|_{S_{\tau'}(\mathsf{R}^N)}$ $(u \in C_0^\infty(\mathsf{R}^N))$ for some $c_{\tau\tau'} < \infty$; and the analogous inequality with $S_\tau(\mathsf{R}^N)$ and $S_\tau(\mathsf{R}^N)$ interchanged.

The first inequality is trivial: let $l > N/2$ be an integer; then, by (3.35)

$$\begin{aligned}\|u\|^2_{S_\tau(\mathbf{R}^N)} &\le \max_{x\in\mathbf{R}^N}((1+|x|^2)^{\tau+l}\sum_{|\mu|\le\tau}|(D^\mu u)(x)|^2)\int_{\mathbf{R}^N}(1+|x|^2)^{-l}\,dx\\ &\le c^2_{\tau\tau'}\|u\|^2_{S_{\tau'}(\mathbf{R}^N)} \qquad (\tau\in\mathrm{T};\ \tau'=\tau+l)\end{aligned}$$

for $u \in C_0^\infty(\mathbf{R}^N)$.

We now show that for each $\tau \in \mathrm{T}$ there is a $\tau' \in \mathrm{T}$ such that

$$\|u\|_{S_\tau(\mathbf{R}^N)} \le c'_{\tau\tau'}\|u\|_{S_{\tau'}(\mathbf{R}^N)} \qquad (u \in C_0^\infty(\mathbf{R}^N))$$

for some $c'_{\tau\tau'} < \infty$. Fix an integer $l > N/2$ and denote by K_x the open ball of radius 1 about a point $x \in \mathbf{R}^N$. According to the imbedding theorems,

$$\|u\|_{C(\tilde{K}_x)} \le c_1\|u\|_{W_2^l(K_x)} \qquad (u \in W_2^l(K_x))$$

with a constant c_1 independent of x. This gives us that for $u \in C_0^\infty(\mathbf{R}^N)$

$$|u(x)| \le c_1\|u \restriction K_x\|_{W_2^l(K_x)} \le c_1\|u\|_{W_2^l(\mathbf{R}^N)} \qquad (x \in \mathbf{R}^N). \tag{3.37}$$

Replacing $u(x)$ by $(1+|x|^2)^{\tau/2}(D^\mu u)(x)$ in (3.37), where $u \in C_0^\infty(\mathbf{R}^N)$ and $|\mu| \le \tau$, we get that

$$\begin{aligned}(1+|x|^2)^{\tau/2}|(D^\mu u)(x)| &\le c_1\|(1+|x|^2)^{\tau/2}(D^\mu u)(x)\|_{W_2^l(\mathbf{R}^N)}\\ &= c_1\left(\sum_{|\nu|\le l}\int_{\mathbf{R}^N}|(D^\nu((1+|x|^2)^{\tau/2}(D^\mu u)))(x)|^2\,dx\right)^{1/2}\\ &= c_1\left(\sum_{|\nu|\le l}\int_{\mathbf{R}^N}\left|\sum_{|\kappa|\le l,|\lambda|\le\tau+l}c_{\mu\nu\kappa\lambda}(D^\kappa(1+|x|^2)^{\tau/2}\right.\right.\\ &\qquad\left.\left.\times (D^\lambda u)(x)\right|^2 dx\right)^{1/2}\end{aligned} \tag{3.38}$$

Here the $c_{\mu\nu\kappa\lambda}$ are coefficients obtained from the Leibniz formula. Since

$$|D^\kappa(1+|x|^2)^{\tau/2}| \le c_2(1+|x|^2)^{\tau/2} \qquad (x \in \mathbf{R}^N;\ |\kappa| \le m),$$

it follows from an upper estimate of the right-hand side of (3.38) by means of elementary inequalities that

$$\begin{aligned}(1+|x|^2)^{\tau/2}|(D^\mu u)(x)| &\le c_3\sum_{|\lambda|\le\tau+l}\left(\int_{\mathbf{R}^N}|(D^\lambda u)(x)|^2(1+|x|^2)^\tau\,dx\right)^{1/2}\\ &\le c_4\|u\|_{S_{\tau+l}(\mathbf{R}^N)}.\end{aligned}$$

Setting $\tau' = \tau + l$, we get by this and (3.36) that

$$\|u\|_{S_\tau(\mathbf{R}^N)} \le c'_{\tau\tau'}\|u\|_{S_{\tau'}(\mathbf{R}^N)} \qquad (u \in C_0^\infty(\mathbf{R}^N)).$$

Accordingly, $\mathcal{S}(\mathbf{R}^N) = \operatorname{pr}\lim_{\tau\in\mathrm{T}} S_\tau(\mathbf{R}^N)$. The fact that $\mathcal{S}(\mathbf{R}^N)$ is a nuclear space follows from Theorem 3.7. ∎

Note that Theorem 1.5 implies the equality

$$\mathcal{S}'(\mathbf{R}^N) = \bigcup_{\tau=0}^{\infty} S_{-\tau}(\mathbf{R}^N), \quad S_{-\tau}(\mathbf{R}^N) = W_2^{-\tau}(\mathbf{R}^N, (1+|x|^2)^\tau\, dx). \tag{3.39}$$

3.10. ***The space $\mathcal{D}(\mathbf{R}^N)$ as a projective limit of Sobolev spaces.*** The well-known space $\mathcal{D}(\mathbf{R}^N)$ ($N = 1, 2, \ldots$) is defined as the collection of functions $C_0^\infty(\mathbf{R}^N)$, with the following convergence: $C_0^\infty(\mathbf{R}^N) \ni \varphi_n \to \varphi \in C_0^\infty(\mathbf{R}^N)$ as $n \to \infty$ if the functions $\varphi_n(x)$ are uniformly compactly supported (i.e., there exists an $r > 0$ depending on $(\varphi_0(x))_1^\infty$ such that $\varphi_n(x) = 0$ for $|x| \geq r$ and all $n = 1, 2, \ldots$), and $(D^\mu \varphi_n)(x) \to (D^\mu \varphi)(x)$ as $n \to \infty$ uniformly for each derivative. Such a convergence arises if the inductive or the projective topology is introduced in $C_0^\infty(\mathbf{R}^N)$. (These topologies are not equivalent.)

In what follows, $\mathcal{D}(\mathbf{R}^N)$ is understood to be $C_0^\infty(\mathbf{R}^N)$ with the projective topology, which is defined as follows. Let T be the collection of all pairs $\tau = (\tau_1, \tau_2(x))$, where $\tau_1 = 0, 1, \ldots$, and $\tau_2 \in C^\infty(\mathbf{R}^N)$ and $\tau_2(x) \geq 1$ ($x \in \mathbf{R}^N$). For each $\tau \in \mathrm{T}$ we introduce a norm on $C_0^\infty(\mathbf{R}^N)$ by setting

$$\|u\|_{\mathcal{D}_\tau(\mathbf{R}^N)} = \max_{x\in\mathbf{R}^N} \left(\tau_2(x) \sum_{|\mu|\leq\tau_1} |(D^\mu u)(x)| \right). \tag{3.40}$$

Let $\mathcal{D}_\tau(\mathbf{R}^N)$ be the completion of $C_0^\infty(\mathbf{R}^N)$ with respect to (3.40). Then

$$\mathcal{D}(\mathbf{R}^N) = \operatorname*{pr\,lim}_{\tau\in\mathrm{T}} \mathcal{D}_\tau(\mathbf{R}^N), \quad \text{i.e.} \quad \mathcal{D}(\mathbf{R}^N) = \bigcap_{\tau\in\mathrm{T}} \mathcal{D}_\tau(\mathbf{R}^N),$$

and a base of neighborhoods of zero is formed by the sets

$$\begin{aligned} &U(0; \tau^{(1)}, \ldots, \tau^{(m)}, \varepsilon_1, \ldots, \varepsilon_m) \\ &\quad = \left\{\varphi \in \mathcal{D}(\mathbf{R}^N) \,|\, \|\varphi\|_{\mathcal{D}_{\tau^{(1)}}(\mathbf{R}^N)} < \varepsilon_1, \ldots, \|\varphi\|_{\mathcal{D}_{\tau^{(m)}}(\mathbf{R}^N)} < \varepsilon_m\right\} \end{aligned}$$

for arbitrary $\tau^{(1)}, \ldots, \tau^{(m)} \in \mathrm{T}$ and $\varepsilon_1, \ldots, \varepsilon_m > 0$ ($m = 1, 2, \ldots$).

The system of norms (3.40) is clearly directed: for each $\tau', \tau'' \in \mathrm{T}$ there is a $\tau''' \in \mathrm{T}$ such that $\mathcal{D}_{\tau'''}(\mathbf{R}^N) \subseteq \mathcal{D}_{\tau'}(\mathbf{R}^N)$ and $\mathcal{D}_{\tau'''}(\mathbf{R}^N) \subseteq \mathcal{D}_{\tau''}(\mathbf{R}^N)$ topologically. (For example, we can set $\tau''' = (\tau_1' + \tau_1'', \tau_2'(x) + \tau_2''(x))$.) Therefore, an equivalent base of neighborhoods of zero is formed by the sets

$$U(0; \tau, \varepsilon) = \{\varphi \in \mathcal{D}(\mathbf{R}^N) \,|\, \|\varphi\|_{\mathcal{D}_\tau(\mathbf{R}^N)} < \varepsilon\}$$

for arbitrary $\tau \in \mathrm{T}$ and $\varepsilon > 0$. It is not hard to show that

$$\operatorname*{pr\,lim}_{\tau\in\mathrm{T}} \mathcal{D}_\tau(\mathbf{R}^N) = C_0^\infty(\mathbf{R}^N),$$

as a set, while the convergence in $\mathcal{D}(\mathbf{R}^N)$ coincides with the convergence described at the beginning of this subsection.

We show that $\mathcal{D}(\mathbf{R}^N)$ can be constructed in a similar way from Sobolev spaces. Let

$$D_\tau(\mathbf{R}^N) = W_2^{\tau_1}(\mathbf{R}^N, \tau_2(x)\,dx) \qquad (\tau = (\tau_1, \tau_2(x)) \in \mathrm{T}). \tag{3.41}$$

Observe that the system of norms of (3.41) is also directed: for each $\tau', \tau'' \in \mathrm{T}$ we can set $\tau''' = (\tau_1' + \tau_1'', \tau_2'(x) + \tau_2''(x)) \in \mathrm{T}$. Then

$$D_{\tau'''}(\mathbf{R}^N) \subseteq D_{\tau'}(\mathbf{R}^N) \quad \text{and} \quad D_{\tau'''}(\mathbf{R}^N) \subseteq D_{\tau''}(\mathbf{R}^N)$$

topologically.

THEOREM 3.9. *The space $\mathcal{D}(\mathbf{R}^N)$ coincides with the projective limit of the spaces $D_\tau(\mathbf{R}^N)$:*

$$\mathcal{D}(\mathbf{R}^N) = \operatorname*{pr\,lim}_{\tau \in \mathrm{T}} D_\tau(\mathbf{R}^N).$$

This space is nuclear.

PROOF. The proof is analogous to that of Theorem 3.8. Let $p(x) \in C^\infty(\mathbf{R}^N)$ be such that $p(x) \geq 1$ $(x \in \mathbf{R}^N)$ and $\int_{\mathbf{R}^N} p^{-1}(x)\,dx < \infty$. Then, by (3.30), for each $\tau \in \mathrm{T}$ and any $u \in C_0^\infty(\mathbf{R}^N)$ we have

$$\begin{aligned} \|u\|^2_{D_\tau(\mathbf{R}^N)} &\leq \max_{x \in \mathbf{R}^N} \left(\tau_2(x) p(x) \sum_{|\mu| \leq \tau_1} |(D^\mu u)(x)|^2 \right) \int_{\mathbf{R}^N} p^{-1}(x)\,dx \\ &\leq c^2_{\tau\tau'} \|u\|^2_{D_{\tau'}(\mathbf{R}^N)} \qquad (\tau' = (\tau_1, (\tau_2(x)p(x))^{1/2}). \end{aligned}$$

Let us prove the opposite inequality. Fix $\tau = (\tau_1, \tau_2(x)) \in \mathrm{T}$ and an integer $l > N/2$ and substitute the function $\tau_2(x)(D^\mu u)(x)$ in place of $u(x)$ in (3.37), where $u \in C_0^\infty(\mathbf{R}^N)$ and $|\mu| \leq \tau_1$. We get the estimate (analogous to (3.38))

$$\begin{aligned} &\tau_2(x)|(D^\mu u)(x)| \\ &\quad \leq c_1 \left(\sum_{|\nu| \leq l} \int_{\mathbf{R}^N} \left| \sum_{|\kappa| \leq l, |\lambda| \leq \tau_1 + l} c_{\mu\nu\kappa\lambda} (D^\kappa \tau_2)(x)(D^\lambda u)(x) \right|^2 dx \right)^{1/2}. \end{aligned} \tag{3.42}$$

Denote by $\tau_2'(x)$ a function in $C^\infty(\mathbf{R}^N)$ such that $|(D^\kappa \tau_2)(x)|^2 \leq \tau_2'(x)$ $(x \in \mathbf{R}^N)$ for all $|\kappa| \leq l$. Getting an upper estimate for the right-hand side of (3.42), we find that

$$\begin{aligned} \tau_2(x)|(D^\mu u)(x)| &\leq c_2 \sum_{|\lambda| \leq \tau_1 + l} \left(\int_{\mathbf{R}^N} |(D^\lambda u)(x)|^2 \, \tau_2'(x)\,dx \right)^{1/2} \\ &\leq c_3 \|u\|_{W_2^{\tau_1 + l}(\mathbf{R}^N, \tau_2'(x)\,dx)}. \end{aligned}$$

This and (3.40) imply that $\|u\|_{D_\tau(\mathbf{R}^N)} \leq c'_{\tau\tau'}\|u\|_{D_{\tau'}(\mathbf{R}^N)}$ $(u \in C_0^\infty(\mathbf{R}^N))$, where $\tau' = (\tau_1 + l, \tau_2'(x)) \in \mathrm{T}$.

Accordingly, $\mathcal{D}(\mathbf{R}^N) = \operatorname{pr\,lim}_{\tau\in\mathrm{T}} D_\tau(\mathbf{R}^N)$. The fact that $\mathcal{D}(\mathbf{R}^N)$ is a nuclear space follows from Theorem 3.6. ■

By Theorem 1.5,

$$\mathcal{D}'(\mathbf{R}^N) = \bigcup_{\tau\in\mathrm{T}} D_{-\tau}(\mathbf{R}^N), \quad D_{-\tau}(\mathbf{R}^N) = W_2^{-\tau_1}(\mathbf{R}^N, \tau_2(x)\,dx). \tag{3.43}$$

In conclusion we remark that §§3.9 and 3.10 and Theorem 3.6 make it clear how to construct Sobolev spaces in such a way that their projective limits are nuclear. Consider a collection of Sobolev spaces $W_2^{\tau_1}(\mathbf{R}^N, \tau_2(x)\,dx)$, where τ_1 runs through a subset of the numbers $\{0, 1, \ldots\}$ and $\tau_2(x)$ through the set of positive functions in $C^\infty(\mathbf{R}^N)$. Let T be the collection of these pairs $\tau = (\tau_1, \tau_2(x))$. For the space $\operatorname{pr\,lim}_{\tau\in\mathrm{T}} W_2^{\tau_1}(\mathbf{R}^N, \tau_2(x)\,dx)$ to be nuclear it suffices that the following condition hold (condition N): for each $\tau = (\tau_1, \tau_2(x)) \in \mathrm{T}$ there is a $\tau = (\tau_1', \tau_2'(x)) \in \mathrm{T}$ such that

$$\int_{\mathbf{R}^N} \frac{\tau_2(x)}{q^2(x)}\,dx < \infty,$$

where $q^2(x) \geq \tau_2(x)$, $\tau_2'(x) \geq q^2_{(l)}(x)$ $(x \in \mathbf{R}^N)$ and $\tau_1' \geq \tau_1 + l$ (here l is the smallest integer greater than $N/2$) for some function $q(x) \in C^l(\mathbf{R}^N)$ (depending on $\tau_2(x)$ and $\tau_2'(x)$). Of course, we could impose less stringent restrictions on the smoothness of τ_2, consider a domain $G \subseteq \mathbf{R}^N$ instead of $\mathbf{R}^N$, etc.

3.11. ***The kernel theorem in spaces of square-integrable functions.*** Let us consider a multilinear form $\bigoplus_1^p H_{0,n} \ni (f^{(1)}, \ldots, f^{(p)}) \mapsto F(f^{(1)}, \ldots, f^{(p)})$, where $H_{0,n} = L_2(G^{(n)})$ and $G^{(n)} \subseteq \mathbf{R}^{N^{(n)}}$ $(N^{(n)} = 1, 2, \ldots)$ is an unbounded regular domain. The general Theorems 2.7 and 2.8 can be applied to such a form or to a corresponding bilinear form in the same way as in §3.5, by taking the chains (2.31) to be, for example,

$$W_2^{-l^{(n)}}(G^{(n)}, q^2_{n,(l^{(n)})}(x)\,dx) \supseteq L_2(G^{(n)}) \supseteq W_2^{l^{(n)}}(G^{(n)}, q^2_{n,(l^{(n)})}(x)\,dx), \tag{3.44}$$

where $l^{(n)} > N^{(n)}/2$ and $\int_{G^{(n)}} q_n^{-2}(x)\,dx < \infty$. (The required quasinuclearity of the imbeddings is satisfied by virtue of Theorem 3.6.)

A bilinear form $L_2(G) \oplus L_2(G) \ni (f, g) \mapsto B(f, g)$, where $G \subseteq \mathbf{R}^N$ is unbounded and regular, also has a representation of the type (3.15)

$$B(u, v) = \int_G \int_G K(x, y)(J^{-1}v)(y)\overline{(J^{-1}u)(x)}\,dx\,dy. \tag{3.45}$$

Here $u, v \in W_2^l(G, q^2_{(l)}(x)\,dx)$, $l > N/2$, $\int_G q^{-2}(x)\,dx < \infty$, and the operator J is connected with the chain

$$W_2^{-l}(G, q^2_{(l)}(x)\,dx) \supseteq L_2(G) \supseteq W_2^l(G, q^2_{(l)}(x)\,dx). \tag{3.46}$$

There is an analogue of Theorem 3.3 in this case: the kernel K in (3.45) is in $C((G \times G)^{\sim})$. This is proved in exactly the same way as Theorem 3.3, except that the continuity of the vector-valued function $\tilde{G} \ni x \mapsto \delta_x \in W_2^{-l}(G, q_{(l)}^2(x)\,dx)$ must be used. The latter follows from Theorem 3.4 and the inclusion dual to (3.32). We remark also that (3.25) and the formula $K(x,y) = \overline{(JA^*\mathbf{J}\delta_x)(y)}$ give us a certain estimate of the growth of this kernel at infinity.

If the chains (3.44) or (3.46) are taken to be the chains

$$S_{-l^{(n)}}(\mathbb{R}^{N^{(n)}}) \supseteq L_2(\mathbb{R}^{N^{(n)}}) \supseteq S_{l^{(n)}}(\mathbb{R}^{N^{(n)}})$$

($l^{(n)} > N^{(n)}/2$), then the corresponding kernel is in the tensor product of the spaces $S_{-l^{(n)}}(\mathbb{R}^{N^{(n)}})$. A fact analogous to Lemma 3.3 (see §4.4) is easy to prove for this product.

We pass to another, "elementary" kind of kernel theorem for bilinear forms defined in $L_2(G)$. (An analogous construction is possible for multilinear forms.) Accordingly, let $G \subseteq \mathbb{R}^N$ be a bounded (or unbounded) domain. Consider the characteristic function of the open parallelepiped in $\mathbb{R}^N$ formed by the coordinate hyperplanes and the hyperplanes parallel to them which pass through some point $x = (x_1, \dots, x_N) \in \mathbb{R}^N$. Let $\omega(x,\xi)$ ($\xi \in G$) denote the product of this function, the characteristic function $\kappa_G(\xi)$ of G, and $(-1)^N \operatorname{sgn} x_1 \cdots \operatorname{sgn} x_N$. Obviously, $\omega(x,\cdot) \in L_2(G)$, and it depends continuously on $x \in \mathbb{R}^N$ as an $L_2(G)$-valued function. Let

$$\mathcal{D} = D_1 \cdots D_N = \frac{\partial^N}{\partial x_1 \cdots \partial x_N}. \tag{3.47}$$

THEOREM 3.10. *Let $L_2(G) \oplus L_2(G) \ni (f,g) \to B(f,g) = (Af,g)_{\mathcal{D}_2(G)}$ be a continuous bilinear form, and let $A\colon L_2(G) \to L_2(G)$ be the corresponding continuous operator. For any $u, v \in C_0^N(G)$ the representation*

$$B(u,v) = \int_G \int_G L(x,y)(\mathcal{D}u)(y)\overline{(\mathcal{D}v)(x)}\,dx\,dy \tag{3.48}$$

is valid, with kernel $L(x,y)$ which depends continuously on $(x,y) \in \mathbb{R}^N \times \mathbb{R}^N$ and has the form

$$L(x,y) = (A\omega(y,\cdot), \omega(x,\cdot))_{L_2(G)} \qquad (x, y \in \mathbb{R}^N). \tag{3.49}$$

Here $C_0^N(G)$, as usual, is the space of functions in $C^N(G)$ with compact supports relative to G and infinity.

PROOF. First of all let us establish the equality

$$\int_G (f, \omega(x,\cdot))_{L_2(G)} \overline{(\mathcal{D}u)(x)}\,dx = (f,u)_{L_2(G)} \qquad (f \in L_2(G),\ u \in C_0^N(G)). \tag{3.50}$$

Indeed, extending the functions f and u by zero outside G and integrating by parts, we get

$$\int_G (f, \omega(x,\cdot))_{L_2(G)} \overline{(\mathcal{D}u)(x)}\, dx$$
$$= (-1)^N \int_{\mathbf{R}^N} \left(\int_0^{x_1} \cdots \int_0^{x_N} f(\xi)\, d\xi_1 \cdots d\xi_N \right) \overline{(\mathcal{D}u)(x)}\, dx$$
$$= \int_{\mathbf{R}^N} \mathcal{D}_x \left(\int_0^{x_1} \cdots \int_0^{x_N} f(\xi)\, d\xi_1 \cdots d\xi_N \right) \overline{u(x)}\, dx = (f, u)_{L_2(G)}.$$

Using (3.50), we prove (3.49):

$$\int_G \int_G (A\omega(y,\cdot), \omega(x,\cdot))_{L_2(G)} (\mathcal{D}u)(y) \overline{(\mathcal{D}v)(x)}\, dx\, dy$$
$$= \int_G \left\{ \int_G (A\omega(y,\cdot), \omega(x,\cdot))_{L_2(G)} \overline{(\mathcal{D}v)(x)}\, dx \right\} (\mathcal{D}u)(y)\, dy$$
$$= \int_G (A\omega(y,\cdot), v)_{L_2(G)} (\mathcal{D}u)(y)\, dy = \overline{\int_G (v, A\omega(y,\cdot))_{L_2(G)} \overline{(\mathcal{D}u)(y)}\, dy}$$
$$= \int_G (A\omega(y,\cdot), v)_{L_2(G)} (\mathcal{D}u)(y)\, dy = \overline{\int_G (v, A\omega(y,\cdot))_{L_2(G)} \overline{(\mathcal{D}u)(y)}\, dy}$$
$$= \overline{\int_G (A^*v, \omega(y,\cdot))_{L_2(G)} \overline{(\mathcal{D}u)(y)}\, dy} = \overline{(A^*v, u)}_{L_2(G)}$$
$$= (Au, v)_{L_2(G)} = B(u,v) \qquad (u, v \in C_0^N(G)).$$

The continuity of the kernel (3.49) follows from that of the vector-valued function $\mathbf{R}^N \ni x \mapsto \omega(x,\cdot) \in L_2(G)$. ■

If the kernel $L(x,y)$ turns out to be sufficiently smooth, then the expressions $\mathcal{D}$ in (3.48) can be transferred to it by integration by parts, and we obtain an integral representation of the form $B(u,v)$. In the general case the derivatives $\mathcal{D}_x \mathcal{D}_y L(x,y)$ exist only in the sense of generalized functions, and the representation (3.48) can be understood as a representation with a generalized (in the Schwartz sense) kernel $\mathcal{D}_x \mathcal{D}_y L$.

Let us establish a connection between the representations (3.48) and (3.15), (3.45). We present some general constructions. The formula (2.48), written in a way similar to that for (3.15), will be used:

$$B(u,v) = (\mathrm{B}, v \otimes u^*)_{H_0 \otimes H_0} = (K, (J^{-1}v) \otimes (J^{-1}u)^*)_{H_0 \otimes H_0}, \qquad (3.51)$$
$$K = (\mathbf{J} \otimes \mathbf{J})\mathrm{B} \in H_0 \otimes H_0 \qquad (u, v \in H_+).$$

Assume that the chain (2.43) has been constructed from the operator D as indicated in §1.8, and assume that an involution $*$ is given in the original space H_0 for which D is real (i.e., $(\mathfrak{D}(D))^* = \mathfrak{D}(D)$ and $Df^* = (Df)^*$, $f \in \mathfrak{D}(D)$). This is clearly a chain with involution $*$, i.e., $(u^*, v^*)_{H_+} = \overline{(u,v)}_{H_0}$ $(u, v \in H_+ = \mathfrak{D}(D))$. In (3.51) we can replace J^{-1} by $J_0^{-1} = \sqrt{D^*D} = UD$, where

U is a unitary operator on H_0 connecting the metrically equal operators D and $\sqrt{D^*D}$. As a result,

$$\begin{aligned}B(u,v) &= (K,(J_0^{-1}v)\otimes(J_0^{-1}u^*))_{H_0\otimes H_0} = (K,(UDv)\otimes(UDu^*))_{H_0\otimes H_0} \\ &= ((U^{-1}\otimes U^{-1})K,(Dv)\otimes(Du)^*)_{H_0}\end{aligned}$$

for $u, v \in \mathfrak{D}(D)$. Accordingly, (3.51) takes the form

$$B(u,v) = (K_1,(Dv)\otimes(Du)^*)_{H_0\otimes H_0}, \quad K_1 = (U^{-1}\otimes U^{-1})K \in H_0\otimes H_0 \\ (u,v\in\mathfrak{D}(D)). \quad (3.52)$$

We now compare (3.48) and (3.52). To do this, we construct a function $\delta(x) \in C(G)$ with $\delta(x) > 0$ $(x \in G)$ such that

$$\int_G \|\omega(x,\cdot)\|^2_{L_2(G)}\delta^2(x)\,dx \le 1. \quad (3.53)$$

($\delta(x)$ can be taken to be a constant in some cases, for instance, for a bounded G.) Let D be the operator defined in $L_2(G)$ as the closure of the operator $C_0^\infty(G) \ni u(x) \mapsto \delta^{-1}(x)(\mathcal{D}u)(x)$. Then *$D$ is suitable for writing the representation* (3.52) *in the case when $H_0 = L_2(G)$ and the involution is $L_2(G) \ni f(x) \mapsto f^*(x) = \overline{f(x)} \in L_2(G)$. The formulas* (3.48) *and* (3.52) *pass one into the other when L is replaced by K_1 and $\mathcal{D}$ by $\delta^{-1}\mathcal{D}$ in* (3.48), *where* $K_1(x,y) = L(x,y)\delta(x)\delta(y)$ $(x, y \in G)$.

Indeed, by (3.50) and with the help of (3.53),

$$\begin{aligned}\|u\|^2_{L_2(G)} &= \int_G (u,\omega(x,\cdot))_{L_2(G)}\overline{(\mathcal{D}u)(x)}\,dx \\ &\le \left(\int_G |(u,\omega(x,\cdot))_{L_2(G)}|^2\delta^2(x)\,dx\right)^{1/2}\|Du\|_{L_2(G)} \\ &\le \|u\|_{L_2(G)}\|Du\|_{L_2(G)} \qquad (u\in C_0^\infty(G)).\end{aligned}$$

This implies that $\|Du\|_{L_2(G)} \ge \|u\|_{L_2(G)}$ $(u \in \mathfrak{D}(D))$. The constructions in §1.8 can be carried out for such a D, and (3.52) can be written because D is real with respect to the involution introduced. On the other hand, it can be concluded from (3.49) that

$$|L(x,y)| \le \|A\|\,\|\omega(x,\cdot)\|_{L_2(G)}\|\omega(y,\cdot)\|_{L_2(G)} \qquad (x,y\in\mathbb{R}^N).$$

Therefore, $L(x,y)\delta(x)\delta(y) \in L_2(G\times G)$. By (3.48) and (3.52),

$$\begin{aligned}&\int_G\int_G L(x,y)\delta(x)\delta(y)(Du)(y)\overline{(Dv)(x)}\,dx\,dy \\ &\qquad = B(u,v) = \int_G\int_G K_1(x,y)(Du)(y)\overline{(Dv)(x)}\,dx\,dy \\ &\qquad\qquad (u,v\in C_0^N(G)). \quad (3.54)\end{aligned}$$

Passing to the limit, we get that this equality is valid also for $u, v \in \mathfrak{D}(D) = H_+$. But D is metrically equal to J_0^{-1}, and so $\mathfrak{R}(D) = \mathfrak{R}(J_0^{-1}) = H_0 = L_2(G)$. Thus, the linear span of the functions $(Du)(y)\overline{(Dv)(x)}$ $(u, v \in \mathfrak{D}(D))$ is dense in $L_2(G \times G)$, and, therefore, (3.54) gives us the equality $K_1(x, y) = L(x, y)\delta(x)\delta(y)$, which can be regarded as satisfied for all $(x, y) \in G \times G$ after redefining K_1 on a set of zero measure. ■

Hence, (3.48) is in essence a representation of the form (3.15), (3.45) constructed from the rigging of $L_2(G)$ introduced with the help of the operator D defined above. It is simple to compute the kernel of the representation (3.48), due to (3.49).

We make some remarks.

Denote by D° the operator D in (3.52), acting from $H_+ = \mathfrak{D}(D)$ to H_0. Then it is easy to see that (3.52) leads to the equality

$$\mathrm{B} = (D^\circ \otimes D^\circ)^+ K_1. \tag{3.55}$$

This equality, together with (3.49) and the procedure for choosing $\delta(x)$, makes it possible to compute the kernel B of the form $B(f, g)$ in the representation (2.48) constructed from the chain $H_- \supseteq L_2(G) \supseteq \mathfrak{D}(D)$. If G is bounded, then for $l \geq N$ the subspace $\overset{\circ}{W}{}_2^l(G)$ is in $\mathfrak{D}(D)$, so the kernel B being computed in this way is essentially also the kernel of the form $B(f, g)$ when the chain $\overset{\circ}{W}{}_2^{-l}(G) \supseteq L_2(G) \supseteq \overset{\circ}{W}{}_2^l(G)$ is used.

We note that (3.49) is connected with the expression for the elements of the matrix $(A_{jk})_{j,k=1}^m$ of A in the m-dimensional space $\mathbb{C}^m$:

$$A_{jk} = (A\delta_k, \delta_j)_{\mathbb{C}^m} \qquad (j, k = 1, \ldots, m). \tag{3.56}$$

Here the $\delta_j = (\delta_{jn})_{n=1}^m$ are the vectors of the orthonormal basis in $\mathbb{C}^m$. We explain this. In passing from $\mathbb{C}^m$ to $L_2(G)$ the role of the δ_j must be played by the δ-functions $\delta_x(\xi)$ $(x \in G)$; however, they are not in $L_2(G)$, and so (3.56) is devoid of meaning. At the same time, we can pass to a transformed basis in (3.56) by setting $\omega_j = \sum_1^j \delta_l$, $j = 1, \ldots, m$, and then form the matrix $L_{jk} = (A\omega_k, \omega_j)_{\mathbb{C}^m}$ $(j, k = 1, \ldots, m)$. It is easy to express the action of the operator with the help of this matrix. The kernel (3.49) is analogous to the matrix $(L_{j,k})_{j,k=1}^m$: for example, if $G = (0, \infty) \subset \mathbb{R}^1$, then we can formally write $\omega_x(\xi) = \int_0^x \delta_z(\xi)\, dz$ $(x \in (0, \infty))$. The generalized kernel $\mathcal{D}_x \mathcal{D}_y L$ is analogous to the matrix (3.56), and the formula (3.48) is analogous to the formula for producing the action of A by the matrix $(L_{j,k})_{j,k=1}^m$.

3.12. ***The chain constructed from a continuous positive-definite kernel.*** We saw in §3.2 that the inner product in $W_2^{-l}(G)$ $(l = 1, 2, \ldots)$, where G is a bounded domain in $\mathbb{R}^N$, is defined with the help of a certain positive-definite kernel: the Green's function of the corresponding elliptic problem of order $2l$. For $l > N/2$ the Green's function is continuous up to the boundary of the domain (Berezanskiĭ [4], Chapter 3, §5). On the other hand, in this case

$W_2^l(G) \subset C(\tilde{G})$, and δ_ξ is in $W_2^{-l}(G)$ and depends continuously on $\xi \in \tilde{G}$, etc. (see subsections 3 and 4). We show that a similar situation is sufficiently general for construction of a chain from a positive-definite kernel (see the example in §1.7).

Accordingly, let G be a bounded domain in $\mathbb{R}^N$ $(N = 1, 2, \ldots)$, and suppose that the kernel $\tilde{G} \times \tilde{G} \ni (x, y) \mapsto K(x, y) \in \mathbb{C}^1$ is continuous and positive-definite. In §1.7 we set $R = G$ and $d\mu(x) = dx$; then $(f, f)_{H_K} \geq 0$, where

$$(f, g)_{H_K} = \int_G \int_G K(x, y) f(y) \overline{g(x)}\, dx\, dy \qquad (f, g \in L_2(G)). \tag{3.57}$$

For simplicity it will be assumed that $K(x, y)$ is nondegenerate: if $(f, f)_{H_K} = 0$ for some $f \in L_2(G)$, then $f(x) = 0$ almost everywhere. Without loss of generality it is assumed that $(f, f)_{H_K} \leq \|f\|^2_{L_2(G)}$ $(f \in L_2(G))$.

According to §1.7, it is possible to construct the chain

$$H_K \supseteq L_2(G) \supseteq H_+, \tag{3.58}$$

where the role of the zero space is played by $L_2(G)$, and that of the negative space is played by the completion H_K of $L_2(G)$ with respect to (3.57).

THEOREM 3.11. *The space H_+ in the chain* (3.58) *consists of continuous functions, and the inclusion $H_+ \subset C(\tilde{G})$ is topological, while the imbedding $H_+ \to L_2(G)$ is quasinuclear. Therefore, $\delta_\xi \in H_K$. Then the vector-valued function $\tilde{G} \ni \xi \mapsto \delta_\xi \in H_K$ is continuous, and*

$$(\delta_\xi, \delta_\eta)_{H_K} = K(\eta, \xi) \qquad (\xi, \eta \in \tilde{G}). \tag{3.59}$$

PROOF. According to the proof of Theorem 1.4, to construct H_+ we must use the relation $(f, g)_{H_K} = (Kf, g)_{L_2(G)}$ $(f, g \in L_2(G))$ to introduce a continuous positive operator $K \colon L_2(G) \to L_2(G)$, and then let

$$(u, v)_{H_+} = (K^{-1}u, v)_{L_2(G)} \qquad (u, v \in \mathfrak{R}(K)). \tag{3.60}$$

The space H_+ coincides with the completion of $\mathfrak{R}(K)$ with respect to (3.60).

Comparing the foregoing with (3.57), we conclude that the operator K has the form

$$(Kf)(x) = \int_G K(x, y) f(y)\, dy \qquad (f \in L_2(G)). \tag{3.61}$$

It follows from (3.61) that $\mathfrak{R}(K) \subseteq C(\tilde{G})$. We need to show that the completion of $\mathfrak{R}(K)$ with respect to (3.60) is also in $C(\tilde{G})$. Due to the difficulty in finding K^{-1} we proceed as follows.

Suppose that $\xi \in \tilde{G}$ and $\varepsilon > 0$. Denote by $\chi_{\xi,\varepsilon}(x)$ the characteristic function of the set $\{x \in \tilde{G} \,|\, |x - \xi| < \varepsilon\}$, divided by its measure. Since

$$(\chi_{\xi,\varepsilon}, \chi_{\eta,\delta})_{H_K} \to (\eta, \xi) \quad \text{as } \varepsilon, \delta \to 0$$

by (3.57), it follows that

$$\|\chi_{\xi,n^{-1}} - \chi_{\xi,m^{-1}}\|^2_{H_K} \to 0 \quad \text{as } n, m \to \infty,$$

i.e., $\chi_{\xi,n^{-1}}(x)$ is a Cauchy sequence in the norm of H_K $(\xi,\eta \in \tilde{G})$. This implies that for each $\xi \in \tilde{G}$ there exists a $\kappa_\xi \in H_K$ such that $\chi_{\xi,\varepsilon} \to \kappa_\xi$ as $\varepsilon \to 0$ in H_K. It is also clear that

$$(\kappa_\xi, \kappa_\eta)_{H_K} = K(\eta, \xi) \qquad (\xi, \eta \in \tilde{G}). \tag{3.62}$$

Consider a $u \in \mathfrak{R}(K)$; since this is a function in H_+, and $\chi_{\xi,\varepsilon}(x)$ converges to κ_ξ in the norm of $H_K = H_-$ as $\varepsilon \to 0$, it follows that

$$(\kappa_\xi, u)_{L_2(G)} = \lim_{\varepsilon\to 0}(\chi_{\xi,\varepsilon}, u)_{L_2(G)}.$$

But the last limit is equal to $\overline{u(\xi)}$ by the continuity of u on $\tilde{G}$. Thus,

$$(\kappa_\xi, u)_{L_2(G)} = \overline{u(\xi)} \qquad (u \in \mathfrak{R}(K);\ \xi \in \tilde{G}). \tag{3.63}$$

This implies that

$$\begin{aligned} |u(\xi)| &\le \|\kappa_\xi\|_{H_K}\, \|u\|_{H_+} = (K(\xi,\xi))^{1/2}\|u\|_{H_+} \\ &\le \max_{\xi\in\tilde{G}}(K(\xi,\xi))^{1/2}\|u\|_{H_+} = c\|u\|_{H_+} \qquad (\xi \in G) \end{aligned}$$

for $u \in \mathfrak{R}(K)$. Thus, $\|u\|_{C(\tilde{G})} \le c\|u\|_{H_+}$ $(u \in \mathfrak{R}(K))$. Passing now to the completion of $\mathfrak{R}(K)$ with respect to $(\cdot,\cdot)_{H_+}$, we conclude that $H_+ \subset C(\tilde{G})$, and the last inequality is preserved. Accordingly, $H_+ \subset C(\tilde{G})$ topologically.

It follows from (3.63) and the denseness of $\mathfrak{R}(K)$ in $H_+ \subset C(\tilde{G})$ that $\kappa_\xi = \delta_\xi$ $(\xi \in \tilde{G})$. Therefore, (3.59) follows from (3.62). The continuity of the vector-valued function $\tilde{G} \ni \xi \mapsto \delta_\xi \in H_K$ follows from (3.59) and the continuity of the kernel $K(x,y)$ on $\tilde{G}\times\tilde{G}$. Finally, a repetition of the computation in (3.7) proves that the imbedding $H_+ \to L_2(G)$ is quasinuclear. ■

The functions in H_+ can be characterized as follows. Since the kernel $K(x,y)$ is positive-definite and nondegenerate, it has a collection of eigenfunctions $(\varphi_n(x))_1^\infty$ and corresponding eigenvalues $(\lambda_n)_1^\infty$ with $\lambda_n > 0$. It follows from (3.60) that $(u,v)_{H_K} = (K^{-1/2}u, K^{-1/2}v)_{L_2(G)}$ for $u, v \in \mathfrak{R}(K)$. The result of completion is that H_+ consists of precisely those $u \in L_2(G)$ such that $K^{-1/2}u \in L_2(G)$. In other words, H_+ consists of precisely those $u \in L_2(G)$ such that

$$\sum_{n=1}^\infty |(u,\varphi_n)_{L_2(G)}|^2\lambda_n^{-1} < \infty;$$

$$(u,v)_{H_+} = \sum_{n=1}^\infty (u,\varphi_n)_{L_2(G)}\overline{(v,\varphi_n)}_{L_2(G)}\lambda_n^{-1} \qquad (u, v \in H_+). \tag{3.64}$$

It is also easy to show that every complex-valued measure of bounded variation defined on the Borel subsets of $\tilde{G}$ is in H_K, and the inner product of two such measures α and β is computed as follows:

$$(\alpha,\beta)_{H_K} = \int_{\tilde{G}}\int_{\tilde{G}} K(x,y)\,d\alpha(x)\,\overline{d\beta(y)}. \tag{3.65}$$

§4. Spaces of functions of infinitely many variables

We shall construct spaces of functions of infinitely many variables mainly as infinite tensor products of spaces of functions of a single variable and projective limits of such products. We first consider the question of constructing riggings.

4.1. ***The space of square-integrable functions with respect to a product measure, and the rigging of it.*** Below we consider functions of points $x \in \mathsf{R}^\infty = \mathsf{R}^1 \times \mathsf{R}^1 \times \cdots$ (no topology is introduced in R^∞); the coordinate notation for such points is $x = (x_n)_1^\infty$ ($x_n \in \mathsf{R}^1$). Let $(p_n(x_n))_1^\infty$ be a fixed sequence of continuous positive probability weights, i.e., $\int_{\mathsf{R}^1} p_n(x_n)\,dx = 1$. The measure on R^∞ given by

$$d\theta(x) = (d\theta_1(x_1)) \otimes (d\theta_2(x_2)) \otimes \cdots = (p_1(x_1)\,dx_1) \otimes (p_2(x_2)\,dx_2) \otimes \cdots$$

is called a (weighted) product measure. According to §2.3, $L_2(\mathsf{R}^\infty, d\theta(x))$ can be understood as the infinite tensor product $\bigotimes_{n=1;e,1}^\infty H_{0,n}$, where $H_{0,n} = L_2(\mathsf{R}^1, d\theta_n, x_n))$, and $e = (e^{(n)}(x_n))_1^\infty$, $e^{(n)}(x_n) \equiv 1$. Therefore, the chains $H_- \supseteq H_0 \supseteq H_+$, where $H_0 = L_2(\mathsf{R}^\infty, d\theta(x))$ and the imbedding $H_+ \to H_0$ has this or that character, can be constructed by the general constructions in §§2.4–2.6.

For example, to construct a chain with a quasinuclear imbedding $H_+ \to H_0$, we first use Theorem 3.6 to construct spaces $H_{+,n} \subseteq H_{0,n}$ such that the imbeddings $H_{+,n} \to H_{0,n}$ are quasinuclear and $\|e^{(n)}\|_{H_{+,n}} = 1$ ($n = 1, 2, \ldots$).

These spaces can be constructed as follows. Consider the space

$$W_2^1(\mathsf{R}^1, p_{+,n}(x_n)\,dx_n),$$

where $p_{+,n}(x_n)$ is a nonnegative function defined on the basis of Theorem 3.6 such that the imbedding $W_2^1(\mathsf{R}^1, p_{+,n}(x_n)\,dx_n) \to H_{0,n}$ is quasinuclear. It is not yet possible to take this space as $H_{+,n}$, because the norm of the function $e^{(n)}$ in it is only greater than or equal to 1. However, by a lemma to be proved now, it can be renormed in the required way. The space $(W_2^1(\mathsf{R}^1, p_{+,n}(x_n)\,dx_n))^\square$ obtained in this way is taken as $H_{+,n}$.

LEMMA 4.1. *Let G and H be two Hilbert spaces, with H dense in G and $\|u\|_G \le \|u\|_H$ ($u \in H$), and let $e \in H$ be such that $\|e\|_G = 1$. Then the inner product*

$$(u,v)_{H^\square} = (u,e)_G\overline{(v,e)_G} + (Pu,Pv)_H \qquad (u,v \in H), \tag{4.1}$$

where P is the projection of G onto the orthogonal complement of e, is equivalent to the original inner product in H, and

$$\|u\|_G \le \|u\|_{H^\square} \quad (u \in H), \qquad \|e\|_{H^\square} = 1. \tag{4.2}$$

The space H, renormed by means of (4.1), is denoted by $H^{\square}$.

PROOF. Since $H \ni u = (u,e)_G e + Pu$ and $e \in H$, it follows that $Pu \in H$; therefore, (4.1) makes sense. The relations (4.2) obviously hold, and so it remains to prove the equivalence of the norms $\|\cdot\|_{H^{\square}}$ and $\|\cdot\|_H$. For $u \in H$

$$\|u\|_H = \|(u,e)_G e + Pu\|_H \leq |(u,e)_G|\, \|e\|_H + \|Pu\|_H \leq \sqrt{2}\|e\|_H\, \|u\|_{H^{\square}}.$$

Conversely, suppose that $u_n \in H$ is such that $\|u_n\|_H \to 0$ as $n \to \infty$. Then $(u_n, e)_G \to 0$ in view of the estimate $|(u_n,e)_G| \leq \|u_n\|_G \leq \|u_n\|_H$; hence also

$$\|Pu_n\|_H = \|u_n - (u_n,e)_G\, e\|_H \leq \|u_n\|_H + |(u_n,e)_G|\|e\|_H \to 0.$$

This implies that $\|u_n\|_{H^{\square}} \to 0$; therefore, $\|u\|_{H^{\square}} \leq c\|u\|_H$ $(u \in H)$ for some $c > 0$. ■

Accordingly, the spaces $H_{+,n}$ have been constructed. By Theorem 2.5, we can now form the chain

$$\bigotimes_{n=1;e,\delta^{-1}}^{\infty} H_{-,n} \supseteq L_2(\mathbb{R}^\infty, d\theta(x)) \supseteq \bigotimes_{n=1;e,\delta}^{\infty} H_{+,n}, \tag{4.3}$$

where $H_{-,n}$ is the negative space with respect to the zero space $H_{0,n} = L_2(\mathbb{R}^1, d\theta_n(x_n))$ and the positive space $H_{+,n} = (W_2^1(\mathbb{R}^1, p_{+,n}(x_n)\, dx_n)^{\square}$ and $\delta = (\delta_n)_1^\infty$ $(\delta_n \geq 1)$ is some weight. If this weight is such that the condition (2.25) holds, where the O_n are the imbedding operators $H_{+,n} \to H_{0,n}$, then the imbedding of the positive space in the zero space in the chain (4.3) is quasinuclear. Of course, this condition can always be achieved.

We remark that a construction analogous to (4.3) can be carried through with the spaces $W_2^1(\mathbb{R}^1, p_{+,n}(x_n)\, dx_n)$ replaced by $W_2^{(1,q+,n)}(\mathbb{R}^1)$ and with the use of Theorem 3.5.

4.2. *The case of a Gaussian measure.* The constructions in subsection 1 can be made more orderly in the case when the measures being multiplied are one-dimensional Gaussian measures, i.e., when they have the form

$$\sqrt{\frac{\varepsilon}{\pi}}e^{-\varepsilon t^2}\, dt = \gamma_\varepsilon(t)\, dt = dg_\varepsilon(t) \qquad (t \in \mathbb{R}^1; \varepsilon > 0). \tag{4.4}$$

(In the case $\varepsilon = 1$, i.e., in the case of a standard Gaussian measure, the index ε is usually omitted.) First of all we get results analogous to those presented in §3.8 for the case $N = 1$ and for Gaussian measure. An assertion of the type of Theorem 3.2 is now valid in place of Theorem 3.6.

We construct the Sobolev space $W_2^l(\mathbb{R}^1, dg(t))$ and denote it by $G^l(\mathbb{R}^1) = G^l(\mathbb{R}^1, 1)$. It is the completion of $C_0^\infty(\mathbb{R}^1)$ in the inner product

$$(u,v)_{G^l(\mathbb{R}^1)} = \sum_{\mu=0}^{l} \int_{\mathbb{R}^1} (D^\mu u)(t)\overline{(D^\mu v)(t)}\, dg(t) \qquad (u, v \in C_0^\infty(\mathbb{R}^1);\ l = 0, 1, \dots). \tag{4.5}$$

THEOREM 4.1. *The imbedding $G^{l+2}(\mathbf{R}^1) \to G^l(\mathbf{R}^1)$ $(l = 0, 1, \ldots)$ is quasinuclear, and its Hilbert-Schmidt norm is bounded above by the number 2^{l+3}.*

LEMMA 4.2. *The closure A of the operator*

$$\begin{aligned} C_0^\infty(\mathbf{R}^1) \ni u \mapsto & -\gamma^{-1}(t)(\gamma(t)u')'(t) + u(t) \\ &= -u''(t) + 2tu'(t) + u(t), \end{aligned} \tag{4.6}$$

in the space $G^0(\mathbf{R}^1) = L_2(\mathbf{R}^1, dg(t))$ is selfadjoint, and its spectrum consists of the set $(2n+1)_{n=0}^\infty$. The Hermite polynomial $h_n(t)$, $n = 0, 1, \ldots,$ is an eigenvector corresponding to the eigenvalue $2n+1$. Moreover, $(u, v)_{G^1(\mathbf{R}^1)} = (Au, v)_{G^0(\mathbf{R}^1)}$ $(u, v \in C_0^\infty(\mathbf{R}^1))$.

PROOF OF THE LEMMA. Consider the closure B (the harmonic oscillator) of the operator $C_0^\infty(\mathbf{R}^1) \ni v \mapsto -v''(t) + t^2 v(t)$ in $L_2(\mathbf{R}^1)$. It is well known that B is selfadjoint, its spectrum consists of the numbers $2n+1$, and the functions $\gamma^{1/2}(t)h_n(t)$ $(n = 0, 1, \ldots)$ are corresponding eigenvectors (see, for example, S. G. Kreĭn [2], Chapter 9, §2.2). The mapping $G^0(\mathbf{R}^1) \ni u(t) \mapsto \gamma^{1/2}(t)u(t) = U(t) \in L_2(\mathbf{R}^1)$ is an isometry between $G^0(\mathbf{R}^1)$ and $L_2(\mathbf{R}^1)$ under which the expression

$$-\gamma^{-1/2}(t)(\gamma(t)(\gamma^{-1/2}(t)U)')'(t) + U(t) = -U''(t) + t^2 U(t)$$

is the image of the differential expression in (4.6). Since $C_0^\infty(\mathbf{R}^1)$ is invariant under this mapping, B is the image of the operator A, and the lemma follows from this. ■

PROOF OF THE THEOREM. We prove it first for the imbedding $G^2(\mathbf{R}^1) \to G^0(\mathbf{R}^1)$. Denote by H_+ the Hilbert space consisting of $\mathfrak{D}(A)$ with the graph inner product $(u, v)_{H_+} = (Au, Av)_{G^0(\mathbf{R}^1)}$ $(u, v \in \mathfrak{D}(A))$. Obviously, $\|u\|_{G^0(\mathbf{R}^1)} \le \|Au\|_{G^0(\mathbf{R}^1)} = \|u\|_{H_+}$ $(u \in H_+)$, and the imbedding $S\colon H_+ \to G^0(\mathbf{R}^1)$ is quasinuclear. The latter follows from the fact that the functions $e_n(t) = (2n+1)^{-1}h_n(t)$ $(n = 0, 1, \ldots)$ form an orthonormal basis in H_+, and hence

$$|S|^2 = \sum_{n=0}^\infty \|e_n\|^2_{G^0(\mathbf{R}^1)} = \sum_{n=0}^\infty (2n+1)^{-2} = \frac{\pi^2}{8} < \infty. \tag{4.7}$$

To prove that the imbedding $G^2(\mathbf{R}^1) \to G^0(\mathbf{R}^1)$ is quasinuclear, it now suffices to establish that the imbedding $G^2(\mathbf{R}^1) \to H_+$ is continuous. According to (4.6),

$$\|Au\|_{G^0(\mathbf{R}^1)} \le \|u''\|_{G^0(\mathbf{R}^1)} + \|u\|_{G^0(\mathbf{R}^1)} + 2\|tu'\|_{G^0(\mathbf{R}^1)}$$

for $u \in C_0^\infty(\mathbf{R}^1)$. We estimate the last term, assuming in addition that the function $u(t)$ is real:

$$\begin{aligned}\|tu'\|^2_{G^0(\mathbf{R}^1)} &= \frac{1}{\sqrt{\pi}}\int_{\mathbf{R}^1} t^2(u'(t))^2 e^{-t^2}\,dt = -\frac{1}{2\sqrt{\pi}}\int_{\mathbf{R}^1} t(u'(t))^2 d(e^{-t^2}) \\ &= \frac{1}{2\sqrt{\pi}}\int_{\mathbf{R}^1}(u'(t))^2 e^{-t^2}\,dt + \frac{1}{\sqrt{\pi}}\int_{\mathbf{R}^1} tu'(t)u''(t)e^{-t^2}\,dt \\ &\le \frac{1}{2}\|u'\|^2_{G^0(\mathbf{R}^1)} + \frac{1}{2}\|tu^2\|^2_{G^0(\mathbf{R}^1)} + \frac{1}{2}\|u''\|^2_{G^0(\mathbf{R}^1)}. \end{aligned} \tag{4.8}$$

In turn we get $\|tu'\|^2_{G^0(\mathbf{R}^1)} \le \|u'\|^2_{G^0(\mathbf{R}^1)} + \|u''\|^2_{G^0(\mathbf{R}^1)}$. This inequality is valid also for a general $u \in C_0^\infty(\mathbf{R}^1)$. (It is necessary to pass to $\operatorname{Re} u(t)$ and $\operatorname{Im} u(t)$ and add the estimates.) In the final analysis we get that

$$\begin{aligned}\|Au\|_{G^0(\mathbf{R}^1)} &\le 3(\|u\|_{G^0(\mathbf{R}^1)} + \|u'\|_{G^0(\mathbf{R}^1)} + \|u''\|_{G^0(\mathbf{R}^1)}) \\ &\le 3\sqrt{3}\|u\|_{G^2(\mathbf{R}^1)} \qquad (u \in C_0^\infty(\mathbf{R}^1)).\end{aligned}$$

This implies the inclusion $G^2(\mathbf{R}^1) \subseteq H_+$ and the inequality

$$\|u\|_{H_+} \le 3\sqrt{3}\|u\|_{G^2(\mathbf{R}^1)} \qquad (u \in G^2(\mathbf{R}^1)).$$

It has thus been proved that the imbedding $G^2(\mathbf{R}^1) \to H_+$ is continuous, and consequently, the imbedding $G^2(\mathbf{R}^1) \to G^0(\mathbf{R}^1)$ is quasinuclear. The last estimate and (4.7) give us the estimate $|O| < 2^3$ for the corresponding imbedding operator.

Now consider the case of the imbedding $G^{(l+2)}(\mathbf{R}^1) \to G^l(\mathbf{R}^1)$ $(l = 0, 1, \ldots)$. Let us use Lemma 3.1 and Remark 1 after it, with $E = C_0^\infty(\mathbf{R}^1)$, $(Tu)(x) = u'(x)$, $H_1 = G^2(\mathbf{R}^1)$ and $G_1 = G^0(\mathbf{R}^1)$. We get that the imbedding $G^{3'}(\mathbf{R}^1) \to G^1(\mathbf{R}^1)$ is quasinuclear, and its Hilbert-Schmidt norm does not exceed $8\sqrt{2}$. (Here $G^{3'}(\mathbf{R}^1)$ denotes the renorming of $G^3(\mathbf{R}^1)$ with the coefficient two in front of the terms with the first and second derivatives in the inner product (4.5) for $G^3(\mathbf{R}^1)$.) The imbedding $G^3(\mathbf{R}^1) \to G^1(\mathbf{R}^1)$ is thus quasinuclear, and its Hilbert-Schmidt norm does not exceed 2^4.

Continuing this process, we take the previous E, T and $H_1 = G^3(\mathbf{R}^1)$, $G_1 = G^1(\mathbf{R}^1)$; then $H_1 = G^4(\mathbf{R}^1), G_1 = G^2(\mathbf{R}^1)$, and so on. As a result, the imbedding $G^{l+2}(\mathbf{R}^1) \to G^l(\mathbf{R}^1)$ is quasinuclear, and its norm does not exceed 2^{l+3}. ■

We can, of course, introduce the spaces $G^l(\mathbf{R}^1, \varepsilon)$ with respect to the general Gaussian measure (4.4), which are analogous to $G^l(\mathbf{R}^1) = G^l(\mathbf{R}^1, 1)$. It is convenient to construct them as the completions of $C_0^\infty(\mathbf{R}^1)$ in the inner product

$$(u, v)_{G^l(\mathbf{R}^1,\varepsilon)} = \sum_{\mu=0}^{l} \varepsilon^{-\mu}\int_{\mathbf{R}} (D_\mu u)(t)\overline{(D^\mu v)(t)}\,dg_\varepsilon(t)$$
$$(u, v \in C_0^\infty(\mathbf{R}^1);\ l = 0, 1, \ldots;\ \varepsilon > 0). \tag{4.9}$$

The mapping $G^l(\mathbb{R}^1) \ni u(t) \mapsto u(\sqrt{\varepsilon}\, t) = U(t) \in G^l(\mathbb{R}^1, \varepsilon)$ is an isometry between $G^l(\mathbb{R}^1)$ and $G^l(\mathbb{R}^1, \varepsilon)$; therefore, the assertion of Theorem 4.1 is preserved for the spaces $G^l(\mathbb{R}^1, \varepsilon)$ $(l = 0, 1, \dots;\ \varepsilon > 0)$. Let $G^{-l}(\mathbb{R}^1, \varepsilon)$ denote the negative space in the chain $H_- \supseteq G^0(\mathbb{R}^1, \varepsilon) \supseteq G^l(\mathbb{R}^1, \varepsilon)$ $(l = 0, 1, \dots;\ \varepsilon > 0)$.

Taking the projective limits of the spaces $G^l(\mathbb{R}^1, \varepsilon)$, we can construct the nuclear spaces

$$\mathcal{G}(\mathbb{R}^1, \varepsilon) = \operatorname*{pr\,lim}_{\tau \in \{0,1,\dots\}} G^\tau(\mathbb{R}^1, \varepsilon) \quad (\varepsilon > 0) \quad \text{and} \quad (\mathcal{G}(\mathbb{R}^1) = \mathcal{G}(\mathbb{R}^1, 1).$$

(It is easy to see that all the conditions ensuring the possibility of such constructions are fulfilled.) The following theorem describes these spaces intrinsically.

THEOREM 4.2. *A function $\varphi(t)$ is in $\mathcal{G}(\mathbb{R}^1)$ if and only if $\varphi(t)e^{-t^2/2} \in S(\mathbb{R}^1)$. The mapping $\mathcal{G}(\mathbb{R}^1) \ni \varphi(t) \mapsto \varphi(t)e^{-t^2/2} = \Phi(t) \in \mathcal{G}(\mathbb{R}^1)$ is a topological isomorphism between the spaces $\mathcal{G}(\mathbb{R}^1)$ and $S(\mathbb{R}^1)$.*

PROOF. It suffices to prove two inequalities: for each $\tau = 0, 1, \dots$ there exist $\tau' = 0, 1, \dots$ and $c < \infty$ such that

$$\|\Phi\|_{S_\tau(\mathbb{R}^1)} \leq c\|\varphi\|_{G^{\tau'}(\mathbb{R}^1)}, \quad \|\varphi\|_{G^\tau(\mathbb{R}^1)} \leq c\|\Phi\|_{S_{\tau'}(\mathbb{R}^1)} \qquad (\varphi \in C_0^\infty(\mathbb{R}^1)). \tag{4.10}$$

To prove the first relation note that (4.8), with u' replaced by φ, leads to the inequality

$$\|t\varphi\|^2_{G^0(\mathbb{R}^1)} \leq \|\varphi\|^2_{G^0(\mathbb{R}^1)} + \|\varphi'\|^2_{G^0(\mathbb{R}^1)},$$

and thereby to the inequality

$$\|t\varphi\|_{G^0(\mathbb{R}^1)} \leq \|\varphi\|_{G^0(\mathbb{R}^1)} + \|\varphi'\|_{G^0(\mathbb{R}^1)} \qquad (\varphi \in C_0^\infty(\mathbb{R}^1)). \tag{4.11}$$

Iterating (4.11), we get that

$$\begin{aligned} \|t^2\varphi\|_{G^0(\mathbb{R}^1)} &= \|t(t\varphi)\|_{G^0(\mathbb{R}^1)} \leq \|t\varphi\|_{G^0(\mathbb{R}^1)} + \|(t\varphi)'\|_{G^0(\mathbb{R}^1)} \\ &\leq 2\|\varphi\|_{G^0(\mathbb{R}^1)} + 2\|\varphi'\|_{G^0(\mathbb{R}^1)} + \|\varphi''\|_{G^0(\mathbb{R}^1)}; \end{aligned}$$

then we estimate $\|t^3\varphi\|_{G^0(\mathbb{R}^1)} = \|t^2(t\varphi)\|_{G^0(\mathbb{R}^1)} \leq \cdots$, and so on. As a result, there exist constants $c_{0k}, \dots, c_{kk}$ such that

$$\|t^k\varphi\|_{G^0(\mathbb{R}^1)} \leq \sum_{j=0}^{k} c_{jk}\|D^j\varphi\|_{G^0(\mathbb{R}^1)} \qquad (\varphi \in C_0^\infty(\mathbb{R}^1);\ k = 0, 1, \dots,). \tag{4.12}$$

Let us get an upper estimate for $\|\Phi\|_{S_\tau(\mathbf{R}^1)}$. According to (3.35), it suffices to get an upper estimate for the expression $\|(1+t^2)^{\tau/2}(D^\mu\Phi)(t)\|_{L_2(\mathbf{R}^1)}$ ($\mu = 0,\dots,\tau$) or, with the substitution $\Phi(t) = \varphi(t)e^{-t^2/2}$ and subsequent differentiation, for the expression

$$\|t^\nu(D^\lambda\varphi)(t)e^{-t^2/2}\|_{L_2(\mathbf{R}^1)} = \|t^\nu(D^\lambda\varphi)(t)\|_{G^0(\mathbf{R}^1)} \qquad (\nu = 0,\dots,2\tau;\ \lambda = 0,\dots,\tau).$$

It follows from (4.12) that these norms can be estimated above by the sum of the norms $\|D^j\varphi\|_{G^0(\mathbf{R}^1)}$ ($j = 0,\dots,3\tau$) with certain coefficients. Thus, the first of the estimates (4.10) is proved, with $\tau' = 3\tau$.

We prove the second estimate. Since $\varphi(t) = \Phi(t)e^{t^2/2}$, an upper estimate of $\|\Phi(t)e^{t^2/2}\|_{G^\tau(\mathbf{R}^1)}$ is needed, and for this it suffices to estimate $\|D^\mu(\Phi(t)e^{t^2/2})\|_{G^0(\mathbf{R}^1)}$ ($\mu = 0,\dots,\tau$) from above, where $\Phi \in C_0^\infty(\mathbf{R}^1)$. Differentiating, we see that these norms can be estimated above by the norm $\|\Phi\|_{S_\tau(\mathbf{R}^1)}$ with some coefficient. Accordingly, the second inequality in (4.10) also holds, and $\tau' = \tau$. ■

In view of the mapping $G^l(\mathbf{R}^1) \ni u(t) \mapsto u(\sqrt{\varepsilon}\,t) = U(t) \in G^l(\mathbf{R}^1,\varepsilon)$, Theorem 4.2 is preserved also for the space $\mathcal{G}(\mathbf{R}^1,\varepsilon)$ ($\varepsilon > 0$) (just replace $e^{-t^2/2}$ by $e^{-\varepsilon t^2/2}$ in its statement).

Let us return to the construction of riggings. Consider the Gaussian measure on $\mathbf{R}^\infty$ given by

$$d\mathfrak{g}_\varepsilon(x) = (dg_{\varepsilon_1}(x_1)) \times (dg_{\varepsilon_2}(x_2)) \times \cdots, \tag{4.13}$$

where $\varepsilon = (\varepsilon_n)_1^\infty$ ($\varepsilon_n > 0$) is a fixed sequence, and form the space

$$L_2(\mathbf{R}^\infty, d\mathfrak{g}_\varepsilon(x)).$$

In the case when $\varepsilon_1 = \varepsilon_2 = \cdots = 1$ we usually write $\mathfrak{g} = \mathfrak{g}_{(1,1,\dots)}$ for the standard Gaussian measure on $\mathbf{R}^\infty$ with the form

$$d\mathfrak{g}(x) = (dg(x_1)) \times (dg(x_2)) \times \cdots.$$

Of course, we could employ the general construction presented in subsection 1 to form a chain of the type (4.3) for this space; however, it is more convenient to use Theorem 4.1 and proceed as follows. Let $H_{+,n} = G^2(\mathbf{R}^1,\varepsilon_n)$ and $H_{-,n} = G^{-2}(\mathbf{R}^1,\varepsilon_n)$, and form the chain

$$\bigotimes_{n=1;e,\delta^{-1}}^{\infty} G^{-2}(\mathbf{R}^1,\varepsilon_n) \supseteq L_2(\mathbf{R}^\infty, d\mathfrak{g}_\varepsilon(x)) \supseteq \bigotimes_{n=1;e,\delta}^{\infty} G^2(\mathbf{R}^1,\varepsilon_n). \tag{4.14}$$

Here, as earlier, $e = (e^{(n)}(x_n))_1^\infty, e^{(n)}(x_n) \equiv 1$, and $\delta = (\delta_n)_1^\infty$ ($\delta_n \geq 1$) is a weight. Since $\|e^{(n)}\|_{G^2(\mathbf{R}^1,\varepsilon_n)} = 1$, the construction is valid. The imbedding of the positive space in the zero space in the chain (4.14) is quasinuclear if δ satisfies the condition (2.25), which holds (by Theorem 4.1) if, for example, $\sum_1^\infty 2^{6n}\delta_n^{-1} < \infty$.

4.3. *Weighted infinite tensor products of nuclear spaces.* We proceed to the construction of spaces of functions of infinitely many variables, with a certain general construction presented first of all. A nuclear space will be constructed as a projective limit of Hilbert spaces in an ambient Hilbert space H_0 (see §1.10); the assumption that H_0 exists is convenient from a purely technical viewpoint (the negative spaces are determined), and the construction below is also possible without this assumption. Namely, let $(H_\tau)_{\tau\in \mathrm{T}}$ be a family of Hilbert spaces, where T is an arbitrary index set. It will be assumed that each H_τ is a dense subset of some ambient Hilbert space H_0 $(0 \in \mathrm{T})$, where the imbedding $H_\tau \to H_0$ is continuous and has norm at most 1. Suppose that $\Phi = \bigcap_{\tau\in\mathrm{T}} H_\tau$ is dense in each H_τ and that for any $\tau', \tau'' \in \mathrm{T}$ there is a $\tau''' \in \mathrm{T}$ such that $H_{\tau'''} \subseteq H_{\tau'}, H_{\tau''}$ and the imbeddings $H_{\tau'''} \to H_{\tau'}$ and $H_{\tau'''} \to H_{\tau''}$ are quasinuclear. Providing Φ with the topology with the collection of neighborhoods

$$U(\psi;\tau,\varepsilon) = \{\varphi \in \Phi \mid \|\varphi - \psi\|_{H_\tau} < \varepsilon\} \qquad (\tau \in \mathrm{T},\ \varepsilon > 0)$$

as a base, we get a nuclear space $\Phi = \operatorname{pr\,lim}_{\tau\in\mathrm{T}} H_\tau$.

Consider now a sequence of nuclear spaces $(\Phi_n)_1^\infty$ so constructed:

$$\Phi_n = \operatorname*{pr\,lim}_{\tau_n\in\mathrm{T}_n} H_{\tau_n},$$

where $(H_{\tau_n})_{\tau_n\in\mathrm{T}_n}$ is a corresponding family of Hilbert spaces, with H_{0_n} $(0_n \in \mathrm{T}_n)$ the space H_0 for this family. Assume that there exists a simultaneous stabilizing sequence $e = (e^{(n)})_1^\infty$, where $e^{(n)} \in \Phi_n$, and $\|e^{(n)}\|_{H_{\tau_n}} = 1$ for each $\tau_n \in \mathrm{T}_n$. In each T_n choose an index τ_n and consider the weighted infinite tensor product $\mathcal{H}_{\boldsymbol{\tau};e,\boldsymbol{\delta}} = \bigotimes_{n=1;e,\boldsymbol{\delta}}^\infty H_{\tau_n}$, where $\boldsymbol{\delta} = (\delta_n)_1^\infty$ $(\delta_n \geq 1)$ is a weight, and $\boldsymbol{\tau} = (\tau_n)_1^\infty \in \times_1^\infty \mathrm{T}_n$ is the chosen sequence. We construct the projective limit of the Hilbert spaces $\mathcal{H}_{\boldsymbol{\tau};e,\delta}$, letting τ_n vary arbitrarily in T_n and taking all possible weights δ of the indicated form; denote it by w. $\bigotimes_{n=1;e}^\infty \Phi_n$ and call it the weighted tensor product of the spaces Φ_n. Thus,

$$\mathrm{T} = \{\tau = (\boldsymbol{\tau},\boldsymbol{\delta}) = ((\tau_n)_{n=1}^\infty, (\delta_n)_{n=1}^\infty) \mid \tau_n \in \mathrm{T}_n,\ \delta_n \geq 1\},$$

$$\mathrm{w.}\bigotimes_{n=1;e}^\infty \Phi_n = \operatorname*{pr\,lim}_{(\boldsymbol{\tau},\delta)\in\mathrm{T}} \mathcal{H}_{\boldsymbol{\tau};e,\delta}. \tag{4.15}$$

The role of $0 \in \mathrm{T}$ is played by $(\mathbf{0},1) = ((0_n)_1^\infty, 1)$.

THEOREM 4.3. *The definition of the weighted infinite tensor product* w. $\bigotimes_{n=1;e}^\infty \Phi_n$ *of nuclear spaces is unambiguous. This product is a nuclear space.*

PROOF. Since each H_{τ_n} is dense in H_{0_n}, clearly $\mathcal{H}_{\tau;e,\delta} = \bigotimes_{n=1;e,\delta}^\infty H_{\tau_n}$ is dense in $\mathcal{H}_{0;e,1} = \bigotimes_{n=1;e,1}^\infty H_{0_n}$. The norm of the corresponding imbedding operator does not exceed 1; this follows from Lemma 2.3. Clearly, $\bigcap_{(\boldsymbol{\tau},\delta)\in\mathrm{T}} \mathcal{H}_{\boldsymbol{\tau};e,\delta}$ is dense in each $H_{\boldsymbol{\tau};e,\delta}$: it contains all possible vectors of the form $\varphi^{(1)} \otimes \cdots \otimes \varphi^{(p)} \otimes e^{(p+1)} \otimes e^{(p+2)} \otimes \cdots$, with $\varphi^{(n)} \in \bigcap_{\tau_n \in \mathrm{T}_n} H_{\tau_n}$ and $p = 1, 2, \ldots$. Let $\tau' = (\boldsymbol{\tau}', \boldsymbol{\delta}') = ((\tau_n')_1^\infty, (\delta_n')_1^\infty)$ and $\tau'' = (\boldsymbol{\tau}'', \boldsymbol{\delta}'') = ((\tau_n'')_1^\infty, (\delta_n'')_1^\infty)$ be given. For each $n = 1, 2, \ldots$ there is a τ_n such that $H_{\tau_n''} \subseteq H_{\tau_n'}, H_{\tau_n''}$ and the imbeddings $H_{\tau_n'''} \to H_{\tau_n'}$ and $H_{\tau_n'''} \to H_{\tau_n''}$ are quasinuclear; this is possible because Φ_n is nuclear. Let $\boldsymbol{\tau}''' = (\tau_n''')_1^\infty$. By the remark in §2.6, there is a weight $\boldsymbol{\delta}''' = (\delta_n''')_1^\infty$ $(\delta_n''' \geq 1)$ such that $\mathcal{H}_{\boldsymbol{\tau}''';e,\boldsymbol{\delta}'''} \subseteq \mathcal{H}_{\boldsymbol{\tau}';e,\boldsymbol{\delta}'}$, $\mathcal{H}_{\boldsymbol{\tau}'';e,\boldsymbol{\delta}''}$ and the imbeddings $\mathcal{H}_{\boldsymbol{\tau}''';e,\boldsymbol{\delta}'''} \to \mathcal{H}_{\boldsymbol{\tau}';e,\boldsymbol{\delta}'}$ and $\mathcal{H}_{\boldsymbol{\tau}''';e,\boldsymbol{\delta}'''} \to \mathcal{H}_{\boldsymbol{\tau}'';e,\boldsymbol{\delta}'}$ are quasinuclear. Accordingly, we can set $\tau''' = (\boldsymbol{\tau}''', \boldsymbol{\delta}''')$. ■

It is not hard to see that the elements of the weighted infinite tensor product $\mathrm{w.}\bigotimes_{n=1;e}^\infty \Phi_n$ are cylindrical in a definite sense. We introduce the corresponding definitions. Fix $p = 1, 2, \ldots$ and let

$$\left(\bigotimes_{n=1}^p \Phi_n\right) \otimes e^{(p+1)} \otimes e^{(p+2)} \otimes \cdots$$
$$= \bigcap_{\tau_1 \in \mathrm{T}_1, \ldots, \tau_p \in \mathrm{T}_p} \left(\left(\bigotimes_{n=1}^p H_{\tau_n}\right) \otimes e^{(p+1)} \otimes e^{(p+2)} \otimes \cdots\right)$$
$$= \operatorname*{pr\,lim}_{\tau_1 \in \mathrm{T}_1, \ldots, \tau_p \in \mathrm{T}_p} \left(\left(\bigotimes_{n=1}^p H_{\tau_n}\right) \otimes e^{(p+1)} \otimes e^{(p+2)} \otimes \cdots\right). \tag{4.16}$$

This space is a subspace of $\mathrm{w.}\bigotimes_{n=1;e}^\infty \Phi_n$. It is natural to say that an element $\varphi \in \mathrm{w.}\bigotimes_{n=1;e}^\infty \Phi_n$ in (4.16) for some p is cylindrical.

THEOREM 4.4. *Each element in the space* $\mathrm{w.}\bigotimes_{n=1;e}^\infty \Phi_n$ *is cylindrical, i.e.,*

$$\mathrm{w.}\bigotimes_{n=1;e}^\infty \Phi_n = \bigcup_{p=1}^\infty \left(\left(\bigotimes_{n=1}^p \Phi_n\right) \otimes e^{(p+1)} \otimes e^{(p+2)} \otimes \cdots\right). \tag{4.17}$$

Convergence of a sequence of vectors in $\mathrm{w.}\bigotimes_{n=1;e}^\infty \Phi_n$ *is equivalent to convergence in the space* $\bigcup_{p=1}^\infty((\bigotimes_{n=1}^p \Phi_n) \otimes e^{(p+1)} \otimes e^{(p+2)} \otimes \cdots)$, *endowed with the topology of the inductive limit of the spaces* $(\bigotimes_{n=1}^p \Phi_n) \otimes e^{(p+1)} \otimes e^{(p+2)} \otimes \cdots$. *In other words, a sequence* $(\varphi^{(k)})_1^\infty$ $(\varphi^{(k)} \in \mathrm{w.}\bigotimes_{n=1;e}^\infty \Phi_n)$ *converges to a* $\varphi \in \mathrm{w.}\bigotimes_{n=1;e}^\infty \Phi_n$, *if and only if it is uniformly cylindrical, i.e.,* $\varphi^{(k)} \in (\bigotimes_{n=1}^p \Phi_n) \otimes e^{(p+1)} \otimes e^{(p+2)} \otimes \cdots$ $(k = 1, 2, \ldots)$ *for some* $p = 1, 2, \ldots$ *and* $\varphi^{(k)} \to \varphi$ *as* $k \to \infty$ *in the topology of* $(\bigotimes_{n=1}^p \Phi_n) \otimes e^{(p+1)} \otimes e^{(p+2)} \otimes \cdots$.

PROOF. Let $\varphi \in \mathrm{w.}\bigotimes_{n=1;e}^\infty \Phi_n$. Fix a sequence $\boldsymbol{\tau}' = (\tau_n')_1^\infty$ with $\tau_n' \in \mathrm{T}_n$. We claim that for some $p' = 1, 2, \ldots$

$$\varphi \in \left(\bigotimes_{n=1}^{p'} H_{\tau_n'}\right) \otimes e^{(p'+1)} \otimes e^{(p'+2)} \otimes \cdots. \tag{4.18}$$

Indeed, assuming the opposite, we find a subsequence $(n_j)_1^\infty$ of $(n)_1^\infty$ such that $\varphi = \sum_{n=1}^\infty \sum_{\alpha\in \mathrm{A}_n} \varphi_\alpha e_\alpha$ and $\varphi_\alpha \not\equiv 0$ for $\alpha \in \mathrm{A}_{n_j}$ $(j = 1, 2, \ldots)$. Suppose that $\sum_{\alpha\in\mathrm{A}_{n_j}} |\varphi_\alpha|^2 = d_{n_j} > 0$. We construct a weight $\delta = (\delta_n)_1^\infty$ by setting $\delta_{n_j} = \max(d_{n_j}^{-1}, 1)$ for $j = 1, 2, \ldots$, and assuming that $\delta_n = 1$ for the remaining n. Then $\sum_{n=1}^\infty (\sum_{\alpha\in\mathrm{A}_n} |\varphi_\alpha|^2)\delta_n = \infty$, which contradicts the relation $\varphi \in \bigotimes_{n=1;e,\delta}^\infty H_{\tau_n'}$.

Consider another sequence $\boldsymbol{\tau}'' = (\tau_n'')_1^\infty$ $(\tau_n' \in \mathrm{T}_n)$. We conclude similarly that it satisfies an inclusion of the form (4.18), with the prime replaced by a double prime. It can be assumed that $p'' \le p'$. This shows that φ is in (4.16), with p replaced by p'. Accordingly, the element φ is cylindrical.

Suppose now that $(\varphi_k)_1^\infty$ is the convergent sequence in the formulation of the theorem, and φ is its limit; it suffices to assume that $\varphi = 0$. To prove that this sequence is uniformly cylindrical, it suffices, as earlier, that all the $\varphi^{(k)}$ are in the space (4.18) for some p'. Assume the opposite. Then there is a subsequence $(\varphi^{(k_j)})_{j=1}^\infty$ such that $\varphi^{(k_j)} = \sum_{n=1}^\infty \sum_{\alpha\in\mathrm{A}_n} \varphi_\alpha^{(k_j)} e_\alpha$ and $\varphi_\alpha^{(k_j)} \not\equiv 0$ for $\alpha \in \mathrm{A}_{n_j}$, where $n_j \to \infty$ as $j \to \infty$. Let $\sum_{\alpha\in\mathrm{A}_{n_j}} |\varphi_\alpha^{(k_j)}|^2 = d_{n_j} > 0$. As earlier, we construct a weight δ from d_{n_j}. Then

$$\|\varphi^{(k_j)}\|^2_{\bigotimes_{n=1;e,\delta}^\infty H_{\tau_n'}} = \sum_{n=1}^\infty \left(\sum_{\alpha\in\mathrm{A}_n} |\varphi_\alpha^{(k_j)}|^2) \right) \delta_n \ge 1,$$

which contradicts the convergence $\varphi^{(k)} \to 0$ in the topology of the space $\mathrm{w}.\bigotimes_{n=1;e}^\infty \Phi_n$, understood as the projective limit of the spaces $\mathcal{H}_{\boldsymbol{\tau};e,\delta}$. It is established that the sequence is uniformly cylindrical. The convergence $\varphi^{(k)} \to 0$ in the space (4.16) follows from the corresponding convergence in $\mathrm{w}.\bigotimes_{n=1;e}^\infty \Phi_n$ and from the definition of the topology in the last space. ■

The procedure for forming a weighted infinite tensor product of nuclear spaces admits an important generalization: instead of considering the whole set T in (4.15) and taking the projective limit with respect to it, we can consider some subset $\Sigma \subseteq \mathrm{T}$, and take the projective limit with respect to it. Moreover, for the construction of this projective limit to be valid and for the projective limit to be nuclear, it is obviously necessary to require that Σ have the following property: for any two pairs $\tau' = (\boldsymbol{\tau}', \delta'), \tau'' = (\boldsymbol{\tau}'', \delta'') \in \Sigma$ there exists a $\tau''' = (\boldsymbol{\tau}''', \delta''') \in \Sigma$ such that $\mathcal{H}_{\boldsymbol{\tau}''';e,\delta'''} \subseteq \mathcal{H}_{\boldsymbol{\tau}';e,\delta'}, \mathcal{H}_{\boldsymbol{\tau}'';e,\delta''}$ and the imbeddings $\mathcal{H}_{\boldsymbol{\tau}''';e,\delta'''} \to \mathcal{H}_{\boldsymbol{\tau}';e,\delta'}$ and $\mathcal{H}_{\boldsymbol{\tau}''';e,\delta'''} \to \mathcal{H}_{\boldsymbol{\tau}'';e,\delta''}$ are quasinuclear. The space $\mathrm{pr}\lim_{(\boldsymbol{\tau},\delta)\in\Sigma} \mathcal{H}_{\boldsymbol{\tau};e,\delta}$ constructed in this way is called the Σ-infinite tensor product of the spaces Φ and is denoted by $\mathrm{w}.\bigotimes_{n=1;e,\Sigma}^\infty \Phi_n$. Of course, such a tensor product need not consist of cylindrical elements.

Especially useful is the particular case when no weighting is taken in general, i.e., the projective limit of some set of Hilbert spaces $\mathcal{H}_{\boldsymbol{\tau};e,1}$ is considered. We formulate the corresponding result as the following simple theorem.

THEOREM 4.5. *Let $(\Phi_n)_1^\infty$ be the previous sequence of nuclear spaces $\Phi_n = \operatorname{pr}\lim_{\tau_n\in T_n} H_{\tau_n}$. Assume that Σ is the set of pairs $(\boldsymbol{\tau}, 1)$ such that for any $((\tau_n')_1^\infty, 1), ((\tau_n'')_1^\infty, 1) \in \Sigma$ there exists a pair $((\tau_n''')_1^\infty, 1) \in \Sigma$ satisfying the condition*

$$\sum_{n=1}^{\infty}(|O_{\tau_n'''\tau_n'}|^2 - 1) < \infty, \quad \sum_{n=1}^{\infty}(|O_{\tau_n'''\tau_n''}|^2 - 1) < \infty. \tag{4.19}$$

Here $H_{\tau_n'''} \subseteq H_{\tau_n'}, H_{\tau_n''}$ and $O_{\tau_n'''\tau_n'}\colon H_{\tau_n'''} \to H_{\tau_n'}$ and $O_{\tau_n'''\tau_n''}\colon H_{\tau_n'''} \to H_{\tau_n''}$ are the corresponding imbedding operators ($n = 1, 2, \ldots$). Then the Σ-infinite tensor product w. $\bigotimes_{n=1;e,\Sigma}^\infty \Phi_n$ *is well defined and is a nuclear space.*

PROOF. The conditions (4.19) in combination with Theorem 2.3 show that the imbeddings $\mathcal{H}_{\boldsymbol{\tau}''';e,1} \to \mathcal{H}_{\boldsymbol{\tau}';e,1}$ and $\mathcal{H}_{\boldsymbol{\tau}''';e,1} \to \mathcal{H}_{\boldsymbol{\tau}'';e,1}$ are quasinuclear, where $\boldsymbol{\tau}' = (\tau_n')_1^\infty, \boldsymbol{\tau}'' = (\tau_n'')_1^\infty$ and $\boldsymbol{\tau}''' = (\tau_n''')_1^\infty$. This gives the assertion. ∎

4.4. *Examples of infinite tensor products of nuclear spaces.*

1. *The space $C^\infty(\times_{n=1}^\infty[0,1])$.* For each $n = 1, 2, \ldots$ let

$$\Phi_n = C^\infty([0,1]) = \operatorname*{pr\,lim}_{\tau\in\{0,1,\ldots\}} C^\tau([0,1]) = \operatorname*{pr\,lim}_{\tau\in\{0,1,\ldots\}} W_2^\tau((0,1)).$$

The fact that these two projective limits coincide follows easily from the imbedding theorems: we have the estimates

$$\|u\|_{C^\tau([0,1])} \le c_\tau'\|u\|_{W_2^{\tau+1}((0,1))} \le c_\tau''\|u\|_{C^{\tau+1}([0,1])}$$
$$(u \in C^{\tau+1}([0,1]);\ c_\tau', c_\tau'' < \infty;\ \tau = 0, 1, \ldots).$$

The fact that $C^\infty([0,1])$ is nuclear follows from Theorem 3.2. The sequence $e = (e^{(n)}(x_n))_1^\infty$, where $e^{(n)}(x_n) \equiv 1$, can be taken as a simultaneous stabilizing sequence. Thus, it is possible to construct the weighted infinite tensor product w. $\bigotimes_{n=1;e}^\infty \Phi_n$. The space (4.16) now coincides with

$$C^\infty\left(\times_{n=1}^{p}[0,1]\right) = \operatorname*{pr\,lim}_{\tau\in\{0,1,\ldots\}} C^\tau\left(\times_{n=1}^{p}[0,1]\right) = \operatorname*{pr\,lim}_{\tau\in\{0,1,\ldots\}} W_2^\tau\left(\times_{n=1}^{p}(0,1)\right),$$

as follows from Lemma 3.3. (The coincidence of the last two spaces can be established as in the case $p = 1$.) With the help of Theorem 4.4 and this fact we conclude that w. $\bigotimes_{n=1;e}^\infty \Phi_n$ consists of all the cylindrical functions

$$\varphi(x) = \varphi(x_1, x_2, \ldots) = \varphi_c(x_1, \ldots, x_p) \in C^\infty\left(\times_{n=1}^{p}[0,1]\right),$$

and $\varphi^{(k)} \to \varphi$ as $k \to \infty$ if and only if the $\varphi^{(k)}$ are uniformly cylindrical and the limit relation $\varphi_c^{(k)}(x_1, \ldots, x_p) \to \varphi_c(x_1, \ldots, x_p)$ holds uniformly on $\times_{n=1}^p[0,1]$ for the corresponding functions along with any derivative. The space w. $\bigotimes_{n=1;e}^\infty \Phi_n$ constructed in this way is denoted by $C^\infty(\times_{n=1}^\infty[0,1])$.

If in this example we consider not all the weights δ (or, what is the same, we carry out not all the possible renormings according to Lemma 2.1), then we get nuclear spaces which may consist not solely of cylindrical functions. We give the most simple of these renormings.

For each $n = 1, 2, \ldots$ let $G_n = L_2((0,1)), H_n = W_2^1((0,1))$ and $e = (e^{(n)}(x_n))_1^\infty$ with $e^{(n)}(x_n) \equiv 1$. Then, by Theorem 2.2,

$$\bigotimes_{n=1;e}^{\infty} H_n \subseteq \bigotimes_{n=1;e}^{\infty} G_n = L_2\left(\bigtimes_{n=1}^{\infty} (0,1), (dx_1) \times (dx_2) \times \cdots\right)$$

and the norm of the corresponding imbedding operator is at most 1. However, this imbedding is not quasinuclear, because condition (2.15) is no longer satisfied. To make it quasinuclear we renorm $W_2^1((0,1))$. It is easy to compute that

$$(u,v)_{(W_2^1((0,1)))^{\#_n}} = c_n (u,v)_{W_2^1((0,1))} + (1 - c_n) \int_0^1 u(x_n)\, dx_n \int_0^1 \overline{v(x_n)}\, dx_n$$

$$(c_n \geq 1;\ n = 1, 2, \ldots;\ u, v \in W_2^1((0,1))).$$

If $\sum_1^\infty c_n^{-1} < \infty$, then the condition (2.15) holds and the imbedding

$$\bigotimes_{n=1;e}^{\infty} (W_2^1((0,1)))^{\#_n} \to L_2\left(\bigtimes_{n=1}^{\infty} (0,1), (dx_1) \times (dx_2) \times \cdots\right)$$

is quasinuclear.

2. *The space* $\mathcal{G}(\mathbb{R}^\infty, \varepsilon)$. Let $\varepsilon = (\varepsilon_n)_1^\infty$ $(\varepsilon_n > 0)$ be a fixed sequence, and let $\Phi_n = \mathcal{G}(\mathbb{R}^1, \varepsilon_n) = \operatorname{pr\,lim}_{\tau \in \{0,1,\ldots\}} G^\tau(\mathbb{R}^1, \varepsilon_n)$ and $e = (e^{(n)}(x_n))_1^\infty$, with $e^{(n)}(x_n) \equiv 1$. Then it is possible to construct the weighted infinite tensor product $\mathcal{G}(\mathbb{R}^\infty, \varepsilon) = \text{w.} \bigotimes_{n=1,e}^\infty \mathcal{G}(\mathbb{R}^1, \varepsilon_n)$ with simultaneous stabilizing sequence e. *The elements of this product are cylindrical functions of the form*

$$\varphi(x) = \varphi(x_1, x_2, \ldots) = \varphi_c(x_1, \ldots, x_p) \exp\left(\sum_{n=1}^{p} \frac{\varepsilon_n}{2} x_n^2\right),$$

where $\varphi_c \in S(\mathbb{R}^p)$. *The relation* $\mathcal{G}(\mathbb{R}^\infty, \varepsilon) \ni \varphi^{(k)} \to \varphi \in \mathcal{G}(\mathbb{R}^\infty, \varepsilon)$ *as* $k \to \infty$ *holds if and only if all the* $\varphi^{(k)}$ *are uniformly cylindrical and* $\varphi_c^{(k)}(x_1, \ldots, x_p) \to \varphi_c(x_1, \ldots, x_p)$ *in* $S(\mathbb{R}^p)$.

Indeed, the space (4.16) coincides with

$$\operatorname*{pr\,lim}_{\tau_1, \ldots, \tau_p \in \{0,1,\ldots\}} \bigotimes_{n=1}^{p} G_{\tau_n}(\mathbb{R}^1, \varepsilon_n),$$

and the last projective limit is equal to

$$\left(\operatorname*{pr\,lim}_{\tau_1, \ldots, \tau_p \in \{0,1,\ldots\}} \bigotimes_{n=1}^{p} S_{\tau_n}(\mathbb{R}^1)\right) \exp\left(\sum_{n=1}^{p} \frac{\varepsilon_n}{2} x_n^2\right)$$

because of (4.10). Just below we prove that the projective limit in the last expression coincides with $S(\mathbb{R}^p)$. The assertion follows from this and Theorem 4.4. ■

LEMMA 4.3.

$$\operatorname*{pr\,lim}_{\tau_1,\ldots\tau_p\in\{0,1,\ldots\}} \bigotimes_{n=1}^{p} S_{\tau_n}(\mathbf{R}^1) = \mathcal{S}(\mathbf{R}^p) \qquad (p = 1, 2, \ldots). \tag{4.20}$$

PROOF. We carry out a proof for the simplest case $p = 2$; the proof is analogous in the general case $p > 2$ (see Theorem 3.8 for $p = 1$). The relation (4.20) with $p = 2$ is an obvious consequence of the following topological inclusion:

$$S_{\min(\tau_1,\tau_2)}(\mathbf{R}^2) \supseteq S_{\tau_1}(\mathbf{R}^1) \otimes S_{\tau_2}(\mathbf{R}^1) \supseteq S_{\tau_1+\tau_2}(\mathbf{R}^2) \qquad (\tau_1, \tau_2 = 0, 1, \ldots). \tag{4.21}$$

This inclusion can be proved by establishing that $\|u\|_1 \le \|u\|_2 \le \|u\|_3$ for functions of the form

$$u(x_1, x_2) = \sum_{n=1}^{q} u_n^{(1)}(x_1)u_n^{(2)}(x_2) \qquad (u_n^{(j)} \in C_0^\infty(\mathbf{R}^1)),$$

where $\|\cdot\|_1, \|\cdot\|_2$, and $\|\cdot\|_3$ are the norms in the first, second, and third spaces of the chain (4.21).

We establish a simple identity. Let $p_1, p_2 \in C(\mathbf{R}^1)$ be two weights, let $x = (x_1, x_2)$, and let $\tau_1, \tau_2 = 0, 1, \ldots$. For the u under consideration

$$\begin{aligned}
&\sum_{\mu_1=0}^{\tau_1}\sum_{\mu_2=0}^{\tau_2} \|D_1^{\mu_1} D_2^{\mu_2} u\|^2_{L_2(\mathbf{R}^2, p_1(x_1)p_2(x_2)\,dx)} \\
&= \sum_{\mu_1=0}^{\tau_1}\sum_{\mu_2=0}^{\tau_2}\sum_{n,m=1}^{q} (D_1^{\mu_1} u_n^{(1)}, D_1^{\mu_1} u_m^{(1)})_{L_2(\mathbf{R}^1, p_1(x_1)\,dx_1)} \\
&\qquad\qquad \times (D_2^{\mu_2} u_n^{(2)}, D_2^{\mu_2} u_m^{(2)})_{L_2(\mathbf{R}^1, p_2(x_2)\,dx_2)} \\
&= \sum_{n,m=1}^{q} (u_n^{(1)}, u_m^{(1)})_{W_2^{\tau_1}(\mathbf{R}^1, p_1(x_1)\,dx_1)} (u_n^{(2)}, u_m^{(2)})_{W_2^{\tau_2}(\mathbf{R}^1, p_2(x_2)\,dx_2)} \\
&= \|u\|^2_{W_2^{\tau_1}(\mathbf{R}^1, p_1(x_1)\,dx_1) \otimes W_2^{\tau_2}(\mathbf{R}^1, p_2(x_2)\,dx_2)}.
\end{aligned} \tag{4.22}$$

Observe also that

$$(1 + |x|^2)^{\min(\tau_1,\tau_2)} \le (1 + x_1^2)^{\tau_1}(1 + x_2^2)^{\tau_2} \le (1 + |x|^2)^{\tau_1+\tau_2} \qquad (x = (x_1, x_2) \in \mathbf{R}^2;\ \tau_1, \tau_2 = 0, 1, \ldots).$$

By (4.22) and this inequality,

$$\begin{aligned}
\|u\|_1^2 &\le \sum_{\mu_1=0}^{\tau_1}\sum_{\mu_2=0}^{\tau_2} \|D_1^{\mu_1} D_2^{\mu_2} u\|^2_{L_2(\mathbf{R}^2, (1+x_1^2)^{\tau_1}(1+x_2^2)^{\tau_2}\,dx)} = \|u\|_2^2 \\
&\le \sum_{\mu_1=0}^{\tau_1}\sum_{\mu_2=0}^{\tau_2} \|D_1^{\mu_1} D_2^{\mu_2} u\|^2_{L_2(\mathbf{R}^2, (1+|x|^2)^{\tau_1+\tau_2}\,dx)} = \|u\|_3^2. \qquad \blacksquare
\end{aligned}$$

We present two more very important examples of infinite tensor products.

4.5. ***The space*** $\mathcal{A}(\mathsf{R}^\infty)$. Consider the space $L_2(\mathsf{R}^1, dg(t)) = G^0(\mathsf{R}^1)$ of functions square-integrable with respect to the standard Gaussian measure $dg(t)$. The Hermite polynomials $(h_j(t))_0^\infty$ form an orthonormal basis in it, where

$$h_j(t) = \frac{(-1)^j}{2^{j/2}\sqrt{j!}} e^{t^2} D^j(e^{-t^2}), \qquad h_0(t) \equiv 1 \qquad \left(D = \frac{d}{dt}\right). \tag{4.23}$$

With each function $u \in L_2(\mathsf{R}^1, dg(t))$ we associate the sequence $(u_j)_0^\infty$ of its Fourier coefficients with respect to the basis $(h_j(t))_0^\infty$. As a result we obtain the isometry

$$L_2(\mathsf{R}^1, dg(t)) \ni u \mapsto (u_j)_{j=0}^\infty \in l_2,$$

which we shall always have in mind.

For each $l = 0, 1, \dots$ consider the Hilbert space

$$A_l(\mathsf{R}^1) = \{u \in L_2(\mathsf{R}^1, dg(t)) | \sum_{j=0}^\infty |u_j|^2 (1+l)^j < \infty\};$$

$$(u, v)_{A_l(\mathsf{R}^1)} = \sum_{j=0}^\infty u_j \overline{v_j} (1+l)^j \qquad (u, v \in A_l(\mathsf{R}^1)). \tag{4.24}$$

Obviously, $\|\cdot\|_{A_l(\mathsf{R}^1)} \le \|\cdot\|_{A_{l+1}(\mathsf{R}^1)}$, $A_{l+1}(\mathsf{R}^1) \subseteq A_l(\mathsf{R}^1)$ $(l = 0, 1, \dots)$, and $A_0(\mathsf{R}^1) = G^0(\mathsf{R}^1)$. The set $\bigcap_0^\infty A_l(\mathsf{R}^1)$ contains all possible finite combinations of Hermite polynomials and, consequently, is dense in each $A_l(\mathsf{R}^1)$. We form the projective limit

$$\operatorname*{pr\,lim}_{l \in \{0,1,\dots\}} A_l(\mathsf{R}^1) = \mathcal{A}(\mathsf{R}^1). \tag{4.25}$$

LEMMA 4.4. *The projective limit* (4.25) *is a nuclear space.*

PROOF. Fix $l, l' = 0, 1, \dots$ with $l' > l$. The vectors $e_j(t) = (1+l')^{-j/2} h_j(t)$ $(j = 0, 1, \dots)$ form an orthonormal basis in the space $A_{l'}(\mathsf{R}^1)$; therefore, by (4.24), we have for the imbedding operator $O_{l'l}\colon A_{l'}(\mathsf{R}^1) \to A_l(\mathsf{R}^1)$ that

$$|O_{l'l}|^2 = \sum_{j=0}^\infty \|(1+l')^{-j/2} h_j(t)\|^2_{A_l(\mathsf{R}^1)} = \sum_{j=0}^\infty \left(\frac{1+l}{1+l'}\right)^j$$
$$= \frac{1+l'}{l'-l} = 1 + \frac{1+l}{l'-l} < \infty. \tag{4.26}$$

Thus, it is even possible to set $l' = l + 1$. ■

According to the general rules in §1.10, we can construct a sequence of negative spaces $A_{-l}(\mathsf{R}^1)$ with $A_0(\mathsf{R}^1) = G^0(\mathsf{R}^1)$ as the zero space and $A_l(\mathsf{R}^1)$ $(l = 1, 2, \dots)$ as the positive space. Then the chain (1.30) takes the form

$$\bigcup_{l=0}^\infty A_{-l}(\mathsf{R}^1) = (\mathcal{A}(\mathsf{R}^1))' \supseteq \dots \supseteq A_{-2}(\mathsf{R}^1) \supseteq A_{-1}(\mathsf{R}^1)$$
$$\supseteq A_0(\mathsf{R}^1) \supseteq A_1(\mathsf{R}^1) \supseteq A_2(\mathsf{R}^1) \supseteq \dots \supseteq \mathcal{A}(\mathsf{R}^1).$$

It follows at once from (4.24) that

$$A_{-l}(\mathsf{R}^1) = \left\{ \xi = \sum_{j=0}^{\infty} \xi_j h_j(t) \middle| \sum_{j=0}^{\infty} |\xi_j|^2 (1+l)^{-j} < \infty \right\}$$

and $(\xi, u)_{A_0(\mathsf{R}^1)} = \sum_{j=0}^{\infty} \xi_j \overline{u_j}$ $(u \in A_l(\mathsf{R}^1),\ \xi \in A_{-l}(\mathsf{R}^1);\ l = 0, 1, \ldots)$.

Let us now consider a countable set of spaces $\mathcal{A}(\mathsf{R}^1)$ and construct a certain Σ-infinite tensor product by taking the simultaneous stabilizing sequence $e = (e^{(n)}(x_n))_1^\infty$, $e^{(n)}(x_n) \equiv 1$. We do not choose a weighting, and take Σ to be all possible pairs $(\boldsymbol{\tau}, 1)$, where $\boldsymbol{\tau} = (\tau_n)_1^\infty \in \mathsf{X}_1^\infty \mathrm{T}_n = \mathrm{T}$, $\mathrm{T}_n = \{0, 1, \ldots\}$. Thus, $(\tau_n)_1^\infty$ is an arbitrary sequence of nonnegative integers. We construct the Hilbert infinite tensor product

$$A_\tau(\mathsf{R}^\infty) = \bigotimes_{n=1;e}^{\infty} A_{\tau_n}(\mathsf{R}^1) \qquad (\tau = (\tau_n)_{n=1}^\infty) \tag{4.27}$$

(it is now convenient to denote the sequence $(\tau_n)_1^\infty$ by the lightface letter τ), and then consider

$$\mathcal{A}(\mathsf{R}^\infty) = \operatorname*{pr\,lim}_{\tau \in \mathrm{T}} A_\tau(\mathsf{R}^\infty). \tag{4.28}$$

The projective limit (4.28) *is defined unambiguously and is a nuclear space.*

Indeed, it suffices to check the conditions in Theorem 4.5. Let $\boldsymbol{\tau}' = (\tau_n')_1^\infty$, $\boldsymbol{\tau}'' = (\tau_n'')_1^\infty \in \mathrm{T}$ be two fixed sequences. By (4.26), the series (4.19) take the form

$$\sum_{n=1}^{\infty} (|O_{\tau_n''' \tau_n'}|^2 - 1) = \sum_{n=1}^{\infty} \frac{1 + \tau_n'}{\tau_n''' - \tau_n'},$$

$$\sum_{n=1}^{\infty} (|O_{\tau_n''' \tau_n''}|^2 - 1) = \sum_{n=1}^{\infty} \frac{1 + \tau_n'}{\tau_n''' - \tau_n''}. \tag{4.29}$$

If $\tau_n''' = (2\tau_n' + 2\tau_n'' + 1)n^2$ $(n = 1, 2, \ldots)$, then

$$\tau_n''' - \tau_n' \geq (\tau_n' + 2\tau_n'' + 1)n^2 \geq (1 + \tau_n')n^2,$$

and so the nth term of the first series in (4.29) is bounded above by n^{-2}. The general term of the second series in (4.29) can be estimated similarly. Thus, both these series converge for the sequence $\boldsymbol{\tau}''' = (\tau''')^\infty_{n\ 1}$ ■

The Hilbert series (4.27) can be regarded as positive spaces with respect to the zero space

$$A_0(\mathsf{R}^\infty) = \bigotimes_{n=1;e}^{\infty} A_0(\mathsf{R}^1) = L_2(\mathsf{R}^\infty, d\mathfrak{g}(x)).$$

According to Theorem 2.4, the corresponding negative space $A_{-\tau}(\mathsf{R}^\infty)$ is equal to $\bigotimes_{n=1;e}^{\infty} A_{-\tau_n}(\mathsf{R}^1)$.

The vectors (4.23) form an orthonormal basis in the space $A_0(\mathsf{R}^1) = L_2(\mathsf{R}^1, dg(t))$; therefore, the vectors

$$e_\alpha = h_\alpha(x) = h_{\alpha_1}(x_1)\cdots h_{\alpha_n}(x_n) \qquad (x = (x_n)_{n=1}^\infty) \tag{4.30}$$

form an orthonormal basis in the infinite tensor product

$$A_0(\mathsf{R}^\infty) = \bigotimes_{n=1;e}^{\infty} A_0(\mathsf{R}^1) = L_2(\mathsf{R}^\infty, d\mathfrak{g}(x)),$$

constructed according to the general rules from the stabilizing sequence consisting of ones; here $\alpha_1, \dots, \alpha_n = 0, 1, \dots$ and $n = 1, 2, \dots$; in other words, α varies over the set A of finite sequences of nonnegative integers.

The vectors $((1+l)^{-j/2}h_j(t))_{j=0}^\infty$ form an orthonormal basis in $A_l(\mathsf{R}^1)$ ($l = 0, 1, \dots$); therefore, the vectors

$$h_\alpha(x)(1+\tau_1)^{-\alpha_1/2}(1+\tau_2)^{-\alpha_2/2}\cdots \qquad (\alpha \in \mathrm{A})$$

form an orthonormal basis for $A_\tau(\mathsf{R}^\infty) = \bigotimes_{n=1;e}^\infty A_{\tau_n}(\mathsf{R}_1)$ according to the same rules. Thus, we can write

$$A_\tau(\mathsf{R}^\infty) = \Big\{ u(x) = \sum_{\alpha\in\mathrm{A}} u_\alpha h_\alpha(x) \in L_2(\mathsf{R}^\infty, d\mathfrak{g}(x)) \,|\, \sum_{\alpha\in\mathrm{A}} |u_\alpha|^2 (1+\tau_1)^{\alpha_1} \times (1+\tau_2)^{\alpha_2}\cdots < \infty \Big\},$$

$$(u, v)_{A_\tau(\mathsf{R}^\infty)} = \sum_{\alpha\in\mathrm{A}} u_\alpha \bar{v}_\alpha (1+\tau_1)^{\alpha_1}(1+\tau_2)^{\alpha_2}\cdots \qquad (u, v \in A_\tau(\mathsf{R}^\infty)). \tag{4.31}$$

The space $A_{-\tau}(\mathsf{R}^\infty)$ of vectors $\xi = \sum_{\alpha\in\mathrm{A}} \xi_\alpha h_\alpha$ is defined by relations analogous to (4.31), with $(1+\tau_n)^{\alpha_n}$ replaced by $(1+\tau_n)^{-\alpha_n}$. (Of course, ξ no longer belongs to $L_2(\mathsf{R}^\infty, d\mathfrak{g}(x))$.) If

$$\xi = \sum_{\alpha\in\mathrm{A}} \xi_\alpha h_\alpha \in A_{-\tau}(\mathsf{R}^\infty) \quad \text{and} \quad u = \sum_{\alpha\in\mathrm{A}} u_\alpha h_\alpha \in A_\tau(\mathsf{R}^\infty),$$

then $(\xi, u)_{A_0(\mathsf{R}^\infty)} = \sum_{\alpha\in\mathrm{A}} \xi_\alpha \bar{u}_\alpha$.

We constructed the space $A_\tau(\mathsf{R}^\infty)$ as a certain infinite tensor product in the space $A_0(\mathsf{R}^\infty) = L_2(\mathsf{R}^\infty, d\mathfrak{g}(x))$. Let us show that if the numbers τ_n increase rapidly enough as $n \to \infty$, then the functions in $A_\tau(\mathsf{R}^\infty)$ are continuous on some Hilbert space in R^∞.

THEOREM 4.6. *Let $\tau = (\tau_n)_1^\infty$ be such that $((1+\tau_n)^{-1})_1^\infty \in l_1$. Denote by $R_\tau \subset \mathsf{R}^\infty$ the weighted Hilbert space of full measure distinguished by the condition*

$$R_\tau = \{x \in \mathsf{R}^\infty | \sum_{n=1}^\infty x_n^2 (1+\tau_n)^{-1} < \infty\};$$

$$(x, y)_{R_\tau} = \sum_{n=1}^\infty x_n y_n (1+\tau_n)^{-1} \qquad (x, y \in R_\tau). \tag{4.32}$$

Then the restriction to R_τ *of each function* $u \in A_\tau(\mathsf{R}^\infty) \subseteq L_2(\mathsf{R}^\infty, d\mathfrak{g}(x))$ *coincides* $\mathfrak{g}$*-almost everywhere with a continuous function of points* $x \in R_\tau$.

PROOF. We first show that for each fixed $x \in R_\tau$ the functional

$$\delta_x = \sum_{\alpha \in \mathrm{A}} h_\alpha(x) h_\alpha \tag{4.33}$$

is in $A_{-\tau}(\mathsf{R}^\infty)$. Using relations of the type (4.31) for $A_{-\tau}(\mathsf{R}^\infty)$ and the formula (see, for example, Erdélyi et al. [1], §10.13 in Chapter 10)

$$\sum_{j=0}^{\infty} z^j h_j^2(t) = (1 - z^2)^{-1/2} \exp\left(2t^2 \left(1 + \frac{1}{z}\right)^{-1}\right)$$
$$(0 < z < 1,\ t \in \mathsf{R}^1),$$

we get

$$\begin{aligned}
\|\delta_x\|^2_{A_{-\tau}(\mathsf{R}^\infty)} &= \sum_{\alpha \in \mathrm{A}} h_\alpha^2(x)(1 + \tau_1)^{-\alpha_1}(1 + \tau_2)^{-\alpha_2} \cdots \\
&= \sum_{n=1}^{\infty} \sum_{\alpha \in \mathrm{A}_n} h_\alpha^2(x)(1 + \tau_1)^{-\alpha_1}(1 + \tau_2)^{-\alpha_2} \cdots \\
&= \lim_{m \to \infty} \sum_{n=1}^{m} \sum_{\alpha \in \mathrm{A}_n} h_\alpha^2(x)(1 + \tau_1)^{-\alpha_1}(1 + \tau_2)^{-\alpha_2} \cdots \\
&= \lim_{m \to \infty} \sum_{\alpha_1, \ldots, \alpha_m = 0}^{\infty} h^2_{(\alpha_1, \ldots, \alpha_m, 0, 0, \ldots)}(x)(1 + \tau_1)^{-\alpha_1} \cdots (1 + \tau_m)^{-\alpha_m} \\
&= \lim_{m \to \infty} \prod_{k=1}^{m} \left(\sum_{\alpha_k = 0}^{\infty} h_{\alpha_k}^2(x_k)(1 + \tau_k)^{-\alpha_k} \right) \\
&= \prod_{k=1}^{\infty} ((1 - (1 + \tau_k)^{-2})^{-1/2} \exp(2x_k^2(2 + \tau_k)^{-1})) \\
&= \left(\prod_{k=1}^{\infty} (1 - (1 + \tau_k)^{-2}) \right)^{-1/2} \exp\left(2 \sum_{k=1}^{\infty} x_k^2 (2 + \tau_k)^{-1} \right) \\
&\leq \left(\prod_{k=1}^{\infty} (1 - (1 + \tau_k)^{-2}) \right)^{-1/2} \exp(2\|x\|^2_{R_\tau}).
\end{aligned} \tag{4.34}$$

Since $((1 + \tau_n)^{-1})_1^\infty \in l_1$, this sequence is a fortiori in l_2, and so the product on the right-hand side of (4.34) converges. Accordingly, $\delta_x \in A_{-\tau}(\mathsf{R}^\infty)$.

The fact that $\mathfrak{g}(R_\tau) = 1$ follows from the theorem of Khintchine and Kolmogorov (see, for example, Shilov and Fan Dyk Tin′ [1], §3.8):

$$\mathfrak{g}\left(\left\{x \in \mathbb{R}^\infty \middle| \sum_{n=1}^{\infty} a_n x_n^2 < \infty,\ a_n \geq 0\right\}\right)$$

is equal to 1 or 0, depending on whether $\sum_1^\infty a_n$ converges or diverges. Thus, for $x \in R_\tau$ and $u \in A_\tau(\mathbb{R}^\infty)$, we can write

$$(\delta_x, u)_{A_0(\mathbb{R}^\infty)} = \sum_{\alpha \in \mathrm{A}} h_\alpha(x)\bar{u}_\alpha = \overline{u(x)}, \tag{4.35}$$

where the last equality holds $\mathfrak{g}$-almost everywhere.

To conclude the proof of the theorem, it remains to show that the left-hand side of (4.35) is continuous with respect to $x \in R_\tau$, and to do this it suffices to see that the vector-valued function $R_\tau \ni x \mapsto \delta_x \in A_{-\tau}(\mathbb{R}^\infty)$ is weakly continuous. Let $(x^{(k)})_1^\infty$ be a sequence of points $x^{(k)} \in R_\tau$ such that $x^{(k)} \to x^{(0)}$ as $k \to \infty$ in R_τ. It is required to prove that $\delta_{x^{(k)}} \to \delta_{x^{(0)}}$ as $k \to \infty$ in the sense of weak convergence. For this it is necessary to prove that 1) $\|\delta_{x^{(k)}}\|_{A_{-\tau}(\mathbb{R}^\infty)} \leq c$ $(k = 1, 2, \ldots)$ and 2) $(\delta_{x^{(k)}}, v)_{A_0(\mathbb{R}^\infty)} \to (\delta_{x^{(0)}}, v)_{A_0(\mathbb{R}^\infty)}$ as $k \to \infty$ for the vectors v in a total subset of $A_\tau(\mathbb{R}^\infty)$. But 1) follows from (4.34), and 2) can be verified immediately: if $\|x^{(k)} - x^{(0)}\|_{R_\tau} \to 0$ as $k \to \infty$, then $x_j^{(k)} \to x_j^{(0)}$ as $k \to \infty$ for each $j = 0, 1, \ldots,$ and thus also $h_\alpha(x^{(k)}) \to h_\alpha(x^{(0)})$ as $k \to \infty$ for each $\alpha \in \mathrm{A}$; the total set can be taken to be $(h_\alpha)_{\alpha \in \mathrm{A}}$. ■

In what follows, a $u \in A_\tau(\mathbb{R}^\infty)$ is assumed to be redefined on a set of $\mathfrak{g}$-measure zero in such a way that $R_\tau \ni x \mapsto u(x) \in \mathbb{C}^1$ is continuous (under the condition that $((1+\tau_n)^{-1})_1^\infty \in l_1$). The relation (4.35) now holds for all $x \in R_\tau$, and we can write (see also (4.34))

$$|u(x)| \leq \|u\|_{A_\tau(\mathbb{R}^\infty)} \|\delta_x\|_{A_{-\tau}(\mathbb{R}^\infty)} \qquad (u \in A_\tau(\mathbb{R}^\infty);\ x \in R_\tau);$$

$$\|\delta_x\|_{A_{-\tau}(\mathbb{R}^\infty)} = d_\tau \exp\left(\sum_{k=1}^{\infty} x_k^2 (2+\tau_k)^{-1}\right),$$

$$d_\tau = \left(\prod_{k=1}^{\infty} (1 - (1+\tau_k)^{-2})\right)^{-1/4}. \tag{4.36}$$

We explain how the operators of differentiation and multiplication by a polynomial act in $A_\tau(\mathbb{R}^\infty)$. In essence, it suffices to consider elementary operators of the following kind: the differentiation $D_k = \partial/\partial x_k$ and the operator X_k of multiplication by x_k $(k = 1, 2, \ldots)$.

The symbol $P_c(\mathbb{R}^\infty)$ will denote the linear set of cylindrical polynomials, i.e., the functions of the form

$$u(x) = \sum_{\alpha \in \mathrm{A}} u_\alpha h_\alpha(x), \tag{4.37}$$

where $u_\alpha = 0$ for multi-indices α such that $\nu(\alpha)+\sum_1^\infty \alpha_n$ exceeds some finite number depending on u (i.e., the summation in (4.37) is actually over a finite set depending on u). Clearly, $P_c(\mathbf{R}^\infty) \subset A_\tau(\mathbf{R}^\infty)$ for any τ, and it is dense in this space. The operators D_k and X_k are defined on $P_c(\mathbf{R}^\infty)$ and carry it into itself:

$$P_c(\mathbf{R}^\infty) \ni u(x) \mapsto (D_k u)(x), \quad (X_k u)(x) = x_k u(x) \in P_c(\mathbf{R}^\infty).$$

LEMMA 4.5. *For any $\tau \in \mathrm{T}$ and $k = 1, 2, \dots$ the operators D_k and X_k extend by continuity from $P_c(\mathbf{R}^\infty)$ to continuous operators acting from the space $A_{\tau'}(\mathbf{R}^\infty)$ to $A_\tau(\mathbf{R}^\infty)$, where $\tau' = (\tau'_n)_1^\infty$, with $\tau'_n = \tau_n$ for $n \neq k$ and $\tau'_k = \tau_k + 1$.*

PROOF. It is known that

$$Dh_j(t) = \sqrt{2j}\,h_{j-1}(t) \qquad (j = 1, 2, \dots); \quad Dh_0 = 0. \tag{4.38}$$

If we take $k = 1$ for simplicity, then, by (4.37) and (4.38), we get that

$$(D_1 u)(x) = \sum_{\alpha\in \mathrm{A}} u_{(\alpha_1,\alpha_2,\dots)}\sqrt{2\alpha_1}\,h_{\alpha_1-1}(x_1)h_{\alpha_2}(x_2)\cdots,$$

for $u \in P_c(\mathbf{R}^\infty)$, which gives us that

$$(D_1 u)_\alpha = \sqrt{2(\alpha_1+1)}\,u_{(\alpha_1+1,\alpha_2,\dots)} \qquad (\alpha \in \mathrm{A}).$$

Therefore,

$$\begin{aligned}
\|D_1 u\|^2_{A_\tau(\mathbf{R}^\infty)} &= \sum_{\alpha\in\mathrm{A}} |(D_1 u)_\alpha|^2 (1+\tau_1)^{\alpha_1}(1+\tau_2)^{\alpha_2}\cdots \\
&= 2\sum_{\alpha\in\mathrm{A}} (\alpha_1+1)|u_{(\alpha_1+1,\alpha_2,\dots)}|^2(1+\tau_1)^{\alpha_1}(1+\tau_2)^{\alpha_2}\cdots \\
&= 2{\sum_{\alpha\in\mathrm{A}}}' |u_\alpha|^2(\alpha_1(1+\tau_1)^{\alpha_1-1})(1+\tau_2)^{\alpha_2}\cdots,
\end{aligned} \tag{4.39}$$

where the prime means that no terms corresponding to $\alpha = (0, \alpha_2, \alpha_3, \dots)$ appear in the sum. Since $2\alpha_1(1+\tau_1)^{\alpha_1-1} \le c_1(\tau_1)(2+\tau_1)^{\alpha_1}$ $(\alpha_1 = 1, 2, \dots)$ for some $c_1(\tau_1)$, the estimate in (4.39) can be continued; we get that

$$\|D_1 u\|^2_{A_\tau(\mathbf{R}^\infty)} \le c_1(\tau_1)\|u\|^2_{A_{\tau'}(\mathbf{R}^\infty)}.$$

The assertion about D_k is proved.

Consider the operator X_k, with $k = 1$ for simplicity as before. It is known that

$$th_j(t) = \sqrt{\frac{j}{2}}\,h_{j-1}(t) + \sqrt{\frac{j+1}{2}}\,h_{j+1}(t) \qquad (j = 0, 1, \dots;\ h_{-1} = 0). \tag{4.40}$$

Again, for $u \in P_c(\mathbf{R}^\infty)$

$$x_1 u(x) = \sum_{\alpha\in\mathrm{A}} u_{(\alpha_1,\alpha_2,\dots)}\left(\sqrt{\frac{\alpha_1}{2}}\,h_{\alpha_1-1}(x_1) + \sqrt{\frac{\alpha_1+1}{2}}\,h_{\alpha_1+1}(x_1)\right)h_{\alpha_2}(x_2)\cdots,$$

and so $x_1 = A + B$, where

$$(Au)_\alpha = \sqrt{\frac{\alpha_1+1}{2}}\, u_{(\alpha_1+1,\alpha_2,\dots)}, \quad (Bu)_\alpha = \sqrt{\frac{\alpha_1}{2}}\, u_{(\alpha_1-1,\alpha_2,\dots)} \quad (\alpha \in \mathrm{A}).$$

As also for D_1, it is easy to obtain an estimate

$$\|Au\|^2_{A_\tau(\mathbb{R}^\infty)} \le c_2(\tau_1)\|u\|^2_{A_{\tau'}(\mathbb{R}^\infty)}$$

with some constant $c_2(\tau_1)$. For B we have

$$\begin{aligned}
\|Bu\|^2_{A_\tau(\mathbb{R}^\infty)} &= \sum_{\alpha\in\mathrm{A}} |(Bu)_\alpha|^2(1+\tau_1)^{\alpha_1}(1+\tau_2)^{\alpha_2}\cdots \\
&= \frac{1}{2}\sum_{\alpha\in\mathrm{A}} \alpha_1|u_{(\alpha_1-1,\alpha_2,\dots)}|^2(1+\tau_1)^{\alpha_1}(1+\tau_2)^{\alpha_2}\cdots \\
&= \frac{1}{2}\sum_{\alpha\in\mathrm{A}} |u_\alpha|^2(\alpha_1+1)(1+\tau_1)^{\alpha_1+1}(1+\tau_2)^{\alpha_2}\cdots \\
&\le c_3(\tau_1)\sum_{\alpha\in\mathrm{A}} |u_\alpha|^2(2+\tau_1)^{\alpha_1}(1+\tau_2)^{\alpha_2}\cdots \\
&= c_3(\tau_1)\|u\|^2_{A_{\tau'}(\mathbb{R}^\infty)},
\end{aligned} \tag{4.41}$$

where the constant $c_3(\tau_1)$ is such that

$$\frac{1}{2}(\alpha_1+1)(1+\tau_1)^{\alpha_1+1} \le c_3(\tau_1)(2+\tau_1)^{\alpha_1} \qquad (\alpha_1 = 0, 1, \dots).$$

It follows from (4.41) and the analogous estimate obtained earlier for A that X_1 acts continuously from $A_{\tau'}(\mathbb{R}^\infty)$ to $A_\tau(\mathbb{R}^\infty)$. ■

Let $\mathfrak{A}_c(X, D)$ be the algebra spanned by the operators X_j and D_k, $j, k = 1, 2, \dots$, i.e., take all possible finite linear combinations of products of these operators, with coefficients in $\mathbb{C}^1$. Considering the obvious commutation rules for X_j and D_k, we can assume that each element Q in this algebra is a differential operator with polynomial coefficients depending only on finitely many variables. It is clear that Q can be written in a minimal way, with the number of operators X_j and D_k in its expression. Of course, $P_c(\mathbb{R}^\infty) \ni u(x) \mapsto (Qu)(x) \in P_c(\mathbb{R}^\infty)$.

THEOREM 4.7. *For any $\tau \in \mathrm{T}$ an operator $Q \in \mathfrak{A}_c(X, D)$ extends by continuity from $P_c(\mathbb{R}^\infty)$ to a continuous operator acting from $A_{\tau(Q)}(\mathbb{R}^\infty)$ to $A_\tau(\mathbb{R}^\infty)$. Here the $\tau(Q) = (\tau_n(Q))_1^\infty \in \mathrm{T}$ are computed as follows: $\tau_n(Q)$ is equal to τ_n plus the number of operators X_n and D_n appearing in a minimal expression for Q.*

PROOF. Lemma 4.5 must be iterated in the proper way. ■

We remark that in the proof of Lemma 4.5 it is easy to clear up the dependence of the constants $c_l(\tau_1)$ $(l = 1, 2, 3)$ on τ_1 and thereby estimate in terms of τ the norms of the operators X_j and D_k, regarded as acting from $A_{\tau'}(\mathbb{R}^\infty)$

to $A_\tau(\mathsf{R}^\infty)$. Having these estimates, we can proceed as follows. Consider the algebra $\mathfrak{A}(X, D)$ of all formal infinite series spanned by finite products of operators X_j and D_k. When the behavior of the coefficients of a $Q \in \mathfrak{A}(X, D)$ has a specific nature, it can be guaranteed that this operator acts continuously from $A_{\tau(Q)}(\mathsf{R}^\infty)$ to $A_\tau(\mathsf{R}^\infty)$ for some $\tau(Q) = (\tau_n(Q))_1^\infty, \tau_n(Q) \geq \tau_n$. On the other hand, the possibility of applying Q to functions in $A_{\tau(Q)}(\mathsf{R}^\infty)$ can be interpreted as a certain smoothness of these functions. This gives us an analogue of the imbedding theorems for the spaces $A_\tau(\mathsf{R}^\infty)$: it can be determined which operators $Q \in \mathfrak{A}(X, D)$ can be applied to a $u \in A_\tau(\mathsf{R}^\infty)$, and in what spaces such operators act. We shall not dwell in more detail on these questions.

Let us establish one more simple fact about $A_\tau(\mathsf{R}^\infty)$ which will be used in what follows. First of all we give the following known formula (see, for example, Vilenkin [1], Chapter 11, §4.4):

$$\int_{\mathsf{R}^1} e^{i\lambda t} h_j(t)\, dg(t) = \frac{(i2^{-1/2}\lambda)^j}{\sqrt{j!}} e^{-\lambda^2/4} \qquad (j = 0, 1, \ldots;\ \lambda \in \mathsf{R}^1). \quad (4.42)$$

LEMMA 4.6. *Each exponential*

$$u^{(k)}(x) = \exp\left(i \sum_{n=1}^{k} \lambda_n x_n\right) \qquad (\lambda_1, \ldots, \lambda_k \in \mathsf{R}^1;\ k = 1, 2, \ldots) \quad (4.43)$$

is in any space $A_\tau(\mathsf{R}^\infty)$ $(\tau \in \mathrm{T})$. *Moreover, let* $\tau \in \mathrm{T}$. *Denote by* $R^\tau \subset \mathsf{R}^\infty$ *the real Hilbert space distinguished by the condition*

$$R^\tau = \left\{ \lambda = (\lambda_n)_{n=1}^\infty \in \mathsf{R}^\infty \mid \sum_{n=1}^{\infty} \lambda_n^2 (1 + \tau_n) < \infty \right\};$$

$$(\lambda, \mu)_{R^\tau} = \sum_{n=1}^{\infty} \lambda_n \mu_n (1 + \tau_n) \quad (\lambda, \mu \in R^\tau). \quad (4.44)$$

Then the vector-valued function

$$R^\tau \ni \lambda \mapsto \exp\left(i \sum_{n=1}^{\infty} \lambda_n x_n\right) = e^{i(\lambda, x)_{l_2}} \in A_\tau(\mathsf{R}^\infty) \quad (4.45)$$

is defined and continuous in the strong sense as the strong limit in the $A_\tau(\mathsf{R}^\infty)$*-norm of the functions* (4.43) *as* $k \to \infty$.

PROOF. With the help of (4.42) we conclude that

$$u_\alpha^{(k)} = \int_{\mathsf{R}^\infty} \exp\left(i \sum_{n=1}^{k} \lambda_n x_n\right) h_\alpha(x)\, d\mathfrak{g}(x)$$

$$= \frac{(i2^{-1/2}\lambda_1)^{\alpha_1} \cdots (i2^{-1/2}\lambda_k)^{\alpha_k}}{\sqrt{\alpha_1! \cdots \alpha_k!}} \exp\left(-\frac{1}{4} \sum_{n=1}^{k} \lambda_n^2\right)$$

$$(\alpha \in \mathrm{A};\ \lambda_1, \ldots, \lambda_k \in \mathsf{R}^1).$$

Therefore, in a way similar to that for (4.34),

$$\begin{aligned}\|u_\alpha^{(k)}\|^2_{A_\tau(\mathsf{R}^\infty)} &= \sum_{\alpha\in\mathrm{A}} |u_\alpha^{(k)}|^2(1+\tau_1)^{\alpha_1}(1+\tau_2)^{\alpha_2}\cdots \\ &= \lim_{m\to\infty} \sum_{\alpha_1,\dots,\alpha_m=0}^{\infty} |u^{(k)}_{(\alpha_1,\dots,\alpha_m,0,0,\dots)}|^2(1+\tau_1)^{\alpha_1}\cdots(1+\tau_m)^{\alpha_m} \\ &= \exp\left(-\frac{1}{2}\sum_{n=1}^{k}\lambda_n^2\right) \sum_{\alpha_1\dots,\alpha_k=0}^{\infty} \frac{(2^{-1}\lambda_1^2)^{\alpha_1}\cdots(2^{-1}\lambda_k^2)^{\alpha_k}}{\alpha_1!\cdots\alpha_k!} \\ &\qquad\times(1+\tau_1)^{\alpha_1}\cdots(1+\tau_k)^{\alpha_k} \\ &= \exp\left(\frac{1}{2}\sum_{n=1}^{k}\lambda_n^2\tau_n\right) < \infty.\end{aligned}$$

The inclusion $u^{(k)} \in A_\tau(\mathsf{R}^\infty)$ follows from this. Similarly,

$$\begin{aligned}&\left(\exp\left(i\sum_{n=1}^{k}\lambda_n x_n\right), \exp\left(i\sum_{n=1}^{k}\mu_n x_n\right)\right)_{A_\tau(\mathsf{R}^\infty)} \\ &\qquad= \exp\left(\frac{1}{2}\sum_{n=1}^{k}\left(\lambda_n\mu_n\tau_n - \frac{1}{2}(\lambda_n-\mu_n)^2\right)\right) \\ &\qquad\qquad(\lambda_1,\dots,\lambda_k,\mu_1,\dots,\mu_k \in \mathsf{R}^1;\ k=1,2,\dots).\end{aligned} \tag{4.46}$$

With the help of (4.46) it is easy to prove that if $\lambda \in R^\tau$, then the functions (4.43) form a Cauchy sequence in $A_\tau(\mathsf{R}^\infty)$; therefore, the vector-valued function (4.45) is defined. Passing to the limit as $k \to \infty$ in (4.46), we get

$$\begin{aligned}(e^{i(\lambda,\cdot)_{l_2}}, e^{i(\mu,\cdot)_{l_2}})_{A_\tau(\mathsf{R}^\infty)} &= \exp\left(\frac{1}{2}\sum_{n=1}^{\infty}\left(\lambda_n\mu_n\tau_n - \frac{1}{2}(\lambda_n-\mu_n)^2\right)\right) \\ &= \exp\left(\frac{1}{4}(2(\lambda,\mu)_{R^\tau} - \|\lambda\|^2_{l_2} - \|\mu\|^2_{l_2})\right) \\ &\qquad (\lambda,\mu \in R^\tau).\end{aligned} \tag{4.47}$$

With the help of (4.47) it is not hard to compute $\|e^{i(\lambda,\cdot)_{l_2}} - e^{i(\mu,\cdot)_{l_2}}\|_{A_\tau(\mathsf{R}^\infty)}$, which implies that (4.45) is continuous. ■

REMARK 1. Suppose that τ is such that $((1+\tau_n)^{-1})_1^\infty \in l_1$. Considering (4.32), we conclude that R_τ is the negative space with respect to the positive space R^τ and the zero space l_2. Accordingly, the following chain of real spaces can be written in the case when $((1+\tau_n)^{-1})_1^\infty \in l_1$:

$$R_\tau \supset l_2 \supset R^\tau. \tag{4.48}$$

REMARK 2. The above facts about $A_\tau(\mathsf{R}^\infty)$ and $A_{-\tau}(\mathsf{R}^\infty)$ imply in an obvious way the corresponding results about the projective limit $\mathcal{A}(\mathsf{R}^\infty)$ and its dual space $\mathcal{A}'(\mathsf{R}^\infty)$. In this connection we mention only the following.

Let $((1+\tau_n)^{-1})_1^\infty \in l_1$. *For* $\lambda \in R^\tau$ *and* $p(x) \in P_c(\mathsf{R}^\infty)$ *the function* $p(x)e^{i(\lambda,x)_{l_2}}$ *is in* $A_\tau(\mathsf{R}^\infty)$ *and depends continuously on* $\lambda \in R^\tau$ *in the norm of this space.* Indeed, denote by Q the operator of multiplication by the polynomial $p(x)$. According to Theorem 4.7, Q acts continuously from $A_{\tau(Q)}(\mathsf{R}^\infty)$ to $A_\tau(\mathsf{R}^\infty)$, and $\tau_n(Q) \geq \tau_n$ $(n = 1, 2, \ldots)$, with $\tau_n(Q) = \tau_n$ for all but finitely many n. Since $\tau_n \to \infty$ as $n \to \infty$, the space $R^{\tau(Q)}$ is equivalent to the space R^τ, and thus, by Lemma 4.6, $e^{i(\lambda,\cdot)_{l_2}} \in A_{\tau(Q)}(\mathsf{R}^\infty)$ and it depends continuously on $\lambda \in R^{\tau(Q)}$, i.e., on $\lambda \in R^\tau$, in the norm of the space. But then $Qp(x)e^{i(\lambda,x)_{l_2}} = (Qe^{i(\lambda,\cdot)_{l_2}})(x)$ has the required properties by continuity. ■

Finally, we note the following generalization of the construction of $A_\tau(\mathsf{R}^\infty)$ and $\mathcal{A}(\mathsf{R}^\infty)$. Let $d = (d_\alpha)_{\alpha\in\mathrm{A}}$ $(d_\alpha \geq 1)$ be a weight. We introduce the space $A_d(\mathsf{R}^\infty)$ according to (4.31), with the product $(1+\tau_1)^{\alpha_1}(1+\tau_2)^{\alpha_2}\cdots$ replaced by d_α. If D is some set of such weights d, then we can consider the projective limit

$$\mathcal{A}_{\mathrm{D}}(\mathsf{R}^\infty) = \operatorname*{pr\,lim}_{d\in\mathrm{D}} A_d(\mathsf{R}^\infty).$$

(Here it is necessary only to require that for each $d', d'' \in \mathrm{D}$ there exist a weight $d''' \in \mathrm{D}$ and constants $c', c'' \in (0, \infty)$ such that $d'_\alpha \leq c'd'''_\alpha$ and $d''_\alpha \leq c''d'''_\alpha$ $(\alpha \in \mathrm{A})$.) If D is taken to be the collection of all weights $d = (d_\alpha)_{\alpha\in\mathrm{A}}$ $(d_\alpha \geq 1)$, then, clearly, $\mathcal{A}_{\mathrm{D}}(\mathsf{R}^\infty) = P_c(\mathsf{R}^\infty)$ (cf., for example, Theorem 4.4). The convergence of the sequence $(p_k)_1^\infty$, $p_k \in P_c(\mathsf{R}^\infty)$, to $p \in P_c(\mathsf{R}^\infty)$ now means that the number of variables in the polynomials $p_k(x)$ and $p(x)$ and the orders of these polynomials are bounded uniformly with respect to k, and the coefficients of the polynomials $p_k(x)$ converge to the corresponding coefficients of the polynomial $p(x)$.

During consideration of the next example it will be shown that *the space* $\mathcal{A}(\mathsf{R}^\infty)$ *consists of cylindrical functions.* At the same time, it is not difficult to construct analogous spaces containing also noncylindrical functions. To do this it is necessary to consider the space $\mathcal{A}_{\mathrm{D}}(\mathsf{R}^\infty)$, where

$$d_\alpha = (1+\tau_1)^{\alpha_1}(1+\tau_2)^{\alpha_2}\cdots;$$

however, the sequences $\tau = (\tau_n)_1^\infty$ are not arbitrary in T, but vary over a certain countable subset T_1 of T having the property that for any $\tau' \in \mathrm{T}_1$ there exists a $\tau'' \in \mathrm{T}_1$ such that $\sum_0^\infty \tau'_n(\tau''_n)^{-1} < \infty$. (The last relation ensures that $\mathcal{A}_{\mathrm{D}}(\mathsf{R}^\infty)$ is a nuclear space; this is checked just as in the case of $\mathcal{A}(\mathsf{R}^\infty)$.)

4.6. ***The space*** $\mathcal{H}(\mathsf{R}^\infty)$**.** In a way similar to that for $\mathcal{A}(\mathsf{R}^\infty)$ we construct an unweighted tensor product with the role of the spaces $A_l(\mathsf{R}^1)$ played, in essence, by the spaces $G^l(\mathsf{R}^1)$. Let us first rewrite the construction of $G^l(\mathsf{R}^1)$ in terms of expansions in the Hermite polynomials (4.23). Take $u \in C^\infty(\mathsf{R}^1)$ and let $u(t) = \sum_0^\infty u_j h_j(t)$ be the expansion of this function in Hermite polynomials. Then, by (4.38),

$$(Du)(t) = \sum_{j=1}^{\infty} u_j \sqrt{2j} h_{j-1}(t) = \sum_{j=0}^{\infty} u_{j+1} \sqrt{2(j+1)} h_j(t) \qquad (t \in \mathbb{R}^1).$$

Iterating this formula, we get that

$$(D^\mu u)(t) = \sum_{j=0}^{\infty} u_{j+\mu} (2^\mu (j+1) \cdots (j+\mu))^{1/2} h_j(t) \qquad (t \in \mathbb{R}^1;\ \mu = 1, 2, \ldots).$$

Thus, (4.5) can be rewritten in the form

$$(u, v)_{G^l(\mathbb{R}^1)} = \sum_{\mu=0}^{l} \left(\sum_{j=0}^{\infty} u_{j+\mu} \bar{v}_{j+\mu} (2^\mu (j+1) \cdots (j+\mu)) \right)$$

$$(u, v \in C_0^\infty(\mathbb{R}^1);\ l = 0, 1, \ldots) \quad (4.49)$$

(the coefficient of $u_j \bar{v}_j$ is absent in the case $\mu = 0$).

For each $\mu = 0, \ldots, l$

$$\sum_{j=0}^{\infty} |u_{j+\mu}|^2 (2^\mu (j+1) \cdots (j+\mu)) \leq 2^l \sum_{j=0}^{\infty} |u_{j+\mu}|^2 (j+\mu+1)^l$$

$$\leq 2^l \sum_{j=0}^{\infty} |u_j|^2 (1+j)^l. \quad (4.50)$$

For $\mu = l$ the left-hand side of (4.50) has a lower estimate in terms of $2^l \sum_{j=l}^{\infty} |u_j|^2 (1+j)^l$ and, moreover,

$$\sum_{j=0}^{l-1} |u_j|^2 (1+j)^l \leq l^l \sum_{j=0}^{\infty} |u_j|^2.$$

From the last two estimates and (4.50) we conclude that there exist constant c_l' and c_l'' such that

$$c_l' \sum_{j=0}^{\infty} |u_j|^2 (1+j)^l \leq \|u\|^2_{G^l(\mathbb{R}^1)} \leq c_l'' \sum_{j=0}^{\infty} |u_j|^2 (1+j)^l$$

$$(l = 0, 1, \ldots). \quad (4.51)$$

For each $l = 0, 1, \ldots$ we introduce the Hilbert space

$$H^l(\mathbb{R}^1) = \{u \in L_2(\mathbb{R}^1, dg(t)) | \sum_{j=0}^{\infty} |u_j|^2 (1+j)^l < \infty\};$$

$$(u, v)_{H^l(\mathbb{R}^1)} = \sum_{j=0}^{\infty} u_j \bar{v}_j (1+j)^l \qquad (u, v \in H^l(\mathbb{R}^1)). \quad (4.52)$$

It follows from (4.51) that $H^l(\mathsf{R}^1)$ is the space $G^l(\mathsf{R}^1)$, suitably renormed, and $H^0(\mathsf{R}^1) = G^0(\mathsf{R}^1) = L_2(\mathsf{R}^1, dg(t))$. Obviously, $\|\cdot\|_{H^l(\mathsf{R}^1)} \leq \|\cdot\|_{H^{l+1}(\mathsf{R}^1)}$, and $H^{l+1}(\mathsf{R}^1) \subseteq H^l(\mathsf{R}^1)$ ($l = 0, 1, \ldots$). The set $\bigcap_0^\infty H^l(\mathsf{R}^1)$ is dense in each $H^l(\mathsf{R}^1)$. Thus, we can form the projective limit

$$\operatorname*{pr\,lim}_{l\in\{0,1,\ldots\}} H^l(\mathsf{R}^1) = \mathcal{H}(\mathsf{R}^1), \tag{4.53}$$

which coincides with

$$\mathcal{G}(\mathsf{R}^1) = \operatorname*{pr\,lim}_{l\in\{0,1,\ldots\}} G^l(\mathsf{R}^1)$$

and is hence a nuclear space.

Comparing (4.24) with (4.52), we conclude that the spaces $A_l(\mathsf{R}^1)$ are much smaller than $H^l(\mathsf{R}^1)$, i.e., than $G^l(\mathsf{R}^1)$. It can be shown that $A_l(\mathsf{R}^1)$ consists of entire functions of a specific growth, and, as we just explained, $H^l(\mathsf{R}^1)$ consists of functions l times differentiable in the mean with respect to Gaussian measure. (A precise result on the functions in $\mathcal{A}(\mathsf{R}^\infty)$ will be formulated below.)

For what follows it will be useful to give a direct proof (similar to that for Lemma 4.4) that $\mathcal{H}(\mathsf{R}^1)$ is nuclear. Fix $l, l' = 0, 1, \ldots$ with $l' > l$. The vectors $e_j(t) = (1+j)^{-l'/2} h_j(t)$ ($j = 0, 1, \ldots$), form an orthonormal basis in the space $A_{l'}(\mathsf{R}^1)$; therefore, for the imbedding operator $O_{l'l}\colon H^{l'}(\mathsf{R}^1) \to H^l(\mathsf{R}^1)$ we have, by (4.52), that

$$|O_{l'l}|^2 = \sum_{j=0}^{\infty} \|(1+j)^{-l'/2} h_j(t)\|^2_{H^l(\mathsf{R}^1)} = \sum_{j=0}^{\infty} (1+j)^{l-l'}. \tag{4.54}$$

Hence, if $l' = l + 2$, then (4.54) converges, and the imbedding $O_{l'l}$ is quasinuclear. (This is, in essence, another proof of Theorem 4.1.)

As in the case of the spaces $A_l(\mathsf{R}^1)$, the chain

$$\bigcup_{l=0}^{\infty} H^{-l}(\mathsf{R}^1) = (\mathcal{H}(\mathsf{R}^1))' \supseteq \cdots \supseteq H^{-2}(\mathsf{R}^1) \supseteq H^{-1}(\mathsf{R}^1)$$
$$\supseteq H^0(\mathsf{R}^1) \supseteq H^1(\mathsf{R}^1) \supseteq H^2(\mathsf{R}^1) \supseteq \cdots \supseteq \mathcal{H}(\mathsf{R}^1)$$

is constructed in the usual way, where $H^{-l}(\mathsf{R}^1)$ is the negative space with respect to the positive space $H^l(\mathsf{R}^1)$ and the zero space $H^0(\mathsf{R}^1)$. Obviously,

$$H^{-l}(\mathsf{R}^1) = \left\{ \xi = \sum_{j=0}^{\infty} \xi_j h_j(t) \,\middle|\, \sum_{j=0}^{\infty} |\xi_j|^2 (1+j)^{-l} < \infty \right\},$$

and

$$(\xi, u)_{H^0(\mathsf{R}^1)} = \sum_{j=0}^{\infty} \xi_j \bar{u}_j \qquad (u \in H^l(\mathsf{R}^1), \xi \in H^{-l}(\mathsf{R}^1);\ l = 0, 1, \ldots).$$

We now construct spaces similar to $A_\tau(\mathsf{R}^\infty)$ and $\mathcal{A}(\mathsf{R}^\infty)$. Take

$$e = (e^{(n)}(x_n))_1^\infty \text{ with } e^{(n)}(x_n) \equiv 1$$

as the simultaneous stabilizing sequence; the role of Σ is played as before by all possible pairs $(\boldsymbol{\tau}, 1)$, where $\boldsymbol{\tau} = (\tau_n)_1^\infty \in \times_1^\infty \mathrm{T}_n = \mathrm{T}$, with $\mathrm{T}_n = \{0, 1, \ldots\}$. Let us construct the Σ-infinite tensor product of a countable set of spaces $\mathcal{H}(\mathsf{R}^1)$. Take an arbitrary sequence $(\tau_n)_1^\infty$ of nonnegative integers, and form the Hilbert infinite tensor product

$$H^{\boldsymbol{\tau}}(\mathsf{R}^\infty) = \bigotimes_{n=1;e}^{\infty} H^{\tau_n}(\mathsf{R}^1) \qquad (\boldsymbol{\tau} = (\tau_n)_{n=1}^\infty) \tag{4.55}$$

and the corresponding projective limit

$$\mathcal{H}(\mathsf{R}^\infty) = \operatorname*{pr\,lim}_{\boldsymbol{\tau} \in \mathrm{T}} H^{\boldsymbol{\tau}}(\mathsf{R}^\infty). \tag{4.56}$$

The projective limit (4.56) *is defined unambiguously and is a nuclear space.*

Indeed, it suffices to verify the conditions in Theorem 4.5. Let $\boldsymbol{\tau}' = (\tau_n')_1^\infty$, $\boldsymbol{\tau}'' = (\tau_n'')_1^\infty \in \mathrm{T}$ be two fixed sequences, and let $\boldsymbol{\tau}''' = (\tau_n''')_1^\infty$ where

$$\tau_n''' = \tau_n' + \tau_n'' + 2n. \tag{4.57}$$

Then, by (4.54), summation of the geometric progression gives us that

$$\begin{aligned} \sum_{n=1}^\infty (|O_{\tau_n''' \tau_n'}|^2 - 1) &= \sum_{n=1}^\infty \sum_{j=1}^\infty (1+j)^{\tau_n' - \tau_n'''} \le \sum_{n=1}^\infty \sum_{j=1}^\infty (1+j)^{-2n} \\ &= \sum_{j=1}^\infty (2j + j^2)^{-1} < \infty. \end{aligned} \tag{4.58}$$

The convergence of the second series in (4.19) can be proved similarly. ∎

As for the spaces $A_\tau(\mathsf{R}^\infty)$, Theorem 2.4 can be used to prove easily that the negative space $H^{-\boldsymbol{\tau}}(\mathsf{R}^\infty)$ with respect to the positive space $H^{\boldsymbol{\tau}}(\mathsf{R}^\infty)$ ($\tau \in \mathrm{T}$) and the zero space $H^0(\mathsf{R}^\infty) = \bigotimes_{n=1;e}^\infty H^0(\mathsf{R}^1) = L_2(\mathsf{R}^\infty, d\mathfrak{g}(x))$ is equal to $\bigotimes_{n=1;e}^\infty H^{-\tau_n}(\mathsf{R}^1)$. For the constructed spaces we have (in a way similar to that for (4.31)) that

$$H^{\boldsymbol{\tau}}(\mathsf{R}^\infty) = \Bigg\{ u(x) = \sum_{\alpha \in \mathrm{A}} u_\alpha h_\alpha(x) \in L_2(\mathsf{R}^\infty, d\mathfrak{g}(x)) \,\Big|\, \sum_{\alpha \in \mathrm{A}} |u_\alpha|^2 (1+\alpha_1)^{\tau_1} \times (1+\alpha_2)^{\tau_2} \cdots < \infty \Bigg\}.$$

$$(u, v)_{H^{\boldsymbol{\tau}}(\mathsf{R}^\infty)} = \sum_{\alpha \in \mathrm{A}} u_\alpha \bar{v}_\alpha (1+\alpha_1)^{\tau_1} (1+\alpha_2)^{\tau_2} \cdots \qquad (u, v \in H^{\boldsymbol{\tau}}(\mathsf{R}^\infty)). \tag{4.59}$$

The space $H^{-\tau}(\mathbb{R}^\infty)$ consists of the vectors $\xi = \sum_{\alpha\in \mathrm{A}} \xi_\alpha h_\alpha$, where

$$\sum_{\alpha\in\mathrm{A}} |\xi_\alpha|^2(1+\alpha_1)^{-\tau_1}(1+\alpha_2)^{-\tau_2}\cdots < \infty; \quad (\xi, u)_{H^0(\mathbb{R}^\infty)} = \sum_{\alpha\in\mathrm{A}} \xi_\alpha \bar{u}_\alpha.$$

Comparing these constructions with the constructions in Example 2 of subsection 4, we see that it is possible to take the infinite tensor product of the nuclear spaces

$$\mathcal{G}(\mathbb{R}^1, 1) = \operatorname*{pr\,lim}_{l\in\{0,1,\ldots\}} G^l(\mathbb{R}^1, 1) = \operatorname*{pr\,lim}_{l\in\{0,1,\ldots\}} G^l(\mathbb{R}^1)$$

not only with a weighting but also without one. Here it is only necessary to properly renorm the generating Hilbert spaces $G^l(\mathbb{R}^1)$, i.e., to pass from these spaces to the spaces $H^l(\mathbb{R}^1)$. As a result we get $\mathcal{H}(\mathbb{R}^\infty)$. It is not hard to show that $\mathcal{H}(\mathbb{R}^\infty)$ *consists of cylindrical functions.* Indeed, by (4.56) and (4.59),

$$u(x) = \sum_{\alpha\in\mathrm{A}} u_\alpha h_\alpha(x), \quad \text{for } u \in \mathcal{H}(\mathbb{R}^\infty),$$

where $\sum_{\alpha\in\mathrm{A}} |u_\alpha|^2(1+\alpha_1)^{\tau_1}(1+\alpha_2)^{\tau_2}\cdots < \infty$ for any sequence $(\tau_n)_1^\infty \in \mathrm{T}$. Consider the decomposition of A into the sets A_n: $\mathrm{A} = \bigcup_{n=1}^\infty \mathrm{A}_n$. It is required to prove that $u_\alpha = 0$ for $\alpha \in \mathrm{A}_n$, beginning with some n. Assume not. Then there is an infinite set of indices $\alpha(n_k) \in \mathrm{A}_{n_k} (n_1 < n_2 < \cdots)$ such that $u_{\alpha(n_k)} \neq 0$. Let $\alpha(n_k) = ((\alpha(n_k))_1, \ldots, (\alpha(n_k))_{n_k}, 0, 0, \ldots)$. Since $\alpha(n_k) \in \mathrm{A}_{n_k}$, it follows that $(\alpha(n_k))_{n_k} > 0$. For each $k = 1, 2, \ldots$ we now choose a $\tau_{n_k} = 0, 1, \ldots$ large enough that

$$|u_{\alpha(n_k)}|^2(1 + (\alpha(n_k))_{n_k})^{\tau_{n_k}} \geq 1$$

and construct the sequence $\tau = (\tau_n)_1^\infty$ with the chosen τ_{n_k} at the n_kth place and zeros elsewhere. Then

$$\begin{aligned}\sum_{\alpha\in\mathrm{A}} &|u_\alpha|^2(1+\alpha_1)^{\tau_1}(1+\alpha_2)^{\tau_2}\cdots \\ &\geq \sum_{k=1}^\infty |u_{\alpha(n_k)}|^2(1+(\alpha(n_k))_1)^{\tau_1}(1+(\alpha(n_k))_2)^{\tau_2}\cdots \\ &\geq \sum_{k=1}^\infty |u_{\alpha(n_k)}|^2(1+(\alpha(n_k))_{n_k})^{\tau_{n_k}} = \infty,\end{aligned}$$

which is absurd. ■

Comparing this result with Example 2 in subsection 4, we conclude that $\mathrm{w.}\bigotimes_{n=1;e}^\infty \mathcal{G}(\mathbb{R}^1, 1)$ and $\mathcal{H}(\mathbb{R}^\infty)$ coincide as sets. It is easy to see that the convergent sequences in these two spaces also coincide (though their topologies are different).

Since $A_l(\mathbb{R}^1) \subset H^l(\mathbb{R}^1)$ $(l = 1, 2, \ldots)$, it follows that $\mathcal{A}(\mathbb{R}^\infty) \subset \mathcal{H}(\mathbb{R}^\infty)$. This and the fact just proved give us that $\mathcal{A}(\mathbb{R}^\infty)$ also consists of cylindrical functions. We have thereby proved the result formulated in the last paragraph of subsection 5.

We describe the nature of the analyticity of the functions in $\mathcal{A}(\mathbb{R}^\infty)$. If $u \in \mathcal{A}(\mathbb{R}^\infty)$, then u is cylindrical: $u(x) = u_c(x_1, \dots, x_p)$. It turns out that $u_c(x_1, \dots, x_p)$ coincides with the restriction to $\mathbb{R}^p$ of an entire function of p complex variables, with order at most 2 and of minimal type (i.e., an entire function $\mathbb{C}^p \ni (z_1, \dots, z_p) \mapsto u_c(z_1, \dots, z_p) \in \mathbb{C}^1$, admitting an estimate

$$|u_c(z_1, \dots, z_p)| \le c_{u_c,\varepsilon} \exp\left(\varepsilon \sum_{n=1}^{p} |z_n|^2\right) \qquad ((z_1, \dots, z_p) \in \mathbb{C}^p)$$

for any $\varepsilon > 0$). Conversely, each such restriction is in $\mathcal{A}(\mathbb{R}^\infty)$. This can be proved by studying the analyticity properties of a function that is expanded in Hermite polynomials with a certain order of decrease of the coefficients.

As in the case of the spaces $A_l(\mathbb{R}^1)$, T could have been taken to be not the collection of all sequences $\tau = (\tau_n)_1^\infty$, where $\tau_n = 0, 1, \dots$, but some subset T_1 of it. The relations (4.57) and (4.58) show that this subset must be sufficiently rich in order to obtain a nuclear space: for example, it is not possible to choose l' and l''' in such a way that the imbedding $O_{\tau_n''' \tau_n'}$ is quasinuclear, where $\tau' = (l', l', \dots)$ and $\tau''' = (l''', l''', \dots)$. (The spaces being multiplied must have smoothness that increases sufficiently rapidly.) It is not hard to formulate a condition on T_1 which ensures that the space is nuclear.

The subsequent results in subsection 5 turn out no longer to hold for the spaces $H^\tau(\mathbb{R}^\infty)$. This has to do with the fact that the δ-function (4.33) can be an element of some space $H^{-\tau}(\mathbb{R}^\infty)$ only for $x \in l_2$, since in the other cases the series $\sum_{\alpha \in \mathrm{A}} h_\alpha^2(x)(1+\alpha_1)^{-\tau_1}(1+\alpha_2)^{-\tau_2} \cdots$ cannot be made convergent by choosing τ_n to be rapidly enough increasing; this follows from the asymptotic formula for the Hermite polynomials $h_j(t)$ as $j \to \infty$.

CHAPTER 2

General Selfadjoint and Normal Operators

§0. Introduction

The main part of this chapter involves proving a theorem on expansion in generalized joint eigenvectors of an arbitrary family of commuting unbounded (in general) normal operators (the "spectral projection theorem"). This theorem singles out "projections" on the generalized eigensubspaces; the spectral integrals themselves are functional integrals. Applications of the theorem are given which illustrate the convenience of this form of spectral theorem for certain questions, as well as some facts about verifying that operators are selfadjoint.

We clarify the foregoing. Let A be a selfadjoint operator defined in a Hilbert space H_0. To it there corresponds a resolution of the identity (RI) E(1), i.e., a projection-valued measure on the σ-algebra $\mathcal{B}(\mathbf{R}^1)$ of Borel subsets of $\mathbf{R}^1$, for which we have the spectral integral representations

$$1 = \int_{\mathbf{R}^1} dE(\lambda), \qquad A = \int_{\mathbf{R}^1} \lambda\, dE(\lambda). \tag{0.1}$$

For an operator A with a discrete spectrum $(\lambda_j)_1^\infty$ we have that $E(\alpha) = \sum_{\lambda_j \in \alpha} P(\lambda_j)$ $(\alpha \in \mathcal{B}(\mathbf{R}^1))$, where $P(\lambda_j)$ is the projection onto the eigensubspace corresponding to λ_j, and the relations (0.1) pass into the equalities $1 = \sum_1^\infty P(\lambda_j)$ and $A = \sum_1^\infty \lambda_j P(\lambda_j)$. In the case of a continuous spectrum such equalities can no longer be written, but it is possible (Berezanskiĭ [2], and Chapter 5 in [4]) to generalize them for separable H_0 as follows on the basis of work of Gel′fand and Kostyuchenko [1], Berezanskiĭ [1], and Kats [1] on generalized eigenfunction expansions.

Let

$$H_- \supseteq H_0 \supseteq H_+ \supseteq D \tag{0.2}$$

be a rigging of H_0 by Hilbert spaces H_+ and H_- of "test" and "generalized" vectors with a quasinuclear (i.e., Hilbert-Schmidt) imbedding $H_+ \subset H_0$, and

(1)The abbreviation RI will be used in §§1–5 of this chapter, where this term is encountered particularly often.

let D be a linear topological space topologically (i.e., densely and continuously) imbedded in H_+. Assume that the operator A is connected in the standard way with the chain (0.2): the domain $\mathfrak{D}(A)$ contains D, and the restriction $A \restriction D$ acts continuously from D to H_+. Then the RI E, understood as an operator-valued measure whose values are operators from H_+ to H_-, can be differentiated with respect to a certain scalar measure $\mathcal{B}(\mathsf{R}^1) \ni \alpha \mapsto \rho(\alpha) \in [0, \infty)$ (the so-called spectral measure): $(dE/d\rho)(\lambda) = P(\lambda)\colon H_+ \to H_-$ exists, and the equalities (0.1) are rewritten in the form

$$1 = \int_{\mathsf{R}^1} P(\lambda)\, d\rho(\lambda), \qquad A = \int_{\mathsf{R}^1} \lambda P(\lambda)\, d\rho(\lambda). \tag{0.3}$$

It can be shown that for ρ-almost all $\lambda \in \mathsf{R}^1$ the range $\mathfrak{R}(P(\lambda))$ consists of generalized eigenvectors $\varphi \in H_-$ corresponding to λ (i.e., $A\varphi = \lambda\varphi$ in the natural generalized sense). Thus, $P(\lambda)$ "projects" onto the corresponding generalized eigensubspace, and the equalities (0.3) generalize the equalities given above in the case of a discrete spectrum and constitute the spectral projection theorem for a single operator.

It is often convenient to consider a somewhat more special situation in which the "quasinuclear" chain (0.2) is replaced by a "nuclear" chain

$$\Phi' \supseteq H_0 \supseteq \Phi, \tag{0.4}$$

where Φ is a nuclear space topologically imbedded in H_0, and Φ' is the dual space of continuous antilinear functionals on Φ. A picture analogous to (0.2) holds for a selfadjoint operator A defined in H_0 and satisfying the more stringent restriction that $\Phi \subseteq \mathfrak{D}(A)$ and $A \restriction \Phi$ acts continuously in Φ. More precisely, the required chain (0.2) can be chosen for the chain (0.4), and the results (0.3) described above are preserved for A.

This is the state of affairs in the case of a single selfadjoint operator. Now consider a family $A = (A_x)_{x \in X}$ of selfadjoint operators A_x acting in H_0, commuting in the sense that their RI's E_x commute, and connected in the standard way with the chain (0.2). If $X = \{1, \dots, p\}$ is finite, then we can construct a joint RI for the family A: a projection-valued measure defined on the σ-algebra $\mathcal{B}(\mathsf{R}^p)$ in the space R^p of points $\lambda = (\lambda_1, \dots, \lambda_p)$. The formulas (0.1) and (0.3) are preserved, except that A must be replaced by A_x and λ by λ_x $(x \in X)$ (Berezanskiĭ [4]).

In this chapter we give analogues of the representations (0.1) and (0.3) in the case when the index set X has arbitrary cardinality (for example, X can coincide with the space $\mathcal{D}(\mathsf{R}^d)$ of test functions—a typical quantum field situation). Difficulties now arise that are especially noticeable when we are dealing with a family of unbounded operators that is more than countable. For example, the joint RI of the family A must naturally be a projection-valued measure on the space R^X of all possible functions $X \ni x \mapsto \lambda(x) \in \mathsf{R}^1$. Such an RI E can be constructed with the help of the Kolmogorov theorem on extending a finite-dimensional distribution to a measure; however, it is defined on a σ-algebra $\mathfrak{N}$ fairly meager in comparison with the σ-algebra $\mathcal{B}(\mathbf{R}^X)$ of Borel sets ($\mathbf{R}^X$ is endowed with the Tychonoff topology). For instance, if X is more than countable, then the singleton sets are not in $\mathfrak{N}$.

This creates a number of difficulties, in particular, when working with the concept of the support $\operatorname{supp} E$ of such a measure. The concept of a support is essential in the questions under discussion. For example, in the integrals (0.1) and (0.3) the domain of integration $\mathbf{R}^1$ can be replaced by the spectrum of the operator A, i.e., by $\operatorname{supp} E$. A similar approach to the spectrum would be natural and useful also for general X (see also below), but if X is more than countable and the operators A_x are unbounded, then it is possible that $\operatorname{supp} E = \varnothing$ (even though $E \neq 0$), and this transition cannot be made.

Further, the role of a generalized eigenvector must now be played by a generalized joint eigenvector, i.e, a vector $\varphi \in H_-$ such that $A_x\varphi = \lambda(x)\varphi$ in the generalized sense for each $x \in X$; the function $\lambda(\cdot) \in \mathbf{R}^X$ corresponding to φ is the "eigenvalue". In the case of a single operator $\mathfrak{R}(P(\lambda))$ consists of generalized eigenvectors for ρ-almost all λ. Analogously, for a particular x there now exists a set $\beta_x \subseteq \mathbf{R}^X$ of full ρ- (or E-) measure such that if $\lambda(\cdot) \in \beta_x$, then $\mathfrak{R}(P(\lambda(\cdot)))$ consists of vectors φ for which $A_x\varphi = \lambda(x)\varphi$. To obtain a similar equality for all $x \in X$ we would like to take the set $\pi = \bigcap_{x\in X} \beta_x$ and consider $P(\lambda(\cdot))$ for $\lambda(\cdot) \in \pi$. If X is not more than countable, then π has full measure, and this procedure gives what is required. However, the "continuum difficulty" arises when X is more than countable: π can fail to be a set of full measure. Overcoming this difficulty involves constructing π in the correct way.

We have already mentioned that it is not completely reasonable to take the support $\operatorname{supp} E$ of the RI as the spectrum of a family A for general families: it is generally impossible for the spectral integrals to be over $\operatorname{supp} E$. The question arises as to what to take as the spectrum of a family from this point of view; more precisely, what is a minimal (in a specific sense) set $g(A) \subseteq \mathbf{R}^X$ such that, for example, the modification

$$A_x = \int_{g(A)} \lambda(x)\, dE(\lambda(\cdot)) \qquad (x \in X) \tag{0.5}$$

of the second formula in (0.1) is valid (the function $\mathbf{R}^X \supseteq g(A) \ni \lambda(\cdot) \mapsto \lambda(x) \in \mathbf{R}^1$ is being integrated)? It turns out that $g(A)$ can be understood as the generalized spectrum of A—the collection of all eigenvalues corresponding to generalized joint eigenvectors (for a fixed chain (0.2) or (0.4)). We shall explain what consequences can be derived, in particular, from spectral representations like (0.5). (Note first that everything said is valid also for normal commuting operators, with $\mathbf{R}^X$ replaced by $\mathbf{C}^X$.)

For example, suppose that the operators A_x are bounded and are connected in a definite way for different $x \in X$. To illustrate, let X be an abelian group and $A_{x+y} = A_xA_y$ $(x, y \in X)$, or let X be a linear space such that

$A_{\alpha x+\beta y} = \alpha A_x + \beta A_y$ $(x, y \in X; \alpha, \beta \in \mathbb{C}^1)$; i.e., the function $X \ni x \mapsto A_x$ realizes a representation of this or that algebraic object—a group, a linear space, an algebra, etc. Moreover, the relations can be of a less traditional type, for instance, $X = \mathbb{R}^1$ and $A_x A_y = \frac{1}{2}(A_{x+y} + A_{x-y})$ $(x, y \in \mathbb{R}^1)$, A_x can satisfy some differential equation with respect to x, and so on. We shall explain how the representation (0.5) is now transformed.

To do this we note that application of the expressed relations to the joint generalized eigenvector φ and consideration of the equality $A_x\varphi = \lambda(x)\varphi$ $(x \in X)$ gives us equations of the form $\lambda(x+y) = \lambda(x)\lambda(y)$, $\lambda(\alpha x + \beta y) = \alpha\lambda(x) + \beta\lambda(y)$, $\lambda(x)\lambda(y) = \frac{1}{2}(\lambda(x+y) + \lambda(x-y))$, and so on, for the corresponding eigenvalues $\lambda(\cdot)$. Thus, the $g(A)$ in (0.5) can be replaced by the solution space of the corresponding scalar equation (a functional equation, differential equation, etc.), and the "multiplicity" of the integral (0.5) is essentially diminished.

These scalar equations can be solved explicitly in a number of cases. For example, $\lambda(x) = e^{i\lambda x}$ $(x \in X = \mathbb{R}^1)$, for Stone's classical theorem, where $\lambda \in \mathbb{R}^1$, and we get the mapping $g(A) \ni \lambda(\cdot) = e^{i\lambda x} \mapsto \lambda \in \mathbb{R}^1$. Transporting the measure $E(\alpha)$ $(\alpha \subseteq g(A))$ to an RI F according to this mapping, we rewrite (0.5) in the classical form $A_x = \int_{\mathbb{R}^1} e^{i\lambda x}\, dF(\lambda)$ $(x \in \mathbb{R}^1)$. It is clear from the foregoing that the method developed makes it possible to obtain broad generalizations of theorems of the Stone type and of the Sz.-Nagy–Hille type (on groups of selfadjoint operators), as well as results of other authors on spectral representation of families of operators. §4 contains these generalizations in part; a more detailed presentation of them, connections with spectral representations of the corresponding function classes (positive-definite functions, the moment problem, etc.), and applications to noncommuting operators are given in the coming second part of the book.

We note also the following point. Our method of obtaining spectral representations for families of operators A_x assumes that these operators are connected in the standard way with the riggings (0.2) or (0.4). It turns out that the converse is also valid: the existence of sufficiently good representations automatically implies also the existence of the riggings (§5.5).

In §1 we construct and investigate a joint RI E from RI's of the operators A_x. Here the constructions call to mind and exploit constructions in measure theory, but they conceal some unexpected features: for example, two abstract commuting RI's cannot always be multiplied. In §2 we present a procedure for differentiating a joint RI, understood as a measure whose values are operators from H_+ to H_-. The projection properties of the resulting derivative $P(\lambda(\cdot))\colon H_+ \to H_-$ are then studied, and the main result of the chapter is proved: the spectral projection theorem.

There is another approach to the theory of expansions in generalized joint eigenvectors for a family $A = (A_x)_{x\in X}$, based on a consideration of the C^*-algebra $\mathcal{A}$ associated with A (Maurin [1]–[4], the "nuclear spectral theorem").

It is effective in the case of bounded operators A_x, when $\mathcal{A}$ is constructed as the closure of the algebraic span of A in the operator norm (although the results are somewhat less sharp even in this case when X is more than countable: with this approach the function $\lambda(\cdot) \mapsto P(\lambda(\cdot))$ is measurable with respect to a more extensive σ-algebra than in reality). It is now possible to identify $\operatorname{supp} E$ with the maximal ideal space of the algebra $\mathcal{A}$. The corresponding facts are presented in §5.6. This approach becomes inconvenient when there is a massive set of unbounded operators among the A_x. (We note that if A_x is connected in the standard way with the chain (0.2) or (0.4), then as a rule a function $F(A_x)$ of it no longer has such a connection.) In his papers [1] and [2] Richter outlines another way, based on Choquet's theorem, of obtaining results about expansions for a family A. In both of these approaches the spectral integrals for general families A are written over a certain space which is mapped into the spectrum, and it is not possible to rewrite them as integrals of the type (0.5) over the generalized spectrum itself in the case of unbounded operators. This makes it difficult to apply such representations, for example, to theorems of the Stone type or the Sz.-Nagy–Hille type.

§5 also contains a theory of expansions for Carleman operators, along with certain other closely related facts.

The remaining two sections of this chapter contain additional information about the theory developed here. Thus, simple and natural connections with the theory of random processes are studied in §3. In §6 we present one sufficiently effective way of proving selfadjointness of operators, based on the fact that if the corresponding evolution equation has unique strong solutions for the Cauchy problem, then, roughly speaking, the operator is selfadjoint.

§1. A joint resolution of the identity

A joint RI can be constructed in a way similar to that for infinite products of probability measures, as an operator-valued measure equal to an infinite product of RI's of commuting normal operators. We need the product of commuting abstractly defined RI's.

1.1. ***A general resolution of the identity.*** Let R be an abstract space, $\mathfrak{R}$ a σ-algebra of subsets of it, and $\mathfrak{R} \ni \alpha \mapsto E(\alpha)$ an operator-valued function on $\mathfrak{R}$ whose values are (orthogonal) projections on a fixed Hilbert space H. Such a function E is called an RI on R if the following two conditions hold:

a) $E(\varnothing) = 0$ and $E(R) = 1$, where 1 is the identity operator on H;

b) the absolute additivity property is valid, i.e., $E(\bigcup_1^\infty \alpha_j) = \sum_1^\infty E(\alpha_j)$ for any sequence $(\alpha_j)_1^\infty$ with $\alpha_j \in \mathfrak{R}$ and $\alpha_j \cap \alpha_k = \varnothing$ for $j \neq k$, where the series converges in the sense of strong operator convergence.

REMARK 1. The orthogonality relation always holds for an RI: $E(\alpha \cap \beta) = E(\alpha)E(\beta)$ $(\alpha, \beta \in \mathfrak{R})$ (and thus the operators $E(\alpha)$ $(\alpha \in \mathfrak{R})$ commute). This is a simple consequence of additivity and the fact that a sum of two projections is a projection only if they are orthogonal.

REMARK 2. Weak convergence of the series in b) can be required instead of strong convergence: it will automatically converge strongly, too, because the terms are orthogonal.

REMARK 3. An RI can also be defined on some algebra $\mathfrak{R}$ (and not on a σ-algebra). The definition stays as before, except that in b) the requirement that $\bigcup_1^\infty \alpha_j \in \mathfrak{R}$ must be added. Remarks 1 and 2 are preserved for an RI on an algebra.

In the case of an RI corresponding to some selfadjoint (normal) operator acting in H we have $R = \mathbb{R}^1$ and $\mathfrak{R} = \mathcal{B}(\mathbb{R}^1)$ ($R = \mathbb{C}^1$ and $\mathfrak{R} = \mathcal{B}(\mathbb{C}^1)$), where $\mathcal{B}(C)$ is the σ-algebra of Borel subsets of a topological space C.

THEOREM 1.1. *If E is an RI on an algebra $\mathfrak{R}$, then there exists an extension of it to an RI E_σ on the σ-algebra $\mathfrak{R}_\sigma$ it generates (i.e., the restriction $E_\sigma \restriction \mathfrak{R}$ is E), and E_σ is uniquely determined by E.*

PROOF. Fix an $f \in H$ and consider the finite measure $\mathfrak{R} \ni \alpha \mapsto \rho_{f,f}(\alpha) = (E(\alpha)f, f)_H \in [0, \infty)$. According to the usual extension theory, there exists a measure $\mathfrak{R}_\sigma \ni \alpha \mapsto \hat{\rho}_{f,f}(\alpha) \in [0, \infty)$ such that $\hat{\rho}_{f,f} \restriction \mathfrak{R} = \rho_{f,f}$. The complex-valued measure (charge) $\mathfrak{R} \ni \alpha \mapsto \rho_{f,g}(\alpha) = (E(\alpha)f, g)_H \in \mathbb{C}^1$ can be expressed as a linear combination of four measures of the form $\rho_{h,h}$ ($h \in H$), by the polarization formula. Extending each of these measures from $\mathfrak{R}$ to $\mathfrak{R}_\sigma$ and taking the corresponding linear combination, we get that the charge $\rho_{f,g}$ also extends to a charge $\mathfrak{R}_\sigma \ni \alpha \mapsto \hat{\rho}_{f,g}(\alpha) \in \mathbb{C}^1$.

For a fixed $\alpha \in \mathfrak{R}_\sigma$ the mapping

$$H \times H \ni (f, g) \mapsto \hat{\rho}_{f,g}(\alpha) \in \mathbb{C}^1 \tag{1.1}$$

is bilinear. Indeed, it is well known that $\mathfrak{R}_\sigma$ coincides with the monotone class generated by $\mathfrak{R}$ and $\hat{\rho}_{f,f}$ (hence also $\hat{\rho}_{f,g}$) is constructed by successive monotone extensions, starting with the sets $\alpha \in \mathfrak{R}$. Bilinearity is preserved for monotone extensions, so the required property follows from the bilinearity of (1.1) when $\alpha \in \mathfrak{R}$.

The mapping (1.1) is continuous: $\hat{\rho}_{f,f}(\alpha) \geq 0$ for $\alpha \in \mathfrak{R}_\sigma$. Therefore,

$$|\hat{\rho}_{f,g}(\alpha)|^2 \leq \hat{\rho}_{f,f}(\alpha)\hat{\rho}_{g,g}(\alpha) \leq \hat{\rho}_{f,f}(R)\hat{\rho}_{g,g}(R) = \|f\|_H^2 \|g\|_H^2 \qquad (f, g \in H)$$

on the strength of the Cauchy-Schwarz-Bunyakovskiĭ inequality. Denote by $E_\sigma(\alpha)$ the bounded operator on H corresponding to the continuous bilinear form (1.1); thus,

$$\hat{\rho}_{f,g}(\alpha) = (E_\sigma(\alpha)f, g)_H \qquad (f, g \in H;\ \alpha \in \mathfrak{R}_\sigma). \tag{1.2}$$

For $\alpha \in \mathfrak{R}$, $E_\sigma(\alpha) = E(\alpha)$ is a projection. The successive monotone extensions $\hat{\rho}_{f,g}$ mean extensions of $E_\sigma(\alpha)$ from $\alpha \in \mathfrak{R}$ to $\alpha \in \mathfrak{R}_\sigma$ by means of weak limits of monotone sequences of projection; therefore, $E_\sigma(\alpha)$ is also a projection for any $\alpha \in \mathfrak{R}_\sigma$. Of course, $E_\sigma(\varnothing) = 0$ and $E_\sigma(R) = 1$. The set function $\mathfrak{R} \ni \alpha \mapsto E_\sigma(\alpha)$ is absolutely additive in the sense of weak convergence of the corresponding series in b): by (1.2), this is a rephrasing of the absolute additivity of the charge $\hat{\rho}_{f,g}$ $(f, g \in H)$. Accordingly, E_σ is a resolution of the identity defined on $\mathfrak{R}_\sigma$. It coincides with E on $\mathfrak{R}$ by construction. The last assertion of the theorem follows from the uniqueness of an extension of a charge from $\mathfrak{R}$ to $\mathfrak{R}_\sigma$. ■

1.2. ***Products of finitely many resolutions of the identity.*** Let R_j be a space, $\mathfrak{R}_j$ a σ-algebra of subsets of it, and $\mathfrak{R}_j \ni \alpha_j \mapsto E_j(\alpha_j)$ an RI $(j = 1, \dots, p)$. Assume that the RI's $E_1, \dots, E_p$ commute, i.e.,

$$E_j(\alpha_j)E_k(\alpha_k) = E_k(\alpha_k)E_j(\alpha_j) \qquad (\alpha_j \in \mathfrak{R}_j,\ \alpha_k \in \mathfrak{R}_k;\ j, k = 1, \dots, p).$$

The question is whether it is possible (in a way similar to that for ordinary measures) to define an RI E on the σ-algebra $\mathfrak{R}$ generated by all the rectangles $\alpha_1 \times \dots \times \alpha_p$ in the direct product $R = R_1 \times \dots \times R_p$ in such a way that

$$E(\alpha_1 \times \dots \times \alpha_p) = E_1(\alpha_1) \cdots E_p(\alpha_p) \qquad (\alpha_1 \in \mathfrak{R}_1, \dots, \alpha_p \in \mathfrak{R}_p).$$

Such an RI E is called the product of the RI's $E_1, \dots, E_p$, or the joint RI generated by these RI's; we use the notation $E = E_1 \times \dots \times E_p = \times_1^p E_j$. As mentioned in the introduction, it does not always exist (see Birman, Vershik, and Solomyak [1]); however, if the RI's to be multiplied are defined on "nice" σ-algebras, then E exists. We present a corresponding result. Of course, it suffices to consider the case $p = 2$.

Let R be a complete separable metric space, and $\mathfrak{R} = \mathcal{B}(R)$ the σ-algebra of Borel subsets of it. An RI defined on $\mathcal{B}(R)$ is called a Borel RI. The joint RI is known to exist if the RI's to be multiplied are Borel RI's. Before proving this, we mention the following important fact: every finite scalar measure $\mathcal{B}(R) \ni \alpha \mapsto \mu(\alpha) \in [0, \infty)$ is automatically regular, i.e.,

$$\mu(\alpha) = \inf_{o \supseteq \alpha} \mu(o) \qquad (\alpha \in \mathcal{B}(R),\ o \text{ open}); \tag{1.3}$$

by passing to complements it is easy to see that (1.3) is equivalent to the relation

$$\mu(\alpha) = \sup_{\varphi \subseteq \alpha} \mu(\varphi) \qquad (\alpha \in \mathcal{B}(R),\ \varphi \text{ closed}); \tag{1.4}$$

moreover, the supremum in (1.4) can be only over the compact subsets $\varphi \subseteq \alpha$ (Ulam's theorem; see Billingsley [1]).

THEOREM 1.2. *Suppose that E_1 and E_2 are two commuting Borel RI's in the respective spaces R_1 and R_2. Then there exists an RI $E = E_1 \times E_2$ on $\mathfrak{R} = \mathcal{B}(R_1 \times R_2)$ uniquely determined by E_1 and E_2.*

PROOF. Denote by $\mathfrak{R}'$ the algebra generated by the collection of all rectangles $\alpha_1 \times \alpha_2$ $(\alpha_1 \in \mathcal{B}(R_1),\ \alpha_2 \in \mathcal{B}(R_2))$. The function

$$\mathcal{B}(R_1) \times \mathcal{B}(R_2) \ni (\alpha_1, \alpha_2) \mapsto E(\alpha_1 \times \alpha_2) = E_1(\alpha_1)E_2(\alpha_2)$$

extends uniquely by additivity to a finitely additive operator-valued function $\mathfrak{R}' \ni \alpha \mapsto E(\alpha)$; this can be proved by the usual argument as in the case of a scalar measure. Each value $E(\alpha)$ ($\alpha \in \mathfrak{R}'$) is a projection, since α is representable as the union of finitely many disjoint rectangles $\alpha_1 \times \alpha_2$, and thus $E(\alpha)$ is equal to a finite sum of projections $E(\alpha_1 \times \alpha_2) = E_1(\alpha_1)E_2(\alpha_2)$ which are mutually orthogonal due to the orthogonality relations for E_1 and E_2. Hence, $\mathfrak{R}' \ni \alpha \mapsto E(\alpha)$ is a projection-valued finitely additive set function on the algebra $\mathfrak{R}'$; also, $E(\varnothing) = 0$ and $E(R_1 \times R_2) = 1$. Let us prove that it is absolutely additive.

Note that for each rectangle $\alpha_1 \times \alpha_2$ ($\alpha_1 \in \mathcal{B}(R_1)$, $\alpha_2 \in \mathcal{B}(R_2)$), $f \in H$, and $\varepsilon > 0$ there are rectangles of the form $o_1 \times o_2$ and $\varphi_1 \times \varphi_2$($o_1 \supseteq \alpha_1$ and $o_2 \supseteq \alpha_2$ open, and $\varphi_1 \subseteq \alpha_1$ and $\varphi_2 \subseteq \alpha_2$ compact), such that

$$\begin{aligned}(E(o_1 \times o_2)f, f)_H - (E(\alpha_1 \times \alpha_2)f, f)_H < \varepsilon,\\ (E(\alpha_1 \times \alpha_2)f, f)_H - (E(\varphi_1 \times \varphi_2)f, f)_H < \varepsilon.\end{aligned} \tag{1.5}$$

Let us establish, for example, the first relation. Using what was mentioned about the regularity of scalar measures on $\mathcal{B}(R_1)$ and $\mathcal{B}(R_2)$, we choose for a $\delta > 0$ two open sets $o_1 \supseteq \alpha_1$ and $o_2 \supseteq \alpha_2$ such that $(E_1(o_1 \backslash \alpha_1)f, f)_H$ and $(E_2(o_2 \backslash \alpha_2)f, f)_H < \delta$. Since

$$(o_1 \times o_2) \backslash (\alpha_1 \times \alpha_2) = ((o_1 \backslash \alpha_1) \times o_2) \cup (\alpha_1 \times (o_2 \backslash \alpha_2)),$$

it follows that

$$\begin{aligned}&(E(o_1 \times o_2)f, f)_H - (E(\alpha_1 \times \alpha_2)f, f)_H\\ &\quad = (E((o_1 \backslash \alpha_1) \times o_2)f, f)_H + (E(\alpha_1 \times (o_2 \backslash \alpha_2))f, f)_H\\ &\quad = (E_1(o_1 \backslash \alpha_1)E_2(o_2)f, f)_H + (E_1(\alpha_1)E_2(o_2 \backslash \alpha_2)f, f)_H\\ &\quad \le \|f\|_H(\|E_1(o_1 \backslash \alpha_1)f\|_H + \|E_2(o_2 \backslash \alpha_2)f\|_H)\\ &\quad = \|f\|_H((E_1(o_1 \backslash \alpha_1)f, f)_H^{1/2} + (E_2(o_2 \backslash \alpha_2)f, f)_H^{1/2})\\ &\quad < 2\|f\|_H \delta^{1/2}.\end{aligned}$$

The required fact follows by taking $\delta > 0$ sufficiently small.

Suppose now that $\alpha_1, \alpha_2, \ldots \in \mathfrak{R}'$ are disjoint and such that $\alpha = \bigcup_1^\infty \alpha_n \in \mathfrak{R}'$. To prove that E is absolutely additive it suffices to see that for any $f \in H$

$$(E(\alpha)f, f)_H \le \sum_{n=1}^{\infty} (E(\alpha_n)f, f)_H. \tag{1.6}$$

(The opposite inequality follows from finite additivity and the monotonicity of the function $\mathfrak{R}' \ni \beta \mapsto (E(\beta)f, f)_H \in [0, \infty)$; therefore, (1.6) means that these expressions are equal.)

Since α is equal to the union of finitely many disjoint rectangles, application to each of them of the second inequality in (1.5) gives us for a specified $\varepsilon > 0$ some compact set $\mathfrak{R}' \ni \varphi \subseteq \alpha$ such that $(E(\alpha)f, f)_H - (E(\varphi)f, f)_H < \varepsilon$. Similarly, with the help of the first inequality in (1.5), it is possible to find for each $n = 1, 2, \dots$ an open set $\mathfrak{R}' \ni o_n \supseteq \alpha$ such that $(E(o_n)f, f)_H - (E(\alpha_n)f, f)_H < 2^{-n}\varepsilon$. The family $(o_n)_1^\infty$ covers φ. Since φ is compact, there exists a $p = 1, 2, \dots$ such that $\bigcup_1^p o_n \supseteq \varphi$. Thus from the monotonicity and finite subadditivity of a scalar measure, we get by the last two inequalities that

$$(E(\alpha)f, f)_H < (E(\varphi)f, f)_H + \varepsilon \leq \left(E\left(\bigcup_{n=1}^{p} o_n\right) f, f\right)_H + \varepsilon$$
$$\leq \sum_{n=1}^{p} (E(o_n)f, f)_H + \varepsilon < \sum_{n=1}^{\infty} (E(\alpha_n)f, f)_H + 2\varepsilon.$$

Passing here to the limit as $\varepsilon \to 0$, we arrive at (1.6). It is proved that E is absolutely additive.

Thus, an RI E has been constructed on the algebra $\mathfrak{R}'$. We extend it by Theorem 1.1 to an RI E_σ on $(\mathfrak{R}')_\sigma = \mathfrak{R}$; this RI is obviously the required one. The equality $\mathfrak{R} = \mathcal{B}(R_1 \times R_2)$ follows from the definition of the topology in $R_1 \times R_2$. ■

1.3. ***Construction of a joint resolution of the identity in the general case.*** Let X be a set of indices x of arbitrary cardinality, and let $(R_x)_{x \in X}$ be a family of complete separable metric spaces R_x. Assume that a Borel RI E_x: $\mathcal{B}(R_x) \ni \alpha \mapsto E_x(\alpha)$ whose values are projections $E_x(\alpha)$ on a fixed Hilbert space H is given on each R_x. The family $(E_x)_{x \in X}$ is assumed to be a commuting family: $E_x(\alpha)E_y(\beta) = E_y(\beta)E_x(\alpha)$ $(\alpha \in \mathcal{B}(R_x), \beta \in \mathcal{B}(R_y); x, y \in X)$. We shall construct a corresponding joint RI $E = E_X = \times_{x \in X} E_x$; E can be called the product of the RI's E_x.

The joint RI E is an RI in the space $R_X = \times_{x \in X} R_x$ consisting of all the mappings $\lambda(\cdot)$ of the form $X \ni x \mapsto \lambda(x) \in R_x$. (A topology is not introduced in R_X at this point.) We construct a σ-algebra of subsets of R_X on which E will be defined. Recall that a set $\mathfrak{C} \subset R_X$ is called a cylindrical set if it is determined by the relation

$$\mathfrak{C} = \mathfrak{C}(x_1, \dots, x_p; \delta) = \{\lambda(\cdot) \in R_X | (\lambda(x_1), \dots, \lambda(x_p)) \in \delta\} \qquad (p = 1, 2, \dots), \quad (1.7)$$

where $x_1, \dots, x_p \in X$ are distinct points (its coordinates) and $\delta \in \mathcal{B}(R_{x_1, \dots, x_p})$ $(R_{x_1, \dots, x_p} = R_{x_1} \times \cdots \times R_{x_p})$ is the base.

In (1.7) the given cylindrical set $\mathfrak{C}$ does not uniquely determine the sequence of points $x_1, \dots, x_p$ nor the base δ. Thus, we can change the order of these points (for example, $\mathfrak{C}(x_1, x_2; \delta_1 \times \delta_2) = \mathfrak{C}(x_2, x_1; \delta_2 \times \delta_1)$ $(\delta_1 \in \mathcal{B}(R_{x_1}), \delta_2 \in \mathcal{B}(R_{x_2})))$ and insert a superfluous point $y \in X$, multiplying the base by R_y:

$$\mathfrak{C}(x_1, \dots, x_p; \delta) = \mathfrak{C}(x_1, \dots, x_p, y; \delta \times R_y)$$

(or eliminate y and the corresponding factor R_y).

The collection of all cylindrical sets form an algebra $C(R_X)$. Indeed, to prove the inclusions $\mathfrak{C}' \cup \mathfrak{C}'', \mathfrak{C}' \backslash \mathfrak{C}'' \in C(R_X)$ for $\mathfrak{C}', \mathfrak{C}'' \in C(R_X)$ it suffices to pass in (1.7) to common coordinates $x_1, \ldots, x_p$ for $\mathfrak{C}'$ and $\mathfrak{C}''$ by adding supplementary points by the above procedure, if necessary. The assertion then follows from the formula

$$\mathfrak{C}' \cup \mathfrak{C}'' = \mathfrak{C}(x_1, \ldots, x_p; \delta') \cup \mathfrak{C}(x_1, \ldots, x_p; \delta'') = \mathfrak{C}(x_1, \ldots, x_p; \delta' \cup \delta'')$$

and the analogous formula for the difference. ■

Denote by $\mathcal{C}_\sigma(R_X)$ the σ-algebra generated by the algebra $\mathcal{C}(R_X)$.

THEOREM 1.3. *Given a family $(E_x)_{x \in X}$ of commuting Borel RI's E_x, it is possible to construct a unique RI $\mathcal{C}_\sigma(R_X) \ni \alpha \mapsto E(\alpha)$ such that*

$$E(\mathfrak{C}(x_1, \ldots, x_p; \delta_1 \times \cdots \times \delta_p)) = E_{x_1}(\delta_1) \cdots E_{x_p}(\delta_p)$$
$$(x_1, \ldots, x_p \in X; \delta_1 \in \mathcal{B}(R_{x_1}), \ldots, \delta_p \in \mathcal{B}(R_{x_p}); p = 1, 2, \ldots); \quad (1.8)$$

E is called the joint RI constructed from $(E_x)_{x \in X}$.

PROOF. We construct according to subsection 2 a joint RI $E_{x_1, \ldots, x_p} = E_{x_1} \times \cdots \times E_{x_p}$ defined on $\mathcal{B}(R_{x_1, \ldots, x_p})$, and let

$$E(\mathfrak{C}(x_1, \ldots, x_p; \delta)) = E_{x_1, \ldots, x_p}(\delta) \qquad (\delta \in \mathcal{B}(R_{x_1, \ldots, x_p}); p = 1, 2, \ldots). \quad (1.9)$$

This definition is unambiguous: an operator $E(\mathfrak{C})$ is associated with each cylindrical set $\mathfrak{C}$. Indeed, each change in the expression for $\mathfrak{C}$ involves a permutation of coordinates, or direct multiplication of δ by an R_y, or removal of such a factor from δ. Since $E_y(R_y) = 1$, the construction of $E_{x_1, \ldots, x_p}(\delta)$ implies that $E(\mathfrak{C})$ is independent of the expression for $\mathfrak{C}$. The set function $\mathcal{C}(R_X) \ni \mathfrak{C} \mapsto E(\mathfrak{C})$ is such that $E(\varnothing) = 0$ and $E(R_X) = 1$, and it is finitely additive. (The latter is obtained from the definition (1.9) by expressing the finite collection of sets in $\mathcal{C}(R_X)$ under consideration by a common collection of coordinates.) Moreover, it satisfies also the absolute additivity condition: if $f \in H$ is fixed, then the nonnegative set function $\mathcal{C}(R_X) \ni \mathfrak{C} \mapsto (E(\mathfrak{C})f, f)_H$ is a consistent system of finite-dimensional distributions, which extends by Kolmogorov's theorem (see Gikhman and Skorokhod [1]) to a measure defined on $\mathcal{C}_\sigma(R_X)$; but then it is in any case absolutely additive on $\mathcal{C}(R_X)$, which is equivalent to b). Thus, $\mathcal{C}(R_X) \ni \mathfrak{C} \mapsto E(\mathfrak{C})$ is an RI defined on the algebra $\mathcal{C}(R_X)$. According to Theorem 1.1, E extends to an RI E_σ on $\mathcal{C}_\sigma(R_X)$, and this is obviously the required RI. ■

REMARK 1. Every RI E defined on $\mathcal{C}_\sigma(R_X)$ can be constructed with the help of Theorem 1.3 from some family $(E_x)_{x \in X}$ of commuting Borel RI's. Indeed, for each $x \in X$ it suffices to set $\mathcal{B}(R_x) \ni \alpha \mapsto E_x(\alpha) = E(\mathfrak{C}(x; \alpha))$ and multiply these RI's.

If $R_x = C$ $(x \in X)$, then we use the notation $R_X = C^X$. Thus, the RI generated by RI's of selfadjoint (normal) operators is defined on $\mathbb{R}^X$ $(\mathbb{C}^X)$. In the case when X is countable we also write $C^X = C^\infty$.

1.4. ***Topologization.*** It is convenient for us to topologize R_X with the Tychonoff topology. For a base of neighborhoods in R_X we take the collection

Σ of all cylindrical sets of the form $\mathfrak{C}_{\mathrm{b}} = \mathfrak{C}(x_1, \ldots, x_p; u_1 \times \cdots \times u_p)$, where u_n is an arbitrary open subset of the space R_{x_n} $(x_1, \ldots, x_p \in X;\ p = 1, 2, \ldots)$. It is well known that this topologization turns R_X into a regular topological space having a countable base if and only if X is at most countable. It is compact if and only if R_x is compact for each $x \in X$.

We fix distinct points $x_1, \ldots, x_p \in X$ $(p = 1, 2, \ldots)$ and consider the "coordinate" mapping $\pi_{x_1,\ldots,x_p}$ given by

$$R_X \ni \lambda(\cdot) \mapsto \pi_{x_1,\ldots,x_p}(\lambda(\cdot)) = (\lambda(x_1), \ldots, \lambda(x_p)) \in R_{x_1,\ldots,x_p}. \tag{1.10}$$

The continuity of $\pi_{x_1,\ldots,x_p}$ follows immediately from the way in which the Tychonoff topology is introduced. It is also open, i.e., it carries open subsets of R_X into open subsets $R_{x_1,\ldots,x_p}$. This follows from the fact that under the mapping $\pi_{x_1,\ldots,x_p}$ each basic neighborhood $\mathfrak{C}_{\mathrm{b}} = \mathfrak{C}(y_1, \ldots, y_q;\ u_1 \times \cdots \times u_q)$ is carried into some neighborhood in $R_{x_1,\ldots,x_p}$ (write $\mathfrak{C}_{\mathrm{b}}$ so that $\{y_1, \ldots, y_q\} \supseteq \{x_1, \ldots, x_p\}$, by possibly adding coordinates y_j).

Each cylindrical set $\mathfrak{C} = \mathfrak{C}(x_1, \ldots, x_p; \delta)$ can be written in the form $\mathfrak{C} = \pi^{-1}_{x_1,\ldots,x_p}(\delta)$, i.e., $\delta = \pi_{x_1,\ldots,x_p}(\mathfrak{C})$. Since $\pi_{x_1,\ldots;x_p}$ is continuous and open, $\mathfrak{C}(x_1, \ldots, x_p, \delta)$ is open in R_X if and only if its base δ is open in $R_{x_1,\ldots,x_p}$. The same assertion can be made with respect to $\mathfrak{C}$ and δ being closed, because $R_X \backslash \mathfrak{C}(x_1, \ldots, x_p; \delta) = \mathfrak{C}(x_1, \ldots, x_p; R_{x_1,\ldots,x_p} \backslash \delta)$.

For particular coordinates $x_1, \ldots, x_p$ the operations of union and difference on cylindrical sets $\mathfrak{C}(x_1, \ldots, x_p; \delta)$ reduce to the same operations on their bases δ. Therefore, $\mathfrak{C}(x_1, \ldots, x_p; \delta)$, where $\delta \in \mathcal{B}(R_{x_1,\ldots,x_p})$, appears in the σ-algebra generated by the basic neighborhoods of the form $\mathfrak{C}_{\mathrm{b}} = \mathfrak{C}(x_1, \ldots, x_p; u_1 \times \cdots \times u_p)$ as the u_n vary. This implies that the σ-algebra $C_\sigma(R_X)$ coincides with the σ-algebra generated by the collection of all basic neighborhoods $\mathfrak{C}_{\mathrm{b}}$. Therefore, $C_\sigma(R_X) \subseteq \mathcal{B}(R_X)$ is always true.

If X is at most countable, then R_X has a countable neighborhood base: the neighborhoods $\hat{\mathfrak{C}}_{\mathrm{b}} = \mathfrak{C}(x_1, \ldots, x_p; v_1 \times \cdots \times v_p)$ $(x_1, \ldots, x_p \in X; p = 1, 2, \ldots)$, where v_n varies over a countable base in R_{x_n}, form a countable system $\hat{\Sigma} \subseteq \Sigma$ giving the previous topology in R_X. In this case each open subset of R_X is a union of neighborhoods of the form $\hat{\mathfrak{C}}_{\mathrm{b}}$, and thus coincides with an at most countable union of different sets $\hat{\mathfrak{C}}_{\mathrm{b}}$ and is hence in $C_\sigma(R_X)$. But then $\mathcal{B}(R_X) \subseteq C_\sigma(R_X)$. Accordingly, if X is at most countable, then $C_\sigma(R_X) = \mathcal{B}(R_X)$.

If X is more than countable, then the σ-algebra $C_\sigma(R_X)$ is fairly meager in comparison with $\mathcal{B}(R_X)$; we now clear up the structure of the sets in it. Given a sequence $(x_n)_1^\infty$ of distinct points $x_n \in X$, let us construct the space $R_{x_1,x_2,\ldots} = \times_1^\infty R_{x_n}$, a particular case of R_X. It can be understood as the collection of all sequences $(\lambda(x_1), \lambda(x_2), \ldots)$ with $\lambda(x_n) \in R_{x_n}$, and is always topologized by the Tychonoff topology. A generalized cylindrical set with coordinates $x_1, x_2, \ldots$ is defined to be a subset of R_X of the form

$$\alpha = \mathfrak{C}(x_1, x_2, \ldots; \delta) = \{\lambda(\cdot) \in R_X | (\lambda(x_1), \lambda(x_2), \ldots) \in \delta\}, \tag{1.11}$$

where the base δ is in $\mathcal{B}(R_{x_1,x_2,\dots}) = \mathcal{C}_\sigma(R_{x_1,x_2,\dots})$. As in the case of cylindrical sets, it is possible to change the order of coordinates in (1.11) and to add (or eliminate) new coordinates $y \in X$ while multiplying the base by R_y.

The σ-algebra $\mathcal{C}_\sigma(R_X)$ coincides with the collection of all generalized cylindrical sets. Indeed, the collection of all sets of the form (1.11) forms a σ-algebra $\mathfrak{R} \supseteq \mathcal{C}(R_X)$—this is proved in a way similar to the proof that $\mathcal{C}(R_X)$ is an algebra, except that in passing to common coordinates it is necessary to "insert in δ" also countable sequences of factors R_y. On the other hand, $\mathfrak{R} \subseteq \mathcal{C}_\sigma(R_X)$. This follows from the fact that operations on the bases δ imply the analogous operations on the sets $\mathfrak{C}(x_1, x_2, \dots; \delta)$ (provided that the coordinates $x_1, x_2, \dots$ of these sets are the same); therefore, the construction of a $\delta \in \mathcal{C}_\sigma(R_{x_1,x_2,\dots})$ from sets in $\mathcal{C}(R_{x_1,x_2,\dots})$ implies the analogous construction of $\mathfrak{C}(x_1, x_2, \dots; \delta)$ from sets $\mathfrak{C}(x_1, x_2, \dots; \gamma)$ with $\gamma \in \mathcal{C}(R_{x_1,x_2,\dots})$, and these sets are in $\mathcal{C}(R_X)$. Thus, $\mathfrak{R} = \mathcal{C}_\sigma(R_X)$. ∎

It follows from what was proved that subsets of R_X "distinguished by uncountably many conditions" can fail to be in $\mathcal{C}_\sigma(R_X)$. For example, if X is more than countable, then a closed subset of R_X of the form $\{\lambda(\cdot) \in R_X | \lambda(x) \in \alpha_x,\ x \in X\} = \times_{x \in X} \alpha_x$, where $\alpha_x \neq R_x$ and $\neq \varnothing$, is closed in R_x and is not in $\mathcal{C}_\sigma(R_X)$. (In particular, $\mathcal{C}_\sigma(R_X)$ does not contain the set consisting of a single point $\lambda(\cdot)$.)

As in (1.10), we can introduce the coordinate mapping $\pi_{x_1,x_2,\dots}$ for particular distinct points $x_1, x_2, \dots \in X$:

$$R_X \ni \lambda(\cdot) \mapsto \pi_{x_1,x_2,\dots}(\lambda(\cdot)) = (\lambda(x_1), \lambda(x_2), \dots) \in R_{x_1,x_2,\dots}. \tag{1.12}$$

This mapping is clearly continuous and open (cf. (1.10)). Each generalized cylindrical set $\alpha = \mathfrak{C}(x_1, x_2, \dots; \delta)$ can be written in the form $\alpha = \pi^{-1}_{x_1,x_2,\dots}(\delta)$, i.e., $\delta = \pi_{x_1,x_2,\dots}(\alpha)$. From the continuity and openness of (1.12) we conclude as before that $\mathfrak{C}(x_1, x_2, \dots; \delta)$ is open (closed) in R_X if and only if its base δ is open (closed) in $R_{x_1,x_2,\dots}$.

Let $X = X_1 \cup X_2$ and $X_1 \cap X_2 = \varnothing$. Then $R_X = R_{X_1} \times R_{X_2}$ if a $\lambda(\cdot) \in R_X$ is identified with the pair $(\lambda(\cdot) \restriction X_1, \lambda(\cdot) \restriction X_2)$. (The topology in R_X coincides with the direct product topology.) In particular, if $x_1, x_2, \dots \in X$ are distinct, then $R_X = R_{x_1,x_2,\dots} \times R_{X_2}$ ($X_2 = X \setminus \{x_1, x_2, \dots\}$) and $\mathfrak{C}(x_1, x_2, \dots; \delta) = \delta \times R_{X_2}$. The σ-algebra $\mathcal{C}_\sigma(R_X)$ can be constructed from $\mathcal{C}_\sigma(R_{X_1})$ and $\mathcal{C}_\sigma(R_{X_2})$ just as $\mathfrak{R}$ is constructed from $\mathfrak{R}_1$ and $\mathfrak{R}_2$ in subsection 2. It is easy to see that $E_X = E_{X_1} \times E_{X_2}$.

1.5. ***Regularity of a joint resolution of the identity.*** In accordance with the definition (1.3), a finite scalar measure $\mathcal{C}_\sigma(R_X) \ni \alpha \mapsto \mu(\alpha) \in [0, \infty)$ is said to be regular if one of the following two equivalent relations holds:

$$\mu(\alpha) = \inf_{o \supseteq \alpha} \mu(o) \qquad (\alpha, o \in \mathcal{C}_\sigma(R_X),\ o \text{ open}),$$
$$\mu(\alpha) = \sup_{\varphi \subseteq \alpha} \mu(\varphi) \qquad (\alpha, \varphi \in \mathcal{C}_\sigma(R_X),\ \varphi \text{ closed}). \tag{1.13}$$

(Generally speaking, it is no longer possible to confine oneself to compact sets $\varphi \in \mathcal{C}_\sigma(R_X)$ in the latter relation.)

THEOREM 1.4. *A joint RI E is always a regular measure, i.e., for any $f \in H$ the finite scalar measure $\mathcal{C}_\sigma(R_X) \ni \alpha \mapsto (E(\alpha)f, f)_H \in [0, \infty)$ is regular.*

PROOF. Fix an $\alpha \in \mathcal{C}_\sigma(R_X)$ and an $f \in H$. Since a measure coincides with the corresponding outer measure on measurable sets, we can write

$$(E(\alpha)f, f)_H = \inf \sum_{n=1}^{\infty} (E(\mathfrak{C}_n)f, f)_H,$$

where the infimum is over all sets $\mathfrak{C}_n \in \mathcal{C}(R_X)$ with $\bigcup_1^\infty \mathfrak{C}_n \supseteq \alpha$. For a given $\varepsilon > 0$ choose sets $\mathfrak{C}_n$ such that

$$\sum_{n=1}^{\infty} (E(\mathfrak{C}_n)f, f)_H < (E(\alpha)f, f)_H + \varepsilon.$$

If $\mathfrak{C}_n = \mathfrak{C}(x_1, \ldots, x_{p_n}; \delta_n)$, then

$$(E(\mathfrak{C}_n)f, f)_H = (E_{x_1, \ldots, x_{p_n}}(\delta_n)f, f)_H,$$

according to (1.9). As mentioned in subsection 2, the finite scalar measure

$$\mathcal{B}(R_{x_1, \ldots, x_{p_n}}) \ni \delta \mapsto (E_{x_1, \ldots, x_{p_n}}(\delta)f, f)_H \in [0, \infty)$$

is automatically regular. Therefore, $R_{x_1, \ldots, x_{p_n}}$ contains an open set $o_n \supseteq \delta_n$ such that

$$(E_{x_1, \ldots, x_{p_n}}(o_n)f, f)_H - (E_{x_1, \ldots, x_{p_n}}(\delta_n)f, f)_H < \varepsilon \cdot 2^{-n}.$$

We construct the open set $\mathfrak{C}'_n = \mathfrak{C}_n(x_1, \ldots, x_{p_n}; o_n)$; then the last inequality can be rewritten in the form

$$(E(\mathfrak{C}'_n)f, f)_H - (E(\mathfrak{C}_n)f, f)_H < \varepsilon \cdot 2^{-n} \qquad (n = 1, 2, \ldots).$$

Summing these inequalities, we get that

$$\sum_{n=1}^{\infty} (E(\mathfrak{C}'_n)f, f)_H < \sum_{n=1}^{\infty} (E(\mathfrak{C}_n)f, f)_H + \varepsilon < (E(\alpha)f, f)_H + 2\varepsilon.$$

As a consequence of the monotonicity of the measure, the absolute subadditivity, and the above inequality, it is possible to write a relation for the open set $\mathcal{C}_\sigma(R_X) \ni o = \bigcup_1^\infty \mathfrak{C}'_n \supseteq \bigcup_1^\infty \mathfrak{C}_n \supseteq \alpha$ which implies (1.13):

$$(E(\alpha)f, f)_H \le (E(o)f, f)_H \le \sum_{n=1}^{\infty} (E(\mathfrak{C}'_n)f, f)_H < (E(\alpha)f, f)_H + 2\varepsilon. \quad \blacksquare$$

1.6. ***The concept of the support of a measure and its properties.*** In this and in §§1.8 and 1.9 we give some simple general constructions from measure theory. Let R be a Hausdorff topological space in which a neighborhood base Σ containing R is fixed, and let $\mathfrak{R}$ be the σ-algebra generated by Σ. Of course, $\mathfrak{R} \subseteq \mathcal{B}(R)$, but not every open set need be in $\mathfrak{R}$, so the last inclusion can be strict. An example of such a situation occurs when $R = R_X$, Σ consists of all the basic neighborhoods $\mathfrak{C}_b$ introduced in subsection 4, and $\mathfrak{R} = \mathcal{C}_\sigma(R_X)$.

Consider an operator-valued (in particular, a finite scalar) measure θ on $\mathfrak{R}$, i.e., an operator-valued function $\mathfrak{R} \ni \alpha \mapsto \theta(\alpha)$, where $\theta(\alpha)$ is a nonnegative bounded operator on H, $\theta(\varnothing) = 0$, and $\theta(R) \neq 0$, that has the absolute additivity property ($\theta(\bigcup_1^\infty \alpha_n) = \sum_1^\infty \theta(\alpha_n)$ if the α_n are in $\mathfrak{R}$ and are disjoint, where the series converges weakly). Of course, such a measure is monotone (if $\alpha', \alpha'' \in \mathfrak{R}$ and $\alpha' \subseteq \alpha''$, then $\theta(\alpha') \leq \theta(\alpha'')$) and absolutely subadditive ($\theta(\bigcup_1^\infty \alpha_n) \leq \sum_1^\infty \theta(\alpha_n)$ for any $\alpha_n \in \mathfrak{R}$ such that the series converges weakly). We often assume that the measure θ is regular, i.e., for each finite measure $\mathfrak{R} \ni \alpha \mapsto \mu(\alpha) = (\theta(\alpha)f, f)_H \in [0, \infty)$ ($f \in H$) one of the two equivalent relations (1.13) holds, with $\mathcal{C}_\sigma(R_X)$ replaced by $\mathfrak{R}$. A joint RI is an example of such a measure.

A set $\alpha \in \mathfrak{R}$ is called a set of full θ-measure if $\theta(\alpha) = \theta(R)$. The support $\operatorname{supp}\theta$ of a measure θ is defined as the intersection of all closed sets φ_ξ in $\mathfrak{R}$ of full measure: $\operatorname{supp}\theta = \bigcap_{\xi \in \Xi} \varphi_\xi$, $\theta(\varphi_\xi) = \theta(R)$. The support always exists and is a closed set (but is not necessarily in $\mathfrak{R}$); there are cases when $\operatorname{supp}\theta = \varnothing$ (see subsection 7).

A measure θ is said to be proper if any open set $\mathfrak{R} \ni o \supseteq \operatorname{supp}\theta$ has full measure; if the measure is also regular, then any set $\mathfrak{R} \ni \alpha \supseteq \operatorname{supp}\theta$ has full measure. We give simple general conditions on R ensuring that an arbitrary measure is proper.

First of all, note that if $u \in \mathfrak{R}$ is open and $u \cap \operatorname{supp}\theta \neq \varnothing$, then $\theta(u) \neq 0$: assuming the contrary, we get that $R \backslash u \in \mathfrak{R}$ is closed and of full measure. Therefore, it must be one of the sets φ_ξ in the definition of $\operatorname{supp}\theta$. But then $u \cap \operatorname{supp}\theta = \varnothing$. ■.

THEOREM 1.5. *Suppose that one of the following two conditions holds: 1) $R = \bigcup_1^\infty R_k$, where each R_k is compact; 2) R has a countable neighborhood base in the system Σ. Then any measure θ is proper, and if case 2) holds, then $\operatorname{supp}\theta$ is in $\mathfrak{R}$ and is of full measure.*

PROOF. Consider case 1). Let $\mathfrak{R} \ni o \subseteq \operatorname{supp}\theta$ be open, and let $\varphi_\xi \in \mathfrak{R}$ be the closed sets in the definition of $\operatorname{supp}\theta$ ($\xi \in \Xi$). The collection of open sets o, $(R \backslash \varphi_\xi)_{\xi \in \Xi}$ covers R and, in particular, R_k ($k = 1, 2, \ldots$). Fix k and choose a finite subcovering from this covering, say the sets o and $(R \backslash \varphi_{\xi_{k,j}})_{j=1}^{n_k}$. Now $R_k \backslash o \subseteq \bigcup_{j=1}^{n_k} (R \backslash \varphi_{\xi_{k,j}})$ and, consequently,

$$R \backslash o = \bigcup_{k=1}^{\infty} (R_k \backslash o) \subseteq \bigcup_{k=1}^{\infty} \bigcup_{j=1}^{n_k} (R \backslash \varphi_{\xi_{k,j}}).$$

Since $\theta(R\backslash\varphi_{\xi_{k,j}}) = 0$, this implies that $\theta(R\backslash o) = 0$ by monotonicity and absolute subadditivity. Accordingly, o has full measure, and this means that θ is proper.

Consider case 2). If the $\varphi_\xi \in \mathfrak{R}$ $(\xi \in \Xi)$ are as before, then

$$R\backslash \operatorname{supp}\theta = \bigcup_{\xi\in\Xi}(R\backslash\varphi_\xi).$$

Denote by $\hat{\Sigma}$ a countable collection of neighborhoods in Σ determining the topology of R. The open set $R\backslash\varphi_\xi$ can be represented in the form $R\backslash\varphi_\xi = \bigcup_{\eta_\xi\in\mathrm{H}_\xi} u_{\eta_\xi}$ $(u_{\eta_\xi} \in \hat{\Sigma})$, and so $R\backslash\operatorname{supp}\theta = \bigcup_{\xi\in\Xi}\bigcup_{\eta_\xi\in\mathrm{H}_\xi} u_{\eta_\xi}$. Since there are at most countably many distinct neighborhoods u_{η_ξ}, we can retain only distinct neighborhoods in the last union and get that $R\backslash\operatorname{supp}\theta = \bigcup_1^\infty u_n$, where u_n is one of the neighborhoods u_{η_ξ} (the union can also be finite). This gives us that

$$\operatorname{supp}\theta = \bigcap_{n=1}^\infty (R\backslash u_n) = \bigcap_{n=1}^\infty \psi_n, \quad \text{where } \psi_n = \bigcap_{k=1}^n (R\backslash u_k) \in \mathfrak{R}.$$

For each $k = 1, 2, \ldots$ there is a η_ξ such that $u_k = u_{\eta_\xi} \subseteq R\backslash\varphi_\xi$; therefore, $\theta(u_k) = 0$ and, consequently, ψ_n is a set of full measure. Since $\psi_1 \supseteq \psi_2 \supseteq \cdots$, it follows that $\operatorname{supp}\theta \in \mathfrak{R}$ and

$$\theta(\operatorname{supp}\theta) = \lim_{n\to\infty}\theta(\psi_n) = \theta(R).$$

All the more so, $\theta(\alpha) = \theta(R)$ for each $\mathfrak{R} \ni \alpha \supseteq \operatorname{supp}\theta$. ■

It follows, of course, from this theorem that every Borel RI is proper and defined on its support. In connection with the joint RI E constructed from a family of commuting RI's on spaces R_x, the theorem gives us that E is proper in two cases: 1) when each R_x is compact, and 2) when X is at most countable. What is more, it is possible to establish that E is proper when all but possibly a countable number of the R_x are compact (see §1.10).

In the general case we have the inclusion

$$\operatorname{supp} E \subseteq \bigtimes_{x\in X} \operatorname{supp} E_x = \{\lambda(\cdot) \in R_X | \lambda(x) \in \operatorname{supp} E_x,\ x \in X\}. \tag{1.14}$$

Indeed, the closed subset $\mathfrak{C}(x; \operatorname{supp} E_x)$ $(x \in X)$ of R_X has full E-measure: by (1.9), $E(\mathfrak{C}(x; \operatorname{supp} E_x)) = E_x(\operatorname{supp} E_x) = 1$. Therefore it is one of the sets φ_ξ, so $\operatorname{supp} E \subseteq \bigcap_{x\in X} \mathfrak{C}(x; \operatorname{supp} E_x)$. The latter set coincides with the set on the right-hand side of (1.14). ■

The inclusion in (1.14) can be strict. (It is not hard to construct an example.) At the same time, equality holds here for scalar measures. More precisely, let $(R_x)_{x\in X}$ be the previous family of spaces, and let $\mathcal{B}(R_x) \ni \alpha \mapsto \mu_x(\alpha) \in [0, 1]$ for $x \in X$ be a scalar probability measure (i.e., $\mu_x(R_x) = 1$). In the usual way we construct the product $\mu = \bigtimes_{x\in X} \mu_x$ of them on R_X; μ is regular, according to Theorem 1.4. It is not hard to prove that μ is a proper measure and

$$\operatorname{supp}\mu = \bigtimes_{x\in X} \operatorname{supp}\mu_x. \tag{1.15}$$

Indeed, let us first prove (1.15). The inclusion $\operatorname{supp}\mu \subseteq \bigtimes_{x\in X}\operatorname{supp}\mu_x = \varphi$ can be established as in (1.14). We prove the opposite inclusion. If the φ_ξ are closed sets of full measure appearing in the definition of $\operatorname{supp}\mu$, then $\varphi \subseteq \varphi_\xi$ ($\xi \in \Xi$). Indeed, if $\varphi \not\subseteq \varphi_\xi$ for some ξ, then there exists a point $\varphi \ni \lambda_0(\cdot) \notin \varphi_\xi$. Hence, some basic neighborhood

$$\mathfrak{C}_{\mathrm{b}} = \mathfrak{C}(x_1, \dots, x_p; u_1 \times \cdots \times u_p)$$

of it does not intersect φ_ξ, and thus $\mu(\mathfrak{C}_{\mathrm{b}}) = 0$. On the other hand, $\mu(\mathfrak{C}_{\mathrm{b}}) = \mu_{x_1}(u_1)\cdots\mu_{x_p}(u_p) > 0$ by (1.8), because $\mu_{x_n}(u_n) > 0$ for each n and because of the fact that $u_n \cap \operatorname{supp}\mu_{x_n} \ni \lambda_0(x_n)$ and, consequently, it is nonempty. Accordingly, $\varphi \subseteq \operatorname{supp}\mu$.

Let us prove that μ is proper. Consider an open set $C_\sigma(R_X) \ni o \supseteq \operatorname{supp}\mu$. According to (1.11), $o = \mathfrak{C}(x_1, x_2, \dots; \delta)$, where δ is open in $R_{x_1,x_2,\dots}$. This generalized cylindrical set can contain $\operatorname{supp}\mu = \bigtimes_{x\in X}\operatorname{supp}\mu_x$ only when the base δ contains $\delta \supseteq \bigtimes_1^\infty \operatorname{supp}\mu_{x_n}$. Therefore, considering the formula $\mu(\mathfrak{C}(x_1, x_2, \dots; \delta)) = (\bigtimes_1^\infty \mu_{x_n})(\delta)$ (see Remark 2 in §1.4; it is, of course, also valid for μ), we get that

$$\begin{aligned}\mu(o) = \mu(\mathfrak{C}(x_1, x_2, \dots; \delta)) &= \left(\bigtimes_{n=1}^{\infty} \mu_{x_n}\right)(\delta) \\ &\geq \left(\bigtimes_{n=1}^{\infty} \mu_{x_n}\right)\left(\bigtimes_{n=1}^{\infty} \operatorname{supp}\mu_{x_n}\right) \\ &= \prod_{n=1}^{\infty} (\mu_{x_n}(\operatorname{supp}\mu_{x_n})) = 1 = \mu(R_X). \qquad \blacksquare\end{aligned}$$

This argument does not work for E, and now only the inclusion (1.14) holds. (It no longer follows from $E_{x_n}(u_n) > 0$ that $E_{x_1}(u_1)\cdots E_{x_p}(u_p) > 0$: the projections can be orthogonal.) It is easy to give an example in which the inclusion in (1.14) is strict.

1.7. *Families of multiplication operators.* We consider an elementary example of a commuting family of RI's of arbitrary cardinality. Let M be an abstract space of points ω, $\mathfrak{M}$ a σ-algebra of subsets of it, and $\mathfrak{M} \ni \alpha \mapsto \mu(\alpha) \in [0, \infty]$ a σ-finite measure. Given an $\mathfrak{M}$-measurable almost-everywhere finite complex-valued function a, we define the operator A of multiplication by a in the space $H = L_2(M, d\mu(\omega))$:

$$H \supseteq \mathfrak{D}(A) \ni f(\omega) \mapsto (Af)(\omega) = a(\omega)f(\omega) \in H,$$
$$\mathfrak{D}(A) = \{f \in H | a(\omega)f(\omega) \in H\}.$$

The domain $\mathfrak{D}(A)$ is dense in H: any function in H vanishing on $\alpha_n = \{\omega \in M \mid |\alpha(\omega)| > n\}$ for some $n = 1, 2, \dots$ is in $\mathfrak{D}(A)$. On the other hand,

$\mathfrak{D}(A) \ni f(\omega)\kappa_{M\setminus\alpha_n}(\omega) \to f(\omega)$ in H as $n \to \infty$ for each $f \in H$, because $\mu(\alpha_n) \to 0$ ($\kappa_\alpha(\omega)$ is the characteristic function of a set α). The operator A is normal, and A^* is constructed similarly from the function $\overline{a(\omega)}$; it is bounded if and only if $a(\omega)$ is essentially bounded. The resolvent R_z is the operator of multiplication by the function $(a(\omega)-z)^{-1}$, where $z \in \mathbb{C}^1$ is such that this function is essentially bounded. This gives us that the resolution of the identity $\mathcal{B}(\mathbb{C}^1) \ni \alpha \mapsto E(\alpha)$ corresponding to A is

$$H \ni f(\omega) \mapsto (E(\alpha)f)(\omega) = \kappa_\alpha(a(\omega))f(\omega) = \kappa_{a^{-1}(\alpha)}(\omega)f(\omega).$$

Let $(a_x)_{x\in X}$ ($X = \{1,\dots,p\}$) be p functions of the indicated form, $(A_x)_{x\in X}$ a corresponding family of normal commuting operators (i.e., their RI's E_x commute), and $\mathcal{B}(\mathbb{C}^p) \ni \alpha \mapsto E(\alpha)$ the joint RI of the family $(E_x)_{x\in X}$. On rectangles it is equal to

$$\begin{aligned} H \ni f(\omega) \mapsto (E(\alpha_1 \times \cdots \times \alpha_p)f)(\omega) &= (E_1(\alpha_1)\cdots E_p(\alpha_p)f)(\omega) \\ &= \kappa_{a_1^{-1}(\alpha_1)}(\omega)\cdots\kappa_{a_p^{-1}(\alpha_p)}(\omega)f(\omega) \\ &= \kappa_{(a_1^{-1}(\alpha_1))\cap\cdots\cap(a_p^{-1}(\alpha_p))}(\omega)f(\omega) \qquad (\alpha_1,\dots,\alpha_p \in \mathcal{B}(\mathbb{C}^1)). \end{aligned}$$

It is not hard to show that the set $\psi = \{(a_1(\omega),\dots,a_p(\omega)) \in \mathbb{C}^p | \omega \in M\}^\sim$ has full E-measure. Indeed, it suffices to see that any point of the open set $\mathbb{C}^p\setminus\psi$ has a neighborhood of E-measure zero. For this it suffices to verify that $E(o_1\times\cdots\times o_p) = 0$ if the o_n are open in $\mathbb{C}^1$ and $(o_1\times\cdots\times o_p)\cap\psi = \varnothing$. Assume not. Then $(a_1^{-1}(o_1))\cap\cdots\cap(a_p^{-1}(o_p)) \neq \varnothing$; let $\omega_0 \in (a_1^{-1}(o_1))\cap\cdots\cap(a_p^{-1}(o_p))$. For each $n = 1,\dots,p$ we have that $a_n(\omega_0) \in o_n$, i.e.,

$$\psi \ni (a_1(\omega_0),\dots,a_p(\omega_0)) \in o_1 \times \cdots \times o_p. \quad \blacksquare$$

Thus, $\operatorname{supp} E \subseteq \psi$. It is easy to give sufficient conditions for equality here. For example, if M is a Hausdorff topological space and $\mathfrak{M} = \mathcal{B}(M)$, the measure μ is positive on open sets, and the functions a_x, $x \in X = \{1,\dots,p\}$, are continuous, then $\operatorname{supp} E = \psi$. This equality no longer holds if X is infinite. Moreover, in the case of uncountable X it can happen that $\operatorname{supp} E = \varnothing$. We give an example, observing first that all the foregoing is transformed in the natural way in the case of real-valued functions a_x.

Suppose that $H = L_2(\mathbb{R}^1, d\omega)$ for Lebesgue measure $d\omega$, and let $X = \{\xi\} \cup \mathbb{R}^1$, where ξ is a particular abstract point; A_ξ and A_x are the operators of multiplication by the functions ω and $(\omega - x)^{-1}$ ($x \in \mathbb{R}^1$), respectively. It is claimed that the joint RI E of this family $(A_x)_{x\in X}$ has empty support. Indeed, consider the joint two-dimensional RI $E_{\xi,x}$ ($x \in \mathbb{R}^1$). Here $a_\xi(\omega) = \omega$ and $a_x(\omega) = (\omega - x)^{-1}$ ($\omega \in \mathbb{R}^1 = M$). The set

$$\psi_x = \{(a_\xi(\omega), a_x(\omega)) \in \mathbb{R}^2 | \omega \in \mathbb{R}^1\} = \{(\omega, (\omega - x)^{-1}) \in \mathbb{R}^2 | \omega \in \mathbb{R}^1\}$$

is the graph of a hyperboloid and is closed in $\mathbb{R}^2$, and $\operatorname{supp} E_{\xi,x} \subseteq \psi_x$ according to the foregoing (it is easy to see that equality actually holds here). The closed cylindrical set $\varphi_x = \mathfrak{C}(\xi, x; \psi_x)$ $(x \in \mathbb{R}^1)$ in $\mathbb{R}^X$ is of full E-measure $(E(\mathfrak{C}(\xi, x; \psi_x)) = E_{\xi,x}(\psi_x) = 1)$; therefore, $\operatorname{supp} E \subseteq \bigcap_{x\in\mathbb{R}^1} \varphi_x = \varphi$. However, the intersection φ is empty. Assume not, and let $\mathbb{R}^X \ni \lambda_0(\cdot) \in \varphi$. Then $\mathbb{R}^1 \supseteq \pi_\xi(\varphi) \ni \pi_\xi(\lambda_0(\cdot)) = \lambda_0(\xi)$. On the other hand, $\varphi \subseteq \varphi_{\lambda_0(\xi)} = \mathfrak{C}(\xi, x; \psi_{\lambda_0(\xi)})$, and so $\pi_\xi(\varphi) \subseteq \pi_\xi(\varphi_{\lambda_0(\xi)}) = \mathbb{R}^1\setminus\{\lambda_0(\xi)\}$. Accordingly, $\varnothing = \varphi \supseteq \operatorname{supp} E$. ■.

REMARK 1. This argument shows that $\operatorname{supp} E = \varnothing$ also when $a_x(\omega)$ is an arbitrary real-valued function that is continuous on $\mathbb{R}^1\setminus\{x\}$ and such that $\lim_{\omega\to x} |a_x(\omega)| = \infty$ $(x \in \mathbb{R}^1;\ a_\xi(\omega) = \omega)$. In particular, it is possible to set $a_x(\omega) = |\omega - x|^{-\varepsilon}$ $(\varepsilon \in (0, \frac{1}{2}))$; now $\bigcap_{x\in X} \mathfrak{D}(A_x) \supseteq H \cap L_\infty(\mathbb{R}^1, d\omega)$ is dense in H, contrary to the example presented. We note that uncountably many unbounded operators appear in all these examples. This circumstance is essential: see Theorem 1.6 below.

1.8. ***Construction of a measure on a larger space from a measure on a smaller space. Compactification.*** As in subsection 6, suppose that R is a Hausdorff topological space, Σ is a neighborhood base for it containing R, and $\mathfrak{R}$ is the σ-algebra generated by Σ. Fix a set $R' \subseteq R$ (not in $\mathfrak{R}$, in general) and topologize it by the relative topology of R: a neighborhood base in R' is formed by the collection Σ' of all sets of the form $u' = u \cap R'$, where $u \in \Sigma$. Let $\mathfrak{R}'$ be the σ-algebra generated by Σ'; the mapping

$$\mathfrak{R} \ni \alpha \mapsto \alpha' = \alpha \cap R' \in \mathfrak{R}' \tag{1.16}$$

obviously maps the whole of $\mathfrak{R}$ onto the whole of $\mathfrak{R}'$ (of course, coalescence is possible here).

We assume that an operator-valued measure $\mathfrak{R}' \ni \alpha' \mapsto \theta'(\alpha')$ is defined on $\mathfrak{R}'$, and we define a measure θ on $\mathfrak{R}$ according to (1.16) by setting

$$\mathfrak{R} \ni \alpha \mapsto \theta(\alpha) = \theta'(\alpha \cap R') \tag{1.17}$$

The following relation is valid:

$$\operatorname{supp} \theta' = (\operatorname{supp} \theta) \cap R'. \tag{1.18}$$

Indeed, let $\operatorname{supp} \theta = \bigcap_{\xi\in\Xi} \varphi_\xi$, where φ_ξ runs through the collection of all closed sets in $\mathfrak{R}$ of full θ-measure. Then $\varphi_\xi \cap R' \in \mathfrak{R}'$, it is closed in the topology of R', and it is of full θ-measure; therefore, it appears in the intersection defining $\operatorname{supp} \theta'$. Consequently,

$$\operatorname{supp} \theta' \subseteq \bigcap_{\xi\in\Xi} (\varphi_\xi \cap R') = (\operatorname{supp} \theta) \cap R'.$$

Conversely, suppose that $(\varphi'_\eta)_{\eta\in\mathrm{H}}$ is the collection of all closed subsets of R' in $\mathfrak{R}'$ of full θ'-measure; $\operatorname{supp} \theta' = \bigcap_{\eta\in\mathrm{H}} \varphi'_\eta$. Assume that the inclusion of the left-hand set in the right-hand set in (1.18) is strict. Then there exists a $\lambda' \in (\operatorname{supp} \theta) \cap R'$, with $\lambda' \notin \operatorname{supp} \theta'$. Thus $\lambda' \notin \varphi'_\eta$ for some $\eta \in \mathrm{H}$. Let $u' = u \cap R'$ be a neighborhood in Σ' of λ' such that $u' \cap \varphi'_\eta = \varnothing$ $(u \in \Sigma)$. Since

$u' \subseteq R' \backslash \varphi'_\eta$ and $\theta'(R' \backslash \varphi'_\eta) = 0$, this and (1.17) give us that $0 = \theta'(u') = \theta(u)$. On the other hand, $\lambda' \in u \cap \operatorname{supp} \theta$, and thus the last set is nonempty. But then $\theta(u) \neq 0$, which is absurd. ■

Let us clear up the relation between integrals with respect to the measures θ and θ'. If $R \ni \lambda \mapsto F(\lambda) \in \mathbb{C}^1$ is measurable with respect to the σ-algebra $\mathfrak{R}$, then its restriction $F \restriction R'$ is measurable with respect to $\mathfrak{R}'$. Let $f \in H$ be fixed. The function F is integrable with respect to the measure $\mathfrak{R} \ni \alpha \mapsto (\theta(\alpha)f, f)_H \in [0, \infty)$ if and only if $F \restriction R'$ is integrable with respect to $\mathfrak{R}' \ni \alpha \mapsto (\theta'(\alpha)f, f)_H \in [0, \infty)$, and

$$\int_R F(\lambda)\, d(\theta(\lambda)f, f)_H = \int_{R'} (F \restriction R')(\lambda')\, d(\theta'(\lambda')f, f)_H, \tag{1.19}$$

Indeed, the assertion will be proved if we establish the relation (1.19) for step functions F. For such functions it is valid if (1.19) holds for the characteristic function $F(\lambda) = \kappa_\alpha(\lambda)$ of a set $\alpha \in \mathfrak{R}$, i.e., if $(\theta(\alpha)f, f)_H = (\theta'(\alpha \cap R')f, f)_H$ (obviously, $\kappa_\alpha \restriction R' = \kappa_{\alpha \cap R'}$). This equality follows at once from (1.17). ■

If $o \in \mathfrak{R}$ is open in R, then $o' = o \cap R' \in \mathfrak{R}'$ is open in R'. This, (1.17) and (1.18) give us that θ' is regular if θ is, and θ is proper if θ' is. The converse implications are false in general.

It will be shown in subsection 7 that a joint RI is not always a proper measure. This leads to a number of difficulties. We present one way of surmounting them which involves compactification of the spaces R_x on which the RI's to be multiplied are defined. For simplicity it is assumed that $R_x = \mathbb{C}^1$ $(x \in X)$ (i.e., RI's of normal operators are considered).

We compactify $\mathbb{C}^1$ by formally adding a point ∞ at infinity, and denote the space obtained by $\mathbf{C}^1 = \mathbb{C}^1 \cup \{\infty\}$; any set $\mathbf{C}^1 \backslash \varphi$ with φ compact in $\mathbb{C}^1$ is regarded as a neighborhood of ∞. The space $\mathbf{C}^1$ is a compact separable complete metric space. The Tychonoff product $\mathbf{C}^X = \times_{x \in X} \mathbf{C}^1$ is a compact set containing $\mathbb{C}^X$. Its points and subsets are denoted by boldface letters. In particular, $X \ni x \mapsto \boldsymbol{\lambda}(x) \in \mathbf{C}^1$ is a point in $\mathbf{C}^X$.

If $\boldsymbol{\mathfrak{C}}_{\mathrm{b}}$ is a basic neighborhood in $\mathbf{C}^X$, then $\boldsymbol{\mathfrak{C}}_{\mathrm{b}} \cap \mathbb{C}^X$ is a basic neighborhood in $\mathbb{C}^X$, and all basic neighborhoods $\mathfrak{C}_{\mathrm{b}}$ in $\mathbb{C}^X$ can be so described. Indeed, if $\boldsymbol{\mathfrak{C}}_{\mathrm{b}} = \boldsymbol{\mathfrak{C}}(x_1, \ldots, x_p; \mathbf{u}_1 \times \cdots \times \mathbf{u}_p)$ is a basic neighborhood in $\mathbf{C}^X$, then its intersection with $\mathbb{C}^X$ consists of all the functions $\lambda(\cdot) \in \mathbb{C}^X$ such that $\lambda(x_n) \in \mathbf{u}_n \backslash \{\infty\} = u_n$ $(n = 1, \ldots, p)$, i.e., it coincides with the cylindrical subset $\mathfrak{C}(x_1, \ldots, x_p; u_1 \times \cdots \times u_p)$ of $\mathbb{C}^X$. Since each $\mathbf{u}_n$ is open in $\mathbf{C}^1$, u_n is open in $\mathbb{C}^1$, and thus the last set is a basic neighborhood $\mathfrak{C}_{\mathrm{b}}$ in $\mathbb{C}^X$. Clearly, the neighborhoods formed in this way run through the whole of the collection Σ of basic neighborhoods in $\mathbb{C}^X$ (taking the u_n to be neighborhoods in $\mathbf{C}^1$ not containing ∞). ■

Thus, the topology in $\mathbb{C}^X = R'$ coincides with the relative topology induced by the topology of $\mathbf{C}^X = R$. Moreover, the basic neighborhoods $\boldsymbol{\mathfrak{C}}_{\mathrm{b}}$ in $\mathbf{C}^X = R$ and $\mathfrak{C}_{\mathrm{b}}$ in $\mathbb{C}^X = R'$ are connected as required at the beginning of this subsection. Note that $\mathbb{C}^p$ is an open set in $\mathbf{C}^p$, and $\mathbf{C}^p \setminus \mathbb{C}^p$ is nowhere dense. The situation is different in the case of infinite X: $\mathbb{C}^X$ is not open in $\mathbf{C}^X$, and both it and $\mathbf{C}^X \setminus \mathbb{C}^X$ are dense in $\mathbf{C}^X$.

We apply the construction at the beginning of this subsection in the case when $R = \mathbf{C}^X$, $R' = \mathbb{C}^X$, $\Sigma = \{\text{arbitrary } \mathfrak{C}_{\mathrm{b}}\}$, and $\Sigma' = \{\text{arbitrary } \mathfrak{C}_{\mathrm{b}}\}$. As already mentioned, $\mathcal{C}_\sigma(R_X)$ coincides with the σ-algebra generated by the corresponding basic neighborhoods. Therefore, here $\mathfrak{R} = \mathcal{C}_\sigma(\mathbf{C}^X)$ and $\mathfrak{R}' = \mathcal{C}_\sigma(\mathbb{C}^X)$ and (1.16) gives the connection between these two σ-algebras. Take θ' to be the joint RI E constructed from the RI's E_x on $\mathbb{C}^1$. Then (1.17) defines a measure $\mathbf{E} = \theta$ on $\mathbf{C}^X$ which is obviously an RI. This $\mathbf{E}$ constructed is called the compactified joint RI. Thus,

$$\mathcal{C}_\sigma(\mathbf{C}^X) \ni \alpha \mapsto \mathbf{E}(\boldsymbol{\alpha}) = E(\alpha \cap \mathbb{C}^X). \tag{1.20}$$

It is also possible to proceed differently. Namely, let $R = \mathbf{C}^1$, $R' = \mathbb{C}^1$, $\Sigma = \mathcal{B}(\mathbf{C}^1)$ and $\Sigma' = \mathcal{B}(\mathbb{C}^1)$. (The connections required at the beginning of this subsection are satisfied, of course.) From $E_x = \theta'$ we define the compactified one-dimensional RI $\mathbf{E}_x = \theta \colon \mathcal{B}(\mathbf{C}^1) \ni \boldsymbol{\alpha} \mapsto \mathbf{E}_x(\alpha) = E_x(\boldsymbol{\alpha} \cap \mathbb{C}^1)$ according to (1.17) (i.e., the construction (1.20) is considered for $X = \{1\}$). The RI's $\mathbf{E}_x$ and $\mathbf{E}_y$ $(x, y \in X)$ obviously commute: therefore, it is possible to construct their joint RI $\mathbf{E} = \times_{x \in X} \mathbf{E}_x$. It is easy to see that this definition of $\mathbf{E}$ coincides with the definition (1.20). Indeed, denote the RI $\mathbf{E}$ constructed here by $\mathbf{F}$. Then, by (1.8) and (1.20),

$$\begin{aligned}
\mathbf{F}(\mathfrak{C}(x_1, \dots, x_p; \boldsymbol{\delta}_1 \times \cdots \times \boldsymbol{\delta}_p)) &= \mathbf{E}_{x_1}(\boldsymbol{\delta}_1) \cdots \mathbf{E}_{x_p}(\boldsymbol{\delta}_p) \\
&= E_{x_1}(\boldsymbol{\delta}_1 \cap \mathbb{C}^1) \cdots E_{x_p}(\boldsymbol{\delta}_p \cap \mathbb{C}^1) \\
&= E(\mathfrak{C}(x_1, \dots, x_p; (\boldsymbol{\delta}_1 \cap \mathbb{C}^1) \times \cdots \times (\boldsymbol{\delta}_p \cap \mathbb{C}^1))) \\
&= E((\mathfrak{C}(x_1, \dots, x_p; \boldsymbol{\delta}_1 \times \cdots \times \boldsymbol{\delta}_p)) \cap \mathbb{C}^X) \\
&= E(\mathfrak{C}(x_1, \dots, x_p; \boldsymbol{\delta}_1 \times \cdots \times \boldsymbol{\delta}_p)) \\
&\qquad (x_1, \dots, x_p \in X;\ \boldsymbol{\delta}_1, \dots, \boldsymbol{\delta}_p \in \mathcal{B}(\mathbf{C}^1)). \quad \blacksquare
\end{aligned}$$

The compactification can clearly be repeated also in the case $R_x = \mathbb{R}^1$ $(x \in X)$; the role of $\mathbf{C}^1$ is now played by the compactified line $\mathbf{R}^1 = \mathbb{R}^1 \cup \{\infty\}$.

1.9. ***Construction of a measure on a smaller space from a measure on a larger space. Modification of a measure.*** Let $R, R', \Sigma, \Sigma', \mathfrak{R}$, and $\mathfrak{R}'$ be as at the beginning of §1.8; as before, (1.16) holds. However, we now assume that the operator-valued measure θ is given, and not θ': $R \ni \alpha \mapsto \theta(\alpha)$. If $R' \in \mathfrak{R}$, then it is possible to consider the restriction of θ to R': $(\theta \restriction R')(\alpha') = \theta(\alpha')$ $(\alpha' \in \mathfrak{R}' = \{\alpha' \in \mathfrak{R} | \alpha' \subseteq R'\})$. However, it is often necessary to define such a restriction also in the case when $R' \notin \mathfrak{R}$. This can be done if R' has the following property: any $\mathfrak{R} \ni \alpha \supseteq R'$ has full θ-measure (we say that R' has full outer θ-measure).

For such an R' we defined a measure θ' on $\mathfrak{R}'$ by (1.17), read from right to left: if $\alpha' \in \mathfrak{R}'$, then by (1.16) there is an $\alpha \in \mathfrak{R}$ such that $\alpha' = \alpha \cap R'$, and we set

$$\mathfrak{R}' \ni \alpha' \mapsto \theta'(\alpha') = \theta'(\alpha \cap R') = \theta(\alpha). \tag{1.21}$$

It is easy to see that the definition (1.21) is unambiguous. Indeed, suppose that along with α there is a $\beta \in \mathfrak{R}$ such that $\alpha' = \beta \cap R'$. It is required to show that $\theta(\alpha) = \theta(\beta)$. Since $\alpha \cap R' = \beta \cap R'$, it follows that $(\alpha\backslash\beta) \cap R' = (\beta\backslash\alpha) \cap R' = \varnothing$, and thus $\mathfrak{R} \ni R\backslash(\alpha\backslash\beta)$ and $R\backslash(\beta\backslash\alpha) \supseteq R'$. Consequently, $\theta(R) = \theta(R\backslash(\alpha\backslash\beta)) = \theta(R\backslash(\beta\backslash\alpha))$, i.e., $\theta(\alpha\backslash\beta) = \theta(\beta\backslash\alpha) = 0$, and $\theta(\alpha) = \theta(\beta)$. It is clear that $\theta'(\varnothing) = 0$ and $\theta'(R') = \theta(R) \neq 0$. We verify that θ' is absolutely additive. Consider a sequence $(\alpha'_j)_1^\infty$, $\alpha'_j \in \mathfrak{R}'$, $\alpha'_j \cap \alpha'_k = \varnothing$ $(j \neq k)$, and choose $\alpha_j \in \mathfrak{R}$ such that $\alpha'_j = \alpha_j \cap R'$ $(j = 1, 2, \ldots)$. Let $\beta_1 = \alpha_1$, $\beta_2 = \alpha_2\backslash\alpha_1$, $\beta_3 = \alpha_3\backslash(\alpha_1 \cup \alpha_2), \ldots$; these sets are in $\mathfrak{R}$, are disjoint, and are such that $\beta_j \cap R' = \alpha'_j$ $(j = 1, 2, \ldots)$ and $(\bigcup_1^\infty \beta_j) \cap R' = \bigcup_1^\infty \alpha'_j$. The absolute additivity follows from the equality

$$\theta'\left(\bigcup_{j=1}^{\infty} \alpha'_j\right) = \theta\left(\bigcup_{j=1}^{\infty} \beta_j\right) = \sum_{j=1}^{\infty} \theta(\beta_j) = \sum_{j=1}^{\infty} \theta'(\alpha'_j). \quad \blacksquare$$

The measure θ' is called the modified measure of θ (with respect to R'). Since θ and θ' are connected by the relation (1.17), it can be assumed that θ is generated by θ', and thus the facts given in subsection 8 are preserved for a measure and its modification. If $R' \in \mathfrak{R}$, then $\theta' = \theta \restriction R'$. A modification of an RI is again an RI, and it is called a modified RI.

Let $(E_x)_{x\in X}$ be a family of commuting Borel RI's, with E_x given on a space R_x. The joint RI E was defined in the space $R_X = \times_{x\in X} R_x$ and was a measure on the σ-algebra $\mathcal{C}_\sigma(R_X)$. However, each E_x could be regarded as given, generally speaking, on a smaller space $R'_x \subseteq R_x$ (for example, on $R'_x = \operatorname{supp} E_x$). In this case a joint RI E' is defined on the σ-algebra $\mathcal{C}_\sigma(R'_X)$ in the space $R'_X = \times_{x\in X} R'_x$. If X is more than countable, then R'_X can fail to belong to $\mathcal{C}_\sigma(R_X)$, and thus we cannot consider that $E' = E \restriction R'_X$. At the same time, it is not hard to see that E' can be understood as E, modified with respect to R_X. More precisely, suppose that $\operatorname{supp} E_x \subseteq R'_x \subseteq R_x$ for each $x \in X$, where R'_x is closed in R_x and is topologized by the relative topology of R_x. The retriction $E'_x = E_x \restriction R'_x$ is a Borel RI on R'_x, and the family $(E'_x)_{x\in X}$ is commutative. We construct from it a joint RI E': $\mathcal{C}_\sigma(R'_X) \ni \alpha' \mapsto E'(\alpha')$ on the space $R'_X = \times_{x\in X} R'_x$.

LEMMA 1.1. *The set R'_X has full outer E-measure, and the RI E' coincides with the modified RI of E with respect to R'_X.*

PROOF. We use the notation at the beginning of §1.4 and let $R = R_X$, $\Sigma = \{\text{arbitrary } \mathfrak{C}_b\}$, and $\mathfrak{R} = \mathcal{C}_\sigma(R_X)$. The intersection of

$$\mathfrak{C}_b = \mathfrak{C}(x_1, \ldots, x_p; u_1 \times \cdots \times u_p)$$

(u_n open in R_{x_n}) with R'_X coincides with the cylindrical subset

$$\mathfrak{C}'_{\mathrm{b}} = \mathfrak{C}(x_1, \dots, x_p; (u_1 \cap R'_{x_1}) \times \cdots \times (u_p \cap R'_{x_p}))$$

of R'_X; the sets $u_n \cap R'_{x_n}$ are open in R'_{x_n}, and so $\mathfrak{C}'_{\mathrm{b}}$ is a neighborhood in R'_X. Clearly, any basic neighborhood in R'_X can be obtained as such an intersection. Thus, $\mathfrak{R}' = \mathcal{C}_\sigma(R'_X)$.

Let us see that R'_X has full outer E-measure: if $\mathcal{C}_\sigma(R_X) \ni \alpha \supseteq R'_X$, then $E(\alpha) = 1$. The set α is a generalized cylindrical set, i.e., $\alpha = \mathfrak{C}(x_1, x_2, \dots; \delta) = \delta \times R_{X_2}$, where $\delta \in \mathcal{B}(R_{x_1, x_2, \dots})$ and $X_2 = X \setminus \{x_1, x_2, \dots\}$. Since

$$R'_X = \left(\bigtimes_{n=1}^{\infty} R'_{x_n} \right) \times \left(\bigtimes_{x \in X_2} R'_x \right),$$

the inclusion $\alpha \supseteq R'_X$ implies the inclusion

$$\delta \supseteq \bigtimes_{n=1}^{\infty} R'_{x_n} \supseteq \bigtimes_{n=1}^{\infty} (\operatorname{supp} E_{x_n}) \supseteq \operatorname{supp} E_{x_1, x_2, \dots}$$

(see (1.14)). Considering what was mentioned at the end of subsection 4 and the fact that $E_{x_1, x_2, \dots}$ is a proper measure, we get that

$$E(\alpha) = E(\mathfrak{C}(x_1, x_2, \dots; \delta)) = E_{x_1, x_2, \dots}(\delta) = 1.$$

Let F' be the modification of E with respect to R'_X; it must be proved that $F' = E'$. This follows from the relation (see (1.21))

$$\begin{aligned} F'(\mathfrak{C}'_\delta) &= F'(\mathfrak{C}(x_1, \dots, x_p; (u_1 \cap R'_{x_1}) \times \cdots \times (u_p \cap R'_{x_p})) \\ &= E(\mathfrak{C}(x_1, \dots, x_p; u_1 \times \cdots \times u_p)) = E_{x_1}(u_1) \cdots E_{x_p}(u_p) \\ &= E'_{x_1}(u_1 \cap R'_{x_1}) \cdots E'_{x_p}(u_p \cap R'_{x_p}) = E'(\mathfrak{C}'_\delta). \quad \blacksquare \end{aligned}$$

Below we also use the ordinary restriction $\theta \restriction R'$, where R' is in $\mathfrak{R}$ (and is possibly not of full θ-measure) and is topologized by the relative topology of R. If we repeat the beginning of the proof of (1.18), we get the inclusion

$$\operatorname{supp}(\theta \restriction R') \subseteq (\operatorname{supp} \theta) \cap R'. \tag{1.22}$$

Simple examples show that equality can fail to hold in (1.22). Regularity of θ implies regularity of $\theta \restriction R'$: if $\mathfrak{R} \ni \alpha' \subseteq R'$, $f \in H$, and the sets $\mathfrak{R} \ni o_n \supseteq \alpha'$ are open in R and such that

$$(\theta(o_n)f, f)_H \to (\theta(\alpha')f, f)_H$$

as $n \to \infty$, then

$$(\theta(\alpha')f, f)_H \leq (\theta(o_n \cap R')f, f)_H \leq (\theta(o_n)f, f)_H$$

and

$$(\theta(o_n \cap R')f, f)_H \to (\theta(\alpha')f, f)_H. \quad \blacksquare$$

1.10. ***Properness of a joint resolution of the identity.*** It was established in subsection 6 that a joint RI E is a proper measure if each R_x is compact or if X is at most countable. What is more, the following general assertion is valid. (The countability condition below is in a certain respect also necessary; see the example above Remark 1 at the end of §1.7.)

THEOREM 1.6. *Let E be the joint RI constructed from a family $(E_x)_{x\in X}$ of commuting Borel RI's E_x on spaces R_x, and let β be a fixed set in $\mathcal{C}_\sigma(R_X)$. If all but at most a countable number of the sets* $\operatorname{supp} E_x$ *are compact, then the restriction $E \upharpoonright \beta$ is a proper measure.*

LEMMA 1.2. *Suppose that R_j is a Hausdorff topological space, $\mathfrak{R}_j$ is the σ-algebra generated by some neighborhood base Σ_j containing R_j in R_j and $\mathfrak{R}_j \ni \alpha_j \mapsto E_j(\alpha_j)$ is an RI $(j = 1, 2)$. Assume that E_1 and E_2 commute and that the joint RI $\mathfrak{R} \ni \alpha \mapsto E(\alpha)$ constructed from E_1 and E_2 exists. Fix a $\beta_1 \in \mathfrak{R}_1$ and consider the restriction $E \upharpoonright (\beta_1 \times R_2)$, which is assumed to be a nontrivial measure (i.e., $E(\beta_1 \times R_2) \neq 0$). Suppose that R_2 is compact. If a set of the form $o_1 \times R_2$ is open in $\beta_1 \times R_2$ and contains* $\operatorname{supp}(E \upharpoonright (\beta_1 \times R_2))$, *then $o_1 \supseteq$* $\operatorname{supp}(E_1 \upharpoonright \beta_1)$.

PROOF. Let $\operatorname{supp}(E \upharpoonright (\beta_1 \times R_2)) = \bigcap_{\xi\in\Xi} \varphi_\xi$, where $\varphi_\xi \subseteq \beta_1 \times R_2$ runs through the collection of all sets in $\mathfrak{R}$ that are of full $E \upharpoonright (\beta_1 \times R_2)$-measure and are closed in $\beta_1 \times R_2$. Assume that what is to be proved is false: $o_1 \not\supseteq \operatorname{supp}(E_1 \upharpoonright \beta_1)$. Then there is a point $\lambda_1^0 \in \operatorname{supp}(E_1 \upharpoonright \beta_1)$ that is not in o_1. If $\lambda_2 \in R_2$, then $(\lambda_1^0, \lambda_2) \notin o_1 \times R_2$; hence

$$(\lambda_1^0, \lambda_2) \notin \operatorname{supp}(E \upharpoonright (\beta_1 \times R_2)) = \bigcap_{\xi\in\Xi} \varphi_\xi.$$

Therefore, for each $\lambda_2 \in R_2$ there is a $\xi(\lambda_2)$ such that $(\lambda_1^0, \lambda_2) \notin \varphi_{\xi(\lambda_2)}$, and thus there is a neighborhood of (λ_1^0, λ_2) in $\beta_1 \times R_2$ disjoint from $\varphi_{\xi(\lambda_2)}$. This neighborhood can be taken in the form $(u_1^{(\lambda_2)} \cap \beta_1) \times u_2(\lambda_2)$ ($u_1^{(\lambda_2)} \in \Sigma_1$ is a neighborhood of the point λ_1^0 and $u_2(\lambda_2) \in \Sigma_2$ is a neighborhood of the point λ_2); its $E \upharpoonright (\beta_1 \times R_2)$-measure is zero, because $\varphi_{\xi(\lambda_2)}$ is of full measure. The neighborhoods $u_2(\lambda_2)$ cover the compact space R_2. Choose a finite subcovering $\bigcup_1^p u_2(\lambda_{2,n}) = R_2$ of this covering and consider the intersection $v_1 = \bigcap_1^p (u_1^{(\lambda_{2,n})} \cap \beta_1) \subseteq \beta_1$. Obviously, $v_1 \in \mathfrak{R}_1$, it is open in β_1, and it is such that

$$v_1 \times R_2 \subseteq \bigcup_{n=1}^{p} ((u_1^{(\lambda_{2,n})} \cap \beta_1) \times u_2(\lambda_{2,n})).$$

This implies that

$$\begin{aligned} 0 &= (E \upharpoonright (\beta_1 \times R_2))(v_1 \times R_2) = E(v_1 \times R_2) \\ &= E_1(v_1)E_2(R_2) = E_1(v_1) = (E_1 \upharpoonright \beta_1)(v_1). \end{aligned}$$

On the other hand, $\lambda_1^0 \in v_1$; hence $v_1 \cap \operatorname{supp}(E_1 \restriction \beta_1) \neq \varnothing$. But then $(E_1 \restriction \beta_1)(v_1) \neq 0$. We have arrived at a contradiction. ■

PROOF OF THE THEOREM. I. Denote by $(x_n)_1^\infty$ a sequence of points $x_n \in X$ such that $\operatorname{supp} E_x$ is compact for $x \neq x_n$. Let $\beta = \mathfrak{C}(y_1, y_2, \ldots; \delta)$ ($\delta \in \mathcal{B}(R_{y_1, y_2, \ldots})$) and $\{z_1, z_2, \ldots\} = \{x_1, x_2, \ldots\} \cup \{y_1, y_2, \ldots\}$; then $\beta = \mathfrak{C}(z_1, z_2, \ldots; \varepsilon)$, where $\varepsilon \in \mathcal{B}(R_{z_1, z_2, \ldots})$ is a new base of β. We set $R'_x = R_{z_n}$ for $x = z_n$ ($n = 1, 2, \ldots$), and $R'_x = \operatorname{supp} E_x$ for the remaining $x \in X$, the latter space being topologized by the relative topology of R_x. According to Lemma 1.1, $R'_X = \times_{x \in X} R'_x$ is of full outer E-measure. Let E' be the modified RI of E with respect to R'_X; E' can also be understood as the joint RI constructed on $\mathcal{C}_\sigma(R'_X)$ in R'_X from the family $(E'_x)_{x \in X}$, where $E'_x = E_x \restriction R'_x$. The cylindrical and generalized cylindrical sets in R'_X will be marked by primes, for example,

$$\mathfrak{C}'(t_1, t_2, \ldots; \delta') = \{\lambda'(\cdot) \in R'_X | (\lambda'(t_1), \lambda'(t_2), \ldots) \in \delta'\}, \qquad \delta' \in \mathcal{B}(R'_{t_1, t_2, \ldots}).$$

II. Since $R'_{z_n} = R_{z_n}$ ($n = 1, 2, \ldots$), it follows that $\mathcal{C}_\sigma(R'_X) \ni \beta' = \beta \cap R'_X = \mathfrak{C}'(z_1, z_2, \ldots; \varepsilon)$. It will be proved just below that the restriction $E' \restriction \beta'$ is a regular proper measure in the space β' (topologized by the relative topology R'_X). Assuming that this fact has been proved, we derive the assertion of the theorem: $E \restriction \beta$ is a proper measure on β.

For a proof, note first of all that β' is of full outer $E \restriction \beta$-measure: if $\mathcal{C}_\sigma(R_X) \ni \alpha \subseteq \beta = \mathfrak{C}(z_1, z_2, \ldots; \varepsilon)$ and $\alpha \supseteq \beta' = \mathfrak{C}'(z_1, z_2, \ldots; \varepsilon)$, then $\mathcal{C}_\sigma(R_X) \ni \beta \backslash \alpha \subseteq R_X \backslash R'_X$ and $0 = E(\beta \backslash \alpha) = E(\beta) - E(\alpha) = (E \restriction \beta)(\beta) - (E \restriction \beta)(\alpha)$. Denote by $(E \restriction \beta')$ the modification of $E \restriction \beta$ with respect to β'; then

$$(E \restriction \beta)' = E' \restriction \beta'. \tag{1.23}$$

Indeed, the measure $(E \restriction \beta)'$ is defined on sets $\alpha' = \alpha \cap \beta'$ with $\mathcal{C}_\sigma(R_X) \ni \alpha \subseteq \beta$ by the equality $(E \restriction \beta)'(\alpha') = (E \restriction \beta)(\alpha) = E(\alpha)$. On the other hand, the measure $E' \restriction \beta'$ is defined on sets $\alpha' \subseteq \beta'$ of the form $\alpha' = \alpha_1 \cap R'_X$ with $\alpha_1 \in \mathcal{C}_\sigma(R_X)$ by the equality $(E' \restriction \beta')(\alpha') = E'(\alpha') = E(\alpha_1)$. Since $\alpha \subseteq \beta$, it follows that

$$\alpha \cap R'_X = (\alpha \cap \beta) \cap R'_X = \alpha \cap \beta' = \alpha' = \alpha_1 \cap R'_X,$$

and thus $\mathcal{C}_\sigma(R_X) \ni \alpha_1 \backslash \alpha,\ \alpha \backslash \alpha_1 \subseteq R_X \backslash R'_X$. Consequently, $E(\alpha_1 \backslash \alpha) = E(\alpha \backslash \alpha_1) = 0$, and $E(\alpha_1) = E(\alpha)$. Thus, (1.23) is valid.

Suppose that $\mathcal{C}_\sigma(R_X) \ni \gamma \subseteq \beta$ and $\gamma \supseteq \operatorname{supp}(E \restriction \beta)$. The properness of $E \restriction \beta$ will be proved if we establish the equality

$$(E \restriction \beta)(\gamma) = (E \restriction \beta)(\beta) = E(\beta).$$

Considering the general relation (1.18) along with (1.23), we get that

$$\gamma \cap \beta' \supseteq (\operatorname{supp}(E \restriction \beta)) \cap \beta' = \operatorname{supp}(E \restriction \beta)' = \operatorname{supp}(E' \restriction \beta').$$

Moreover, $\gamma \cap \beta' = \gamma \cap (\beta \cap R'_X) = \gamma \cap R'_X$; therefore, the set $\gamma \cap \beta'$ is in the σ-algebra on which the measure $E' \restriction \beta'$ is defined. Since this measure is regular and proper by assumption, $(E' \restriction \beta')(\gamma \cap \beta') = (E' \restriction \beta')(\beta')$. But

$$(E' \restriction \beta')(\gamma \cap \beta') = E'(\gamma \cap \beta') = E'(\gamma \cap R'_X) = E(\gamma) = (E \restriction \beta)(\gamma),$$
$$(E' \restriction \beta')(\beta') = E'(\beta') = E'(\beta \cap R'_X) = E(\beta) = (E \restriction \beta)(\beta).$$

Thus, $(E \restriction \beta)(\gamma) = (E \restriction \beta)(\beta)$, and it is established that $E \restriction \beta$ is proper.

III. To prove the theorem it remains to establish that $E' \restriction \beta'$ is a regular proper measure on β'. Its regularity follows from that of E' (Theorem 1.4) and what was noted in subsection 8. We proceed to prove that $E' \restriction \beta'$ is proper.

Suppose that $\mathcal{C}_\sigma(R'_X) \ni o' \subseteq \beta'$ is open in β' and such that

$$o' \supseteq \operatorname{supp}(E' \restriction \beta');$$

it must be proved that $(E' \restriction \beta')(o') = (E' \restriction \beta')(\beta')$, i.e., $E'(o') = E'(\beta')$. Of course, it suffices to assume that the measure $E' \restriction \beta'$ is nontrivial: $E'(\beta') \neq 0$. Suppose that $o' = \mathfrak{C}'(t_1, t_2, \dots; \kappa')$, where $\kappa' \in \mathcal{B}(R'_{t_1,t_2,\dots})$. We consider $\{s_1, s_2, \dots\} = \{z_1, z_2, \dots\} \cup \{t_1, t_2, \dots\}$, and rewrite β' and o' with the help of the coordinates $(s_n)_1^\infty$: $\beta' = \mathfrak{C}'(s_1, s_2, \dots; \beta_1)$ and $o' = \mathfrak{C}'(s_1, s_2, \dots; o_1)$, where $\beta_1, o_1 \in \mathcal{B}(R'_{s_1,s_2,\dots})$ are the bases ε and κ', suitably modified.

In accordance with what was said at the end of subsection 4, we let $X = X_1 \cup X_2$, where $X_1 = \{s_1, s_2, \dots\}$ and $X_2 = X \backslash X_1$, and represent R'_X as the direct product $R'_X = R_1 \times R_2$ ($R_1 = R'_{X_1}$, $R_2 = R'_{X_2}$). The joint RI $E' = E'_X$ is equal to the joint RI constructed from the RI's $E_1 = E'_{X_1} = \times_{x \in X_1} E'_x$ and $E_2 = E'_{X_2} = \times_{x \in X} E'_x$ in the respective spaces R_1 and R_2. Since $\beta' = \mathfrak{C}'(s_1, s_2, \dots; \beta_1) = \beta_1 \times R_2$, it follows that $E'(\beta_1 \times R_2) = E'(\beta') \neq 0$; therefore, the restriction $E' \restriction (\beta_1 \times R_2)$ is a nontrivial measure. The space $R_2 = R'_{X_2} = \times_{x \in X_2} R'_x$ is compact ($X_2 \subseteq X \backslash \{x_1, x_2, \dots\}$, and hence the spaces $R'_x = \operatorname{supp} E_x$ being multiplied are compact). Thus, we find ourselves under the conditions of applicability of Lemma 1.2. The set $o' = \mathfrak{C}(s_1, s_2, \dots; o_1) = o_1 \times R_2$ is open by definition in $\beta' = \beta_1 \times R_2$ and contains the support of the measure $E' \restriction \beta' = E' \restriction (\beta_1 \times R_2)$. By this lemma, $o_1 \supseteq \operatorname{supp}(E_1 \restriction \beta_1)$.

But X_1 is countable; therefore, the system of basic neighborhoods $\mathfrak{C}_\mathrm{b}$ in the space $R_1 = R'_{X_1}$ contains a countable subsystem of neighborhoods $\hat{\mathfrak{C}}_\mathrm{b}$ which give the previous topology in R_1 (see subsection 4). The collection of all intersections $\mathfrak{C}_\mathrm{b} \cap \beta_1$ serves as a basic system in β_1 ($E_1 \restriction \beta_1$ is defined on the σ-algebra it generates), and it contains the countable subsystem of all $\hat{\mathfrak{C}}_\mathrm{b} \cap \beta_1$, which determines the same topology. Consequently, by 2) in Theorem 1.5, the measure $E_1 \restriction \beta_1$ is proper. The measure E_1 is defined on $\mathcal{C}_\sigma(R'_{X_1}) = \mathcal{B}(R'_{X_1}) = \mathcal{B}(R_1)$ (see subsection 4), and the measure $E_1 \restriction \beta_1$ on $\mathcal{B}(\beta_1)$. Since $o_1 \times R_2$ is open in $\beta_1 \times R_2$, o_1 is open in β_1 and is thus in $\mathcal{B}(\beta_1)$. Since $E_1 \restriction \beta_1$ is proper, we have that $(E_1 \restriction \beta_1)(o_1) = (E_1 \restriction \beta_1)(\beta_1)$, i.e.,

$E_1(o_1) = E_1(\beta_1)$, which implies that

$$E'(o') = E'(o_1 \times R_2) = E_1(o_1)E_2(R_2)$$
$$= E_1(\beta_1)E_2(R_2) = E'(\beta_1 \times R_2) = E'(\beta_1). \quad \blacksquare$$

1.11. ***Spectral representation of a family of commuting normal operators.*** Let $A = (A_x)_{x \in X}$ be a family of normal operators A_x acting in the Hilbert space H; each A_x has its RI $\mathcal{B}(\mathbb{C}^1) \ni \alpha \mapsto E_x(\alpha)$. The operators A_x are assumed to commute, which means that the family of RI's $(E_x)_{x \in X}$ is commutative. Let $\mathcal{C}_\sigma(\mathbb{C}^X) \ni \alpha \mapsto E(\alpha) = (\times_{x \in X} E_x)(\alpha)$ be the joint RI constructed according to Theorem 1.3; it is called the joint RI of the family A.

The spectral integral

$$F = \int_R F(\lambda)\, dE(\lambda), \qquad \mathfrak{D}(F) = \left\{ f \in H \mid \int_R |F(\lambda)|^2\, d(E(\lambda)f, f)_H < \infty \right\} \tag{1.24}$$

is defined in the usual way for a general RI $\mathfrak{R} \ni \alpha \mapsto E(\alpha)$ given on a σ-algebra $\mathfrak{R}$ and described in §1.1. Here $F(\lambda)$ is a complex-valued function defined almost everywhere with respect to the measure E on R and measurable with respect to $\mathfrak{R}$; the first of the integrals in (1.24) converges in the sense of strong convergence in H. The construction of the integral (1.24) is exactly the same as in the classical case of an RI corresponding to a single selfadjoint operator.

In particular, the integral (1.24) can be written with respect to the joint RI $E = \times_{x \in X} E_x$ of the family A. Take the function $\mathbb{C}^X \ni \lambda(\cdot) \mapsto F(\lambda(\cdot)) \in \mathbb{C}^1$ to be a cylindrical function, i.e., a function constructed from a function $\mathbb{C}^p \ni \lambda = (\lambda_1, \dots, \lambda_p) \mapsto G(\lambda) = G(\lambda_1, \dots, \lambda_p) \in \mathbb{C}^1$ and particular distinct points $x_1, \dots, x_p \in X$ by the formula $F(\lambda(\cdot)) = G(\lambda(x_1), \dots, \lambda(x_p))$. Each function $\mathbb{C}^X \ni \lambda(\cdot) \mapsto \lambda(x) \in \mathbb{C}^1$ $(x \in X)$ is measurable with respect to the σ-algebra $\mathcal{C}_\sigma(\mathbb{C}^X)$; therefore, if G is a Borel function, then F is measurable with respect to $\mathcal{C}_\sigma(\mathbb{C}^X)$.

It is easy to see that

$$F = \int_{\mathbb{C}^X} F(\lambda(\cdot))\, dE(\lambda(\cdot)) = \int_{\mathbb{C}^X} G(\lambda(x_1), \dots, \lambda(x_p))\, dE(\lambda(\cdot))$$
$$= \int_{\mathbb{C}^p} G(\lambda_1, \dots, \lambda_p)\, dE_{x_1, \dots, x_p}(\lambda) = G(A_{x_1}, \dots, A_{x_p}). \tag{1.25}$$

(It suffices to verify this for the characteristic function $\kappa_\delta(\lambda)$ of a set $\delta \in \mathcal{B}(\mathbb{C}^p)$, but in this case the equality becomes (1.9).) Thus, the left-hand integral in (1.25) gives a spectral representation of a function G of the normal commuting operators $A_{x_1}, \dots, A_{x_p}$; its domain is given by the right-hand part of (1.24).

In the case $\mathbb{C}^1 \ni \lambda \mapsto G(\lambda) = \lambda \in \mathbb{C}^1$ formulas (1.24) and (1.25) give a representation of an arbitrary operator A_x in the family under consideration in the form of the "functional" spectral integral of the function $\mathbb{C}^X \ni \lambda(\cdot) \mapsto \lambda(x) \in \mathbb{C}^1$:

$$A_x = \int_{\mathbb{C}^X} \lambda(x)\, dE(\lambda(\cdot)),$$

$$\mathfrak{D}(A_x) = \{f \in H \mid \int_{\mathbb{C}^X} |\lambda(x)|^2\, d(E(\lambda)f, f)_H < \infty\} \qquad (x \in X). \quad (1.26)$$

The adjoint operator A_x^* can, of course, be represented as the first integral, with $\lambda(x)$ replaced by $\overline{\lambda(x)}$, and $\mathfrak{D}(A_x^*) = \mathfrak{D}(A_x)$.

REMARK 1. The representation (1.26) can also be written with E replaced by the compactified joint RI $\mathcal{C}_\sigma(\mathbf{C}^X) \ni \boldsymbol{\alpha} \mapsto \mathbf{E}(\boldsymbol{\alpha}), \mathbb{C}^X$ replaced by $\mathbf{C}^X$, and $\lambda(x)$ replaced by $\boldsymbol{\lambda}(x)$. Indeed,

$$\mathbf{E}(\{\boldsymbol{\lambda}(\cdot) \in \mathbf{C}^X | \boldsymbol{\lambda}(x) = \infty\}) = \mathbf{E}(\mathfrak{C}(x; \{\infty\})) = \mathbf{E}_x(\{\infty\}) = 0;$$

therefore, the mapping $\mathbf{C}^X \ni \boldsymbol{\lambda}(\cdot) \mapsto \boldsymbol{\lambda}(x) \in \mathbf{C}^1$ takes values in $\mathbb{C}^1$ almost everywhere with respect to $\mathbf{E}$, and the integrals in (1.26) make sense with the indicated change. The equalities (1.26) hold because of (1.20) and (1.19).

What has been said in this subsection is also valid for a family of commuting selfadjoint operators; $\mathbb{C}^X$ and $\mathbf{C}^X$ must now be replaced by $\mathbb{R}^X$ and $\mathbf{R}^X$.

§2. The spectral projection theorem

In subsections 1–4 of the section we prove a theorem of Radon-Nikodým type on differentiation of an operator-valued measure with respect to its trace ρ, along with a corollary about differentiation of an RI E. This fact is used to establish the main result of Chapter 2, which is a proof that the "generalized projection" $P(\lambda(\cdot)) = (dE/d\rho)(\lambda)$ for the joint RI E of a family $(A_x)_{x\in X}$ of commuting normal operators can actually be regarded as a certain "projection on the subspace consisting of the generalized joint eigenvectors corresponding to the eigenvalue $\lambda(\cdot)$". The proof of this assertion is first presented for the simple situation when X is at most countable and then considered for the general case. At the end of the section we construct a decomposition of the Hilbert space into a direct integral with respect to these "eigensubspaces".

2.1. ***Differentiation of an operator-valued measure with respect to its trace.*** Fix a chain

$$H_- \supseteq H_0 \supseteq H_+, \quad (2.1)$$

whose spaces are all separable (for this it is, of course, sufficient for H_+ to be separable). Recall (see §1.5 in Chapter 1) that an operator $A: H_+ \to H_-$ is said to be nonnegative if $(Au, u)_{H_0} \geq 0$ $(u \in H_+)$; the trace of a nonnegative operator A is by definition equal to

$$\operatorname{Tr}(A) = \sum_{j=1}^{\infty} (Ae_j, e_j)_{H_0} \leq +\infty,$$

where $(e_j)_1^\infty$ is an orthonormal basis in H_+. The quantity $\mathrm{Tr}(A)$ does not depend on the choice of the basis: if $\mathbf{I}$ is the isometry connected with (2.1), then we conclude from the relation $(\alpha, u)_{H_0} = (\mathbf{I}\alpha, u)_{H_+}$ $(\alpha \in H_-,\ u \in H_+)$ that A is nonnegative if and only if $\mathbf{I}A\colon H_+ \to H_+$ is nonnegative, and $\mathrm{Tr}(A) = \mathrm{Tr}(\mathbf{I}A)$.

Let R be an abstract space, in which it is unnecessary to introduce a topology, and let $\mathfrak{R}$ be a σ-algebra of subsets of R. A function $\mathfrak{R} \ni \alpha \mapsto \theta(\alpha)$ is called an operator-valued measure with finite trace if the following requirements hold:

a) $\theta(\alpha)$ is a nonnegative operator from H_+ to H_-, with $\theta(\varnothing) = 0$ and $\mathrm{Tr}(\theta(R)) < \infty$;

b) the absolute additivity property holds, namely, if $\alpha_j \in \mathfrak{R}$ $(j = 1, 2, \ldots)$ are disjoint, then $\theta(\bigcup_1^\infty \alpha_j) = \sum_1^\infty \theta(\alpha_j)$, where the series converges in the weak sense.

The additivity and nonnegativity of θ imply its monotonicity: if $\alpha' \subseteq \alpha''$, then $\theta(\alpha') \le \theta(\alpha'')$. Therefore, $\theta(\alpha) \le \theta(R)$, and $\mathrm{Tr}(\theta(\alpha)) \le \mathrm{Tr}(\theta(R))$ $(\alpha \in \mathfrak{R})$.

We introduce the nonnegative numerical set function $\mathfrak{R} \ni \alpha \mapsto \rho(\alpha) = \mathrm{Tr}(\theta(\alpha))$. If $\alpha_j \in \mathfrak{R}$ $(j = 1, 2, \ldots)$ are disjoint, then, by b) and the nonnegativity of the terms,

$$\begin{aligned}\rho\left(\bigcup_{j=1}^\infty \alpha_j\right) &= \mathrm{Tr}\left(\theta\left(\bigcup_{j=1}^\infty \alpha_j\right)\right) = \mathrm{Tr}\left(\sum_{j=1}^\infty \theta(\alpha_j)\right) \\ &= \sum_{k=1}^\infty \left(\left(\sum_{j=1}^\infty \theta(\alpha_j)\right) e_k, e_k\right)_{H_0} = \sum_{k=1}^\infty \sum_{j=1}^\infty (\theta(\alpha_j)e_k, e_k)_{H_0} \\ &= \sum_{j=1}^\infty \sum_{k=1}^\infty (\theta(\alpha_j)e_k, e_k)_{H_0} = \sum_{j=1}^\infty \mathrm{Tr}(\theta(\alpha_j)) = \sum_{j=1}^\infty \rho(\alpha_j).\end{aligned}$$

Thus, $\mathfrak{R} \ni \alpha \mapsto \rho(\alpha)$ is a finite nonnegative numerical measure, which will be called the trace measure for the measure θ.

THEOREM 2.1. *An operator-valued measure θ with finite trace can be differentiated with respect to its trace measure ρ: this means that there exists a weakly $\mathfrak{R}$-measurable operator-valued function $\psi(\lambda)\colon H_+ \to H_-$ defined for ρ-almost all $\lambda \in R$ such that $\psi(\lambda) \ge 0$, $|\psi(\lambda)| \le \mathrm{Tr}(\psi(\lambda)) = 1$, and*

$$\theta(\alpha) = \int_\alpha \psi(\lambda)\, d\rho(\lambda) \qquad (\alpha \in \mathfrak{R}) \tag{2.2}$$

(the integral converges with respect to the Hilbert-Schmidt norm). The function $\psi(\lambda)$ is uniquely determined to within its values on a set of ρ-measure zero and is called the Radon-Nikodým derivative $(d\theta/d\rho)(\lambda) = \psi(\lambda)$.

We note that convergence of the integral in (2.2) in the Hilbert-Schmidt norm means its convergence in the Bochner sense if $\psi(\lambda)$ is understood as a vector-valued function with values in the space of Hilbert-Schmidt operators from H_+ to H_-.

PROOF. Let $(e_j)_1^\infty$ be a fixed orthonormal basis in H_+. The measure θ is absolutely continuous with respect to ρ, i.e., if $\rho(\alpha) = 0$, then $\theta(\alpha) = 0$ $(\alpha \in \mathfrak{R})$:

$$\begin{aligned}|(\theta(\alpha)e_j, e_k)_{H_0}|^2 &\leq (\theta(\alpha)e_j, e_j)_{H_0}(\theta(\alpha)e_k, e_k)_{H_0} \\ &\leq \rho^2(\alpha) = 0 \qquad (j, k = 1, 2, \ldots).\end{aligned}$$

This implies that for any particular u, $v \in H_+$ the complex-valued measure $\mathfrak{R} \ni \alpha \mapsto (\theta(\alpha)u, v)_{H_0} \in \mathbb{C}^1$ is also absolutely continuous with respect to ρ, and, by the ordinary Radon-Nikodým theorem,

$$(\theta(\alpha)u, v)_{H_0} = \int_\alpha \psi(\lambda; u, v)\, d\rho(\lambda) \qquad (\alpha \in \mathfrak{R};\ u, v \in H_+), \tag{2.3}$$

where the derivative $\psi(\lambda; u, v)$ is defined on a set $R_{u,v} \subseteq R$ of full ρ-measure, is $\mathfrak{R}$-measurable, and is integrable; for $u = v$ it is nonnegative. Denote by L the linear span of the vectors $(e_j)_1^\infty$ with rational complex coefficients; $\tilde{L} = H_+$. Since L is countable, the set $\bigcap_{u,v\in L} R_{u,v}$ is also of full measure; all the functions $\psi(\lambda; u, v)$ $(u, v \in L)$ are defined for this set, and $\psi(\lambda; u, u) \geq 0$ $(u \in L)$.

The fact that the derivative is uniquely determined to within its values on a set of measure zero gives us that the bilinearity of the left-hand side of (2.3) with respect to u, v implies the bilinearity of $\psi(\lambda; u, v)$. More precisely, there exists a set $Q \subseteq \bigcap_{u,v\in L} R_{u,v}$ of full measure such that if $\lambda \in Q$, then

$$\begin{aligned}\psi(\lambda; \alpha_1 u_1 + \alpha_2 u_2, \beta_1 v_1 + \beta_2 v_2) &= \alpha_1\overline{\beta}_1\psi(\lambda; u_1, v_1) + \alpha_1\overline{\beta}_2\psi(\lambda; u_1, v_2) \\ &\quad + \alpha_2\overline{\beta}_1\psi(\lambda; u_2, v_1) + \alpha_2\overline{\beta}_2\psi(\lambda; u_2, v_2)\end{aligned}$$

for any $u_1, u_2, v_1, v_2 \in L$ and any complex rational numbers $\alpha_1, \alpha_2, \beta_1, \beta_2$. For a proof we first conclude from the bilinearity of $(\theta(\alpha)u, v)_{H_0}$ and the arbitrariness of the $\alpha \in \mathfrak{R}$ in (2.3) that the indicated equality is valid for λ in a set

$$Q_{\alpha_1,\alpha_2,\beta_1,\beta_2,u_1,u_2,v_1,v_2} \subseteq \bigcap_{u,v\in L} R_{u,v}$$

of full measure, and then take Q to be the (countable) intersection of all possible such sets. Moreover, as noted, $\psi(\lambda; u, u) \geq 0$ $(u \in L)$ for such λ. In the usual way this bilinearity and nonnegativity leads to the Cauchy-Schwarz-Bunyakovskiĭ inequality

$$|\psi(\lambda; u, v)|^2 \leq \psi(\lambda; u, u)\psi(\lambda; v, v) \qquad (\lambda \in Q;\ u, v \in L). \tag{2.4}$$

With the help of Fubini's theorem, we get that

$$\rho(\alpha) = \int_{\alpha} \left(\sum_{j=1}^{\infty} \psi(\lambda; e_j, e_j) \right) d\rho(\lambda) \qquad (\alpha \in \mathfrak{R})$$

by setting $u = v = e_j$ in (2.3) and summing over j. This implies that for almost all $\lambda \in Q$

$$\sum_{j=1}^{\infty} \psi(\lambda; e_j, e_j) = 1. \tag{2.5}$$

By making the set Q somewhat smaller (if necessary) it can be assumed that (2.5) is valid for all $\lambda \in Q$. From (2.4) and (2.5),

$$\sum_{j,k=1}^{\infty} |\psi(\lambda; e_j, e_k)|^2 \le \sum_{j,k=1}^{\infty} \psi(\lambda; e_j, e_j)\psi(\lambda; e_k, e_k) = 1 \qquad (\lambda \in Q). \tag{2.6}$$

Fix $\lambda \in Q$ and denote by $A(\lambda)$ the operator defined in H_+ corresponding to the matrix $(a_{jk}(\lambda))_{j,k=1}^{\infty}$ in the basis $(e_j)_1^{\infty}$, where $a_{jk}(\lambda) = \psi(\lambda; e_k, e_j)$. By (2.6), this operator is well defined and is a Hilbert-Schmidt operator. The fact that each function $Q \ni \lambda \mapsto \psi(\lambda; e_k, e_j)$ $(j, k = 1, 2, \ldots)$ is measurable implies that the operator-valued function $Q \ni \lambda \mapsto A(\lambda)$ is weakly measurable. We introduce the continuous operator $\psi(\lambda) = I^{-1}A(\lambda)\colon H_+ \to H_-$ and show that it is the one desired.

Since $A(\lambda)$ is measurable, $Q \ni \lambda \mapsto \psi(\lambda)$ is weakly measurable. Further, for $u = \sum_1^{\infty} \alpha_k e_k$ and $v = \sum_1^{\infty} \beta_j e_j \in L$

$$\begin{aligned}(\psi(\lambda)u, v)_{H_0} &= (I^{-1}A(\lambda)u, v)_{H_0} = (A(\lambda)u, v)_{H_+} \\ &= \sum_{j,k=1}^{\infty} \psi(\lambda; e_k, e_j)\alpha_k \overline{\beta}_j = \psi(\lambda; u, v). \end{aligned} \tag{2.7}$$

In particular, $(\psi(\lambda)u, u)_{H_0} = \psi(\lambda; u, u) \ge 0$; by passing to the limit, we conclude that this inequality is preserved also for arbitrary $u \in H_+$, i.e., $\psi(\lambda) \ge 0$. According to (2.5) $\mathrm{Tr}(\psi(\lambda)) = \mathrm{Tr}(A(\lambda)) = 1$ $(\lambda \in Q)$. Thus, $|\psi(\lambda)| \le 1$, and $\psi(\lambda)$ $(\lambda \in Q)$ is weakly measurable. Therefore, the integral $\int_\alpha \psi(\lambda)\, d\rho(\lambda)$ $(\alpha \in \mathfrak{R})$ is convergent in the Hilbert-Schmidt norm. According to (2.7) and (2.3), for $u, v \in L$ and $\alpha \in \mathfrak{R}$

$$\begin{aligned}\left(\left(\int_\alpha \psi(\lambda)\, d\rho(\lambda)\right) u, v\right)_{H_0} &= \int_\alpha (\psi(\lambda)u, v)_{H_0}\, d\rho(\lambda) \\ &= \int_\alpha \psi(\lambda; u, v)\, d\rho(\lambda) = (\theta(\alpha)u, v)_{H_0},\end{aligned}$$

i.e., (2.2) holds.

Finally, we establish that $\psi(\lambda)$ is unique. Suppose that along with $\psi(\lambda)$ there is an operator-valued function $\psi_1(\lambda)$ of the same type such that

$$\int_\alpha \psi(\lambda)\,d\rho(\lambda) = \int_\alpha \psi_1(\lambda)\,d\rho(\lambda) \qquad (\alpha \in \mathfrak{R}).$$

This implies that $(\psi(\lambda)u, v)_{H_0} = (\psi_1(\lambda), u, v)_{H_0}$ for each $u, v \in L$ for λ in a set $R_{1;u,v} \subseteq R$ of full measure. But then $(\psi(\lambda)u, v)_{H_0} = (\psi_1(\lambda)u, v)_{H_0}$ for any $u, v \in L$ for λ in the set $Q_1 = \bigcap_{u,v\in L} R_{1;u,v}$ of full measure. This and the continuity of the operators $\psi(\lambda)$ and $\psi_1(\lambda)$ for each particular $\lambda \in Q_1$ leads us to conclude that $\psi(\lambda) = \psi_1(\lambda)$ $(\lambda \in Q_1)$. ■.

In the case when, for example, $R = \mathsf{R}^p$ and $\mathfrak{R} = \mathcal{B}(\mathsf{R}^1)$ $(p = 1, 2, \dots)$ the proof simplifies somewhat; moreover, a useful formula can be obtained for $\psi(\lambda)$. Namely, recall that if ω and ρ are two numerical measures on $\mathcal{B}(\mathsf{R}^p)$ with ρ nonnegative and ω absolutely continuous with respect to ρ, then the Radon-Nikodým derivative $(d\omega/d\rho)(\lambda)$ can be obtained by the following formula, true for ρ-almost all λ:

$$\left(\frac{d\omega}{d\rho}\right)(\lambda) = \lim_{\delta_n \to \lambda} \frac{\omega(\delta_n)}{\rho(\delta_n)}. \tag{2.8}$$

Here δ_n is a rectangle in an nth partition of R^p into rectangles of the form $\times_1^p[a_j, b_j)$ whose diameters tend to zero as $n \to \infty$; $\delta_n \to \lambda$ means that the limit is taken for a sequence of rectangles $\delta_n \ni \lambda$ tending to λ. (We shall not dwell on the use of formulas of the type (2.8), which is true in more general settings (Gikhman and Skorokhod [2], Chapter 2, §2).)

Theorem 2.1 can be proved as follows for the indicated R and $\mathfrak{R}$. We obtain the representation (2.3) as before. Applying (2.8), we conclude that

$$\psi(\lambda; u, v) = \lim_{\delta_n \to \lambda} \rho^{-1}(\delta_n)(\theta(\delta_n)u, v)_{H_0}$$

for λ in a set $R_{2;u,v} \subseteq R$ of full measure. Let L be an arbitrary countable dense subset of H_+ containing the basis $(e_j)_1^\infty$. Then the last equality is valid for all λ in the set $Q_2 = \bigcap_{u,v\in L} R_{2;u,v}$ of full measure and any $u, v \in L$. It is now easy to see that the limit

$$\lim_{\delta_n \to \lambda} \rho^{-1}(\delta_n)\theta(\delta_n) \tag{2.9}$$

exists for $\lambda \in Q_2$ in the sense of weak convergence of operators from H_+ to H_-.

Indeed, we have established that the limit

$$\lim_{\delta_n \to \lambda} (\rho^{-1}(\delta_n)\theta(\delta_n)u, v)_{H_0}$$

exists; therefore, to see that the limit in (2.9) exists it suffices to see that $\|\rho^{-1}(\delta_n)\theta(\delta_n)\| \leq c$ $(n = 1, 2, \dots)$. But

$$\|\rho^{-1}(\delta_n)\theta(\delta_n)\| \leq |\rho^{-1}(\delta_n)\theta(\delta_n)| \leq \mathrm{Tr}(\rho^{-1}(\delta_n)\theta(\delta_n)) = 1,$$

and so the limit in (2.9) exists; denote it by $\psi(\lambda): H_+ \to H_-$. It is nonnegative, since the operators $\rho^{-1}(\delta_n)\theta(\delta_n)$ are nonnegative. Obviously, $(\psi(\lambda)u, v)_{H_0} = \psi(\lambda; u, v)$ $(u, v \in L)$. Further, by (2.3) and Fubini's theorem,

$$\rho(\alpha) = \mathrm{Tr}(\theta(\alpha)) = \sum_{j=1}^{\infty}(\theta(\alpha)e_j, e_j)_{H_0} = \sum_{j=1}^{\infty}\int_\alpha \psi(\lambda; e_j, e_j)\, d\rho(\lambda)$$
$$= \sum_{j=1}^{\infty}\int_\alpha (\psi(\lambda)e_j, e_j)_{H_0}\, d\rho(\lambda) = \int_\alpha \sum_{j=1}^{\infty}(\psi(\lambda)e_j, e_j)_{H_0}\, d\rho(\lambda)$$

for any $\alpha \in \mathfrak{R}$, and this implies that $\mathrm{Tr}(\psi(\lambda)) = 1$ for almost all $\lambda \in Q_2$. It is also clear that (2.2) holds for the $\psi(\lambda)$ constructed.

Accordingly, the main part of Theorem 2.1 has been proved in the particular case under discussion. We have also established that for ρ-almost all λ

$$\psi(\lambda) = \lim_{\delta_n \to \lambda} \rho^{-1}(\delta_n)\theta(\delta_n) \tag{2.10}$$

in the sense of weak convergence of operators.

REMARK 1. It is possible to consider an operator-valued measure θ with σ-finite trace. This means that there exists a sequence $(R_k)_1^\infty$, $R_k \in \mathfrak{R}$, such that $R_1 \subseteq R_2 \subseteq \cdots$, $\bigcup_1^\infty R_k = R$, and $\mathrm{Tr}(\theta(R_k)) < \infty$ $(k = 1, 2, \ldots)$. The formulation and proof of Theorem 2.1 are preserved, except that the trace measure is not finite but σ-finite, and the representation (2.2) holds for each $\mathfrak{R} \ni \alpha \subseteq R_k$ for some $k = 1, 2, \ldots$.

REMARK 2. Theorem 2.1 can be made to approximate the usual Radon-Nikodým theorem in its formulation. Namely, let θ be an operator-valued measure with a σ-finite trace and let $\mathfrak{R} \ni \alpha \mapsto \rho(\alpha)$ be a σ-finite nonnegative numerical measure with respect to which θ is absolutely continuous: if $\rho(\alpha) = 0$ for some $\alpha \in \mathfrak{R}$, then $\theta(\alpha) = 0$. The the representation (2.2) is valid, where $\mathfrak{R} \ni \alpha \subseteq R_k$ $(k = 1, 2, \ldots)$ and $\psi(\lambda)$ is a weakly measurable operator-valued function defined for ρ-almost all $\lambda \in R$ and with nonnegative operators from H_+ to H_- as values, each of finite trace and integrable over R_k $(k = 1, 2, \ldots)$ with respect to ρ. (Write the representation (2.2) with the trace measure, and then differentiate it with respect to ρ in the representation.)

2.2. ***Differentiation of a resolution of the identity. The spectral measure.*** Let R be an abstract space, $\mathfrak{R}$ a σ-algebra of subsets of R, and $\mathfrak{R} \ni \alpha \mapsto E(\alpha)$ a general RI acting in the space H_0. As a rule, the measure E does not have a finite or σ-finite trace; therefore, the direct application of Theorem 2.1 is not possible here. However, it is convenient to proceed as follows.

Assume that we have a rigging (2.1) of the space H_0. The function

$$\mathfrak{R} \ni \alpha \mapsto \theta(\alpha) = O^+E(\alpha)O, \tag{2.11}$$

with continuous operators from H_+ to H_- as values, is clearly an operator-valued measure ($O^+E(\alpha)O \geq 0$, since $(O^+E(\alpha)Ou, u)_{H_0} = (E(\alpha)Ou, Ou)_{H_0} \geq 0$ for $u \in H_+$).

LEMMA 2.1. *If the rigging* (2.1) *is quasinuclear, then the operator-valued measure* (2.11) *has finite trace.*

Observe first that if $A: H_0 \to H_0$ is nonnegative, then so is $O^+AO: H_+ \to H_-$, and

$$\mathrm{Tr}(O^+AO) \leq \|A\| \, |O|^2. \tag{2.12}$$

Indeed, the inequality $O^+AO \geq 0$ was just explained by the example $A = E(\alpha)$. Further, let $(e_j)_1^\infty$ be an orthonormal basis in H_+; then

$$\mathrm{Tr}(O^+AO) = \sum_{j=1}^{\infty}(O^+AOe_j, e_j)_{H_0} = \sum_{j=1}^{\infty}(AOe_j, Oe_j)_{H_0}$$
$$\leq \|A\| \sum_{j=1}^{\infty} \|Oe_j\|_{H_0}^2 = \|A\| \, |O|^2. \quad \blacksquare$$

PROOF OF THE LEMMA. According to (2.12),

$$\mathrm{Tr}(\theta(R)) = \mathrm{Tr}(O^+E(R)O) \leq |O|^2 < \infty. \quad \blacksquare$$

Fix a quasinuclear rigging (2.1). The nonnegative finite measure $\mathfrak{R} \ni \alpha \mapsto \rho(\alpha) = \mathrm{Tr}(O^+E(\alpha)O) \in [0, \infty)$ is called the spectral measure of the RI E. Clearly, E and ρ are absolutely continuous with respect to each other: the equalities $E(\alpha) = 0$ and $\rho(\alpha) = 0$ for some $\alpha \in \mathfrak{R}$ are equivalent. Applying Theorem 2.1 to (2.11) and ρ, we get the following assertion.

THEOREM 2.2. *Let $\mathfrak{R} \ni \alpha \mapsto E(\alpha)$ be an RI acting in the space H_0,* (2.1) *a fixed quasinuclear rigging, and $\mathfrak{R} \ni \alpha \mapsto \rho(\alpha) \in [0, \infty)$ the corresponding spectral measure. Then the representation*

$$O^+E(\alpha)O = \int_\alpha P(\lambda)\, d\rho(\lambda) \qquad (\alpha \in \mathfrak{R}) \tag{2.13}$$

is valid, where the integral converges in the Hilbert-Schmidt norm. Here $P(\lambda): H_+ \to H_-$ is a weakly $\mathfrak{R}$-measurable operator-valued function defined for ρ-almost all $\lambda \in R$ and such that $P(\lambda) \geq 0$ and $|P(\lambda)| \leq \mathrm{Tr}(P(\lambda)) = 1$; $P(\lambda)$ is called a generalized projection.

We observe that in the case of the RI E of a single selfadjoint operator acting in H_0 with discrete spectrum $(\lambda_j)_1^\infty$ the equality $E(\alpha) = \sum_{\lambda_j \in \alpha} P(\lambda_j)$ ($\alpha \in \mathcal{B}(\mathbf{R}^1)$) is valid, where $P(\lambda_j)$ is the projection onto the eigensubspace of A corresponding to the eigenvalue λ_j. It was the comparison of this formula with (2.13) that determined the term "generalized projection".

REMARK 1. According to Remark 2 in subsection 1, it is also possible to introduce the concept of a general spectral measure corresponding to an RI E: such a measure is defined to be a σ-finite nonnegative measure $\mathfrak{R} \ni \alpha \mapsto \rho(\alpha) \ni [0, \infty]$ with the property that E and ρ are mutually absolutely continuous. The representation (2.13) is preserved also for a general spectral measure, except that the operator $P(\lambda)$ acquires a scalar factor.

2.3. *Differentiation of a joint resolution of the identity.* Let $(E_x)_{x\in X}$ be a family of commuting RI's acting in H_0 on the spaces R_x, and let $\mathcal{C}_\sigma(R_X) \ni \alpha \mapsto E(\alpha)$ be its joint RI. The general Theorem 2.2 is applicable to E, of course. We dwell on some of the features connected with the fact that we are now considering $E = \times_{x\in X} E_x$ (a quasinuclear rigging (2.1) is fixed).

THEOREM 2.3. *The spectral measure ρ of the joint RI E is always regular, and* $\operatorname{supp}\rho = \operatorname{supp} E$. *The measure ρ is proper if and only if E is.*

PROOF. Let us establish that ρ is regular. It suffices to prove that for any $\alpha \ni \mathcal{C}_\sigma(R_X)$ and any $\varepsilon > 0$ there is an open set $\mathcal{C}_\sigma(R_X) \ni o \supseteq \alpha$ such that $\rho(o\backslash\alpha) < \varepsilon$. Let $(e_j)_1^\infty$ be an orthonormal basis in H_+. Since $\sum_1^\infty \|Oe_j\|_{H_0}^2 = |O|^2 < \infty$, there is an $n = 1, 2, \dots$ large enough that $\sum_{n+1}^\infty \|Oe_j\|_{H_0}^2 < \varepsilon/2$. The measure E is regular (Theorem 1.4); therefore, for each $j = 1, \dots, n$, there is an open set $\mathcal{C}_\sigma(R_X) \ni o_j \supseteq \alpha$ such that $(E(o_j\backslash\alpha)Oe_j, Oe_j)_{H_0} < \varepsilon/2n$. Then $o = \bigcap_1^n o_j \in \mathcal{C}_\sigma(R_X)$ is open, it contains α, and

$$\rho(o\backslash\alpha) = \operatorname{Tr}(O^+E(o\backslash\alpha)O) = \sum_{j=1}^\infty (O^+E(o\backslash\alpha)Oe_j, e_j)_{H_0}$$

$$\leq \sum_{j=1}^n (E(o_j\backslash\alpha)Oe_j, Oe_j)_{H_0} + \sum_{j=n+1}^\infty \|Oe_j\|_{H_0}^2 < \varepsilon.$$

The regularity of ρ is proved.

The remaining assertions of the theorem follow from the mutual absolute continuity of the measures E and ρ. ∎

Let $(A_x)_{x\in X}$ be a family of commuting normal operators acting in a space H_0, let $\mathcal{C}_\sigma(\mathbb{C}^X) \ni \alpha \mapsto E(\alpha)$ and $\rho(\alpha)$ be the corresponding joint RI and spectral measure, and let $P(\lambda(\cdot))$ be the generalized projection. We can also define the analogous "compactified" objects: $\mathcal{C}_\sigma(\mathbf{C}^X) \ni \boldsymbol{\alpha} \mapsto \mathbf{E}(\boldsymbol{\alpha})$, $\operatorname{Tr}(O^+\mathbf{E}(\boldsymbol{\alpha})O) = \boldsymbol{\rho}(\boldsymbol{\alpha})$ and $\mathbf{P}(\boldsymbol{\lambda}(\cdot))$; according to (1.20), $\mathbf{E}(\boldsymbol{\alpha}) = E(\boldsymbol{\alpha} \cap \mathbb{C}^X)$, and thus $\boldsymbol{\rho}(\boldsymbol{\alpha}) = \rho(\boldsymbol{\alpha} \cap \mathbb{C}^X)$. We have the relation

$$P(\lambda(\cdot)) = (\mathbf{P} \restriction \mathbb{C}^X)(\lambda(\cdot)) \qquad (\lambda(\cdot) \in \mathbb{C}^X). \tag{2.14}$$

Indeed, denote by $P_1(\lambda(\cdot))$ the operator-valued function on the right-hand side of (2.14). By (1.19) and (2.13), $\mathbf{E}$ and E satisfy

$$\begin{aligned}\int_{\boldsymbol{\alpha}\cap\mathbb{C}^X} (P_1(\lambda(\cdot))u, v)_{H_0}\, d\rho(\lambda(\cdot)) &= \int_{\boldsymbol{\alpha}} (\mathbf{P}(\boldsymbol{\lambda}(\cdot))u, v)_{H_0}\, d\boldsymbol{\rho}(\boldsymbol{\lambda}(\cdot)) \\ &= (O^+\mathbf{E}(\boldsymbol{\alpha})Ou, v)_{H_0} \\ &= (O^+E(\boldsymbol{\alpha}\cap\mathbb{C}^X)Ou, v)_{H_0} \\ &= \int_{\boldsymbol{\alpha}\cap\mathbb{C}^X} (P(\lambda(\cdot))u, v)_{H_0}\, d\rho(\lambda(\cdot))\end{aligned}$$

for any $u, v \in H_+$ and $\boldsymbol{\alpha} \in \mathcal{C}_\sigma(\mathbf{C}^X)$. Since $\boldsymbol{\alpha} \cap \mathbb{C}^X$ runs through the whole of the σ-algebra $\mathcal{C}_\sigma(\mathbb{C}^X)$, this implies (2.14). ∎

In the preceding part we introduced the concept of a general spectral measure ρ corresponding to an RI E. We mention an important case of the joint RI generated by the family $A = (A_x)_{x\in X}$ under consideration here when the role of ρ can be played by a measure different from $\mathrm{Tr}(O^+E(\alpha)O)$ and constructed by a standard procedure.

Consider the product $A_{x_1}^{m_1}\cdots A_{x_p}^{m_p}$, where $x_1,\dots,x_p$ are distinct points in X, and $m_1,\dots,m_p$ are nonnegative integers; $p = 1,2,\dots$. A unit vector $\Omega\in H_0$ is called a cyclic vector of the family $(A_x)_{x\in X}$ (or a vacuum) if there exist products $A_{x_1}^{m_1}\cdots A_{x_p}^{m_p}$ such that $\Omega\in\mathfrak{D}(A_{x_1}^{m_1}\cdots A_{x_p}^{m_p})$ and the linear span of all the vectors $A_{x_1}^{m_1}\cdots A_{x_p}^{m_p}\Omega$ is dense in H_0 (in particular, $x_j\in X$, and the $m_j = 0,1,\dots$ can be arbitrary).

THEOREM 2.4. *Assume that a family $(A_x)_{x\in X}$ of commuting normal operators with joint RI E has a cyclic vector Ω. Then the finite nonnegative measure $\mathcal{C}_\sigma(\mathbf{C}^X)\ni\alpha\mapsto\rho(\alpha) = (E(\alpha)\Omega,\Omega)_{H_0}\in[0,\infty)$ is one of the spectral measures for this family.*

PROOF. If $E(\alpha) = 0$ for some $\alpha\in\mathcal{C}_\sigma(\mathbf{C}^X)$, then $\rho(\alpha) = 0$. We prove the opposite implication. Suppose that $0 = \rho(\alpha) = (E(\alpha)\Omega,\Omega)_{H_0} = \|E(\alpha)\Omega\|_{H_0}^2$, i.e., $E(\alpha)\Omega = 0$. Then for any product under consideration

$$0 = A_{x_1}^{m_1}\cdots A_{x_p}^{m_p}E(\alpha)\Omega = E(\alpha)A_{x_1}^{m_1}\cdots A_{x_p}^{m_p}\Omega,$$

and hence also $E(\alpha)f = 0$, where f is in the linear span of the vectors $A_{x_1}^{m_1}\cdots A_{x_p}^{m_p}\Omega$. But this span is dense in H_0 by assumption, so $E(\alpha) = 0$. ∎

Of course, the last two assertions hold also for a family of commuting selfadjoint operators.

2.4. *The case of a nuclear rigging.* Results analogous to those given in §§2.1–2.3 can be obtained if instead of a quasinuclear rigging (2.1) of the Hilbert space H_0 we use a nuclear rigging of it:

$$\Phi'\supseteq H_0\supseteq\Phi. \tag{2.15}$$

(Recall that a rigging (2.15) is called a nuclear rigging if $\Phi = \mathrm{pr\,lim}_{\tau\in\mathrm{T}} H_\tau$ is a nuclear space; see Chapter 1, §1.10.) More precisely, the case of the chain (2.15) reduces to the chain (2.1). We show this.

Consider the rigging (2.15) and denote by O and O^+ the respective imbedding operators $\Phi\subseteq H_0$ and $H_0\subseteq\Phi'$. We shall consider continuous operators $A:\Phi\to\Phi'$. Such an operator is said to be nonnegative ($A\geq 0$) if $(A\varphi,\varphi)_{H_0}\geq 0$ ($\varphi\in\Phi$). In particular, if $\mathfrak{R}\ni\alpha\mapsto E(\alpha)$ is the RI in subsection 2, then $O^+E(\alpha)O$: $\Phi\to\Phi'$, and it is nonnegative. The function $\mathfrak{R}\ni\alpha\mapsto\theta(\alpha) = O^+E(\alpha)O$ is a measure similar to that considered in subsections 1 and 2, but with values in $\mathcal{L}(\Phi,\Phi')$.

THEOREM 2.5. *Suppose that $\mathfrak{R}\ni\alpha\mapsto E(\alpha)$ and $\rho(\alpha)$ are an RI acting in a space H_0 and a spectral measure for it, and* (2.15) *is a fixed nuclear rigging. Then the representation* (2.13) *is valid in the form of a weakly convergent* *integral, where* $0 \le P(\lambda)\colon \Phi \to \Phi'$ *(a generalized projection) is a weakly* $\mathfrak{R}$-*measurable operator-valued function defined for* ρ-*almost all* $\lambda \in R$.

PROOF. As explained in §1.10 of Chapter 1, each H_τ ($\tau \in \mathrm{T}$) is topologically imbedded in H_0, and $0 \in \mathrm{T}$. Choose a τ such that the imbedding $H_\tau \subseteq H_0$ is quasinuclear; this is possible because Φ is nuclear. As a result, we get a chain

$$\Phi' \supseteq H_{-\tau} \supseteq H_0 \supseteq H_\tau \supseteq \Phi \tag{2.16}$$

with topological imbeddings. Now take (2.1) to be the middle part of the chain (2.16), and introduce the imbedding operators $O_1\colon H_\tau \to H_0$ and $O_1^+\colon H_0 \to H_{-\tau}$. Then, by Theorem 2.2 and Remark 1 in Part 2,

$$O_1^+ E(\alpha) O_1 = \int_\alpha P_1(\lambda)\, d\rho(\lambda) \qquad (\alpha \in \mathfrak{R}), \tag{2.17}$$

where $P_1(\lambda)\colon H_\tau \to H_{-\tau}$ is the corresponding generalized projection.

Consider the imbeddings $O_2\colon \Phi \to H_\tau$ and $O_3\colon H_{-\tau} \to \Phi'$. Multiplying (2.17) from the right and the left by O_2 and O_3, respectively, we get the required equality (2.13), where $P(\lambda) = O_3 P_1(\lambda) O_2$. Convergence of the integral (2.17) in the Hilbert-Schmidt norm (from H_τ to $H_{-\tau}$) obviously implies weak convergence of the integral (2.13). ■

2.5. *The concept of a generalized joint eigenvector and the spectrum of a family of operators.* Let $A = (A_x)_{x \in X}$ be a family of commuting normal operators acting in a Hilbert space H_0, and let (2.1) be a rigging of H_0. Assume that there exists a linear topological space D topologically imbedded in H_+ such that $D \subseteq \mathfrak{D}(A_x)$ and the restrictions $A_x \restriction D$ and $A_x^* \restriction D$ act continuously from D to H_+ ($x \in X$). Thus, instead of (2.1) there now appears the rigging (chain)

$$H_- \supseteq H_0 \supseteq H_+ \supseteq D. \tag{2.18}$$

We say that the family A of operators having the described properties is connected in a standard way with the rigging (2.18) (or that A admits (2.18)). The chain (2.18) is called an extension of (2.1). As before, (2.18) is quasinuclear by definition if the imbedding of the (separable) space H_+ in H_0 is quasinuclear.

A nonzero vector $\varphi \in H_-$ is called a generalized joint eigenvector of the family A with eigenvalue $\lambda(\cdot) \in \mathbb{C}^X$ if for each $x \in X$

$$(\varphi, A_x^* u)_{H_0} = \lambda(x)(\varphi, u)_{H_0}, \qquad (\varphi, A_x u)_{H_0} = \overline{\lambda(x)}(\varphi, u)_{H_0} \qquad (u \in D). \tag{2.19}$$

(We emphasize that an eigenvalue is understood to be a function $X \ni x \mapsto \lambda(x) \in \mathbb{C}^1$.) If $\varphi \in \mathfrak{D}(A_x)$, then in (2.19) the operators A_x^* and A_x can be transposed to φ, and as a result we get that $A_x\varphi = \lambda(x)\varphi$ and $A_x^*\varphi = \overline{\lambda(x)}\varphi$ ($x \in X$), i.e.,. φ is an ordinary joint eigenvector of A.

The collection of all eigenvalues $\lambda(\cdot)$ of the family A is called its generalized spectrum $g(A) \subseteq \mathbb{C}^X$. The spectrum $s(A)$ of this family is defined to be the support of its joint RI E: $s(A) = \operatorname{supp} E \subseteq \mathbf{C}^X$. Of course, the definition of $g(A)$ depends essentially on the choice of the rigging (2.18).

The subsets $s(A)$ and $g(A)$ of $\mathbf{C}^1$ are generally distinct even in the case of a family consisting of a single operator $A_1 = A$. For example, let $H_0 = L_2(\mathbf{R}^1, dt)$ with Lebesgue measure dt, and let A be the minimal operator generated by the expression $(\mathcal{L}u)(t) = -iu'(t)$, i.e., the closure of the operator $H_0 \supset C_0^\infty(\mathbf{R}^1) \ni u \mapsto -iu' \in H_0$, where $C_0^\infty(\mathbf{R}^1)$ is the collection of infinitely differentiable functions with compact support in $\mathbf{R}^1$; A is a selfadjoint operator acting in H_0, and $s(A) = \mathbf{R}^1$. Let $H_+ = L_2(\mathbf{R}^1, e^{t^2}\,dt)$ and $D = \mathcal{D}(\mathbf{R}^1)$, where $\mathcal{D}(\mathbf{R}^1)$ is the classical space $C_0^\infty(\mathbf{R}^1)$ of test functions (endowed, for example, with the projective topology). The requirements connected with the rigging (2.18) are satisfied, and $H_- = L_2(\mathbf{R}^1, e^{-t^2}\,dt)$. The function $\varphi(t) = e^{i\lambda t}$ ($t \in \mathbf{R}^1$) is in H_- and satisfies (2.19) for each $\lambda \in \mathbf{C}^1$. Thus, here we have that $g(A) = \mathbf{C}^1 \neq \mathbf{R}^1 = s(A)$ and $g(A) \supset s(A)$.

Consider the same operator A, but let $H_+ = H_0$. Now $H_- = H_0$. There do not exist vectors $H_- \ni \varphi \neq 0$ satisfying (2.19): each $\varphi \in \mathcal{D}'(\mathbf{R}^1) \supset H_0$ satisfying (2.19) is a generalized solution of the equation $-i\varphi' = \lambda\varphi$; hence $\varphi(t) = Ce^{i\lambda t}$ ($t \in \mathbf{R}^1$, $\mathbf{C}^1 \ni C \neq 0$), but this function is not in $H_0 = H_-$ for any $\lambda \in \mathbf{C}^1$. Accordingly, $g(A) = \varnothing \neq \mathbf{R}^1 = s(A)$ and $g(A) \subset s(A)$ here.

If $A = (A_x)_{x \in X}$ is a family of commuting selfadjoint operators, then (2.19) reduces to the relation

$$(\varphi, A_x u)_{H_0} = \lambda(x)(\varphi, u)_{H_0} \qquad (x \in X,\ u \in D). \tag{2.20}$$

It is natural to understand the generalized spectrum $g(A)$ as the collection of all $\lambda(\cdot) \in \mathbf{R}^X$ appearing in (2.20). Thus, here $s(A), g(A) \subseteq \mathbf{R}^X$.

Let us rephrase these definitions in terms of a rigging of H_0 by linear topological spaces. We say that a family A of commuting normal operators is connected with (2.15) in the standard way (or admits (2.15)) if $\Phi \subseteq \mathfrak{D}(A_x)$, and $A_x \restriction \Phi$ and $A_x^* \restriction \Phi$ act continuously in Φ ($x \in X$). Then $\varphi \in \Phi'$ and $u \in \Phi$ in (2.19). The definitions of the spectra are as before. Analogous changes apply to (2.20). Note that if A is connected with (2.15) in the standard way and $\Phi = \operatorname{pr\,lim}_{t \in \mathrm{T}} H_\tau$, then A is connected in the same way also with any chain $H_{-\tau} \supseteq H_0 \supseteq H_\tau \supseteq \Phi = D$ of the form (2.18). If Φ is nuclear, then the rigging (chain) (2.15) is said to be nuclear.

2.6. *The spectral theorem. The case of at most countably many operators.* We consider a family $A = (A_x)_{x \in X}$ of commuting normal operators acting in the space H_0; $\mathcal{C}_\sigma(\mathbf{C}^X) \ni \alpha \mapsto E(\alpha)$ is its joint RI. Fix a quasinuclear rigging (2.18) connected in the standard way with A, and let $\mathcal{C}_\sigma(\mathbf{C}^X) \ni \alpha \mapsto \rho(\alpha) = \operatorname{Tr}(O^+E(\alpha)O) \in [0, \infty)$ be the corresponding spectral measure. According to Theorem 2.2, we have the representation (2.13):

$$O^+E(\alpha)O = \int_\alpha P(\lambda(\cdot))\,d\rho(\lambda(\cdot)) \qquad (\alpha \in \mathcal{C}_\sigma(\mathbf{C}^X)), \tag{2.21}$$

where $P(\lambda(\cdot))$: $H_+ \to H_-$ is a generalized projection.

Suppose that X is at most countable. Then $\mathcal{C}_\sigma(\mathbb{C}^X) = \mathcal{B}(\mathbb{C}^X)$, and the measures E and ρ are defined on the Borel subsets of $\mathbb{C}^X$. The proof of the spectral theorem here is not essentially different from the case of a single operator, and does not use most of the constructions in §1.

THEOREM 2.6. *Suppose that X is at most countable, and assume that the space D in the quasinuclear chain* (2.18) *is separable. There exists a set $\mathcal{B}(\mathbb{C}^X) \ni \pi \subseteq s(A) \cap g(A)$ of full ρ-spectral measure such that for each $\lambda(\cdot) \in \pi$ the range $\mathfrak{R}(P(\lambda(\cdot)))$ consists of generalized joint eigenvectors of A corresponding to the eigenvalue $\lambda(\cdot)$, and* $\operatorname{Tr}(P(\lambda(\cdot))) = 1$.

In the spectral representation (1.25) the domain of integration can be replaced by π, and if A_x and A_x^* are understood as the restrictions to $\mathfrak{D}(A_x) \cap H_+$, then $dE(\lambda(\cdot))$ can be replaced by $P(\lambda(\cdot))\, d\rho(\lambda(\cdot))$.

We make some remarks about the case of a general X.

The first assertion of the theorem amounts to a proof of the relations

$$(P(\lambda(\cdot))u, A_x^* v)_{H_0} = \lambda(x)(P(\lambda(\cdot))u, v)_{H_0},$$
$$(P(\lambda(\cdot))u, A_x v)_{H_0} = \overline{\lambda(x)}(P(\lambda(\cdot))u, v)_{H_0} \qquad (u \in H_+,\ v \in D) \quad (2.22)$$

for $\lambda(\cdot)$ in a suitably chosen set π and $x \in X$.

REMARK 1. It suffices to establish (2.22) for the indicated $\lambda(\cdot)$ and x when u and v vary over dense subsets of H_+ and D, respectively. This follows from the fact that in (2.22), written for u and v in these subsets, we can then pass to the limit thanks to the continuity of the $P(\lambda(\cdot))$: $H_+ \to H_-$ and $A_x^* \restriction D$, $A_x \restriction D$: $D \to H_+$.

LEMMA 2.2. *For each $x \in X$ there exists a set $\beta_x \in \mathcal{C}_\sigma(\mathbb{C}^X)$ of full ρ-measure such that* (2.22) *holds and* $\operatorname{Tr}(P(\lambda(\cdot))) = 1$ *for each $\lambda(\cdot) \in \beta_x$.*

PROOF. Fix $u, v \in D$. By (2.21)

$$\begin{aligned}\int_\alpha (P(\lambda(\cdot))u, A_x^* v)_{H_0}\, d\rho(\lambda(\cdot)) &= \left(\left(\int_\alpha P(\lambda(\cdot))\, d\rho(\lambda(\cdot))\right) u, A_x^* v\right)_{H_0} \\ &= (O^+E(\alpha)Ou, A_x^* v)_{H_0} = A_x E(\alpha)u, v)_{H_0} \\ &= \left(\left(\int_\alpha \lambda(x)\, dE(\lambda(\cdot))\right) u, v\right)_{H_0} \\ &= \int_\alpha \lambda(x)\, d(O^+E(\lambda(\cdot))Ou, v)_{H_0} \qquad (\alpha \in \mathcal{C}_\sigma(\mathbb{C}^X)). \end{aligned} \quad (2.23)$$

Here we have used the relation

$$A_x E(\alpha) = \int_\alpha \lambda(x)\, dE(\lambda(\cdot)) \qquad (\alpha \in \mathcal{C}_\sigma(\mathbb{C}^X)),$$

which follows from (1.25). Next, replacing the differential in the last integral in (2.23) according to (2.21), we find that

$$\int_\alpha (P(\lambda(\cdot))u, A_x^* v)_{H_0}\, d\rho(\lambda(\cdot)) = \int_\alpha \lambda(x)(P(\lambda(\cdot))u, v)_{H_0}\, d\rho(\lambda(\cdot))$$

$$(\alpha \in \mathcal{C}_\sigma(\mathbb{C}^X)).$$

Since α is arbitrary, this implies that there exists a set $\beta_{u,v,x} \in \mathcal{C}_\sigma(\mathbb{C}^X)$ of full ρ-measure such that the first of the relations in (2.22) holds for $\lambda(\cdot) \in \beta_{u,v,x}$.

Suppose now that L is a countable dense subset of the space D, and hence of H_+. Consider the countable intersection $\bigcap_{u,v \in L} \beta_{u,v,x} = \beta'_x$. This is a set of full ρ-measure, and for $\lambda(\cdot) \in \beta'_x$ the first of the relations in (2.22) holds for all $u, v \in L$. According to Remark 1, this implies that it holds also for $\lambda(\cdot) \in \beta'_x$, $u \in H_+$ and $v \in D$.

Replacing A_x^* by A_x in (2.23) and using the corresponding expression for $A_x^* E(\alpha)$, we establish in a completely analogous way that the second of the relations in (2.22) holds for $\lambda(\cdot) \in \beta''_x$, $u \in H_+$ and $v \in D$, where $\beta''_x \in \mathcal{C}_\sigma(\mathbb{C}^X)$ is a set of full ρ-measure. The lemma is proved if we set

$$\beta_x = \beta'_x \cap \beta''_x \cap \{\lambda(\cdot) \in \mathbb{C}^X | \operatorname{Tr}(P(\lambda(\cdot))) = 1\}. \quad \blacksquare$$

PROOF OF THE THEOREM. Suppose that X is at most countable, and let β_x be the set in Lemma 2.2. The measure ρ is now defined on $s(A) = \operatorname{supp} E$, and the latter is of full measure. The set $\pi = (\bigcap_{x \in X} \beta_x) \cap s(A) \in \mathcal{B}(\mathbb{C}^X)$ is also of full ρ-measure. For $\lambda(\cdot) \in \pi$ the relations in (2.22) hold for all $x \in X$, i.e., $\mathfrak{R}(P(\lambda(\cdot)))$ has the required properties. Next, it is necessary to replace $\mathbb{C}^X$ in (1.25) by π (π is also of full E-measure) and use the equality $O^+\, dE(\lambda(\cdot))O = P(\lambda(\cdot))\, d\rho(\lambda(\cdot))$, which is a corollary of (2.21). $\blacksquare$

2.7. *The spectral theorem. The general case.* As already mentioned in the Introduction, an attempt to prove the spectral projection theorem in a way similar to that in subsection 6 for an arbitrary index set X leads to the following "continuum difficulty": the intersection $\pi = \bigcap_{x \in X} \beta_x$ can fail to be of full E-measure. Therefore, the preceding simple considerations no longer work. The difficulty will be overcome here, but the set of full outer E-measure constructed (in a different way) turns out not to be in $\mathcal{C}_\sigma(\mathbb{C}^X)$ in general, and it is thus impossible to integrate over it.

This makes it necessary to modify the RI E and the spectral measure ρ with respect to π. Let $\tau \subseteq \mathbb{C}^X$ be a set of full outer ρ-measure or, what is the same, full outer E-measure. Denote the modified RI with respect to τ by E_τ. Recall that E_τ is defined on the σ-algebra $\mathcal{C}_\sigma(\tau) = \{\alpha' = \alpha \cap \tau | \alpha \in \mathcal{C}_\sigma(\mathbb{C}^X)\}$ by the equality $\mathcal{C}_\sigma(\tau) \ni \alpha \cap \tau \mapsto E_\tau(\alpha \cap \tau) = E(\alpha)$; if $\tau \in \mathcal{C}_\sigma(\mathbb{C}^X)$, then $E_\tau = E \restriction \tau$. The modification of the spectral measure ρ is defined similarly; $E_{\mathbb{C}^X} = E$ and $\rho_{\mathbb{C}^X} = \rho$.

THEOREM 2.7. *Let X be arbitrary, and suppose that the space D in the quasinuclear chain* (2.18) *is a separable projective limit of Hilbert spaces.*

There exists a subset π of full outer ρ-measure in the generalized spectrum $g(A)$ such that for each $\lambda(\cdot) \in \pi$ the range $\mathfrak{R}(P(\lambda(\cdot)))$ consists of generalized joint eigenvectors of A corresponding to the eigenvalue $\lambda(\cdot)$, and $\mathrm{Tr}(P(\lambda(\cdot))) = 1$.

The following representation as a functional spectral integral is valid:

$$A_x = \int_\pi \lambda(x)\, dE_\pi(\lambda(\cdot)),$$

$$\mathfrak{D}(A_x) = \left\{ f \in H_0 \middle| \int_\pi |\lambda(x)|^2\, d(E_\pi(\lambda(\cdot))f, f)_{H_0} < \infty \right\} \qquad (x \in X). \quad (2.24)$$

(A_x^* is representable as the integral of $\overline{\lambda(x)}$.) If A_x and A_x^* are understood as the restrictions to $\mathfrak{D}(A_x) \cap H_+$, then the $dE_\pi(\lambda(\cdot))$ in (2.24) can be replaced by $P(\lambda(\cdot))\, d\rho_\pi(\lambda(\cdot))$.

REMARK 1. Fix an arbitrary set $\tau \subseteq \mathbb{C}^X$ containing π. The representation (2.24) and the last result of the theorem are valid with π replaced by τ. Indeed, τ is of full outer ρ-measure; therefore, E_τ and ρ_τ make sense. It is easy to see that the measures E_τ and ρ_τ can be modified with respect to $\pi \subseteq \tau$, and E_π and ρ_π are obtained again as a result—this follows from the formulas defining the modified measures with respect to E and ρ. Writing the integrals in (2.24) for E_τ and ρ_τ (including also the integrals with $P(\lambda(\cdot))\, d\rho_\tau(\lambda(\cdot))$) and using the rule (1.19), we get that they are equal to the same integrals with τ replaced by π.

REMARK 2. The measure ρ_τ ($\pi \subseteq \tau \subseteq \mathbb{C}^X$) above can be understood as the spectral measure of the RI E:

$$\mathrm{Tr}(O^+ E_\tau(\alpha \cap \tau) O) = \mathrm{Tr}(O^+ E(\alpha) O) = \rho(\alpha) = \rho_\tau(\alpha \cap \tau) \qquad (\alpha \in \mathcal{C}_\sigma(\mathbb{C}^X)).$$

The set π has full outer ρ_τ-measure.

Let us establish some lemmas. We shall use the compactified objects; see §§1.8 and 2.3. According to Theorem 2.2, along with (2.21) the following representation is valid:

$$O^+ \mathbf{E}(\boldsymbol{\alpha}) O = \int_{\boldsymbol{\alpha}} \mathbf{P}(\boldsymbol{\lambda}(\cdot)) d\boldsymbol{\rho}(\boldsymbol{\lambda}(\cdot)) \qquad (\boldsymbol{\alpha} \in \mathcal{C}_{\boldsymbol{\sigma}}(\mathbf{C}^X)). \quad (2.25)$$

LEMMA 2.3. *For each $x \in X$ there exists a set $\boldsymbol{\beta}_x \in \mathcal{C}_\sigma(\mathbf{C}^X)$ of full $\boldsymbol{\rho}$-measure such that for each $\boldsymbol{\lambda}(\cdot) \in \boldsymbol{\beta}_x$ the value of $\boldsymbol{\lambda}(x)$ is $\neq \infty$, and*

$$(\mathbf{P}(\boldsymbol{\lambda}(\cdot))u, A_x^* v)_{H_0} = \boldsymbol{\lambda}(x)(\mathbf{P}(\boldsymbol{\lambda}(\cdot))u, v)_{H_0},$$
$$(\mathbf{P}(\boldsymbol{\lambda}(\cdot))u, A_x v)_{H_0} = \overline{\boldsymbol{\lambda}(x)}(\mathbf{P}(\boldsymbol{\lambda}(\cdot))u, v)_{H_0} \qquad (u \in H_+,\ v \in D),$$
$$\mathrm{Tr}(\mathbf{P}(\boldsymbol{\lambda}(\cdot))) = 1. \quad (2.26)$$

The proof repeats the arguments in the proof of Lemma 2.2, except that (2.25) and the representation (1.25) with E replaced by $\mathbf{E}$ must be used instead of (2.21) and (1.25) (see Remark 1 in §1.11). The α in equalities of the type (2.23) or (2.24) is replaced by $\mathcal{C}_\sigma(\mathbf{C}^X) \ni \boldsymbol{\alpha} \subseteq \mathbf{C}^X \setminus \{\boldsymbol{\lambda}(\cdot) \in \mathbf{C}^X | \boldsymbol{\lambda}(x) = \infty\}$ (recall that the $\boldsymbol{\rho}$-measure of the set subtracted is equal to zero).

We use the following general "separability lemma", which it is convenient to prove a little later.

LEMMA 2.4. *Let Φ be a separable linear topological space, and $\Phi_0 = (\varphi_j)_1^\infty$ a fixed countable dense subset of it. Assume that the dual space Φ' is also separable with respect to the weak topology. Consider an arbitrary set $\mathrm{A} \subseteq \mathcal{L}(\Phi, H)$ of continuous linear operators acting from Φ into some separable Hilbert space H. Then* A *is always separable in the following sense: there exists a sequence $(A_n)_1^\infty$ $(A_n \in \mathrm{A})$ of operators such that for any $A \in \mathrm{A}$ there is a subsequence $(A_{n_k})_1^\infty$ with $\|(A - A_{n_k})\varphi_j\|_H \to 0$ as $k \to \infty$ for each $j = 1, 2, \ldots$.*

The main point in the proof of the theorem is

LEMMA 2.5. *Suppose that the conditions of Theorem 2.7 hold. There exist a set $\mathcal{B}(\mathbf{C}^X) \ni \boldsymbol{\gamma} \subseteq \operatorname{supp} \boldsymbol{\rho}$ of full outer $\boldsymbol{\rho}$-measure and a mapping $\mathbf{C}^X \supseteq \boldsymbol{\gamma} \ni \boldsymbol{\lambda}(\cdot) \mapsto (\operatorname{reg} \boldsymbol{\lambda}(\cdot))(\cdot) \in \mathbb{C}^X$ such that if $\boldsymbol{\lambda}(\cdot) \in \boldsymbol{\gamma}$, then $\mathfrak{R}(P(\boldsymbol{\lambda}(\cdot)))$ consists of generalized joint eigenvectors of A corresponding to the eigenvalue $(\operatorname{reg} \boldsymbol{\lambda}(\cdot))(\cdot)$, and $\operatorname{Tr}(\mathbf{P}(\boldsymbol{\lambda}(\cdot))) = 1$.*

The mapping reg *has the following form. A sequence $(y_n)_1^\infty$ of points $y_n \in X$ $(\boldsymbol{\lambda}(y_n) \neq \infty,\ \boldsymbol{\lambda}(\cdot) \in \boldsymbol{\gamma})$ depending on A is constructed which for each $x \in X$ has a subsequence $(y_{n_k(x)})_1^\infty$ with $(\operatorname{reg} \boldsymbol{\lambda}(\cdot))(x) = \lim_{k\to\infty} \boldsymbol{\lambda}(y_{n_k(x)})$; moreover, $(\operatorname{reg} \boldsymbol{\lambda}(\cdot))(x) = \boldsymbol{\lambda}(x)$ for each $x \in X$ with $\boldsymbol{\lambda}(x) \neq \infty$.*

PROOF. I. Consider the direct sum $\Phi = D \dotplus D$. Like D, Φ is a separable projective limit of Hilbert spaces: $\Phi = \operatorname{pr\,lim}_{\tau \in \mathrm{T}} F_\tau$ is topologically imbedded in $F_0 = H_0 \oplus H_0$ $(0 \in \mathrm{T})$. The space Φ' is separable in the weak topology. Indeed, let Ψ be a countable dense subset of Φ, hence also of each F_τ, and let $\alpha \in \Phi'$. According to (2.4), α is some $F_{-\tau_0}$, i.e., $\alpha(\varphi) = (a_\alpha, \varphi)_{F_{\tau_0}}$, where $a_\alpha \in F_{\tau_0}$ $(\varphi \in \Phi)$. Let the sequence $(\psi_n)_1^\infty \subseteq \Psi$ be such that $\psi_n \to a_\alpha$ in F_{τ_0} as $n \to \infty$; then $(\psi_n, \varphi)_{F_{\tau_0}} \to \alpha(\varphi)$ for each $\varphi \in \Phi$. In other words, the vectors ψ_n, understood as functionals in Φ', approximate α weakly.

II. Let us use the orthogonalization process to construct an orthonormal basis $(e_j)_1^\infty$ in H_+ consisting of vectors in D, and then supplement it in an arbitrary way to form a countable dense subset L of D. We apply Lemma 2.4 with $\Phi = D \dotplus D$, $\Phi_0 = L \dotplus L$, $H = H_+ \oplus H_+$, and $\mathrm{A} = (B_x)_{x \in X} \subseteq \mathcal{L}(\Phi, H)$, where $\Phi = D \dotplus D \ni (\varphi, \psi) \mapsto B_x(\varphi, \psi) = (A'_x\varphi, A''_x\psi) \in H_+ \oplus H_+ = H$, $A'_x = A^*_x \upharpoonright D \in \mathcal{L}(D, H_+)$ and $A''_x = A_x \upharpoonright D \in \mathcal{L}(D, H_+)$ $(x \in X)$; application of the lemma is possible by virtue of I. According to this lemma, there is a sequence $(y_n)_1^\infty$ $(y_n \in X)$ which for each $x \in X$ has a subsequence $(y_{n_k})_{k=1}^\infty$ such that $(A'_{y_{n_k}}\varphi, A''_{y_{n_k}}\psi) \to (A'_x\varphi, A''_x\psi)$ $(\varphi, \psi \in L)$ in $H_+ \oplus H_+$ as $k \to \infty$, i.e.,

$$\|A'_{y_{n_k}}\varphi - A'_x\varphi\|_{H_+}, \|A''_{y_{n_k}}\varphi - A''_x\varphi\|_{H_+} \to 0.$$

In what follows we fix the sequence $(y_n)_1^\infty$ and for each $x \in X$ some such subsequence $(y_{n_k(x)})_{k=1}^\infty$ of it.

III. We construct the set γ. Let $\boldsymbol{\beta} = \bigcap_1^\infty \boldsymbol{\beta}_{y_n} \in \mathcal{C}_\sigma(\mathbf{C}^X)$, where the y_n are the points constructed in II, and the $\boldsymbol{\beta}_{y_n}$ are the sets in Lemma 2.3. It is clear that $\boldsymbol{\beta}$ is of full $\boldsymbol{\rho}$-measure, and (2.26) holds for $\boldsymbol{\lambda}(\cdot) \in \boldsymbol{\beta}$ with $x = y_n$ $(n = 1, 2, \ldots)$. The equality

$$\sum_{j=1}^{\infty} (\mathbf{P}(\boldsymbol{\lambda}(\cdot))e_j, e_j)_{H_0} = \operatorname{Tr}(\mathbf{P}(\boldsymbol{\lambda}(\cdot))) = 1 \qquad (\boldsymbol{\lambda}(\cdot) \in \boldsymbol{\beta})$$

gives us the decomposition

$$\boldsymbol{\beta} = \bigcap_{j,m=1}^{\infty} \{\boldsymbol{\lambda}(\cdot) \in \boldsymbol{\beta} | (\mathbf{P}(\boldsymbol{\lambda}(\cdot))e_j, e_j)_{H_0} \geq 1/m\}. \tag{2.27}$$

Let $(\boldsymbol{\beta}_i')_1^\infty$ be some enumeration of the countable family of sets after the union sign in (2.27). We consider the disjoint sets $\boldsymbol{\beta}_1', \boldsymbol{\beta}_2' \backslash \boldsymbol{\beta}_1', \boldsymbol{\beta}_3' \backslash (\boldsymbol{\beta}_1' \cup \boldsymbol{\beta}_2'), \ldots$ and retain those of them with nonzero $\boldsymbol{\rho}$-measure; denote them by $\boldsymbol{\beta}_1, \boldsymbol{\beta}_2, \ldots$ (the sequence $(\boldsymbol{\beta}_l)_1^\infty$ can be finite). Let $\boldsymbol{\beta}_0$ be the union of the rest of these sets with $\mathbf{C}^X \backslash \boldsymbol{\beta}$. Thus, the $\boldsymbol{\beta}_l$ are in $\mathcal{C}_\sigma(\mathbf{C}^X)$ and are disjoint $(l = 0, 1, \ldots)$, $\boldsymbol{\rho}(\boldsymbol{\beta}_0) = 0$, $\boldsymbol{\rho}(\boldsymbol{\beta}_l) > 0$ $(l = 1, 2, \ldots)$ and $\mathbf{C}^X = \bigcup_0^\infty \boldsymbol{\beta}_l$. Each $\boldsymbol{\beta}_l$ is in some $\boldsymbol{\beta}_{i_l}'$; therefore, by (2.27) there exist $j_l, m_l = 1, 2, \ldots$ such that

$$(\mathbf{P}(\boldsymbol{\lambda}(\cdot))e_{j_l}, e_{j_l})_{H_0} \geq 1/m_l \qquad (\boldsymbol{\lambda}(\cdot) \in \boldsymbol{\beta}_l;\ l = 1, 2, \ldots). \tag{2.28}$$

According to Theorem 2.3, the measure $\boldsymbol{\rho}$ is regular; hence for each $\boldsymbol{\beta}_l$ there exists a sequence $(\boldsymbol{\varphi}_{l,p})_{p=1}^\infty$ of closed subsets $\mathcal{C}_\sigma(\mathbf{C}^X) \ni \boldsymbol{\varphi}_{l,p} \subseteq \boldsymbol{\beta}_l$ of $\mathbf{C}^X$ such that $\boldsymbol{\rho}(\boldsymbol{\varphi}_{l,p}) > 0$, and

$$\lim_{p\to\infty} \boldsymbol{\rho}(\boldsymbol{\varphi}_{l,p}) = \boldsymbol{\rho}(\boldsymbol{\beta}_l) \qquad (l = 1, 2, \ldots). \tag{2.29}$$

We consider the restriction $\boldsymbol{\rho} \restriction \boldsymbol{\varphi}_{l,p}$ and define $\boldsymbol{\psi}_{l,p} = \operatorname{supp}(\boldsymbol{\rho} \restriction \boldsymbol{\varphi}_{l,p})$ $(l, p = 1, 2, \ldots)$. Let

$$\mathcal{B}(\mathbf{C}^X) \ni \boldsymbol{\gamma} = \bigcap_{l,p=1}^{\infty} \boldsymbol{\psi}_{l,p} \subseteq \bigcup_{l=1}^{\infty} \boldsymbol{\beta}_l \subseteq \boldsymbol{\beta}. \tag{2.30}$$

IV. On the basis of Theorem 2.3 and the general remark at the end of §1.9, $\boldsymbol{\rho} \restriction \boldsymbol{\varphi}_{l,p}$ is regular. It is proper by 1) in Theorem 1.5: $\boldsymbol{\varphi}_{l,p}$ is a closed subset of the compact space $\mathbf{C}^X$ and is thus compact. By (1.22), $\boldsymbol{\psi}_{l,p} \subseteq \operatorname{supp} \boldsymbol{\rho}$ $(l, p = 1, 2, \ldots)$; consequently, $\boldsymbol{\gamma} \subseteq \operatorname{supp} \boldsymbol{\rho}$.

V. The set (2.30) is of full outer $\boldsymbol{\rho}$-measure. Indeed, suppose that $\mathcal{C}_\sigma(\mathbf{C}^X) \ni \boldsymbol{\alpha} \supseteq \boldsymbol{\gamma}$. Obviously, $\boldsymbol{\rho}(\boldsymbol{\alpha}) = \bigcup_p^\infty \boldsymbol{\rho}(\boldsymbol{\alpha} \cap \boldsymbol{\beta}_l)$, and it thus suffices to establish that $\boldsymbol{\rho}(\boldsymbol{\alpha} \cap \boldsymbol{\beta}_l) = \boldsymbol{\rho}(\boldsymbol{\beta}_l)$ $(l = 1, 2, \ldots)$. But $\boldsymbol{\rho}(\boldsymbol{\alpha} \cap \boldsymbol{\varphi}_{l,p}) = \boldsymbol{\rho}(\boldsymbol{\varphi}_{l,p})$ $(\boldsymbol{\rho} = 1, 2, \ldots)$: this equality is equivalent to

$$(\boldsymbol{\rho} \restriction \boldsymbol{\varphi}_{l,p})(\boldsymbol{\alpha} \cap \boldsymbol{\varphi}_{l,p}) = (\boldsymbol{\rho} \restriction \boldsymbol{\varphi}_{l,p})(\boldsymbol{\varphi}_{l,p}),$$

and since $\boldsymbol{\alpha} \cap \boldsymbol{\varphi}_{l,p} \supseteq \boldsymbol{\psi}_{l,p} = \operatorname{supp}(\boldsymbol{\rho} \restriction \boldsymbol{\varphi}_{l,p})$, the last relation holds because the measure $\boldsymbol{\rho} \restriction \boldsymbol{\varphi}_{l,p}$ is regular and proper. Passing to the limit as $p \to \infty$, we find by (2.29) that $\boldsymbol{\rho}(\boldsymbol{\alpha} \cap \boldsymbol{\beta}_l) = \boldsymbol{\rho}(\boldsymbol{\beta}_l)$.

VI. We define the mapping $\mathbf{C}^X \supseteq \boldsymbol{\gamma} \ni \boldsymbol{\lambda}(\cdot) \mapsto (\operatorname{reg}\boldsymbol{\lambda}(\cdot))(\cdot) \in \mathbb{C}^X$. Let $(y_n)_1^\infty$ be the sequence constructed in II, and let $\boldsymbol{\lambda}(\cdot) \in \boldsymbol{\gamma}$ and $x \in X$. If $\boldsymbol{\lambda}(x) \neq \infty$, then define $(\operatorname{reg}\boldsymbol{\lambda}(\cdot))(x) = \boldsymbol{\lambda}(x)$; if $\boldsymbol{\lambda}(x) = \infty$, then $(\operatorname{reg}\boldsymbol{\lambda}(\cdot))(x) = \lim_{k\to\infty} \boldsymbol{\lambda}(y_{n_k(x)})$ by definition.

Let us prove that $\boldsymbol{\lambda}(y_{n_k(x)}) \neq \infty$ $(k = 1, 2, \dots)$ and that the indicated limit exists and is finite in the second case. Indeed, the inequality follows from the inclusion $\boldsymbol{\gamma} \subseteq \boldsymbol{\beta} \subseteq \boldsymbol{\beta}_{y_{n_k(x)}}$ $(k = 1, 2, \dots)$. Further, $\boldsymbol{\lambda}(\cdot) \in \boldsymbol{\psi}_{l,p} \subseteq \boldsymbol{\beta}_l$ for some l and p by (2.30); therefore, (2.28) holds.

For $u = v = e_{j_l}$ and $x = y_{n_k(x)}$ we get from the first equality in (2.26) that

$$\begin{aligned}\boldsymbol{\lambda}(y_{n_k(x)}) &= (\mathbf{P}(\boldsymbol{\lambda}(\cdot))e_{j_l}, A'_{y_{n_k(x)}} e_{j_l})_{H_0} (\mathbf{P}(\boldsymbol{\lambda}(\cdot))e_{j_l}, e_{j_l})_{H_0}^{-1} \\ &\underset{k\to\infty}{\longrightarrow} (\mathbf{P}(\boldsymbol{\lambda}(\cdot))e_{j_l}, A'_x e_{j_l})_{H_0} (\mathbf{P}(\boldsymbol{\lambda}(\cdot))e_{j_l}, e_{j_l})_{H_0}^{-1}. \qquad (2.31)\end{aligned}$$

VII. It must be verified that $\mathfrak{R}(\mathbf{P}(\boldsymbol{\lambda}(\cdot)))$ $(\boldsymbol{\lambda}(\cdot) \in \boldsymbol{\gamma})$ consists of generalized joint eigenvectors of A corresponding to the eigenvalue $(\operatorname{reg}\boldsymbol{\lambda}(\cdot))(\cdot)$. In VII and VIII we carry out some preliminary constructions.

Let us see first of all that for each $x \in X$ and $l = 1, 2, \dots$ the restrictions $(\pi_{y_{n_k(x)}} \restriction \boldsymbol{\beta}_l)(\boldsymbol{\lambda}(\cdot))$ converge uniformly with respect to $\boldsymbol{\lambda}(\cdot) \in \boldsymbol{\beta}_l$ as $k \to \infty$ to some function $f_{x,l}(\boldsymbol{\lambda}(\cdot))$ $(\boldsymbol{\lambda}(\cdot) \in \boldsymbol{\beta}_l)$ which is continuous on $\boldsymbol{\beta}_l$ (endowed with the topology induced from $\mathbf{C}^X$) and measurable with respect to the σ-algebra $\{\boldsymbol{\alpha} \in \mathcal{C}_\sigma(\mathbf{C}^X) | \boldsymbol{\alpha} \subseteq \boldsymbol{\beta}_l\}$. (Recall that

$$\mathbf{C}^X \ni \boldsymbol{\lambda}(\cdot) \mapsto \pi_y(\boldsymbol{\lambda}(\cdot)) = \boldsymbol{\lambda}(y) \in \mathbf{C}^1 \text{ and } (\pi_{y_{n_k(x)}} \restriction \boldsymbol{\beta}_l)(\boldsymbol{\lambda}(\cdot)) = \boldsymbol{\lambda}(y_{n_k(x)}) \neq \infty$$

according to VI.)

Indeed, from the first relation in (2.26), which holds for $x = y_{n_k(x)}$ and $\boldsymbol{\lambda}(\cdot) \in \boldsymbol{\beta}_l \subseteq \boldsymbol{\beta}$, and from (2.28),

$$\begin{aligned}(\pi_{y_{n_k(x)}} \restriction \boldsymbol{\beta}_l)(\boldsymbol{\lambda}(\cdot)) &= \boldsymbol{\lambda}(y_{n_k(x)}) \\ &= (\mathbf{P}(\boldsymbol{\lambda}(\cdot))e_{j_l}, A''_{y_{n_k(x)}} e_{j_l})_{H_0} (\mathbf{P}(\boldsymbol{\lambda}(\cdot))e_{j_l}, e_{j_l})_{H_0}^{-1} \\ &\qquad (k = 1, 2, \dots). \qquad (2.32)\end{aligned}$$

We conclude with the help of (2.32) and (2.28) that the following estimate holds uniformly with respect to $\boldsymbol{\lambda}(\cdot) \in \boldsymbol{\beta}_l$:

$$\begin{aligned}&|(\pi_{y_{n_k(x)}} \restriction \boldsymbol{\beta}_l)(\boldsymbol{\lambda}(\cdot)) - (\pi_{y_{n_i(x)}} \restriction \boldsymbol{\beta}_l)(\boldsymbol{\lambda}(\cdot))| \\ &\quad = |(\mathbf{P}(\boldsymbol{\lambda}(\cdot))e_{j_l}, (A'_{y_{n_k(x)}} - A'_{y_{n_i(x)}})e_{j_l})_{H_0}| (\mathbf{P}(\boldsymbol{\lambda}(\cdot))e_{j_l}, e_{j_l})_{H_0}^{-1} \\ &\quad \le m_l \|\mathbf{P}(\boldsymbol{\lambda}(\cdot))e_{j_l}\|_{H_-} \|(A'_{y_{n_k(x)}} - A'_{y_{n_i(x)}})e_{j_l}\|_{H_+} \\ &\quad \le m_l \|(A'_{y_{n_k(x)}} - A'_{y_{n_i(x)}})e_{j_l}\|_{H_+} \qquad (k, i = 1, 2, \dots). \qquad (2.33)\end{aligned}$$

(We have used the fact that

$$\|\mathbf{P}(\boldsymbol{\lambda}(\cdot))e_{j_l}\|_{H_-} \le \|\mathbf{P}(\boldsymbol{\lambda}(\cdot))\| \, \|e_{j_l}\|_{H_+} \le |\mathbf{P}(\boldsymbol{\lambda}(\cdot))| \le \operatorname{Tr}(\mathbf{P}(\boldsymbol{\lambda}(\cdot))) = 1$$

for $\boldsymbol{\lambda}(\cdot) \in \boldsymbol{\beta}$.) The required convergence of the restrictions $\pi_{y_{n_k(x)}} \upharpoonright \boldsymbol{\beta}_l$ follows from (2.33); the properties of the limit are consequences of the properties of π_y.

VIII. We define a function $f_x(\boldsymbol{\lambda}(\cdot)), \boldsymbol{\lambda}(\cdot) \in \bigcup_1^\infty \boldsymbol{\beta}_l \supseteq \boldsymbol{\gamma}$, by setting $f_x(\boldsymbol{\lambda}(\cdot)) = f_{x,l}(\boldsymbol{\lambda}(\cdot))$ for $\boldsymbol{\lambda}(\cdot) \in \boldsymbol{\beta}_l$ $(x \in X)$. We claim that

$$(f_x \upharpoonright \boldsymbol{\gamma})(\boldsymbol{\lambda}(\cdot)) = (\operatorname{reg} \boldsymbol{\lambda}(\cdot))(x) \qquad (\boldsymbol{\lambda}(\cdot) \in \boldsymbol{\gamma}). \tag{2.34}$$

Indeed, fix an $x \in X$ and a $\boldsymbol{\lambda}_0(\cdot) \in \boldsymbol{\gamma}$. If $\boldsymbol{\lambda}_0(x) = \infty$, then, according to VI, $(\operatorname{reg} \boldsymbol{\lambda}_0(\cdot))(x)$ is defined by (2.31), and $(f_x \upharpoonright \boldsymbol{\gamma})(\boldsymbol{\lambda}(\cdot))$ is always constructed by passing to the limit in (2.32), i.e., in exactly the same way. Consequently, (2.34) is now valid.

Assume that $\boldsymbol{\lambda}_0(x) \neq \infty$; according to VI, $(\operatorname{reg} \boldsymbol{\lambda}_0(\cdot))(x) = \boldsymbol{\lambda}_0(x)$. By (2.30), there exist an l and a p such that $\boldsymbol{\lambda}_0(\cdot) \in \boldsymbol{\psi}_{l,p}$. Let us consider the corresponding $\boldsymbol{\varphi}_{l,p} \supseteq \boldsymbol{\psi}_{l,p}$ and $\boldsymbol{\rho} \upharpoonright \boldsymbol{\varphi}_{l,p}$. In view of (1.25) for the operators A_y, written with the RI $\mathbf{E}$, along with (2.25) and (2.28), we get that

$$\begin{aligned}
&\|(A'_{y_{n_k(x)}} - A'_x)e_{j_l}\|^2_{H_+} \\
&\quad \geq \|(A^*_{y_{n_k(x)}} - A^*_x)e_{j_l}\|^2_{H_0} \\
&\quad = \int_{\mathbf{C}^X \setminus \boldsymbol{\delta}} |\boldsymbol{\lambda}(y_{n_k(x)}) - \boldsymbol{\lambda}(x)|^2 \, d(\mathbf{E}(\boldsymbol{\lambda}(\cdot))e_{j_l}, e_{j_l})_{H_0} \\
&\quad = \int_{\mathbf{C}^X \setminus \boldsymbol{\delta}} |\pi_{y_{n_k(x)}}(\boldsymbol{\lambda}(\cdot)) - \pi_x(\boldsymbol{\lambda}(\cdot))|^2 (\mathbf{P}(\boldsymbol{\lambda}(\cdot))e_{j_l}, e_{j_l})_{H_0} \, d\boldsymbol{\rho}(\boldsymbol{\lambda}(\cdot)) \\
&\quad \geq \frac{1}{m_l} \int_{\boldsymbol{\varphi}_{l,p} \setminus \boldsymbol{\delta}} |(\pi_{y_{n_k(x)}} \upharpoonright \boldsymbol{\varphi}_{l,p})(\boldsymbol{\lambda}(\cdot)) - (\pi_x \upharpoonright \boldsymbol{\varphi}_{l,p})(\boldsymbol{\lambda}(\cdot))|^2 \\
&\qquad\qquad \times d(\boldsymbol{\rho} \upharpoonright \boldsymbol{\varphi}_{l,p})(\boldsymbol{\lambda}(\cdot)).
\end{aligned} \tag{2.35}$$

Here $\boldsymbol{\delta} \in \mathcal{C}_\sigma(\mathbf{C}^X)$ is the union of the sets $\{\boldsymbol{\lambda}(\cdot) \in \mathbf{C}^X | \boldsymbol{\lambda}(y_{n_k(x)}) = \infty\}$ $(k = 1, 2, \ldots)$ and $\{\boldsymbol{\lambda}(\cdot) \in \mathbf{C}^X | \boldsymbol{\lambda}(x) = \infty\}$, and $\boldsymbol{\rho}(\boldsymbol{\delta}) = 0$.

We pass to the limit as $k \to \infty$ in (2.35). According to II and VII, the left-hand side of this relation tends to zero, and $\pi_{y_{n_k(x)}} \upharpoonright \boldsymbol{\beta}_l$ tends uniformly to $f_{x,l} = f_x \upharpoonright \boldsymbol{\beta}_l$. Therefore, passage to the limit gives us that

$$0 = \int_{\boldsymbol{\varphi}_{l,p} \setminus \boldsymbol{\delta}} |(f_x \upharpoonright \boldsymbol{\varphi}_{l,p})(\boldsymbol{\lambda}(\cdot)) - (\pi_x \upharpoonright \boldsymbol{\varphi}_{l,p})(\boldsymbol{\lambda}(\cdot))|^2 \, d(\boldsymbol{\rho} \upharpoonright \boldsymbol{\varphi}_{l,p})(\boldsymbol{\lambda}(\cdot)). \tag{2.36}$$

The integrand in (2.36), which we denote by $g(\boldsymbol{\lambda}(\cdot))$, is defined and continuous on $\boldsymbol{\beta}_l$; its restriction to $\boldsymbol{\varphi}_{l,p} \setminus \boldsymbol{\delta}$ is continuous in the topology induced by $\mathbf{C}^X$. Then $\boldsymbol{\lambda}_0(\cdot)$ is in $\boldsymbol{\varphi}_{l,p} \setminus \boldsymbol{\delta}$, because $\boldsymbol{\lambda}_0(x) \neq \infty$. The relation (2.34) now means that $g(\boldsymbol{\lambda}_0(\cdot)) = 0$. Assume that this equality does not hold; then there exist an $\varepsilon > 0$ and a neighborhood of $\boldsymbol{\lambda}_0(\cdot)$ in $\boldsymbol{\varphi}_{l,p} \setminus \boldsymbol{\delta}$, i.e., a set of the form $\mathbf{u} \setminus \boldsymbol{\delta}$, where $\mathbf{u}$ is a neighborhood in $\boldsymbol{\varphi}_{l,p}$, such that $g(\boldsymbol{\lambda}(\cdot)) \geq \varepsilon$ for $\boldsymbol{\lambda}(\cdot) \in \mathbf{u} \setminus \boldsymbol{\delta}$.

Since $\mathbf{u} \cap \operatorname{supp}(\boldsymbol{\rho} \restriction \varphi_{l,p}) \neq \varnothing$ (this intersection contains $\boldsymbol{\lambda}_0(\cdot)$), it follows that $(\boldsymbol{\rho} \restriction \varphi_{l,p})(\mathbf{u}) > 0$. Therefore,

$$\int_{\mathbf{u}\setminus\delta} g(\boldsymbol{\lambda}(\cdot))\, d(\boldsymbol{\rho} \restriction \varphi_{l,p})(\boldsymbol{\lambda}(\cdot)) \geq \varepsilon(\boldsymbol{\rho} \restriction \varphi_{l,p})(\mathbf{u}\setminus\delta) = \varepsilon(\boldsymbol{\rho} \restriction \varphi_{l,p})(\mathbf{u}) > 0,$$

which is absurd, because the integral in (2.36) is bounded below by the last integral.

The relation (2.34) is now proved. In passing we have established that $(\operatorname{reg} \boldsymbol{\lambda}(\cdot))(x)$ can also be defined by the limit $\lim_{k\to\infty} \boldsymbol{\lambda}(y_{n_k(x)})$ in the case when $\boldsymbol{\lambda}(x) \neq \infty$: this is the way the function f_x was defined. The definition of reg given in the formulation of the theorem now becomes clear.

IX. Our goal is to prove the relations

$$\begin{aligned} (\mathbf{P}(\boldsymbol{\lambda}(\cdot))u, A_x^* v)_{H_0} &= (\operatorname{reg} \boldsymbol{\lambda}(\cdot))(x)(\mathbf{P}(\boldsymbol{\lambda}(\cdot))u, v)_{H_0}, \\ (\mathbf{P}(\boldsymbol{\lambda}(\cdot))u, A_x v)_{H_0} &= \overline{(\operatorname{reg} \boldsymbol{\lambda}(\cdot))(x)}(\mathbf{P}(\boldsymbol{\lambda}(\cdot))u, v)_{H_0} \\ &\qquad (\boldsymbol{\lambda}(\cdot) \in \gamma, x \in X; u \in H_+,\ v \in D). \qquad (2.37) \end{aligned}$$

It follows from Remark 1 in §2.6, which is also valid for (2.37), that it suffices to verify them for $u, v \in L$.

The proof of the first of them is easy. Indeed, write the first of the equalities in (2.26) for x of the form $y_{n_k(x)}$, $\boldsymbol{\lambda}(\cdot) \in \gamma$, and $u, v \in L$ (this is possible by III):

$$(\mathbf{P}(\boldsymbol{\lambda}(\cdot))u, A'_{y_{n_k(x)}} v)_{H_0} = (\pi_{y_{n_k(x)}} \restriction \gamma)(\boldsymbol{\lambda}(\cdot))(\mathbf{P}(\boldsymbol{\lambda}(\cdot))u, v)_{H_0}. \qquad (2.38)$$

Let us pass to the limit here as $k \to \infty$. The left-hand side tends to

$$(\mathbf{P}(\boldsymbol{\lambda}(\cdot))u, A'_x v)_{H_0} = (\mathbf{P}(\boldsymbol{\lambda}(\cdot))u, A_x^* v)_{H_0},$$

due to the choice of $y_{n_k(x)}$. The factor $(\pi_{y_{n_k(x)}} \restriction \gamma)(\boldsymbol{\lambda}(\cdot))$ tends to $(f_x \restriction \gamma)(\boldsymbol{\lambda}(\cdot))$ by VII, because $\boldsymbol{\lambda}(\cdot) \in \boldsymbol{\psi}_{l,p} \subseteq \boldsymbol{\beta}_l$ for some $l, p = 1, 2, \ldots$ whenever $\boldsymbol{\lambda}(\cdot) \in \gamma$. But $(f_x \restriction \gamma)(\boldsymbol{\lambda}(\cdot)) = (\operatorname{reg} \boldsymbol{\lambda}(\cdot))(x)$ by (2.34). Thus, we have arrived at the first equality in (2.37).

To prove the second equality we write (by III) the second of the equalities in (2.26) for $x = y_{n_k(x)}$, $\boldsymbol{\lambda}(\cdot) \in \gamma$, and $u, v \in L$. The result is a relation analogous to (2.38) in which $A'_{y_{n_k(x)}}$ is replaced by $A''_{y_{n_k(x)}}$ and $(\pi_{y_{n_k(x)}} \restriction \gamma)(\boldsymbol{\lambda}(\cdot))$ appears with complex overbar. Passing to the limit as $k \to \infty$ and using the relation (see II) $\|A''_{y_{n_k(x)}} \varphi - A''_x \varphi\|_{H_+} \to 0$ ($\varphi \in L$), we get what is required.

Finally, the equality $\operatorname{Tr}(\mathbf{P}(\boldsymbol{\lambda}(\cdot))) = 1$ ($\boldsymbol{\lambda}(\cdot) \in \gamma$) follows from the inclusion $\gamma \subseteq \boldsymbol{\beta}$ and III. ∎

PROOF OF THE THEOREM. Let $\gamma = \operatorname{reg} \gamma \subseteq \mathbb{C}^X$, where reg and γ are taken from Lemma 2.5; we make certain that the representations (2.24) are valid in I–III, with π replaced by γ. (We need only get the formulas for A_x and $\mathfrak{D}(A_x)$; the formula for A_x^* is obtained automatically.)

I. Write (1.25) with the RI $\mathbf{E}$ (see Remark 1 in §1.11), and then modify $\mathbf{E}$ with respect to γ. For the resulting RI $\mathbf{E}_\gamma$ on the σ-algebra $\mathcal{C}_\sigma(\gamma) = \{\boldsymbol{\alpha}' = \boldsymbol{\alpha} \cap \gamma | \boldsymbol{\alpha} \in \mathcal{C}_\sigma(\mathbf{C}^X)\}$ we get

$$A_x = \int_\gamma \boldsymbol{\lambda}(x)\, d\mathbf{E}_\gamma(\boldsymbol{\lambda}(\cdot)),$$

$$\mathfrak{D}(A_x) = \{f \in H_0 | \int_\gamma |\boldsymbol{\lambda}(x)|^2\, d(\mathbf{E}_\gamma(\boldsymbol{\lambda}(\cdot))f, f)_{H_0} < \infty\} \qquad (x \in X). \quad (2.39)$$

We transform the measure $\mathbf{E}_\gamma$ by means of the mapping $\mathbf{C}^X \supseteq \gamma \ni \boldsymbol{\lambda}(\cdot) \mapsto (\operatorname{reg} \boldsymbol{\lambda}(\cdot))(\cdot) \in \gamma \subseteq \mathbb{C}^X$ to a measure E' (which is an RI) on the space γ; E' is defined on the σ-algebra $\mathfrak{R}' = \{\boldsymbol{\alpha} \subseteq \gamma | \operatorname{reg}^{-1} \boldsymbol{\alpha} \in \mathcal{C}_\sigma(\gamma)\}$ by the equality $E'(\boldsymbol{\alpha}) = \mathbf{E}_\gamma(\operatorname{reg}^{-1} \boldsymbol{\alpha})$.

II. To express the representation (2.39) in terms of E', note that for each particular $x \in X$ the function $\gamma \ni \lambda(\cdot) \mapsto \lambda(x) \in \mathbb{C}^1$ is $\mathfrak{R}'$-measurable. Indeed, by the construction of $\mathfrak{R}'$ it is necessary to show that the function $\gamma \ni \boldsymbol{\lambda}(\cdot) \mapsto (\operatorname{reg} \boldsymbol{\lambda}(\cdot))(x) \in \mathbb{C}^1$ is $\mathcal{C}_\sigma(\gamma)$-measurable. According to the definition of reg, there exists a sequence $(y_{n_k(x)})_1^\infty$ such that if $\boldsymbol{\lambda}(\cdot) \in \gamma$, then $\boldsymbol{\lambda}(y_{n_k(x)}) \neq \infty$ and $\boldsymbol{\lambda}(y_{n_k(x)}) \to (\operatorname{reg} \boldsymbol{\lambda}(\cdot))(x)$ as $k \to \infty$, i.e.,

$$(\pi_{y_{n_k(x)}} \restriction \gamma)(\boldsymbol{\lambda}(\cdot)) \to (\operatorname{reg} \boldsymbol{\lambda}(\cdot))(x) \qquad (\boldsymbol{\lambda}(\cdot) \in \gamma).$$

The $\mathcal{C}_\sigma(\gamma)$-measurability of the functions before taking the limit, hence that of the limit function, follows from the $\mathcal{C}_\sigma(\mathbf{C}^X)$-measurability of the function of $\boldsymbol{\lambda}(\cdot) \in \mathbf{C}^X$ equal to $\pi_{y_{n_k(x)}}(\boldsymbol{\lambda}(\cdot))$ if $\boldsymbol{\lambda}(y_{n_k(x)}) \neq \infty$ and equal to 0 otherwise.

Thus, the integral $\int_\gamma \lambda(x)\, dE'(\lambda(\cdot))$ makes sense, and this integral is equal to $\int_\gamma (\operatorname{reg} \boldsymbol{\lambda}(\cdot))(x)\, d\mathbf{E}_\gamma(\boldsymbol{\lambda}(\cdot))$, by the rules for transforming integrals under mappings of measures. By definition,

$$\operatorname{reg}\{\boldsymbol{\lambda}(\cdot) \in \gamma | (\operatorname{reg} \boldsymbol{\lambda}(\cdot))(x) \neq \boldsymbol{\lambda}(x)\} \subseteq \{\boldsymbol{\lambda}(\cdot) \in \mathbf{C}^X | \boldsymbol{\lambda}(x) = \infty\},$$

and, as was just proved, the first of these sets is in $\mathcal{C}_\sigma(\gamma)$. The $\mathbf{E}$-measure of the second set is equal to zero, but then the $\mathbf{E}_\gamma$-measure of the first is also zero. Therefore, the last integral is equal to $\int_\gamma \boldsymbol{\lambda}(x)\, d\mathbf{E}_\gamma(\boldsymbol{\lambda}(\cdot)) = A_x$ (see (2.39)). The second integral in (2.39) is transformed similarly. Accordingly,

$$A_x = \int_\gamma \lambda(x)\, dE'(\lambda(\cdot)),$$

$$\mathfrak{D}(A_x) = \left\{f \in H_0 | \int_\gamma |\lambda(x)|^2\, d(E'(\lambda(\cdot))f, f)_{H_0} < \infty\right\} \qquad (x \in X). \quad (2.40)$$

III. Let us show that (2.40) can be rewritten in the form (2.24) with $\pi = \gamma$. To do this, note that $\mathcal{C}_\sigma(\gamma) \subseteq \mathfrak{R}'$. Indeed, it suffices to establish that $\alpha' = \mathfrak{C}(x_1, \ldots, x_p; \kappa) \cap \gamma \in \mathfrak{R}'$ $(x_1, \ldots, x_p \in X; \kappa \in \mathcal{B}(\mathbb{C}^p); p = 1, 2, \ldots)$. But $\alpha' = \{\lambda(\cdot) \in \gamma | (\lambda(x_1), \ldots, \lambda(x_p)) \in \kappa\}$, i.e., it is the full inverse image of κ under the mapping $\gamma \ni \lambda(\cdot) \mapsto (\lambda(x_1), \ldots, \lambda(x_p)) \in \mathbb{C}^p$. Such a mapping is

$\mathfrak{R}'$-measurable, because each of the coordinates $\gamma \ni \lambda(\cdot) \mapsto \lambda(x_n) \in \mathbb{C}^1$ is measurable in this sense, according to II. Consequently, $\alpha' \in \mathfrak{R}'$, which is what was required.

Consider the restriction E'' of the RI E' to $\mathcal{C}_\sigma(\gamma)$. Since the function $\mathbb{C}^X \supseteq \gamma \ni \lambda(\cdot) \mapsto \lambda(x) \in \mathbb{C}^1$ is measurable with respect to this σ-algebra, E' can be replaced by E'' in the integrals in (2.40) with preservation of these formulas. It is easy to see that E'' is equal to the modification of the original RI E with respect to γ.

Thus, fix an $x \in X$. The mapping $\mathbb{C}^X \supseteq \gamma \ni \lambda(\cdot) \mapsto \pi_x(\lambda(\cdot)) = \lambda(x) \in \mathbb{C}^1$ is $\mathcal{C}_\sigma(\gamma)$-measurable and carries E'' into the RI

$$\mathcal{B}(\mathbb{C}^1) \ni \kappa \mapsto E''_x(\kappa) = E''(\pi_x^{-1}(\kappa)) = E''(\mathfrak{C}(x;\kappa) \cap \gamma),$$

and the first integral in (2.40) (with E' replaced by E'') passes into the integral $\int_{\mathbb{C}^1} \lambda \, dE''_x(\lambda)$. This implies that E''_x is the usual RI E_x of the operator A_x, i.e., $E''(\mathfrak{C}(x;\kappa) \cap \gamma) = E(\mathfrak{C}(x;\kappa))$ $(\kappa \in \mathcal{B}(\mathbb{C}^1))$. We conclude from this equality in the standard way that

$$E''(\mathfrak{C}(x_1, \dots, x_p; \kappa) \cap \gamma) = E(\mathfrak{C}(x_1, \dots, x_p; \kappa)) \qquad (x_1, \dots, x_p \in X; \kappa \in \mathcal{B}(\mathbb{C}^p);\ p = 1, 2, \dots),$$

hence also $E''(\alpha \cap \gamma) = E(\alpha)$ $(\alpha \in \mathcal{C}_\sigma(\mathbb{C}^X))$. (First take κ to be a rectangle, and use the orthogonality of the RI.) If $\mathcal{C}_\sigma(\mathbb{C}^X) \ni \alpha \supseteq \gamma$ in the last equality, then we get that $E(\alpha) = E''(\gamma) = 1$, i.e., γ is of full outer E-measure. The equality itself means that E'' is the modification of E with respect to γ: $E'' = E_\gamma$ (see (1.21)).

Thus, the representation (2.40) is valid with E' replaced by E_γ, i.e.,

$$A_x = \int_\gamma \lambda(x) \, dE_\gamma(\lambda(\cdot)),$$
$$\mathfrak{D}(A_x) = \left\{ f \in H_0 \middle| \int_\gamma |\lambda(x)|^2 \, d(E_\gamma(\lambda(\cdot))f, f)_{H_0} < \infty \right\} \qquad (x \in X). \quad (2.41)$$

IV. Let $\pi \in \mathcal{C}_\sigma(\gamma)$ be a set such that $E_\gamma(\pi) = 1$. Then it is clear that π is of full outer E-measure; analogous representations follow from (2.41), with γ replaced by π, i.e., (2.24). The inclusion $\pi \subseteq g(A)$ is valid: if $\lambda(\cdot) \in \pi \subseteq \gamma$, then $\lambda(\cdot) = (\operatorname{reg} \boldsymbol{\lambda}(\cdot))(\cdot)$ for some $\boldsymbol{\lambda}(\cdot) \in \boldsymbol{\gamma}$, and thus (by Lemma 2.5) $\lambda(\cdot)$ is an eigenvalue for any generalized joint eigenvector $\mathbf{P}(\boldsymbol{\lambda}(\cdot))u$ $(u \in H_+)$.

V. We prove in VI–VIII below that there exists a $\pi \in \mathcal{C}_\sigma(\gamma)$ with $E_\gamma(\pi) = 1$ such that if $\lambda(\cdot) \in \pi$, then $\mathfrak{R}(P(\lambda(\cdot)))$ consists of generalized joint eigenvectors of A corresponding to $\lambda(\cdot)$, and $\operatorname{Tr}(P(\lambda(\cdot))) = 1$. In view of IV, this proves the theorem. (Arguments are necessary only for its last assertion.)

VI. With the help of (1.21) and (1.19) we get that

$$O^+E_\gamma(\alpha')O = O^+E(\alpha)O = \int_\alpha P(\lambda(\cdot))\,d\rho(\lambda(\cdot))$$
$$= \int_{\alpha'} (P \restriction \gamma)(\lambda(\cdot))\,d\rho_\gamma(\lambda(\cdot))$$
$$(\mathcal{C}_\sigma(\gamma) \ni \alpha' = \alpha \cap \gamma,\ \alpha \in \mathcal{C}_\sigma(\mathbb{C}^X)). \quad (2.42)$$

Since $\mathcal{C}_\sigma(\gamma) \subseteq \mathfrak{R}'$ and the measure E_γ is $E' \restriction \mathcal{C}_\sigma(\gamma)$, the inverse image $\hat{\mathbf{E}}$ of E_γ under the mapping $\mathbf{C}^X \supseteq \boldsymbol{\gamma} \ni \boldsymbol{\lambda}(\cdot) \mapsto (\operatorname{reg}\boldsymbol{\lambda}(\cdot))(\cdot) \in \gamma \subseteq \mathbb{C}^X$ satisfies $\hat{\mathbf{E}} = \mathbf{E}_\gamma \restriction \hat{\mathfrak{R}}$, where $\hat{\mathfrak{R}}$ is the σ-algebra $\{\hat{\boldsymbol{\alpha}} = \operatorname{reg}^{-1}\alpha' | \alpha' \in \mathcal{C}_\sigma(\gamma)\} \subseteq \mathcal{C}_\sigma(\boldsymbol{\gamma})$. Passing to these inverse images in (2.42), we find that

$$O^+\hat{\mathbf{E}}(\hat{\boldsymbol{\alpha}})O = \int_{\hat{\boldsymbol{\alpha}}} (\mathbf{P} \restriction \boldsymbol{\gamma})((\operatorname{reg}\boldsymbol{\lambda}(\cdot))(\cdot))\,d\hat{\boldsymbol{\rho}}(\boldsymbol{\lambda}(\cdot)) \qquad (\hat{\boldsymbol{\alpha}} \in \hat{\mathfrak{R}}) \qquad (2.43)$$

Here $\hat{\boldsymbol{\rho}}$ is the inverse image of ρ_γ. At the same time, it is easy to see that $\hat{\boldsymbol{\rho}} = \boldsymbol{\rho}_\gamma \restriction \hat{\mathfrak{R}}$, where $\boldsymbol{\rho}_\gamma$ is the modification of $\boldsymbol{\rho}$ with respect to $\boldsymbol{\gamma}$, and thus $\hat{\boldsymbol{\rho}}(\hat{\boldsymbol{\alpha}}) = \operatorname{Tr}(O^+\hat{\mathbf{E}}(\hat{\boldsymbol{\alpha}})O)$ ($\hat{\boldsymbol{\alpha}} \in \hat{\mathfrak{R}}$). From Lemma 2.1, Theorem 2.1, and (2.43) it follows that the derivative $(d(O^+\hat{\mathbf{E}}O)/d\hat{\boldsymbol{\rho}})(\boldsymbol{\lambda}(\cdot)) = \hat{\mathbf{P}}(\boldsymbol{\lambda}(\cdot))$ exists, it is uniquely determined for $\hat{\boldsymbol{\rho}}$-almost all $\boldsymbol{\lambda}(\cdot) \in \boldsymbol{\gamma}$, and it coincides with $(P \restriction \gamma)((\operatorname{reg}\boldsymbol{\lambda}(\cdot))(\cdot))$.

In VII and VIII it will be proved that there exists a $\hat{\boldsymbol{\pi}} \in \hat{\mathfrak{R}}$ with $\hat{\mathbf{E}}(\hat{\boldsymbol{\pi}}) = 1$ such that if $\boldsymbol{\lambda}(\cdot) \in \hat{\boldsymbol{\pi}}$, then $\mathfrak{R}(\hat{\mathbf{P}}(\boldsymbol{\lambda}(\cdot)))$ consists of generalized joint eigenvectors of A corresponding to $(\operatorname{reg}\boldsymbol{\lambda}(\cdot))(\cdot)$, and $\operatorname{Tr}(\hat{\mathbf{P}}(\boldsymbol{\lambda}(\cdot))) = 1$. Let $\pi = \operatorname{reg}\hat{\pi} \in \mathcal{C}_\sigma(\gamma)$; then $E_\gamma(\pi) = \hat{\mathbf{E}}(\hat{\boldsymbol{\pi}}) = 1$. Since $(P \restriction \gamma)((\operatorname{reg}\boldsymbol{\lambda}(\cdot))(\cdot))$ can be chosen to coincide with $\hat{\mathbf{P}}(\boldsymbol{\lambda}(\cdot))$, this implies the assertion in V.

VII. The assertion we have formulated calls to mind Lemma 2.5, but differs from it in that the RI being differentiated is defined not on the σ-algebra $\mathcal{C}_\sigma(\mathbf{C}^X)$ or $\mathcal{C}_\sigma(\boldsymbol{\gamma})$, but on the σ-subalgebra $\hat{\mathfrak{R}} \subseteq \mathcal{C}_\sigma(\boldsymbol{\gamma})$. Therefore, the corresponding derivative $\hat{\mathbf{P}}(\boldsymbol{\lambda}(\cdot))$ is equal in a certain sense to the "conditional mathematical expectation" of $\mathbf{P}(\boldsymbol{\lambda}(\cdot))$.

Passing in the integrals (2.41) to the inverse image of E_γ, we get that

$$A_x = \int_{\boldsymbol{\gamma}} (\operatorname{reg}\boldsymbol{\lambda}(\cdot))(x)\,d\hat{\mathbf{E}}(\boldsymbol{\lambda}(\cdot)),$$
$$\mathfrak{D}(A_x) = \left\{ f \in H_0 \middle| \int_{\boldsymbol{\gamma}} |(\operatorname{reg}\boldsymbol{\lambda}(\cdot))(x)|^2\,d(\hat{\mathbf{E}}(\boldsymbol{\lambda}(\cdot))f, f)_{H_0} < \infty \right\}$$
$$(x \in X). \qquad (2.44)$$

This can be used in a way similar to that for Lemma 2.2 to prove that for each $x \in X$ there exists a set $\hat{\boldsymbol{\beta}}_x \in \hat{\mathfrak{R}}$ of full $\hat{\boldsymbol{\rho}}$-measure such that $\operatorname{Tr}(\hat{\mathbf{P}}(\boldsymbol{\lambda}(\cdot))) = 1$

for each $\boldsymbol{\lambda}(\cdot) \in \hat{\boldsymbol{\beta}}_x$, and

$$(\hat{\mathbf{P}}(\boldsymbol{\lambda}(\cdot))u, A_x^* v)_{H_0} = (\operatorname{reg} \boldsymbol{\lambda}(\cdot))(x)(\hat{\mathbf{P}}(\boldsymbol{\lambda}(\cdot))u, v)_{H_0},$$
$$(\hat{\mathbf{P}}(\boldsymbol{\lambda}(\cdot))u, A_x v)_{H_0} = \overline{(\operatorname{reg} \boldsymbol{\lambda}(\cdot))(x)}(\hat{\mathbf{P}}(\boldsymbol{\lambda}(\cdot))u, v)_{H_0}$$
$$(u \in H_+, \ v \in D). \qquad (2.45)$$

Indeed, on the basis of the equalities

$$O^+\hat{\mathbf{E}}(\hat{\boldsymbol{\alpha}})O = \int_{\hat{\boldsymbol{\alpha}}} \hat{\mathbf{P}}(\boldsymbol{\lambda}(\cdot))\, d\hat{\boldsymbol{\rho}}(\boldsymbol{\lambda}(\cdot))$$

and

$$A_x\hat{\mathbf{E}}(\hat{\boldsymbol{\alpha}}) = \int_{\hat{\boldsymbol{\alpha}}} (\operatorname{reg} \boldsymbol{\lambda}(\cdot))(x)\, d\hat{\mathbf{E}}(\boldsymbol{\lambda}(\cdot)) \qquad (\hat{\boldsymbol{\alpha}} \in \hat{\mathfrak{R}};\ x \in X)$$

(the latter is a consequence of (2.44)), we conclude as in the case of (2.23) that

$$\int_{\hat{\boldsymbol{\alpha}}} (\hat{\mathbf{P}}(\boldsymbol{\lambda}(\cdot))u, A_x^* v)_{H_0}\, d\hat{\boldsymbol{\rho}}(\boldsymbol{\lambda}(\cdot)) = \int_{\hat{\boldsymbol{\alpha}}} (\operatorname{reg} \boldsymbol{\lambda}(\cdot))(x)(\hat{\mathbf{P}}(\boldsymbol{\lambda}(\cdot))u, v)_{H_0}\, d\hat{\boldsymbol{\rho}}(\boldsymbol{\lambda}(\cdot))$$
$$(\hat{\boldsymbol{\alpha}} \in \hat{\mathfrak{R}};\ u \in H_+,\ v \in D).$$

This and the analogous formula for A_x^* give us (2.45) as in Lemma 2.2.

VIII. Let $\hat{\boldsymbol{\pi}} = \bigcap_1^\infty \hat{\boldsymbol{\beta}}_{y_n} \in \hat{\mathfrak{R}}$, where the y_n are the points in Lemma 2.5. It is clear that $\hat{\mathbf{E}}(\hat{\boldsymbol{\pi}}) = 1$ and $\operatorname{Tr}(\hat{\mathbf{P}}(\boldsymbol{\lambda}(\cdot))) = 1$ $(\boldsymbol{\lambda}(\cdot) \in \hat{\boldsymbol{\pi}})$. The equalities (2.45) are valid for $x = y_n$, and $(\operatorname{reg} \boldsymbol{\lambda}(\cdot))(y_n) = \boldsymbol{\lambda}(y_n) = (\pi_{y_n} \restriction \gamma)(\boldsymbol{\lambda}(\cdot))$ $(n = 1, 2, \ldots)$. We fix $x \in X$ and write (2.45) for $\boldsymbol{\lambda}(\cdot) \in \hat{\boldsymbol{\pi}}$ and $y_n = y_{n_k(x)}$, where $(y_{n_k(x)})_1^\infty$ was introduced in II of the proof of Lemma 2.5.

According to VII–IX in that proof, for each $\boldsymbol{\lambda}(\cdot)$

$$(\pi_{y_{n_k(x)}} \restriction \gamma)(\boldsymbol{\lambda}(\cdot)) \to (\operatorname{reg} \boldsymbol{\lambda}(\cdot))(x),$$
$$(\hat{\mathbf{P}}(\boldsymbol{\lambda}(\cdot))u, A_{y_{n_k(x)}}^* v)_{H_0} \to (\hat{\mathbf{P}}(\boldsymbol{\lambda}(\cdot))u, A_x^* v)_{H_0},$$
$$(\hat{\mathbf{P}}(\boldsymbol{\lambda}(\cdot))u, A_{y_{n_k(x)}} v)_{H_0} \to (\hat{\mathbf{P}}(\boldsymbol{\lambda}(\cdot))u, A_x v)_{H_0} \qquad (u, v \in L)$$

as $k \to \infty$. (The last two relations follow from the fact that $A'_{y_{n_k(x)}}\varphi \to A'_x\varphi$ and $A''_{y_{n_k(x)}}\varphi \to A''_x\varphi$ $(\varphi \in L)$ in H_+ and from the estimate $\left|\hat{\mathbf{P}}(\boldsymbol{\lambda}(\cdot))\right| \le 1$. But $\hat{\boldsymbol{\pi}} \subseteq \gamma$: therefore, the last three limit relations enable us to pass to the limit as $k \to \infty$ in (2.45) when $u, v \in L$. As a result we get (2.45) with arbitrary $\boldsymbol{\lambda}(\cdot) \in \hat{\boldsymbol{\pi}}$, $x \in X$, and $u, v \in L$, and hence also with arbitrary $u \in H_+$ and $v \in D$, by Remark I in subsection 6.

Accordingly, we have proved the fact formulated at the end of VI. By V, the theorem is proved except for its last assertion.

IX. The last assertion of the theorem means that

$$O^+E_\pi(\alpha')O = \int_{\alpha'} (P \restriction \pi)(\lambda(\cdot))\, d\rho_\pi(\lambda(\cdot))$$
$$(\mathcal{C}_\sigma(\pi) \ni \alpha' = \alpha \cap \pi,\ \alpha \in \mathcal{C}_\sigma(\mathbb{C}^X)). \quad (2.46)$$

To derive (2.46), write $(O^+E_\pi(\alpha')Ou, v)_{H_0} = (O^+E(\alpha)Ou, v)_{H_0}$ $(u, v \in H_+)$ according to (2.21) as an integral with respect to $d\rho(\lambda(\cdot))$ and apply (1.19) to it, with

$$R = \mathbb{C}^X, \quad R' = \pi \quad \text{and} \quad F(\lambda) = \kappa_\alpha(\lambda(\cdot))(P(\lambda(\cdot))u, v)_{H_0}. \quad \blacksquare$$

PROOF OF LEMMA 2.4. Denote by $(\beta_m)_1^\infty$ a fixed dense subset of Φ', and by $(e_j)_1^\infty$ an orthonormal basis in H. Consider the collection B of operators from Φ to H of the form

$$\Phi \ni \varphi \mapsto B\varphi = \sum_{j=1}^{n} \overline{\beta_{m_j}(\varphi)} e_j \in H,$$

where the m_j are distinct $(n = 1, 2, \ldots)$. Each such operator is continuous, and there are countably many of them; let $\mathrm{B} = (B_p)_{p=1}^\infty \subseteq \mathcal{L}(\Phi, H)$.

Fix $p, q = 1, 2, \ldots$ and consider the set $\{A \in \mathrm{A} \mid \|(A - B_p)\varphi_k\|_H < q^{-1}, k = 1, \ldots, q\}$. If it is nonempty, then we choose some operator $A_{p,q}$ in it. Varying $p, q = 1, 2, \ldots$ arbitrarily, we obtain an at most countable set of operators in A. Let us show that it is the required family of operators. (It will be clear from what follows that A is nonempty.)

We first show that for any $A \in \mathrm{A}$ and $q = 1, 2, \ldots$ there is an operator $B_{n_q} \in \mathrm{B}$ such that

$$\|(A - B_{n_q})\varphi_k\|_H < q^{-1} \qquad (k = 1, \ldots, q). \tag{2.47}$$

The operator A is in $\mathcal{L}(\Phi, H)$ and can thus be represented in the form

$$A\varphi = \sum_{j=1}^{\infty} \overline{\alpha_j(\varphi)} e_j \qquad (\varphi \in \Phi),$$

where $(\alpha_j)_1^\infty$ is a sequence of functionals in Φ' determined by A. Since $\|A\varphi_k\|_H^2 = \sum_{j=1}^\infty |\alpha_j(\varphi_k)|^2 < \infty$, there exists for each $q = 1, 2, \ldots$ an index $l(q) = 1, 2, \ldots$ large enough that $\sum_{j=l(q)+1}^\infty |\alpha_j(\varphi_k)|^2 < (2q^2)^{-1}$ $(k = 1, \ldots, q)$. Since $(\beta_m)_1^\infty$ is dense in Φ', for any $j = 1, \ldots, l(q)$ this sequence contains functionals $\beta_{m(j,q)}$ such that

$$|(\alpha_j - \beta_{m(j,q)})(\varphi_k)|^2 < (2q^2 l(q))^{-1} \qquad (k = 1, \ldots, q);$$

it is clear that the indices $m(1, q), \ldots, m(l(q), q)$ can be assumed to be distinct. The operator

$$\Phi \ni \varphi \mapsto B\varphi = \sum_{j=1}^{l(q)} \overline{\beta_{m(j,q)}(\varphi)} e_j \in H$$

belongs to B and thus has some index n_q, i.e., $B = B_{n_q}$. The estimate (2.47) holds for it:

$$\|(A - B_{n_q})\varphi_k\|_H^2 = \sum_{j=1}^{l(q)} |(\alpha_j - \beta_{m(j,q)})(\varphi_k)|^2 + \sum_{j=l(q)+1}^{\infty} |\alpha_j(\varphi_k)|^2 < q^{-2}$$
$$(k = 1, \ldots, q).$$

Accordingly, (2.47) is established. From the definition of the operators $A_{p,q}$ and from (2.47) it follows that for each $q = 1, 2, \ldots$ there exists an operator $A_{n_q,q}$. (These operators can coincide for different q.) We show that the sequence $(A_{n_q,q})_{q=1}^{\infty}$ approximates A in the required sense. Indeed, the definition of $A_{p,q}$ gives us that $\|(A_{n_q,q} - B_{n_q})\varphi_k\|_H < q^{-1}$. From this and (2.47) we conclude that $\|(A - A_{n_q,q})\varphi_k\|_H < 2q^{-1}$ ($k = 1, \ldots, q$; $q = 1, 2, \ldots$). Thus, $\|(A - A_{n_q,q})\varphi_j\|_H \to 0$ as $q \to \infty$ for each $j = 1, 2, \ldots$. ■

2.8. *The spectral theorem. The case of at most countably many unbounded operators.* Theorem 2.7 becomes somewhat more refined now, and its proof essentially simpler. Namely, the following result holds.

THEOREM 2.8. *Suppose that the assumptions of Theorem 2.7 hold. If at most countably many of the operators A_x in the family $A = (A_x)_{x \in X}$ are unbounded, then the set π in Theorem 2.7 can be chosen in such a way that $\mathcal{B}(\mathbb{C}^X) \ni \pi \subseteq s(A) \cap g(A)$.*

PROOF. We repeat the proof of Lemma 2.5. Steps I and II remain as before. In III we use the generalized projection $P(\lambda(\cdot))$ and the spectral measure ρ instead of the compactified objects, and Lemma 2.2 instead of Lemma 2.3. Let $\beta = \bigcap_{n=1}^{\infty} \beta_{y_n} \in \mathcal{C}_\sigma(\mathbb{C}^X)$ and define $\mathcal{B}(\mathbb{C}^X) \ni \gamma \subseteq \operatorname{supp} E = s(A)$ according to (2.30): $\gamma = \bigcup_{l,p=1}^{\infty} \psi_{l,p}$, where $\psi_{l,p} = \operatorname{supp}(\rho \restriction \varphi_{l,p})$ and the $\varphi_{l,p}$ are constructed in a way similar to that for $\boldsymbol{\varphi}_{l,p}$. (We could make one more simplification: after (2.28) let $\psi_l = \operatorname{supp}(\rho \restriction \beta_l)$ ($l = 1, 2, \ldots$) and take $\gamma = \bigcup_1^{\infty} \psi_l$; the role of $\rho \restriction \varphi_{l,p}$ is played by $\rho \restriction \beta_l$.) We take $\pi = \gamma$ and check that all the required assertions of the theorem hold.

As in IV, we conclude that $\rho \restriction \varphi_{l,p}$ is regular. It is proper on the basis of Theorem 1.6; the possibility of applying this theorem also gives us a simplification of the proof as compared with Theorem 2.7. It can be verified in a way similar to that in V that γ is of full outer ρ-measure. Further, we see as in VII that for each $x \in X$ and each $l = 1, 2, \ldots$ the restrictions $(\pi_{y_{n_k(x)}} \restriction \beta_l)(\lambda(\cdot))$ converge uniformly with respect to $\lambda(\cdot) \in \beta_l$ as $k \to \infty$ to some function $f_{x,l}(\lambda(\cdot))$ ($\lambda(\cdot) \in \beta_l$) continuous on β_l and measurable with respect to the σ-algebra $\{\alpha \in \mathcal{C}_\sigma(\mathbb{C}^X) | \alpha \subseteq \beta_l\}$. From $f_{x,l}$ we construct a function $f_x(\lambda(\cdot))$ on $\bigcup_1^{\infty} \beta_l \supseteq \gamma$ in a way similar to that in VIII, and then prove that

$$(f_x \restriction \gamma)(\lambda(\cdot)) = (\pi_x \restriction \gamma)(\lambda(\cdot)) = \lambda(x) \qquad (\lambda(\cdot) \in \gamma); \tag{2.48}$$

this last is done by repeating the arguments in VIII, with the fact that $\psi_{l,p} = \operatorname{supp}(\rho \restriction \varphi_{l,p})$ used in an essential way. As in IX, we now verify in an elementary way that the relations (2.22) hold for $x \in X$, $\lambda(\cdot) \in \gamma$, $u \in H_+$ and $v \in D$: first write them with x replaced by $y_{n_k(x)}$ and with $u, v \in L$, then let $k \to \infty$ and use (2.48) and Remark 1 in §2.6.

Thus, the first part of Theorem 2.8, which corresponds to the first part of Theorem 2.7, is established. To prove the second part of it, modify the RI E with respect to $\pi = \gamma$ and apply (1.19) to (1.25). The last part can be verified as in Theorem 2.7 (step IX in its proof). ■

Note that if all the operators A_x in the family $A = (A_x)_{x \in X}$ are bounded, then the proof of Theorem 2.8 simplifies further. Thus, as explained in §1.9, the compact space $\times_{x \in X} \operatorname{supp} E_x$ can now be considered instead of $\mathbb{C}^X$. Therefore, the elementary part 1) of Theorem 1.5 can be used in the proof presented here instead of Theorem 1.6. In this case most of the constructions in §1 are no longer needed.

2.9. *Continuity and smoothness of eigenvalues.* Under a certain simply formulated assumption, the functions $X \ni x \mapsto \lambda(x) \in \mathbb{C}^1$ in the generalized spectrum $g(A)$ are continuous or even smooth. In this case the set $\tau \supseteq g(A)$ in Remark 1 after Theorem 2.7 can automatically be chosen to be continuous (smooth) functions, and the integrals in (2.24) actually become "functional" integrals.

Thus, assume that the index set X is a topological space, and the operators A_x depend continuously on $x \in X$ in the following sense: for each $u \in D$ the vector-valued function

$$X \ni x \mapsto A_x^* u \in H_+ \tag{2.49}$$

is weakly continuous. In this case each eigenvalue $\lambda(\cdot) \in g(A)$ is a continuous function. Indeed, let $\varphi \in H_-$ be a generalized eigenvector of A with eigenvalue $\lambda(\cdot)$. Choose $u \in D$ such that $(\varphi, u)_{H_0} \neq 0$. Then $\lambda(x) = (\varphi, A_x^* u)_{H_0} (\varphi, u)_{H_0}^{-1}$ on the basis of the first equality in (2.19), and our assumption (2.49) gives us that $\lambda(\cdot)$ is continuous. ■

It can also be proved in the same way that if X is a differentiable manifold and (2.49) is k times weakly continuously differentiable, then a $\lambda(\cdot) \in g(A)$ is k times continuously differentiable.

Instead of weak continuity (differentiability) of the function (2.49), we can require the same property with A_x^* replaced by A_x. The conclusions remain valid—use the second equality in (2.19).

2.10. *Supplementary remarks.*

REMARK 1. Theorems 2.7 and 2.8 are preserved if D is a separable linear topological space such that D' is also separable. This follows from the remarks in steps I and II of the proof of Lemma 2.5.

REMARK 2. Assume that the family of commuting normal operators is connected in the standard way with a rigging (2.18) that is not quasinuclear but nuclear. The results in Theorems 2.6–2.8 and §2.9 are preserved if ρ is understood to be some spectral measure for the RI E and the corresponding definition of a generalized joint eigenvector is used (see the end of subsection 5). Indeed, the situation now reduces to the case of a rigging (2.18) in the same way as explained in the proof of Theorem 2.5.

REMARK 3. Suppose that the family $A = (A_x)_{x \in X}$ of commuting normal operators is connected in the standard way with the rigging (2.18) only in the sense that $A_x \restriction D \in \mathcal{L}(D, H_+)$ $(x \in X)$ (or $A_x^* \restriction D \in \mathcal{L}(D, H_+)$ $(x \in X)$). Then the results are preserved, except that in the definition (2.19) only the second (first) equality holds. An analogous situation arises in the case of a nuclear rigging.

REMARK 4. In the case of a family of commuting selfadjoint operators $\mathbb{C}^X$ can be replaced by $\mathbb{R}^X$ in the results of this section.

2.11. ***The Fourier transformation. The direct integral of Hilbert spaces.*** Assume conditions ensuring the validity of Theorem 2.7, and choose $\mathbb{C}^X \supseteq \tau \supseteq \pi$. By (2.24) (see Remark 1 in subsection 7),

$$(E_\tau(\alpha), u, v)_{H_0} = \int_\alpha (P(\lambda(\cdot))u, v)_{H_0}\, d\rho_\tau(\lambda(\cdot)) \qquad (u, v \in H_+;\ \alpha \in \mathcal{C}_\sigma(\tau)). \tag{2.50}$$

In particular, $E_\tau(\tau) = 1$ for $\alpha = \tau$, and (2.50) gives a decomposition of $(u,v)_{H_0}$ into an integral of $(P(\lambda(\cdot))u, v)_{H_0}$. The last expression defines an inner product, and (2.50) becomes a decomposition of H_0 into a direct integral of corresponding Hilbert spaces. Let us dwell on this in more detail.

Fix a $\lambda(\cdot) \in \tau$ such that $\mathrm{Tr}(P(\lambda(\cdot))) = 1$. Such functions $\lambda(\cdot)$ clearly form a set of full ρ_τ-measure. By the relations $P(\lambda(\cdot)) \geq 0$ and $|P(\lambda(\cdot))| \leq \mathrm{Tr}(P(\lambda(\cdot))) = 1$, the operator $\mathbf{J}P(\lambda(\cdot))J \colon H_0 \to H_0$ is a nonnegative Hilbert-Schmidt operator (see (2.1) and (2.2)). Let $\psi_\gamma(\lambda(\cdot)) \in H_0$ ($\gamma = 1, 2, \ldots, N_{\lambda(\cdot)} \leq \infty$) be an orthonormal sequence of eigenvectors of $\mathbf{J}P(\lambda(\cdot))J$ corresponding to the eigenvalues $\nu_\gamma(\lambda(\cdot)) > 0$. Since the dependence of $\mathbf{J}P(\lambda(\cdot))J$ on the parameter $\lambda(\cdot)$ is weakly $\mathcal{C}_\sigma(\tau)$-measurable, $\psi_\gamma(\lambda(\cdot))$ can (as is known) be assumed to be (weakly) measurable and $\nu_\gamma(\lambda(\cdot))$ ($\gamma = 1, 2, \ldots, N_{\lambda(\cdot)}$) to be measurable. Then

$$\begin{aligned}(P(\lambda(\cdot))Jf, Jg)_{H_0} &= (\mathbf{J}P(\lambda(\cdot))Jf, g)_{H_0} \\ &= \sum_{\gamma=1}^{N_{\lambda(\cdot)}} \nu_\gamma(\lambda(\cdot))(f, \psi_\gamma(\lambda(\cdot)))_{H_0}\overline{(g, \psi_\gamma(\lambda(\cdot)))}_{H_0} \\ &= \sum_{\gamma=1}^{N_{\lambda(\cdot)}} (Jf, \varphi_\gamma(\lambda(\cdot)))_{H_0}\overline{(Jg, \varphi_\gamma(\lambda(\cdot)))}_{H_0} \qquad (f, g \in H_0),\end{aligned} \tag{2.51}$$

where

$$\begin{aligned}\varphi_\gamma(\lambda(\cdot)) &= \sqrt{\nu_\gamma(\lambda(\cdot))}\,\mathbf{J}^{-1}\psi_\gamma(\lambda(\cdot)) \\ &= P(\lambda(\cdot))((\nu_\gamma(\lambda(\cdot)))^{-1/2}J\psi_\gamma(\lambda(\cdot))) \subseteq \mathfrak{R}(P(\lambda(\cdot))) \subseteq H_-.\end{aligned}$$

The vectors $\varphi_\gamma(\lambda(\cdot))$ ($\lambda(\cdot) \in \tau; \gamma = 1, 2, \ldots, N_{\lambda(\cdot)} \leq \infty$) are "individual" generalized joint eigenvectors of the family $A = (A_x)_{x \in X}$.

It follows from (2.51) that

$$(P(\lambda(\cdot))u, v)_{H_0} = \sum_{\gamma=1}^{N_{\lambda(\cdot)}} (u, \varphi_\gamma(\lambda(\cdot)))_{H_0} \overline{(v, \varphi_\gamma(\lambda(\cdot)))}_{H_0} \qquad (u, v \in H_+) \tag{2.52}$$

for ρ_τ-almost every $\lambda(\cdot) \in \tau$.

Let $l_2(\infty) = l_2$ and $l_2(N) = \mathbf{C}^N$ $(N < \infty)$, where the latter space is regarded as imbedded in l_2 (all the coordinates of the vector beginning with the $(N+1)$st are zero). The Fourier transformation corresponding to the family A is defined to be the mapping

$$\begin{gathered} H_+ \ni u \mapsto \tilde{u}(\lambda(\cdot)) = (\tilde{u}_1(\lambda(\cdot)), \tilde{u}_2(\lambda(\cdot)), \ldots) \in l_2(N_{\lambda(\cdot)}), \\ \tilde{u}_\gamma(\lambda(\cdot)) = (u, \varphi_\gamma(\lambda(\cdot)))_{H_0} \qquad (\gamma = 1, 2, \ldots, N_{\lambda(\cdot)}). \end{gathered} \tag{2.53}$$

(The inclusion in $l_2(N_{\lambda(\cdot)})$ follows from (2.52) with $v = u$.) The Fourier transform $\tilde{u}(\lambda(\cdot))$ of a vector u is defined for ρ_τ-almost all $\lambda(\cdot) \in \tau$, and each coordinate $\tilde{u}_\gamma(\lambda(\cdot))$ of it is $\mathcal{C}_\sigma(\tau)$-measurable. Substituting (2.52) in (2.50) and using the notation (2.53), we get Parseval's equality for the Fourier transformation:

$$(E_\tau(\alpha)u, v)_{H_0} = \int_\alpha (\tilde{u}(\lambda(\cdot)), \tilde{v}(\lambda(\cdot)))_{l_2(N_{\lambda(\cdot)})}\, d\rho_\tau(\lambda(\cdot)) \qquad (\alpha \in \mathcal{C}_\sigma(\tau);\ u, v \in H_+). \tag{2.54}$$

For a particular $\lambda(\cdot)$ the set $\{\tilde{u}(\lambda(\cdot)) | u \in H_+\}$ coincides with $l_2(N_{\lambda(\cdot)})$ for $N_{\lambda(\cdot)} < \infty$ and contains all the vectors in l_2 that are finitely nonzero sequences when $N_{\lambda(\cdot)} = \infty$. Indeed, it suffices to show that every vector $(0, \ldots, 0, 1, 0, \ldots)$ (the 1 is at the ςth place) is in this set. If

$$u = (\nu_\varsigma(\lambda(\cdot)))^{-1/2} J\psi_\varsigma(\lambda(\cdot)) \in H_+,$$

then

$$\begin{aligned} (u, \varphi_\gamma(\lambda(\cdot)))_{H_0} &= (\nu_\varsigma(\lambda(\cdot)))^{-1/2} (J\psi_\varsigma(\lambda(\cdot)), \varphi_\gamma(\lambda(\cdot)))_{H_0} \\ &= (\nu_\varsigma(\lambda(\cdot)))^{-1/2} (\psi_\varsigma(\lambda(\cdot)), J\varphi_\gamma(\lambda(\cdot)))_{H_0} \\ &= (\psi_\varsigma(\lambda(\cdot)), \psi_\gamma(\lambda(\cdot)))_{H_0} = \delta_{\varsigma\gamma} \\ &\qquad (\gamma = 1, 2, \ldots, N_{\lambda(\cdot)}). \qquad \blacksquare \end{aligned}$$

We consider the direct integral of the Hilbert spaces $l_2(N_{\lambda(\cdot)})$ over τ with respect to the measure ρ_τ:

$$L_2 = \int_\tau \bigoplus l_2(N_{\lambda(\cdot)})\, d\rho_\tau(\lambda(\cdot)).$$

It is defined as the collection of all vector-valued functions $\tau \ni \lambda(\cdot) \mapsto F(\lambda(\cdot)) \in l_2(N_{\lambda(\cdot)})$ defined for ρ_τ-almost all $\lambda(\cdot)$, $\mathcal{C}_\sigma(\tau)$-measurable in the sense that each coordinate $F_\gamma(\lambda(\cdot))$ $(\gamma = 1, 2, \ldots, N_{\lambda(\cdot)})$ is measurable, and such that

$$\int_\tau \|F(\lambda(\cdot))\|^2_{l_2(N_{\lambda(\cdot)})}\, d\rho_\tau(\lambda(\cdot)) < \infty.$$

The direct integral is a Hilbert space with the inner product

$$(F(\cdot), G(\cdot))_{L_2} = \int_\tau (F(\lambda(\cdot)), G(\lambda(\cdot)))_{l_2(N_{\lambda(\cdot)})}\, d\rho_\tau(\lambda(\cdot)) \qquad (F(\cdot), G(\cdot) \in L_2). \quad (2.55)$$

Comparing (2.54) and (2.55), we conclude that for $\alpha = \tau$ the expression on the right-hand side of (2.54) defines the inner product in the direct integral; therefore, Parseval's equality can be written in the form $(u, v)_{H_0} = (\tilde{u}(\cdot), \tilde{v}(\cdot))_{L_2}$ $(u, v \in H_+)$. Extending this equality by continuity to the whole of H_0, we obtain it in the form

$$(f, g)_{H_0} = \int_\tau (\tilde{f}(\lambda(\cdot)), \tilde{g}(\lambda(\cdot)))_{l_2(N_{\lambda(\cdot)})}\, d\rho_\tau(\lambda(\cdot)) \qquad (f, g \in H_0), \quad (2.56)$$

where the Fourier transform $\tilde{f}(\lambda(\cdot))$ of the vector f is understood as the limit of the Fourier transforms (2.53) in the norm of the direct integral. (For $\tilde{f}_\gamma(\lambda(\cdot))$ it is no longer possible, of course, to write the last formula in (2.53).)

THEOREM 2.9. *If D is a core for each operator A_x $(x \in X)$, then the Fourier transforms $\tilde{u}(\lambda(\cdot))$ $(x \in H_+)$ are dense in the direct integral, and the Fourier transformation $H_0 \ni f \mapsto \tilde{f}(\lambda(\cdot)) \in L_2$ thus implements an isomorphism between the spaces H_0 and L_2.*

Accordingly, in this sense H_0 can be regarded as decomposed into the direct integral

$$H_0 = \int_\tau \bigoplus l_2(N_{\lambda(\cdot)})\, d\rho_\tau(\lambda(\cdot)). \quad (2.57)$$

We first establish the following. Under the isomorphism (2.57) each operator $A_x \restriction D$ passes into the operator of multiplication by $\lambda(x)$: for ρ_τ-almost all $\lambda(\cdot) \in \tau$

$$(A_x, u)^\sim(\lambda(\cdot)) = \lambda(x)\tilde{u}(\lambda(\cdot)) = (\lambda(x)\tilde{u}_1(\lambda(\cdot)), \lambda(x)\tilde{u}_2(\lambda(\cdot)), \ldots) \qquad (x \in X,\ u \in D). \quad (2.58)$$

Indeed, it must be proved that

$$f(x, u, \gamma; \lambda(\cdot)) = (A_x u)^\sim_\gamma(\lambda(\cdot)) - \lambda(x)\tilde{u}_\gamma(\lambda(\cdot)) = 0$$

for ρ_τ-almost all $\lambda(\cdot) \in \tau$ for each $x \in X$, $u \in D$, and $\gamma = 1, 2, \ldots$. Assuming the opposite and using the fact that this function is $\mathcal{C}_\sigma(\tau)$-measurable, we find x_0, u_0, and γ_0 such that $\rho_\tau(\alpha) > 0$, where

$$\alpha = \{\lambda(\cdot) \in \tau | f(x_0, u_0, \gamma_0; \lambda(\cdot)) \neq 0\} \in \mathcal{C}_\sigma(\tau).$$

Let $\lambda(\cdot) \in \pi$, where π is the set in Theorem 2.7. According to (2.53), the inclusion $\varphi_{\gamma_0}(\lambda(\cdot)) \in \mathfrak{R}(P(\lambda(\cdot)))$, Theorem 2.7, and the second of the equalities in (2.19), we have that

$$\begin{aligned}(A_{x_0}u_0)^{\sim}_{\gamma_0}(\lambda(\cdot)) &= (A_{x_0}u_0, \varphi_{\gamma_0}(\lambda(\cdot)))_{H_0} \\ &= \lambda(x_0)(u_0, \varphi_{\gamma_0}(\lambda(\cdot)))_{H_0} = \lambda(x_0)(\tilde{u}_0)_{\gamma_0}(\lambda(\cdot)),\end{aligned}$$

which implies that $\lambda(\cdot) \notin \alpha$. Thus, $\pi \subseteq \tau\backslash\alpha$; but this contradicts the fact that π is of full outer ρ_τ-measure. ■

PROOF OF THE THEOREM. Without loss of generality it can be assumed that the operators A_x are selfadjoint and that $\tau \subseteq \mathbf{R}^X$. (Consider instead of $(A_x)_{x\in X}$ the family

$$\left(\frac{1}{2}(A_x + A_x^*), \frac{1}{2i}(A_x - A_x^*)\right)_{x\in X},$$

of selfadjoint commuting operators; the corresponding direct integral is isomorphic to (2.57) for this family.) Fix an $x \in X$ and a nonreal $z \in \mathbf{C}^1$. It is claimed that

$$(\lambda(x) - z)^{-1}\tilde{u}(\lambda(\cdot)) \in \tilde{H}_0 \subseteq L_2$$

for each $u \in H_+$: since $(\lambda(x) - z)^{-1}$ is bounded as a function of a point $\lambda(\cdot) \in \tau \subseteq \mathbf{R}^X$, the operator of multiplication by it is continuous in L_2, and hence, by (2.58) and (2.56), for $v \in D$

$$\begin{aligned}\|(\lambda(x) - z)^{-1}\tilde{u}(\lambda(\cdot)) - \tilde{v}(\lambda(\cdot))\|_{L_2} &\le c\|\tilde{u}(\lambda(\cdot)) - (\lambda(x) - z)\tilde{v}(\lambda(\cdot))\|_{L_2} \\ &= c\|\tilde{u}(\lambda(\cdot)) - ((A_x - z1)v)^{\sim}(\lambda(\cdot))\|_{L_2} \\ &= c\|u - (A_x - z1)v\|_{H_0};\end{aligned}$$

since D forms a core for the selfadjoint operator A_x, the right-hand side of this estimate can be made as small as desired.

Suppose that $F(\lambda(\cdot)) \in L_2$ is orthogonal to $\tilde{H}_0$ in L_2. Then, in particular, for a nonreal $z \in \mathbf{C}^1$ and a $u \in H_+$ chosen so that $\tilde{u}(\lambda(\cdot)) = (1, 0, 0, \ldots)$ (this is possible; see above) we get that

$$\begin{aligned}0 &= \int_\tau \left(F(\lambda(\cdot)), \frac{1}{\lambda(x) - z}\tilde{u}(\lambda(\cdot))\right)_{l_2(N_{\lambda(\cdot)})} d\rho_\tau(\lambda(\cdot)) \\ &= \int_\tau \frac{1}{\lambda(x) - \bar{z}} F_1(\lambda(\cdot))\, d\rho_\tau(\lambda(\cdot)) = \int_{\mathbf{R}^1} \frac{1}{\lambda - \bar{z}}\, d\omega_x(\lambda),\end{aligned}$$

where the complex-valued measure is given by

$$\omega_x(\delta) = \int_{\mathfrak{C}(x;\delta)\cap\tau} F_1(\lambda(\cdot))\, d\rho_\tau(\lambda(\cdot)) \qquad (\delta \in \mathcal{B}(\mathbf{R}^1)).$$

We conclude from the arbitrariness of z that $\omega_x = 0$, and from the arbitrariness of x that $F_1(\lambda(\cdot)) = 0$ ρ_τ-almost everywhere. An analogous argument works with $F_2(\lambda(\cdot))$, and so on. In the final analysis we get that $F(\lambda(\cdot)) = 0$ in L_2, which implies that $\tilde{H}_0 = L_2$. ■

It follows from the preceding that $N_{\lambda(\cdot)} = \dim \mathfrak{R}(P(\lambda(\cdot)))$, i.e., it is the "multiplicity" of the eigenvalue $\lambda(\cdot)$. In the case when $N_{\lambda(\cdot)} = 1$ for ρ_τ-almost all $\lambda(\cdot)$, i.e., when "the spectrum is simple", we have that $\tilde{u}(\lambda(\cdot)) = \tilde{u}_1(\lambda(\cdot)) = (u, \varphi_1(\lambda(\cdot)))_{H_0} \in \mathbb{C}^1$ $(u \in H_+)$, and (2.57) gives a decomposition of H_0 into a direct integral of complex planes $\mathbb{C}^1 = l_2(1)$.

THEOREM 2.10. *Assume that the family $A = (A_x)_{x \in X}$ has a cyclic vector $\Omega \in D$ such that $A_{x_1}^{m_1} \cdots A_{x_p}^{m_p} \Omega \in D$ for all products $A_{x_1}^{m_1} \cdots A_{x_p}^{m_p}$ of operators in the definition of Ω, and the linear span of these vectors is dense not only in H_0 but also in H_+. Then its spectrum is simple.*

Such a cyclic vector is called a *strong* cyclic vector.

In the case of bounded operators A_x $(x \in X)$ the spectrum is simple if the indicated linear span is dense in H_0; see §5.6.

PROOF. Assume that $N_{\lambda(\cdot)} > 1$ for a set of $\lambda(\cdot)$ of positive ρ_τ-measure. Then there is a $\lambda(\cdot) \in \tau$ such that $N_{\lambda(\cdot)} > 1$, $\operatorname{Tr}(P(\lambda(\cdot))) = 1$, and $\mathfrak{R}(P(\lambda(\cdot)))$ consists of generalized joint eigenvectors corresponding to $\lambda(\cdot)$. For $f \in H_0$ and $g = J^{-1}\Omega$ we get from (2.51) that

$$\begin{aligned} (P(\lambda(\cdot))Jf, \Omega)_{H_0} &= \sum_{\gamma=1}^{N_{\lambda(\cdot)}} \nu_\gamma(\lambda(\cdot))(f, \psi_\gamma(\lambda(\cdot)))_{H_0} \overline{(J^{-1}\Omega, \psi_\gamma(\lambda(\cdot)))}_{H_0} \\ &= \sum_{\gamma=1}^{N_{\lambda(\cdot)}} (f, \psi_\gamma(\lambda(\cdot)))_{H_0} \bar{\xi}_\gamma. \end{aligned} \tag{2.59}$$

Since $(\psi_\gamma(\lambda(\cdot)))_{\gamma=1}^{N_{\lambda(\cdot)}}$ is an orthonormal sequence in H_0, the vectors $\eta(f) = ((f, \psi_1(\lambda(\cdot)))_{H_0}, (f, \psi_2(\lambda(\cdot)))_{H_0}, \ldots) \in l_2(N_{\lambda(\cdot)})$ run through the whole of $l_2(N_{\lambda(\cdot)})$ as f runs through H_0. This gives us that for $N_{\lambda(\cdot)} > 1$ there exists an $f_0 \in H_0$ such that $\eta(f_0) \neq 0$ and is orthogonal in $l_2(N_{\lambda(\cdot)})$ to the vector $\xi = (\xi_1, \xi_2, \ldots)$ introduced in (2.59). ($\xi \in l_2(N_{\lambda(\cdot)})$, because $J^{-1}\Omega \in H_0$ and the factors $\nu_\gamma(\lambda(\cdot))$ are bounded.) Setting $f = f_0$ in (2.59), we get that $(P(\lambda(\cdot))Jf_0, \Omega)_{H_0} = 0$.

Since $P(\lambda(\cdot))Jf_0$ is a generalized joint eigenvector of A corresponding to $\lambda(\cdot)$, successive use of the second equality in (2.18) gives us that

$$\begin{aligned} &(P(\lambda(\cdot))Jf_0, A_{x_1}^{m_1} \cdots A_{x_p}^{m_p}\Omega)_{H_0} \\ &\quad = (\overline{\lambda(x_1)})^{m_1} \cdots (\overline{\lambda(x_p)})^{m_p} (P(\lambda(\cdot))Jf_0, \Omega)_{H_0} = 0 \end{aligned}$$

for all $x_j \in X$ and all $m_j = 1, 2, \ldots$ in the definition of the cyclic vector Ω. The linear span of the vectors $A_{x_1}^{m_1} \cdots A_{x_p}^{m_p}\Omega$ is dense in H_+, so we conclude from this that $P(\lambda(\cdot))Jf_0 = 0$.

Setting $f = f_0$ in (2.51) gives us that

$$0 = \sum_{\gamma=1}^{N_{\lambda(\cdot)}} \nu_\gamma(\lambda(\cdot))\eta_\gamma(f_0)\overline{\eta_\gamma(g)} \qquad (g \in H_0).$$

The vectors $\eta(g) \in l_2(N_{\lambda(\cdot)})$ run through the whole of $l_2(N_{\lambda(\cdot)})$ as g varies over H_0; therefore, it follows from the last equality and the relations $0 < \nu_\gamma(\lambda(\cdot)) \leq c < \infty$ that $\eta(f_0) = 0$. We have arrived at a contradiction. ∎

Everything in this part carries over in a natural way to a family of commuting selfadjoint operators, with $\mathbb{C}^X$ replaced by $\mathbb{R}^X$. (We have already used this fact.)

§3. Connections with random processes and commutative algebras of operators. Quantum processes.

A connection is established between the theory in §§1 and 2 and random processes with the help of the Gel′fand theory of commutative normed algebras. Here we confine ourselves to an explanation of the general situation.

3.1. ***The concept of a random process.*** We first recall some well-known definitions (see, for example, Gikhman and Skorokhod [2], Chapter 1, or Venttsel′ [1]). Consider a fixed abstract space M of points ω (the probability space or the space of elementary events ω) and a σ-algebra $\mathfrak{M}$ of subsets (events) of it; a nonnegative measure $\mathfrak{M} \ni \alpha \mapsto \mu(\alpha) > 0$ such that $\mu(M) = 1$ is assumed to be given on $\mathfrak{M}$ (a probability measure; $\mu(\alpha)$ is the probability of an event α). A (complex-valued) random variable φ is defined to be a function $M \ni \omega \mapsto \varphi(\omega) \in \mathbb{C}^1$ which is measurable with respect to $\mathfrak{M}$; its distribution μ_φ is the φ-image of the measure μ under the mapping $M \ni \omega \mapsto \varphi(\omega) \in \mathbb{C}^1$, where the σ-algebra of Borel sets is fixed in $\mathbb{C}^1$ (see §7 of the Introduction). In other words, the distribution of a random variable φ is the measure μ_φ on $\mathcal{B}(\mathbb{C}^1)$ defined by the relation $\mathcal{B}(\mathbb{C}^1) \ni \delta \mapsto \mu_\varphi(\delta) = \mu(\varphi^{-1}(\delta))$.

A vector-valued random variable $\varphi = (\varphi_1, \ldots, \varphi_p)$ is defined to be a mapping $M \ni \omega \mapsto \varphi(\omega) = (\varphi_1(\omega), \ldots, \varphi_p(\omega)) \in \mathbb{C}^p$ which is measurable with respect to $\mathfrak{M}$ (i.e., each coordinate $M \ni \omega \mapsto \varphi_n(\omega) \in \mathbb{C}^1$ is measurable with respect to $\mathfrak{M}$); its distribution μ_φ is the measure $\mathcal{B}(\mathbb{C}^p) \ni \delta \mapsto \mu_\varphi(\delta) = \mu(\varphi^{-1}(\delta))$. Clearly, $\mu_\varphi(\mathbb{C}^p) = 1$ $(p = 1, 2, \ldots)$. If $\delta = \times_1^p \delta_n$ is a rectangle, then $\varphi^{-1}(\delta) = \bigcap_1^p \varphi_n^{-1}(\delta_n)$, and so

$$\mu_\varphi\left(\mathop{\times}_{n=1}^{p} \delta_n\right) = \mu\left(\varphi^{-1}\left(\mathop{\times}_{n=1}^{p} \delta_n\right)\right) = \mu\left(\bigcap_{n=1}^{p} \varphi_n^{-1}(\delta_n)\right)$$
$$= \int_M \left(\prod_{n=1}^{p} \kappa_{\varphi_n^{-1}(\delta_n)}(\omega)\right) d\mu(\omega) \qquad (\delta_n \in \mathcal{B}(\mathbb{C}^1)). \quad (3.1)$$

This trivial equality is essential for what follows.

Let Φ be a set of random variables φ_α: $\Phi = \bigcup_{\alpha \in \mathrm{A}} \varphi_\alpha$. These random variables are said to be independent if the distribution function of any vector-valued random variable $\varphi(\omega) = (\varphi_{\alpha_1}(\omega), \dots, \varphi_{\alpha_p}(\omega))$ ($\alpha_1, \dots, \alpha_p \in \mathrm{A}$ distinct, $p = 1, 2, \dots$) is $\mu_\varphi = \mu_{\varphi_{\alpha_1}} \times \cdots \times \mu_{\varphi_{\alpha_p}}$.

A random process indexed by $x \in X$ is defined to be a family $(\varphi_x)_{x \in X}$ of random variables φ_x. In other words, giving a random process is equivalent to giving a function

$$M \times X \ni (\omega, x) \mapsto \varphi_x(\omega) \in \mathbb{C}^1 \tag{3.2}$$

such that the function $\varphi_x(\cdot)$ is measurable with respect to the σ-algebra $\mathfrak{M}$ for each x.

Each fixed random process $(\varphi_x)_{x \in X}$ gives rise to a probability measure m on the space $\mathbb{C}^X$. More precisely, along with the M, $\mathfrak{M}$, and μ given earlier we can consider a new probability space $\mathbb{C}^X$, where the points $\lambda(\cdot) \in \mathbb{C}^X$ (i.e., the functions $X \ni x \mapsto \lambda(x) \in \mathbb{C}^1$) serve as the elementary events, and the sets α in the σ-algebra $\mathcal{C}_\sigma(\mathbb{C}^X)$ as the events. The probability measure m is constructed in a way analogous to that for a joint RI (see §1.3). Consider first the algebra $\mathcal{C}(\mathbb{C}^X)$ of cylindrical sets $\mathfrak{C}(x_1, \dots, x_p; \delta)$, where $\delta \in \mathcal{B}(\mathbb{C}^p)$. On the sets in this algebra let

$$m(\mathfrak{C}(x_1, \dots, x_p; \delta)) = \mu_\varphi(\delta) = \mu_{(\varphi_{x_1}, \dots, \varphi_{x_p})}(\delta) = \mu(\varphi^{-1}(\delta)), \tag{3.3}$$

where $\varphi = (\varphi_{x_1}, \dots, \varphi_{x_p})$ is a vector-valued random variable. In (3.3) the points $x_1, \dots, x_p \in X$ are distinct, and $\delta \in \mathcal{B}(\mathbb{C}^p)$. As in the case of a joint RI, it can be proved that this definition is unambiguous: the value $m(\mathfrak{C}(x_1, \dots, x_p; \delta))$ does not depend on how the cylindrical set is expressed in choosing a set of coordinates. This has to do with the fact that $\mu_{\varphi_x}(\mathbb{C}^1) = 1$ ($x \in X$); therefore, the addition or removal of coordinates and the corresponding multiplication by or cancellation of a direct factor $\mathbb{C}^1$ in δ does not affect the quantity (3.3). With the help of Kolmogorov's theorem we then extend the set function m from $\mathcal{C}(\mathbb{C}^X)$ to a measure m on $\mathcal{C}_\sigma(\mathbb{C}^X)$. This is the desired probability measure.

By repeating the proof of Theorem 1.2 it is easy to see that m is always regular.

Thus, for a given triple M, $\mathfrak{M}$, μ and a particular random process $(\varphi_x)_{x \in X}$ we have constructed a triple $\mathbb{C}^X$, $\mathcal{C}_\sigma(\mathbb{C}^X)$, m. Corresponding to that triple it is again possible to consider random variables: functions $\mathbb{C}^X \ni \lambda(\cdot) \mapsto F(\lambda(\cdot)) \in \mathbb{C}^1$ measurable with respect to $\mathcal{C}_\sigma(\mathbb{C}^X)$. In particular, the function $\mathbb{C}^X \ni \lambda(\cdot) \mapsto \pi_x(\lambda(\cdot)) = \lambda(x) \in \mathbb{C}^1$ is such a variable for a given $x \in X$. Its distribution function is

$$m_{\pi_x}(\delta) = m(\pi_x^{-1}(\delta)) = m(\mathfrak{C}(x; \delta)) = \mu_{\varphi_x}(\delta) = \mu(\varphi_x^{-1}(\delta)) \qquad (\delta \in \mathcal{B}(\mathbb{C}^1)).$$

3.2. ***Construction of a family of commuting normal operators from a random process.*** The construction of a family of commuting operators from a process is very simple: each random variable φ can be interpreted as the operator of multiplication by the function $\varphi(\omega)$ in the space $L_2(M, d\mu(\omega))$. Then to a process $(\varphi_x)_{x\in X}$ there corresponds the family of such operators, which of course commute.

More precisely, each measurable function $M \ni \omega \mapsto a(\omega) \in \mathbb{C}^1$ determines the operator A of multiplication by it in the space $H = L_2(M, d\mu(\omega))$:

$$H \supseteq \mathfrak{D}(A) \ni f(\omega) \mapsto (Af)(\omega) = a(\omega)f(\omega) \in H,$$
$$\mathfrak{D}(A) = \{f \in H | a(\cdot)f(\cdot) \in H\}. \tag{3.4}$$

The domain $\mathfrak{D}(A)$ is dense in H (see §1.7). The operator A is normal; recall that the resolution $\mathcal{B}(\mathbb{C}^1) \ni \delta \mapsto E(\delta)$ of the identity corresponding to A has the form

$$H \ni f(\omega) \mapsto (E(\delta)f)(\omega) = \kappa_\delta(a(\omega))f(\omega) = \kappa_{a^{-1}(\delta)}(\omega)f(\omega), \tag{3.5}$$

where κ_α is the characteristic function of a set α.

We now fix a random process $(a_x)_{x\in X}$ of the form (3.2) and associate with each random variable a_x the operator A_x by formula (3.4). This gives a family $(A_x)_{x\in X}$ of commuting normal operators. Let $\Omega(\omega) = 1$ $(\omega \in M)$. It is not hard to see that *the joint RI* $\mathcal{C}_\sigma(\mathbb{C}^X) \ni \alpha \mapsto E(\alpha)$ *of the constructed family* $(A_x)_{x\in X}$ *is connected with the measure* $\mathcal{C}_\sigma(\mathbb{C}^X) \ni \alpha \mapsto m(\alpha) \geq 0$ *generated by the process* $(a_x)_{x\in X}$ *by the formula*

$$(E(\alpha)\Omega, \Omega)_{L_2(M,d\mu(\omega))} = m(\alpha) \qquad (\alpha \in \mathcal{C}_\sigma(\mathbb{C}^X)). \tag{3.6}$$

Indeed, since the left-hand side of (3.6) is a nonnegative scalar measure, to prove (3.6) it suffices to establish the equality for $\alpha = \mathfrak{C}(x_1, \dots, x_p; \delta)$, i.e., to prove that

$$(E_{x_1,\dots,x_p}(\delta)\Omega, \Omega)_H = \mu_{(a_{x_1},\dots,a_{x_p})}(\delta)$$
$$(x_1, \dots, x_p \in X;\ \delta \in \mathcal{B}(\mathbb{C}^p);\ p = 1, 2, \dots).$$

The last relation will be proved if it is established for $\delta = \times_1^p \delta_n$ $(\delta_n \in \mathcal{B}(\mathbb{C}^1))$. By the definition of $E_{x_1,\dots,x_p}$, (3.5), and (3.1),

$$\begin{aligned}(E_{x_1,\dots,x_p}(\delta)\Omega, \Omega)_H &= (E_{x_1}(\delta_1)\cdots E_{x_p}(\delta_p)\Omega, \Omega)_H \\ &= \left(\prod_{n=1}^p \kappa_{a_{x_n}^{-1}(\delta_n)}(\omega)\Omega, \Omega\right)_H \\ &= \int_M \left(\prod_{n=1}^p \kappa_{a_{x_n}^{-1}(\delta_n)}(\omega)\right) d\mu(\omega) \\ &= \mu_{(a_{x_1},\dots,a_{x_p})}\left(\mathop{\times}_{n=1}^p \delta_n\right). \quad \blacksquare\end{aligned} \tag{3.7}$$

It follows from (3.6) that m is absolutely continuous with respect to E. If Ω is a cyclic vector for the family $(A_x)_{x\in X}$, then, by Theorem 2.4, E is absolutely continuous with respect to m. In this case a number of the facts in §1 involving the concept of a support remain valid for m. For example, Theorem 1.6 leads to the following. Let $(a_x)_{x\in X}$ be a random process, with at most countably many of the functions a_x essentially unbounded. If for some $x_j \in X$ and some positive powers m_j the functions $a_{x_1}^{m_1}(\omega), \dots, a_{x_p}^{m_p}(\omega)$ are in $L_2(M, d\mu(\omega))$ and their c.l.s. coincides with $L_2(M, d\mu(\omega))$, then the measure m generated by this process is proper. (The last condition is a reformulation of what it means for Ω to be a cyclic vector.)

3.3. *Construction of a random process from a family of commuting normal operators. Connection with commutative algebras of operators.* Roughly speaking, we show that the above-described family $(A_x)_{x\in X}$ generated by a random process is a sufficiently general model to within a Hilbert space isomorphism. In large part the arguments below amount to a suitable variant of the familiar proof of von Neumann's spectral theorem for von Neumann algebras of bounded operators, based on the Gel′fand theory of commutative normed algebras (see, for example, Naĭmark [1], Chapter 8, or Maurin [3], Chapter 1).

Let $(A_x)_{x\in X}$ be a family of commuting normal operators acting in a Hilbert space H, and suppose that it has a cyclic vector Ω. With it we associate (though not in a very reasonable way) a certain commutative normed algebra $\mathcal{A}$ whose maximal ideal space M can be taken as a space of elementary events. ("Not reasonable" here means that this algebra is too broad; it is true that we do not impose any additional conditions on the family $(A_x)_{x\in X}$.) Later, after proving Theorems 3.1 and 3.2, we shall describe an improvement of the construction.

Denote by $(E_x)_{x\in X}$ the family of resolutions of the identity corresponding to $(A_x)_{x\in X}$, and consider the normed algebra $\mathcal{A} \subseteq \mathcal{L}(H \to H)$ spanned by the commuting projections $E_x(\delta)$ ($x \in X$, $\delta \in \mathcal{B}(\mathbb{C}^1)$) (i.e., take all possible linear combinations and products of the operators $E_x(\delta)$ and close this set in the operator norm). It is clear that $\mathcal{A}$ is a commutative C^*-algebra with identity. Let M be the compact space of its maximal ideals ω. By the Gel′fand-Naĭmark theorem, $\mathcal{A}$ is isometrically isomorphic to the algebra $C(M)$ of all complex-valued continuous functions on M with the usual algebraic operations and the uniform norm $\|\cdot\|_{C(M)}$ (Gel′fand, Raĭkov and Shilov [1] (the Supplement), or Maurin [3], Chapter 1, §3). Denote this isomorphism by g: $\mathcal{A} \ni A \mapsto g(A) = a \in C(M)$; $g(1) = e$ ($e(\omega) = 1$, $\omega \in M$).

Consider the linear functional l defined on $C(M)$ by

$$l(a) = (g^{-1}(a)\Omega, \Omega)_H \qquad (a \in C(M)). \tag{3.8}$$

It is continuous because of the estimate $|l(a)| \leq \|g^{-1}(a)\| \, \|\Omega\|_H^2 = \|a\|_{C(M)}$ ($a \in C(M)$), $l(e) = 1$, and l is nonnegative: if $a(\omega) \geq 0$ ($\omega \in M$), then $g^{-1}(a)$ is a nonnegative operator in $\mathcal{A}$ and $l(a) \geq 0$. By the Riesz theorem, we have the representation

$$l(a) = \int_M a(\omega)\, d\mu(\omega) \qquad (a \in C(M)), \tag{3.9}$$

where $\mathcal{B}(M) \ni \alpha \mapsto \mu(\alpha) \geq 0$ is a probability measure ($\mu(M) = l(e) = 1$) on the σ-algebra of Borel subsets of M and is uniquely determined by l.

We introduce the space $L_2(M, d\mu(\omega))$ and show that g induces a natural isometry between H and $L_2(M, d\mu(\omega))$. This isometry is based on the following formula, which is a consequence of (3.8) and (3.9):

$$\begin{aligned}(A_1\Omega, A_2\Omega)_H &= (A_2^* A_1\Omega, \Omega)_H = l(g(A_2^* A_1)) = l(\overline{g(A_2)}g(A_1)) \\ &= \int_M (g(A_1))(\omega)\overline{(g(A_2))(\omega)}\, d\mu(\omega) \\ &= (g(A_1), g(A_2))_{L_2(M, d\mu(\omega))} \qquad (A_1, A_2 \in \mathcal{A}). \end{aligned}\tag{3.10}$$

LEMMA 3.1. *There exists an isometry $H \ni f \mapsto Gf = \hat{f} \in L_2(M, d\mu(\omega))$ carrying the whole space H into the whole of $L_2(M, d\mu(\omega))$ and having the property that*

$$G(A\Omega) = g(A) \qquad (A \in \mathcal{A}); \tag{3.11}$$

moreover, the linear set $\{A\Omega | A \in \mathcal{A}\}$ is dense in H.

PROOF. Let us first establish that the indicated linear set L is dense in H. We fix $f \in H$ and $\varepsilon > 0$ and find, by the definition of a cyclic vector, points $x_1, \ldots, x_p \in X$ and a linear combination (the m_j below are positive integers)

$$f_1 = \sum_{m_1=0}^{l_1} \cdots \sum_{m_p=0}^{l_p} c_{m_1,\ldots,m_p} A_{x_1}^{m_1} \cdots A_{x_p}^{m_p}\Omega, \tag{3.12}$$

such that $\|f - f_1\|_H < \varepsilon$. Suppose that $\mathbb{C}^N \ni (\lambda_1, \ldots, \lambda_p) = \lambda \mapsto F(\lambda) \in \mathbb{C}^1$ is continuous and has compact support. Denote by $E_{x_1,\ldots,x_p}$ the joint RI of the family $(A_{x_n})_1^p$ of operators; then, by (3.12), (1.1), and (1.2),

$$\begin{aligned}&\|f_1 - F(A_{x_1}, \ldots, A_{x_p})\Omega\|_H^2 \\ &= \int_{\mathbb{C}^p} \left| \sum_{m_1=0}^{l_1} \cdots \sum_{m_p=0}^{l_p} c_{m_1,\ldots,m_p} \lambda_1^{m_1} \cdots \lambda_p^{m_p} - F(\lambda_1, \ldots, \lambda_p) \right|^2 \\ &\qquad\qquad \times d(E_{x_1,\ldots,x_p}(\lambda)\Omega, \Omega)_H. \end{aligned}\tag{3.13}$$

(The integral in (3.13) converges because Ω is in the domain of all the operators appearing in (3.12).) The F can be chosen so that the integral in (3.13) is less than ε^2. To prove that L is dense in H it suffices to see that $F(A_{x_1}, \ldots, A_{x_p}) \in \mathcal{A}$.

Suppose that the F chosen is zero off the rectangle $\delta = \times_1^p \delta_n \subset \mathbb{C}^p$, where δ_n is a closed square in $\mathbb{C}^1$. Setting $\lambda_n = \sigma_n + i\tau_n$ ($\sigma_n, \tau_n \in \mathbb{R}^1$; $n = 1, \ldots, p$) and regarding δ as a compact subset of $\mathbb{R}^{2p}$, we get that $F(\sigma_1 + i\tau_1, \ldots, \sigma_p + i\tau_p)$

can, by the Weierstrass theorem, be approximated uniformly on δ arbitrarily well by a polynomial in $\sigma_1, \dots, \sigma_p, \tau_1, \dots, \tau_p$, i.e., by a polynomial $P(\lambda, \overline{\lambda})$ in $\lambda_1, \dots \lambda_p, \overline{\lambda}_1, \dots, \overline{\lambda}_p$. For any $h \in H$

$$\begin{aligned}\left\|(F(A_{x_1}, \dots, A_{x_p}) - \int_\delta P(\lambda, \overline{\lambda})\, dE_{x_1,\dots,x_p}(\lambda))h\right\|_H^2 \\ = \left\|\int_\delta (F(\lambda) - P(\lambda, \overline{\lambda}))\, dE_{x_1,\dots,x_p} h\right\|_H^2 \\ \le \max_{\lambda \in \delta} |F(\lambda) - P(\lambda, \overline{\lambda})|^2 \|h\|_H^2,\end{aligned}$$

so it can be asserted that $F(A_{x_1}, \dots, A_{x_p})$ can be approximated arbitrarily well in the operator norm by the operator $\int_\delta P(\lambda, \overline{\lambda})\, dE_{x_1,\dots,x_p}(\lambda)$. In turn, the last operator is a polynomial in $E_{x_n}(\delta_n)$, $\int_{\delta_n} z\, dE_{x_n}(z)$, and $\int_{\delta_n} \overline{z}\, dE_{x_n}(z)$ $(n = 1, \dots, p)$, and these operators are obviously in $\mathcal{A}$. (It is necessary to approximate the function z uniformly on δ_n by a piecewise constant function.) Accordingly, $F(A_{x_1}, \dots, A_{x_p}) \in \mathcal{A}$, which is what was required.

The proof of the lemma is now almost obvious. Indeed, define G on L by (3.11). From (3.10) it follows that this definition is unambiguous ($g(A)$ depends only on $A\Omega$) and that G is isometric; it is clear that G is linear. Since L and $G(L) = C(M)$ are dense in H and $L_2(M, d\mu(\omega))$, respectively, extension of G by continuity to the whole of H gives us the required isometry. ■

LEMMA 3.2. *The image under the isometry G of an operator $A \in \mathcal{A}$ is the operator of multiplication in $L_2(M, d\mu(\omega))$ by the continuous function $(g(A))(\omega)$, i.e., in $L_2(M, d\mu(\omega))$*

$$(G(Af))(\omega) = (g(A))(\omega)(Gf)(\omega) \qquad (f \in H,\ A \in \mathcal{A}). \tag{3.14}$$

PROOF. Fix an $f \in H$. Let $(A_n)_1^\infty$ be a sequence of operators $A_n \in \mathcal{A}$ such that $A_n\Omega \to f$ as $n \to \infty$ in H. Then (3.11) gives us that, in the sense of convergence in $L_2(M, d\mu(\omega))$,

$$\begin{aligned}(G(Af))(\omega) &= \lim_{n\to\infty} (G(AA_n\Omega))(\omega) = \lim_{n\to\infty} (g(AA_n))(\omega) \\ &= \lim_{n\to\infty} ((g(A))(\omega)(g(A_n))(\omega)) = (g(A))(\omega) \lim_{n\to\infty} (g(A_n))(\omega) \\ &= (g(A))(\omega) \lim_{n\to\infty} (G(A_n\Omega))(\omega) = (g(A))(\omega)(Gf)(\omega).\end{aligned}$$ ■

Since $g(\mathcal{A}) = C(M)$, any operator of multiplication by a function in $C(M)$ acting in $L_2(M, d\mu(\omega))$ can be obtained as the image (3.14) of some $A \in \mathcal{A}$; clearly, A^* passes into the operator of multiplication by $\overline{(g(A))(\omega)}$.

Let us extend the relation (3.14) to broader classes of operators A. Denote by $L_\infty(\mathcal{A})$ the collection of all bounded operators B on H obtained as strong limits of operators in $\mathcal{A}$ (i.e., $L_\infty(\mathcal{A})$ is the closure of $\mathcal{A}$ in the strong operator topology). On the other hand, let $L_\infty(M, d\mu(\omega))$ be the space of all μ-essentially bounded measurable functions $M \ni \omega \mapsto b(\omega) \in \mathbb{C}^1$.

LEMMA 3.3. *Under the isometry G the image of each operator in the collection $L_\infty(\mathcal{A})$ is the operator of multiplication in $L_2(M, d\mu(\omega))$ by some function in $L_\infty(M, d\mu(\omega))$, and any function in the latter space can be so obtained. More precisely, there exists a one-to-one correspondence $L_\infty(\mathcal{A}) \ni B \mapsto b \in L_\infty(M, d\mu(\omega))$ such that in $L_2(M, d\mu(\omega))$*

$$(G(Bf))(\omega) = b(\omega)(Gf)(\omega) \qquad (f \in H,\ B \in L_\infty(\mathcal{A})). \tag{3.15}$$

PROOF. Fix a $B \in L_\infty(\mathcal{A})$ and consider a sequence $(A_n)_1^\infty$ $(A_n \in \mathcal{A})$ such that $Bf = \lim_{n\to\infty} A_n f$ $(f \in H)$. Using Lemma 3.2, we conclude that in $L_2(M, d\mu(\omega))$

$$\begin{aligned}(G(Bf))(\omega) &= \lim_{n\to\infty} (G(A_n f))(\omega) \\ &= \lim_{n\to\infty} (g(A_n))(\omega)(Gf)(\omega) \qquad (f \in H).\end{aligned} \tag{3.16}$$

Let $f = \Omega$ here. Since $(G\Omega)(\omega) = 1$ $(\omega \in M)$, it can be concluded that the limit $\lim_{n\to\infty} g(A_n) = G(B\Omega)$ exists in $L_2(M, d\mu(\omega))$. Taking a subsequence of $((g(A_n))(\omega))_1^\infty$ converging almost everywhere and passing to the limit in (3.16) for this subsequence, we get that

$$(G(Bf))(\omega) = (G(B\Omega))(\omega)(Gf)(\omega)$$

almost everywhere. But

$$\begin{aligned}\|(G(B\Omega))(\cdot)(Gf)(\cdot)\|_{L_2(M,d\mu(\omega))} &= \|G(Bf)\|_{L_2(M,d\mu(\omega))} \\ &= \|Bf\|_H \le \|B\|\,\|f\|_H \\ &= \|B\|\,\|Gf\|_{L_2(M,d\mu(\omega))} \qquad (f \in H).\end{aligned} \tag{3.17}$$

Since the Gf in (3.17) runs through the whole of $L_2(M, d\mu(\omega))$, this implies that $(G(B\Omega))(\omega)$ is essentially bounded. Denoting this function by $b(\omega)$, we arrive at (3.15). It follows from (3.15) that b is uniquely determined by B and can be generated by only one operator $B \in L_\infty(\mathcal{A})$.

It remains to see that for any $b \in L_\infty(M, d\mu(\omega))$ there is an operator $B \in L_\infty(\mathcal{A})$ such that $B \mapsto b$. To prove this, consider the bounded operator $\hat{B}$ of multiplication in $L_2(M, d\mu(\omega))$ by the function $b(\omega)$. We construct a sequence $(a_n(\omega))_1^\infty$ $(a_n \in C(M))$ of functions such that $a_n(\omega) \to b(\omega)$ as $n \to \infty$ in the sense of convergence in μ-measure and the $a_n(\omega)$ are uniformly bounded: $|a_n(\omega)| \le C$ $(C = \operatorname{ess\,sup}_{\omega\in M}|b(\omega)|;\ \omega \in M;\ n = 1, 2, \ldots)$. This can always be done, because $b(\omega)$ is measurable and essentially bounded. Let $\hat{A}_n$ be the operator of multiplication by $a_n(\omega)$ in $L_2(M, d\mu(\omega))$ $(n = 1, 2, \ldots)$.

It is easy to see that $\hat{A}_n\hat{f} \to \hat{B}\hat{f}$ as $n \to \infty$ in the L_2-norm for each function $\hat{f} \in L_2(M, d\mu(\omega))$. Indeed, for any $\sigma > 0$

$$\begin{aligned}
&\|\hat{B}\hat{f} - \hat{A}_n\hat{f}\|^2_{L_2(M,d\mu(\omega))} \\
&\quad = \int_M |b(\omega) - a_n(\omega)|^2 |\hat{f}(\omega)|^2 \, d\mu(\omega) \\
&\quad \leq \sigma^2 \int_M |\hat{f}(\omega)|^2 \, d\mu(\omega) \\
&\qquad + 4C^2 \int_{\{\omega \in M \mid |b(\omega) - a_n(\omega)| \geq \sigma\}} |\hat{f}(\omega)|^2 \, d\mu(\omega). \qquad (3.18)
\end{aligned}$$

Since

$$\mu(\{\omega \in M \mid |b(\omega) - a_n(\omega)| \geq \sigma\}) \to 0 \quad \text{as } n \to \infty,$$

the right-hand side of (3.18) can be made arbitrarily small for n large enough.

To conclude the proof of the lemma it remains to set $B = G^{-1}\hat{B}G$ and $A_n = G^{-1}A_nG$ ($n = 1, 2, \ldots$). Then each A_n is in $\mathcal{A}$, and $Bf = \lim_{n\to\infty} A_nf$ ($f \in H$) in H, i.e., $B \in L_\infty(\mathcal{A})$. According to (3.15), $B \mapsto b$. ■

We look at $L_\infty(M, d\mu(\omega))$ as a commutative algebra of bounded operators $\hat{B}$ of multiplication in $L_2(M, d\mu(\omega))$ by functions $b \in L_\infty(M, d\mu(\omega))$, and $C(M)$ as a subalgebra of it. Then, according to Lemma 3.3, the operators $B = G^{-1}\hat{B}G$ ($\hat{B} \in L_\infty(M, d\mu(\omega))$) run through the whole of $L_\infty(\mathcal{A})$, i.e., the last set is turned into a commutative algebra of bounded operators on H, and it contains the adjoint operator of each of its members (since the last property is enjoyed by $L_\infty(M, d\mu(\omega))$). Summing up the foregoing and Lemmas 3.2 and 3.3, we arrive at the following well-known result.

The collection $L_\infty(\mathcal{A}) \supseteq \mathcal{A}$ is a strongly closed commutative algebra of operators in $\mathcal{L}(H \to H)$ containing the adjoint operator of each of its members (a commutative von Neumann algebra of operators). The isometry

$$H \ni f \mapsto Gf = \hat{f} \in L_2(M, d\mu(\omega))$$

converts $L_\infty(\mathcal{A})$ into the whole algebra $L_\infty(M, d\mu(\omega))$, with the algebra $\mathcal{A}$ passing into the algebra $C(M)$.

Let us now see what are the images under G of certain unbounded operators associated in a definite way with the algebra $\mathcal{A}$. Let C be a normal operator with domain $\mathfrak{D}(C)$ in H. We say that it is affiliated with $\mathcal{A}$ if there exist sequences $(B_n)_1^\infty$ and $(D_n)_1^\infty$ ($B_n, D_n \in L_\infty(\mathcal{A})$) such that 1) $Cf = \lim_{n\to\infty} B_nf$ for any $f \in \mathfrak{D}(C)$ and 2) $\Omega = \lim_{n\to\infty} D_n\Omega$ and $D_n\Omega \in \mathfrak{D}(C)$ ($n = 1, 2, \ldots$). The collection of all operators affiliated with $\mathcal{A}$ is denoted by $L_0(\mathcal{A})$.

It is clear that $L_\infty(\mathcal{A}) \subseteq L_0(\mathcal{A})$. It is essential for us that *each operator A_x ($x \in X$) is affiliated with $\mathcal{A}$*. Indeed, let $\delta_n = \{z \in \mathbb{C}^1 \mid |z| \leq n\}$ and define $B_n = \int_{\delta_n} z \, dE_x(z)$ and $D_n = E_x(\delta_n)$ ($n = 1, 2, \ldots$). Clearly, $B_n, D_n \in \mathcal{A} \subseteq L_\infty(\mathcal{A})$, and these operators have the required properties. ■

Denote by $L_0(M, d\mu(\omega))$ the collection of all functions $M \ni \omega \mapsto c(\omega) \in \mathbb{C}^1$ which are finite μ-almost everywhere and measurable with respect to $\mathcal{B}(M)$; of course, functions in $L_0(M, d\mu(\omega))$ are regarded as equal if they are equal μ-almost everywhere.

THEOREM 3.1. *Under the isometry G the image of each operator affiliated with $\mathcal{A}$ is the operator of multiplication in $L_2(M, d\mu(\omega))$ by some function in $L_0(M, d\mu(\omega))$, and each function in this collection can be so obtained.*

More precisely, there exists a one-to-one correspondence $L_0(\mathcal{A}) \ni C \mapsto c \in L_0(M, d\mu(\omega))$ such that in $L_2(M, d\mu(\omega))$

$$(G(Cf))(\omega) = c(\omega)(Gf)(\omega),$$
$$f \in \mathfrak{D}(C) = \{f \in H | c(\cdot)(Gf)(\cdot) \in L_2(M, d\mu(\omega))\}. \tag{3.19}$$

PROOF. Fix a $C \in L_0(\mathcal{A})$ and consider a sequence of operators $D_n \in L_\infty(\mathcal{A})$ associated with it. Since $D_n\Omega \to \Omega$ as $n \to \infty$ in H, Lemma 3.3 gives us that $d_n(\omega) \to 1$ as $n \to \infty$ in $L_2(M, d\mu(\omega))$, where $d_n \in L_\infty(M, d\mu(\omega))$ is the function corresponding to the operator D_n. Let $(d_{n_k})_{k=1}^\infty$ be a subsequnce of $(d_n)_1^\infty$ converging μ-almost everywhere to 1. Denote by $\beta \in \mathcal{B}(M)$ a set of full measure at whose points this convergence is valid, and let $e_k = d_{n_k}$ $(k = 1, 2, \ldots)$.

Let $(B_n)_1^\infty$ be a sequence of operators $B \in L_\infty(\mathcal{A})$ associated with C according to the definition of $L_0(\mathcal{A})$. For each $f \in \mathfrak{D}(C)$ we have that $B_n f \to Cf$ as $n \to \infty$ in H. Using Lemma 3.3 and denoting by $b_n \in L_\infty(M, d\mu(\omega))$ the function corresponding to the operator B_n, we get that

$$b_n(\omega)(Gf)(\omega) \underset{n\to\infty}{\longrightarrow} (C(Gf))(\omega) \qquad (f \in \mathfrak{D}(C)) \tag{3.20}$$

in the sense of convergence in $L_2(M, d\mu(\omega))$.

Let us show by a diagonal process that the sequence $(b_n)_1^\infty$ of essentially bounded functions on M has a subsequence $(b_{n_l})_{l=1}^\infty$ converging almost everywhere to some measurable function $c(\omega)$.

Take $f = D_{n_1}\Omega$ in (3.20), i.e., $Gf = e_1$, and select from the sequence $(b_n(\omega)e_1(\omega))_1^\infty$, which converges to $(Ce_1)(\omega)$ in $L_2(M, d\mu(\omega))$, a subsequence $(b_{1n}(\omega)e_1(\omega))_1^\infty$ that converges to the same function almost everywhere. Then take $f = D_{n_2}\Omega$ in (3.20), i.e., $Gf = e_2$, and select from the sequence $(b_{1n}(\omega)e_2(\omega))_1^\infty$ a subsequence $(b_{2n}(\omega)e_2(\omega))_1^\infty$ that converges to $(Ce_2)(\omega)$ almost everywhere. Continue this procedure, setting $f = D_{n_3}\Omega$, and so on, and then consider the diagonal sequence $(b_{nn}(\omega))_1^\infty$. It has the property that for each $k = 1, 2, \ldots$ the sequence $(b_{nn}(\omega)e_k(\omega))_1^\infty$ converges almost everywhere; thus, it is possible to construct a set $\gamma \in \mathcal{B}(M)$ of full measure at whose points ω the indicated sequence converges for each $k = 1, 2, \ldots$.

It is now easy to see that at each point ω of the set $\beta \cap \gamma$ of full measure the limit $c(\omega) = \lim_{n\to\infty} b_{nn}(\omega)$ exists. Indeed, since $\omega \in \beta$, it follows that $e_k(\omega) \to 1$ as $k \to \infty$; therefore, there is a k large enough that $e_k(\omega) \neq 0$. Fix this k and consider the sequence $(b_{nn}(\omega)e_k(\omega))_{n=1}^\infty$. Since $\omega \in \gamma$, it converges, and hence, by the inequality $e_k(\omega) \neq 0$, the sequence $(b_{nn}(\omega))_1^\infty$ also converges. Accordingly, $c(\omega)$ has been constructed for $\omega \in \beta \cap \gamma$, and outside this set we can take $c(\omega)$ to be, say, zero. It remains to let $b_{ll} = b_{n_l}$ $(l = 1, 2, \ldots)$.

We now replace n by n_l in (3.20) and pass to the limit as $l \to \infty$. As a result, $(G(Cf))(\omega) = c(\omega)(Gf)(\omega)$ for almost all $\omega \in M$. Thus, we have constructed a function $c(\omega)$ corresponding to C. Moreover, it has been proved that

$$\mathfrak{D}(C) \subseteq \{f \in H | c(\cdot)(Gf)(\cdot) \in L_2(M, d\mu(\omega))\}. \tag{3.21}$$

Note that it has not yet been used anywhere that C is normal.

Let us prove the inclusion opposite to (3.21). Let h be in the right-hand set in (3.21). Then, by what has been proved, it is possible to write

$$\begin{aligned}(Cf, h)_H &= (G(Cf), Gh)_{L_2(M,d\mu(\omega))} = (c(\cdot)(Gf)(\cdot), (Gh)(\cdot))_{L_2(M,d\mu(\omega))} \\ &= ((Gf)(\cdot), \hat{k}(\cdot))_{L_2(M,d\mu(\omega))} = (f, G^{-1}(\hat{k}(\cdot)))_H \qquad (f \in \mathfrak{D}(C)),\end{aligned}$$

where

$$\hat{k}(\omega) = \overline{c(\omega)}(Gh)(\omega) \in L_2(M, d\mu(\omega)).$$

This implies that $h \in \mathfrak{D}(C^*) = \mathfrak{D}(C)$ (because C is normal), which is what was required. Accordingly, (3.19) is established.

Since $G(\mathfrak{D}(C))$ is dense in $L_2(M, d\mu(\omega))$, the function c corresponding to C according to (3.19) is uniquely determined. Further, if C_1 and C_2 are two operators in $L_0(\mathcal{A})$ corresponding to one and the same c according to (3.19), then their domains coincide and their actions are the same, i.e., $C_1 = C_2$.

It remains to see that for any $c \in L_0(M, d\mu(\omega))$ there is an operator $C \in L_0(\mathcal{A})$ such that $C \mapsto c$. To do this, consider the operator $\hat{C}$ of multiplication in $L_2(M, d\mu(\omega))$ by the function $c(\omega)$. Let $b_n(\omega) = c(\omega)$ on the set $\{\omega \in M | \, |c(\omega)| \leq n\}$ and let $b_n(\omega) = 0$ for the remaining ω in M. Take $\hat{B}_n$ to be the corresponding multiplication operator $(n = 1, 2, \ldots)$. For each $\hat{f} \in \mathfrak{D}(\hat{C})$ we have that $\hat{B}_n \hat{f} \to \hat{C}\hat{f}$ as $n \to \infty$ in the norm of $L_2(M, d\mu(\omega))$. Indeed,

$$\|\hat{C}\hat{f} - \hat{B}_n\hat{f}\|^2_{L_2(M,d\mu(\omega))} = \int_{\{\omega \in M | \, |c(\omega)| > n\}} |c(\omega)\hat{f}(\omega)|^2 \, d\mu(\omega) \underset{n\to\infty}{\longrightarrow} 0,$$

since

$$\mu(\{\omega \in M | \, |c(\omega)| > n\}) \underset{n\to\infty}{\longrightarrow} 0$$

because of the fact that $c(\omega)$ is finite almost everywhere, and $c(\cdot)\hat{f}(\cdot) \in L_2(M, d\mu(\omega))$ $(\hat{f} \in \mathfrak{D}(\hat{C}))$. Now let $C = G^{-1}\hat{C}G$ and $B_n = G^{-1}\hat{B}_n G$ $(n = 1, 2, \ldots)$. Then, by Lemma 3.3, each B_n is in $L_\infty(\mathcal{A})$, and in H

$$Cf = \lim_{n\to\infty} B_n f \qquad (f \in G^{-1}\hat{f} \in \mathfrak{D}(C) = G^{-1}\mathfrak{D}(\hat{C})). \tag{3.22}$$

We now construct the required sequence $(D_n)_1^\infty$. Let $d_n(\omega) = 1$ on the set $\{\omega \in M | \, |c(\omega)| > n\}$, and $d_n(\omega) = 0$ for the remaining ω in M. Take $\hat{D}_n$

to be the corresponding multiplication operator acting in $L_2(M, d\mu(\omega))$ ($n = 1, 2, \ldots$). It is clear that $\hat{D}_n e \to e$ as $n \to \infty$ in this space, and $\hat{D}_n e \in \mathfrak{D}(\hat{C})$ ($n = 1, 2, \ldots$). Again passing to H via the isometry G, we get operators

$$D_n = G^{-1}\hat{D}_n G \in L_\infty(\mathcal{A})$$

such that $D_n\Omega \to \Omega$ as $n \to \infty$ in H and $D_n\Omega \in \mathfrak{D}(C)$ ($n = 1, 2, \ldots$).

It follows from (3.22) and what was just proved that C is affiliated with $\mathcal{A}$, i.e., $C \in L_0(\mathcal{A})$. According to (3.19), $C \mapsto c$. ■

REMARK 1. The collection $L_0(M, d\mu(\omega))$ is an algebra with respect to the ordinary algebraic operations, and is closed with respect to convergence almost everywhere. Due to Theorem 3.1, the corresponding algebraic operations and convergence can thereby be introduced in $L_0(\mathcal{A})$. We do not dwell in detail on these questions.

REMARK 2. The selfadjoint operators in $L_0(\mathcal{A})$ correspond to the real-valued functions in $L_0(M, d\mu(\omega))$.

In parallel with Theorem 3.1 we have proved the following result.

THEOREM 3.2. *Let $(A_x)_{x\in X}$ be a family of commuting normal operators that has a cyclic vector. The isometry G constructed in Theorem 3.1 transforms this family into a family of operators of multiplication in $L_2(M, d\mu(\omega))$ by measurable functions that are finite μ-almost everywhere. If*

$$a_x(\cdot) \in L_0(M, d\mu(\omega))$$

is the function corresponding to A_x, then $(a_x)_{x\in X}$ is a random process, with the role of the probability space played by M, the σ-algebra $\mathcal{B}(M)$ of events, and the probability measure μ.

This theorem follows from Theorem 3.1 and the remark above about the fact that $a_x \in L_0(M, d\mu(\omega))$ ($x \in X$). Note that if $\Omega \in \mathfrak{D}(A_x)$, then $a_x = G(A_x\Omega)$.

REMARK 3. As already mentioned, the algebra $\mathcal{A}$ and the space M in Theorems 3.1 and 3.2 are too broad, in general. It is easy to see what additional restriction to impose on the family $(A_x)_{x\in X}$ and the unit vector Ω in order to constrict these formations. Assume that there exists a commutative C^*-algebra $\mathcal{A}_1$ with identity composed of bounded operators in $\mathcal{L}(H \to H)$ and having the following properties:

a) the linear set $\{A\Omega | A \in \mathcal{A}_1\}$ is dense in H;

b) if $L_0(\mathcal{A}_1)$ denotes the collection of all normal operators acting in H and affiliated with $\mathcal{A}_1$ (the definition of being affiliated is as before, except with $\mathcal{A}$ replaced by $\mathcal{A}_1$), then each operator A_x ($x \in X$) is affiliated with $\mathcal{A}_1$.

By repeating the proofs given above with $\mathcal{A}$ replaced by $\mathcal{A}_1$ (the main part of the proof of Lemma 3.1 now simply coincides with the requirement a)), it is easy to establish Theorems 3.1 and 3.2, with $\mathcal{A}$ replaced by $\mathcal{A}_1$, M by the maximal ideal space M_1 of $\mathcal{A}_1$, etc.

3.4. *The concept of a quantum process.* We introduce the following definition. A (general) quantum process indexed by $x \in X$ is defined to be a family $(A_x)_{x \in X}$ of commuting normal operators acting in H which has a cyclic vector Ω.(2) If all the operators are selfadjoint or unitary, then the process is called a selfadjoint or unitary process.

Theorem 3.2 shows that every quantum process can be replaced by an ordinary random process (which will be called a realization of the quantum process) to within an isomorphism of the Hilbert space H. It follows from §3.2 that an ordinary random process can also be regarded as a quantum process, provided only that $\Omega(\cdot) = 1$ is a cyclic vector (see the beginning of §3.3). Despite the apparent equivalence of these two concepts, there are essential differences between them, having to do with the fact that natural conditions formulated in terms of the Hilbert space H often become unnatural for its realization as $L_2(M, d\mu(\omega))$, and conversely. We emphasize also that it is often convenient to introduce the realization itself in various ways (see §3.3).

Each of the operators A_x $(x \in X)$ can be perceived as an "unrealized" random variable. By means of the formula

$$\mathcal{B}(\mathbb{C}^1) \ni \delta \mapsto (E_x(\delta)\Omega, \Omega)_H$$

the resolution E_x of the identity gives the distribution of the corresponding realized random variable a_x, independently of the method of realization (see (3.6), where $\alpha = \mathfrak{C}(x; \delta)$; the term "unrealized" is omitted below). We shall not give a consequent presentation of the corresponding part of the theory of random processes from an "invariant" point of view, but confine ourselves to isolated remarks.

1. Let $(A_x)_{x \in X'}$ $(X' \subseteq X)$ be a set of random variables. These variables are said to be independent if

$$(E_{x_1, \ldots, x_p}(\cdot)\Omega, \Omega)_H = (E_{x_1}(\cdot)\Omega, \Omega)_H \times \cdots \times (E_{x_p}(\cdot)\Omega, \Omega)_H$$

$(x_1, \ldots, x_p \in X'$ distinct, $p = 1, 2, \ldots)$. The expectation E and the variance Var of the random variable A_x are defined by $\mathsf{E}A_x = (A_x\Omega, \Omega)_H$ (if $\Omega \in \mathfrak{D}(A_x)$) and $\operatorname{Var} A_x = \|(A_x - (\mathsf{E}A_x)1)\Omega\|_H^2 \leq \infty$. If $\Omega \notin \mathfrak{D}(A_x)$, then we can set $\mathsf{E}A_x = \lim_{n \to \infty}(A_x D_n \Omega, D_n \Omega)_H$ if the last limit exists (the D_n are operators in $L_\infty(\mathcal{A})$ associated with $A_x \in L_0(\mathcal{A})$). The covariance of the random variables A_x and A_y is defined, if it exists, by the relation

$$\operatorname{cov}(A_x, A_y) = (A_x - (\mathsf{E}A_x)1)\Omega, (A_y - (\mathsf{E}A_y)1\Omega)_H \qquad (x, y \in X).$$

2. A selfadjoint quantum process is said to be Gaussian (with zero mean, and nondegenerate) if for any distinct $x_1, \ldots, x_p \in X$ $(p = 1, 2, \ldots)$ the

(2)In quantum field theory the situation can arise when the operators A_x do not always commute. We shall not touch upon such situations in this book.

following representation holds in terms of the integral with respect to Lebesgue measure:

$$(E_{x_1,\dots,x_p}(\delta)\Omega,\Omega)_H$$
$$= (2\pi)^{-p/2}(\mathrm{Det}(b_{jk})_{j,k=1}^p)^{1/2}\int_\delta \exp\left(-\frac{1}{2}\sum_{j,k=1}^p b_{jk}\xi_j\xi_k\right)d\xi$$
$$(\delta\in\mathcal{B}(\mathsf{R}^p)).\quad (3.23)$$

Here $(b_{jk})_{j,k=1}^p = ((a_{jk})_{j,k=1}^p)^{-1}$, where $(a_{jk})_{j,k=1}^p$, $a_{jk} = (A_{x_j}\Omega, A_{x_k}\Omega)_H$, is the covariance matrix. (It is assumed that $\Omega\in\mathfrak{D}(A_x)$ $(x\in X)$ and that each such matrix is invertible, i.e., the vectors $A_{x_1}\Omega,\dots,A_{x_p}\Omega$ are linearly independent.) We could also define a general Gaussian process (with nonzero mean, or degenerate), for example, by means of the so-called characteristic function of the process.

3. Theorems 2.6–2.8 on expansions in joint generalized eigenvectors can be interpreted as results on choosing functions $X\ni x\mapsto a_x(\omega)\in\mathbb{C}^1$ for a particular ω in the case of an ordinary random process $(a_x)_{x\in X}$, i.e., choosing sample paths of the process. Each such function is an eigenvalue corresponding to the joint generalized eigenvector δ_ω of the operators of multiplication by $a_x(\cdot)$. (Of course, M and μ must be such that δ_ω can be defined in a reasonable way.) The meaning of the theorems from this point of view amounts to the possibility of choosing a full (with respect to the measure m corresponding to the process) supply of these functions by working only with A_x and H, without a preliminary realization of the process.

4. Let us see what the example of a joint RI with empty support (see §1.7) leads to from the point of view of random processes. We apply Theorem 3.2 or (in order to express M and μ explicitly) Remark 3 in §3.3 to this family of selfadjoint operators $(A_x)_{x\in X}$ $(X=(\xi)\cup\mathsf{R}^1)$. Consider all possible operators of multiplication in the space $H = L_2(\mathsf{R}^1, dt)$ by functions $a(\cdot)\in C(\mathsf{R}^1)$ having finite equal limits as $t\to\pm\infty$. They form a commutative C^*-algebra $\mathcal{A}_1$ whose space M of maximal ideals ω is the compactified line $\mathbf{R}^1$. Let $\Omega = \pi^{-1/4}e^{-t^2/2}$. It is easy to see that $\{A\Omega | A\in\mathcal{A}_1\}$ is dense in $L_2(\mathsf{R}^1, dt)$. The algebra $L_\infty(\mathcal{A}_1)$ consists of all the operators of multiplication in $L_2(\mathsf{R}^1, dt)$ by bounded measurable functions, and $L_0(\mathcal{A}_1)$ consists of the operators of multiplication by measurable functions. The action of the operator A_x $(x\in\mathsf{R}^1)$ reduces to multiplication by the function $a_x(t) = (t-x)^{-1}$; therefore, each such operator is in $L_0(\mathcal{A}_1)$; also, $A_\xi\in L_0(\mathcal{A}_1)$. Comparing (3.8) and (3.9), we conclude that $d\mu(\omega) = \pi^{-1/2}e^{-\omega^2}\,d\omega$ here. Accordingly, the following random process is a realization of the quantum process under discussion: $M=\mathsf{R}^1$, $\mathfrak{M} = \mathcal{B}(\mathsf{R}^1)$, $d\mu(\omega) = \pi^{-1/2}e^{-\omega^2}\,d\omega$, $(a_x(\omega))_{x\in X}$, where $a_\xi(\omega)\equiv\omega$, and

$a_x(\omega)$ has the form indicated above for $x \in \mathbb{R}^1$. The probability measure $\mathcal{C}_\sigma(\mathbb{R}^X) \ni \alpha \mapsto m(\alpha) \geq 0$ corresponding to this process has the property that $\operatorname{supp} m = \varnothing$.

§4. Representation of algebraic structures by commuting operators

If a family $(A_x)_{x\in X}$ of commuting normal operators is arbitrary, then an arbitrary RI E on $\mathbb{C}^X$ can serve as its joint RI. However, if there are definite algebraic relations among the operators A_x, then E is no longer arbitrary. We consider certain types of such relations and use the spectral projection theorem to determine where E must be concentrated in these cases. It will be convenient to use the concept of a representation of an algebraic structure X: the mapping $X \ni x \mapsto A_x$ must be such that it translates the algebraic operations in X into the corresponding operations on the operators A_x. For simplicity of the formulations we use mainly a scheme with a nuclear rigging.

4.1. *Representations of a group.* Let X be an abelian group with respect to addition. A generalized character of it is defined to be a complex-valued function $X \ni x \mapsto \chi(x) \in \mathbb{C}^1$ such that $\chi(x+y) = \chi(x)\chi(y)$ $(x, y \in X)$; if in addition $|\chi(x)| = 1$ $(x \in X)$, then χ is a character. The collection of all generalized characters (characters) is denoted by X_g (X); we have $\mathrm{X}_g \subseteq \mathrm{X} \subseteq \mathbb{C}^X$. In the case of a topological group X it is natural to consider characters that are in addition continuous functions; let $\mathrm{X}_{g,c} = \mathrm{X}_g \cap C(X)$ and $\mathrm{X}_c = \mathrm{X} \cap C(X)$. ($C(X)$ is the collection of all complex-valued continuous functions on X.) Let $\mathrm{X}_{g,\mathrm{Re}}$ and $\mathrm{X}_{g,c,\mathrm{Re}}$ be the corresponding collections of real-valued characters.

A function $X \ni x \mapsto A_x$ whose values are normal (in general bounded) operators acting in a separable Hilbert space H_0 is called a representation of X if $A_{x+y} = A_x A_y$ $(x, y \in X)$. Such a representation is said to be nuclear if there exists a nuclear rigging $\Phi' \supseteq H_0 \supseteq \Phi$ that is connected in the standard way with $A = (A_x)_{x\in X}$ (i.e., $\Phi \subseteq \mathfrak{D}(A_x)$ and $A_x \restriction \varphi \in \mathcal{L}(\varphi, \varphi)$ $(x \in X)$), and it is said to be a nuclear continuous representation if in addition the vector-valued function $X \ni x \mapsto A_x\varphi \in \varphi$ is weakly continuous for each $\varphi \in \Phi$.

THEOREM 4.1. *Let X be an abelian group, and $X \ni x \mapsto A_x$ a nuclear representation of it by normal operators. Then the following spectral integral representation is valid:*

$$A_x = \int_{\mathrm{X}_g} \chi(x)\, dE(\chi),$$

$$\mathfrak{D}(A_x) = \left\{ f \in H_0 \mid \int_{\mathrm{X}_g} |\chi(x)|^2\, d(E(\chi)f, f)_{H_0} < \infty \right\} \qquad (x \in X), \quad (4.1)$$

where E is an RI on the σ-algebra $\mathcal{C}_\sigma(\mathrm{X}_g) = \{\alpha \cap \mathrm{X}_g | \alpha \in \mathcal{C}_\sigma(\mathbb{C}^X)\}$. If the operators A_x are unitary (selfadjoint), then X_g is replaced above by X $(\mathrm{X}_{g,\mathrm{Re}})$; if X is a topological group and the representation is nuclear and continuous, *then* X_g *is replaced by* $X_{g,c}$, *and if in addition the* A_x *are unitary* (*selfadjoint*), *then* X_g *is replaced by* X_c ($X_{g,c,\mathrm{Re}}$).

PROOF. We consider the main case. The operators A_x commute: $A_x A_y = A_{x+y} = A_y A_x$ ($x, y \in X$), and it is known (Devinatz and Nussbaum [1]) that if A, B, and C are normal operators acting in a Hilbert space and $AB = BA = C$, then the RI's of A and B commute. Next, consider Remark 3 in §2.10 concerning the definition of the standard connection between a rigging and a family $A = (A_x)_{x\in X}$: below, the generalized spectrum $g(A)$ is understood as the collection of all $\lambda(\cdot) \in \mathbb{C}^X$ such that only the second equality in (2.19) holds for some $\varphi \in \Phi'$ and all $u \in \Phi$. We have the inclusion

$$g(A) \subseteq X_g. \tag{4.2}$$

Indeed, let $\lambda(\cdot) \in g(A)$, and let $\varphi \in \Phi'$ be a corresponding generalized joint eigenvector. With the help of the second equality in (2.19) and the inclusion $\mathfrak{R}(A_y \restriction \Phi) \subseteq \Phi$, we get that for each $u \in \Phi$

$$\begin{aligned}\overline{\lambda(x+y)}(\varphi, u)_{H_0} &= (\varphi, A_{x+y}u)_{H_0} = (\varphi, A_x A_y u)_{H_0} = \overline{\lambda(x)}(\varphi, A_y u)_{H_0} \\ &= \overline{\lambda(x)\lambda(y)}(\varphi, u)_{H_0} \qquad (x, y \in X).\end{aligned} \tag{4.3}$$

Choosing u such that $(\varphi, u)_{H_0} \neq 0$, we find from (4.3) that $\lambda(x+y) = \lambda(x)\lambda(y)$ ($x, y \in X$), i.e., $\lambda(\cdot) \in X_g$, and (4.2) is valid.

We apply Theorem 2.7 to A; more precisely (considering Remark 2 in §2.10), we apply Remark 1 after this theorem. Let $\tau = X_g$; by (4.2), $\tau \supseteq g(A) \supseteq \pi$, and hence it is possible to write the representation (2.24) with π replaced by $\tau = X_g$. Denoting E_τ by E, we get (4.1).

The remaining representations are obtained analogously, except that (4.2) is modified on the basis of the additional information available about the operators A_x. For example, if the A_x are selfadjoint, then $g(A)$ consists of real-valued functions (see §2.5), and thus (4.2) becomes $g(A) \subseteq X_{g,\mathrm{Re}}$. We then proceed as before, but with $\mathbb{C}^X$ replaced by $\mathbb{R}^X$ (see Remark 4 in §2.10). If X is a topological group and the representation is nuclear and continuous, then, by §2.9, each eigenvalue $\lambda(\cdot)$ is in $C(X)$, so $g(A) \subseteq X_{g,c}$. The required formula of the form (4.1) follows from this. We proceed similarly in the case of a nuclear continuous representation by selfadjoint operators.

Suppose now that the representation is unitary, i.e., A_x ($x \in X$) is unitary; then, by Theorem 2.8, (1.14), and (4.2), the set π can be chosen so that $\pi \subseteq s(A) \cap g(A) \subseteq \{z \in \mathbb{C}^1 \mid |z| = 1\}^X \cap X_g = X$; for $\tau = X$ this gives us the corresponding formula of type (4.1). The case of X_c is handled similarly. ■

4.2. ***Generalization of Stone's theorem.*** The conditions of Theorem 4.1 have the nature of sufficient conditions: they imply a formula of type (4.1). The determination of necessary and sufficient conditions for such representations to be valid is a more subtle question; here one needs to go from nuclear chains to quasinuclear chains. We present a result of this kind: Stone's theorem for a representation of a Hilbert space as an additive group. The second part of the proof of this theorem (like that of Theorem 4.4; see below) is closely connected to the constructions in §5.5.

Accordingly, consider the chain

$$H_0 \supseteq H_+ \supseteq D, \tag{4.4}$$

where H_0 and H_+ are Hilbert spaces, H_+ is imbedded by a quasinuclear mapping as a dense subspace of H_0, and D is a separable projective limit of Hilbert spaces that is topologically imbedded in H_+.

THEOREM 4.2. *A unitary representation $X \ni x \mapsto A_x \in \mathcal{L}(H_0, H_0)$ of a real separable Hilbert space X has the form*

$$A_x = \int_X e^{i(\lambda,x)_X}\, dE(\lambda) \qquad (x \in X), \tag{4.5}$$

where E is an RI on the σ-algebra $\mathcal{B}(X)$ of Borel subsets of X, if and only if there exists a rigging (4.4) such that 1) $A_x \restriction D \in \mathcal{L}(D, H_+)$ $(x \in X)$, and 2) for each $u \in D$ the vector-valued function $X \ni x \mapsto A_x u \in H_+$ is weakly continuous. The RI in (4.5) is uniquely determined by $(A_x)_{x\in X}$.

PROOF. The family $A = (A_x)_{x\in X}$ of commuting unitary operators is connected in the standard way with the chain (4.4); according to §2.5, we must prove in addition that $A_x^* \restriction D \in \mathcal{L}(D, H_+)$; but this follows from the assumption that $A_x \restriction D \in \mathcal{L}(D, H_+)$ and the equality $A_x^* = A_x^{-1} = A_{-x}$ $(x \in X)$. (This simplification of the definition of a standard connection in the case of unitary A_x is also valid in Theorem 4.1, of course.) However, (4.3) is now false, because D is not invariant under A_x. We proceed as follows.

If $(u_j)_1^\infty$ ($(x_l)_1^\infty \ni 0$) is a countable set of vectors dense in D (in X), then $H_+ \supset (A_{x_l} u_j)_{j,l=1}^\infty \supseteq (u_j)_1^\infty$ $(A_0 = 1)$ is dense in H_+. Take in $(A_{x_l} u_j)_{j,l=1}^\infty$ an arbitrary sequence of linearly independent vectors whose linear span D_1 contains this set. Each vector in D has the form $u = \sum_{k=1}^\infty \xi_k A_{x_{l_k}} u_{j_k}$, where $(\xi_k)_1^\infty$, $\xi_k \in \mathbf{C}^1$, is an arbitrary finitely nonzero sequence ($\xi_k = 0$ for all large k) of coordinates of u. We turn D_1 into a linear topological space by introducing the convergence: $D_1 \ni u^{(n)} \to u \in D_1$ as $n \to \infty$ if the corresponding coordinates $\xi_k^{(n)}$ are uniformly finitely nonzero and $\xi_k^{(n)} \to \xi_k$ for each $k = 1, 2, \ldots$. It is easy to see that D_1 is a separable projective limit of Hilbert spaces with topological imbeddings (cf. step IV in the proof that the rigging exists). The quasinuclear chain $H_- \supseteq H_0 \supseteq H_+ \supseteq D_1$ is connected in the standard way with A: if $(\xi_k)_1^\infty$ is finitely nonzero, then

$$A_x u = A_x \left(\sum_{k=1}^\infty \xi_k A_{x_{l_k}} u_{j_k} \right) = \sum_{k=1}^\infty \xi_k A_{x + x_{l_k}} u_{j_k} \in H_+;$$

this implies that $A_x \restriction D_1 \in \mathcal{L}(D_1, H_+)$, and the analogous relations are valid for $A_x^* = A_{-x}$ $(x \in X)$. It is also clear that the function $X \ni x \mapsto A_x u \in H_+$ is weakly continuous for each $u \in D_1$. According to §2.9, $g(A) \subseteq C(X)$, where $g(A)$ is understood to be the generalized spectrum connected with the quasinuclear chain constructed.

We show that $g(A) \subseteq \mathrm{X}_{g,c}$. Let $\lambda(\cdot) \in g(A)$, and let $\varphi \in H_-$ be a corresponding generalized joint eigenvector. The second equality in (2.19) gives us that for any $x \in X$

$$\overline{\lambda(x+x_l)}(\varphi, u_j)_{H_0} = (\varphi, A_{x+x_l} u_j)_{H_0} = (\varphi, A_x A_{x_l} u_j)_{H_0}$$
$$= \overline{\lambda(x)}(\varphi, A_{x_l} u_j)_{H_0} = \overline{\lambda(x)\lambda(x_l)}(\varphi, u_j)_{H_0}$$
$$(j, l = 1, 2, \ldots).$$

(We have used the fact that $u_j, A_{x_l} u_j \in D_1$.) It follows from this equality, the continuity of $\lambda(\cdot)$, and the denseness of the vectors x_l in X that $\lambda(x+y) = \lambda(x)\lambda(y)$ $(x, y \in X)$, i.e., $g(A) \subseteq \mathrm{X}_{g,c}$.

By Theorem 2.8, (1.14), and what was just proved, the set π can be chosen in such a way that $\pi \subseteq s(A) \cap g(A) \subseteq \mathrm{X}_c$. We let $\tau = \mathrm{X}_c$ and write the corresponding formula of type (4.1). Each continuous character $\chi(x)$ on X is easily seen to equal $e^{i(\lambda,x)_X}$ for some vector $\lambda \in X$. This gives us a bijection $\mathrm{X}_c \ni \chi(\cdot) = e^{i(\lambda,\cdot)_X} \mapsto \lambda \in X$. Denoting the image of E under this mapping again by E, we pass from the formula (4.1) just written to (4.5).

It is easy to see that the RI E thus obtained is defined on $\mathcal{C}_\sigma(X) = \{\alpha \cap X \mid \alpha \in \mathcal{C}_\sigma(\mathbb{C}^X)\}$ $(x = x')$, and this σ-algebra coincides with $\mathcal{B}(X)$. Indeed, since X is separable, it has a countable neighborhood base in the weak topology. This means that $\mathcal{C}_\sigma(X)$ coincides with the σ-algebra of Borel (with respect to the weak topology) subsets of X, and that σ-algebra coincides with the σ-algebra $\mathcal{B}(X)$ of Borel subsets in the norm-topology of X. (Here it should just be explained that each ball $\{x \in X \mid \|x\|_X \leq 1\}$ can be obtained as the intersection of countably many half-spaces $\{x \in X \mid (x, y_n)_X \leq 1\}$, where $(y_n)_1^\infty$ is a dense sequence of points in its surface.)

We proceed to prove the converse assertion of the theorem; results from §4.5 in Chapter 1 will be exploited here. (The notation from that subsection is also used.)

Consider a sequence $\tau = (\tau_n)_1^\infty \in \mathrm{T}$ such that $((1+\tau_n)^{-1})_1^\infty \in l_1$, and define the space R^τ according to (4.44). Since X is isomorphic to R^τ (any two separable real Hilbert spaces are isomorphic), it can be assumed that the representation (4.5) in the theorem to be proved is valid with $X = R^\tau$. For what follows, it is convenient to identify X' not with X but with the negative space R_τ in the chain (4.48) of Chapter 1. (Use the relation $(x, y)_{R^\tau} = (\mathbf{I}^{-1}x, y)_{l_2}$ $(x, y \in R^\tau)$, where the operator $\mathbf{I}$ is connected with this chain.) Let F be the image of the RI E under the mapping $R^\tau \ni x \mapsto \mathbf{I}_x^{-1} \in R_\tau$, i.e., F is defined by $\mathcal{B}(R_\tau) \ni \alpha \mapsto F(\alpha) = E(\mathbf{I}(\alpha))$. It is clear that F is a resolution of the identity in R_τ. The equality (4.5) can now be rewritten in the form

$$A_x = \int_{R^\tau} e^{i(\lambda,x)_{l_2}}\, dF(\lambda) \qquad (x \in R^\tau). \tag{4.6}$$

Accordingly, it must be proved that the existence of the representation (4.6) implies that a rigging (4.4) with the required properties can be constructed. A proof is given in several steps.

I. Fix a vector $l_1 \in H_0$, $\|l_1\|_{H_0} = 1$, and consider the linear mapping

$$A_\tau(\mathsf{R}^\infty) \ni a \mapsto Q_1 a = \int_{R_\tau} a(\lambda)\|\delta_\lambda\|^{-1}_{A_{-\tau}(\mathsf{R}^\infty)}\, dF(\lambda) l_1 \in H_0. \tag{4.7}$$

Note that $a(\lambda)$ is continuous for $\lambda \in R_\tau$ (see Theorem 4.6 in Chapter 1). By (4.36) in Chapter 1, the function under the integral sign in (4.7) is also continuous, and it has modulus at most $\|a\|_{A_\tau(\mathsf{R}^\infty)}$; therefore, the integral exists and admits the estimate

$$\|Q_1 a\|^2_{H_0} = \int_{R^\tau} |a(\lambda)|^2 \|\delta_\lambda\|^2_{A_{-\tau}(\mathsf{R}^\infty)}\, d(F(\lambda) l_1, l_1)_{H_0} \le \|a\|^2_{A_\tau(\mathsf{R}^\infty)}. \tag{4.8}$$

Thus, Q_1 is continuous and $\|Q_1\| \le 1$. Moreover, Q_1 is a Hilbert-Schmidt operator and $|Q_1| = 1$. Indeed, consider the orthonormal basis $e_\alpha(\lambda) = (1+\tau_1)^{-\alpha_1/2}(1+\tau_2)^{-\alpha_2/2}\cdots h_\alpha(\lambda)$ ($\alpha \in \mathrm{A}$) in the space $A_\tau(\mathsf{R}^\infty)$. By (4.15) and (4.34) in Chapter 1,

$$\begin{aligned} |Q_1|^2 &= \sum_{\alpha\in\mathrm{A}} \|Q_1 e_\alpha\|^2_{H_0} \\ &= \sum_{\alpha\in\mathrm{A}} \int_{R^\tau} (h^2_\alpha(\lambda)(1+\tau_1)^{-\alpha_1}(1+\tau_2)^{-\alpha_2}\cdots) \\ &\qquad\qquad \times \|\delta_\lambda\|^2_{A_{-\tau}(\mathsf{R}^\infty)}\, d(F(\lambda) l_1, l_1)_{H_0} \\ &= 1. \end{aligned}$$

II. Consider the kernel of the operator Q_1, i.e., the subspace $\{a \in A_\tau(\mathsf{R}^\infty) \mid Q_1 a = 0\}$, and denote by $\hat{A}_\tau(\mathsf{R}^\infty)$ the orthogonal complement of it. Let $\hat{Q}_1$ be the restriction of Q_1 to $\hat{A}_\tau(\mathsf{R}^\infty)$. Choosing an orthonormal basis in $A_\tau(\mathsf{R}^\infty)$ whose vectors are contained in $\hat{A}_\tau(\mathsf{R}^\infty)$ and in this kernel, we get that $|\hat{Q}_1| = |Q_1| = 1$. The inverse operator $\hat{Q}_1^{-1}$ exists on $\mathfrak{R}(\hat{Q}_1) = \mathfrak{R}(Q_1)$, and it is closed (since $\hat{Q}_1$ is continuous) and satisfies

$$\|\hat{Q}_1^{-1} f\|_{A_\tau(\mathsf{R}^\infty)} \ge \|f\|_{H_0} \qquad (f \in \mathfrak{R}(Q_1)). \tag{4.9}$$

We now turn $\mathfrak{R}(Q_1) \subseteq H_0$ into a Hilbert space $H_{+,1}$ by setting

$$(f, g)_{H_{+,1}} = (\hat{Q}_1^{-1} f, \hat{Q}_1^{-1} g)_{A_\tau(\mathsf{R}^\infty)} \qquad (f, g \in \mathfrak{R}(Q_1)).$$

Since $\hat{Q}_1^{-1}$ is closed and (4.9) holds, $H_{+,1} = \mathfrak{R}(Q_1)$ is complete. The imbedding O_1: $H_{+,1} \to H_0$ is quasinuclear, and $|Q_1| = 1$. Indeed, if $(a_n)_1^\infty$ is an orthonormal basis in $\hat{A}_\tau(\mathsf{R}^\infty)$, then $(\hat{Q}_1 a_n)_1^\infty$ is an orthonormal basis in $H_{+,1}$. For it,

$$|O_1|^2 = \sum_{n=1}^\infty \|\hat{Q}_1 a_n\|^2_{H_0} = |\hat{Q}_1|^2 = 1.$$

Let $D_1 = Q_1 P_c(\mathbb{R}^\infty) \subset H_{+,1}$, where $P_c(\mathbb{R}^\infty)$ is the collection of all cylindrical polynomials. In other words, D_1 consists of all the vectors of the form $u = \sum_{\alpha \in \mathrm{A}} u_\alpha Q_1 e_\alpha$, where the function $\mathrm{A} \ni \alpha \mapsto u_\alpha \in \mathbb{C}^1$ is finitely nonzero. Hence, we have obtained a chain

$$H_0 \supseteq H_{+,1} \supseteq D_1, \tag{4.10}$$

with the imbedding $O_1 \colon H_{+,1} \to H_0$ quasinuclear and D_1 dense in $H_{+,1}$.

III. If $H_{+,1}$ is dense in H_0, then the construction is finished. If not, then choose a vector $l_2 \in H_0$, $\|l_2\|_{H_0} = 1$, orthogonal to $H_{+,1}$ and repeat the construction of I–III with l_1 replaced by l_2. This gives a chain

$$H_0 \supseteq H_{+,2} \supseteq D_2 \tag{4.11}$$

with quasilinear imbedding $O_2 \colon H_{+,2} \to H_0$ and D_2 dense in $H_{+,2}$. Let Q_2 denote the operator (4.7) with l_1 replaced by l_2.

The space $H_{+,2}$ is orthogonal to $H_{+,1}$ in the inner product $(\cdot,\cdot)_{H_0}$. Indeed, for $a, b \in A_\tau(\mathbb{R}^\infty)$

$$(Q_1 a, Q_2 b)_{H_0} = \int_{R_\tau} a(\lambda)\overline{b(\lambda)} \|\delta_\lambda\|^{-2}_{A_{-\tau}(\mathbb{R}^\infty)}\, d(F(\lambda)l_1, l_2)_{H_0}.$$

This implies that it suffices to prove that

$$\int_\alpha \|\delta_\lambda\|^{-1}_{A_{-\tau}(\mathbb{R}^\infty)}\, d(F(\lambda)l_1, l_2)_{H_0} = 0$$

for any $\alpha \in \mathcal{B}(R_\tau) = \{\beta \cap R_\tau \mid \beta \in \mathcal{C}_\sigma(\mathbb{R}^\infty)\}$, hence for any $\alpha \in \{\beta \cap R_\tau \mid \beta \in \mathcal{C}_\sigma(\mathbb{R}^\infty)\}$. The last set has the form $\alpha = \{\lambda \in R_\tau \mid (\lambda_1, \ldots, \lambda_p) \in \delta\}$, where $\delta \in \mathcal{B}(\mathbb{R}^p)$ for some $p = 1, 2, \ldots$. Therefore, it suffices to establish that the complex-valued finite measure

$$\mathcal{B}(\mathbb{R}^p) \ni \gamma \mapsto \omega(\gamma) = \int_{\{\lambda \in R_\tau \mid (\lambda_1, \ldots, \lambda_p) \in \gamma\}} d_\tau^{-1} \exp\left(- \sum_{k=p+1}^{\infty} \lambda_k^2 (2 + \tau_k)^{-1} \right) \times d(F(\lambda)l_1, l_2)_{H_0}$$

is zero: by (4.36) in Chapter 1,

$$\int_{\{\lambda \in R_\tau \mid (\lambda_1, \ldots, \lambda_p) \in \gamma\}} \|\delta_\lambda\|^{-1}_{A_{-\tau}(\mathbb{R}^\infty)}\, d(F(\lambda)l_1, l_2)_{H_0} = \int_\gamma \exp\left(- \sum_{k=1}^{p} \lambda_k^2 (2 + \tau_k)^{-1} \right) d\omega(\lambda_1, \ldots, \lambda_p).$$

For a proof, note that, by construction, the vector l_2 is orthogonal in H_0 to the space $H_{+,1}$, hence to all the vectors $Q_1 e_\alpha$ ($\alpha \in \mathrm{A}$). In particular, it is orthogonal to all the vectors $Q_1(h_{\alpha_1}(\lambda_1) \cdots h_{\alpha_p}(\lambda_p))$ $(\alpha_1, \ldots, \alpha_p = 0, 1, \ldots)$, i.e.,

$$\begin{aligned}
0 &= (Q_1(h_{\alpha_1}(\lambda_1)\cdots h_{\alpha_p}(\lambda_p)), l_2)_{H_0} \\
&= \int_{R_\tau} h_{\alpha_1}(\lambda_1)\cdots h_{\alpha_p}(\lambda_p)\|\delta_\lambda\|^{-1}_{A_\tau(\mathbf{R}^\infty)}\, d(F(\lambda)l_1, l_2)_{H_0} \\
&= \int_{\mathbf{R}^p} h_{\alpha_1}(\lambda_1)\cdots h_{\alpha_p}(\lambda_p)\exp\left(-\sum_{k=1}^{p}\lambda_k^2(2+\tau_k)^{-1}\right) d\omega(\lambda_1,\ldots,\lambda_p) \\
&\qquad\qquad (\alpha_1,\ldots,\alpha_p = 0,1,\ldots). \qquad (4.12)
\end{aligned}$$

Below we prove Lemma 4.1, from which it follows that (4.12) implies $\omega = 0$. Thus, $H_{+,2}$ is orthogonal in H_0 to $H_{+,1}$.

IV. By (4.10) and (4.11),

$$H_0 \supseteq H_{+,1} \oplus H_{+,2} \supseteq D_1 \oplus D_2, \tag{4.13}$$

where $\oplus$ denotes the orthogonal sum in H_0. If $H_{+,1} \oplus H_{+,2}$ is dense in H_0, then the construction is finished. If not, then we choose a vector $l_3 \in H_0$, $\|l_3\|_{H_0} = 1$, orthogonal to $H_{+,1} \oplus H_{+,2}$ and construct a chain analogous to (4.10) and (4.11), and so on. As a result we obtain the chain

$$H_0 \supseteq \bigoplus_{n=1}^{\infty} H_{+,n} \supseteq \bigoplus_{n=1}^{\infty} D_n, \tag{4.14}$$

where the orthogonal summation is understood in H_0, and $\bigoplus_1^\infty H_{+,n}$ is now dense in H_0. (There can also be finitely many terms; for definiteness we consider the more complicated case when there are infinitely many.)

Let $\bigoplus_1^\infty H_{+,n}$ now be replaced in (4.14) by the weighted orthogonal sum $\bigoplus_{n=1;\delta}^\infty H_{+,n}$ of these spaces, where $\delta = (\delta_n)_1^\infty$ ($\delta_n > 0$) is some weight (see Chapter 1, §1.9). As before, $\bigoplus_{n=1;\delta}^\infty H_{+,n}$ is dense in H_0, the corresponding imbedding operator is equal to $O = \bigoplus_1^\infty O_n$, and

$$|O|^2 = \sum_{n=1}^{\infty} \delta_n^{-1} |O_n|^2 = \sum_{n=1}^{\infty} \delta_n^{-1}. \tag{4.15}$$

We choose δ_n to decrease rapidly enough that the series (4.15) converges, and let $H_+ = \bigoplus_{n=1;\delta}^\infty H_{+,n}$. Thus, the imbedding $H_+ \to H_0$ is now quasinuclear.

Take D to be the collection of all finitely nonzero sequences $u = (u_n)_1^\infty$, $u_n \in D_n$. We topologize D by regarding $u^{(k)} \in D$ as convergent to a $u \in D$ as $k \to \infty$ if the sequences $u^{(k)}$ are finitely nonzero uniformly with respect to k, $u_n^{(k)} \to u_n$ for each n in the sense of convergence of the corresponding coordinates $(u_n^{(k)})(\alpha)$ when the vector $u_n^{(k)}$ is expanded in the basis vectors $(Q_n e_\alpha)_{\alpha\in \mathrm{A}}$ (here Q_n denotes the operator Q_1 for the nth step), and the functions $\mathrm{A} \ni \alpha \mapsto u_n^{(k)}(\alpha) \in \mathbf{C}^1$ are finitely nonzero uniformly with respect to k. It is easy to understand that this topologization of D can be performed more accurately by specifying a neighborhood system. To do this, consider the Hilbert space

$$G_{n,\sigma_n} = \left\{ \sum_{\alpha\in A} f_\alpha(Q_n e_\alpha) \,\middle|\, \sum_{\alpha\in A} |f_\alpha|^2 \sigma_n(\alpha) < \infty \right\}$$

$$(\sigma_n = (\sigma_n(\alpha))_{\alpha\in A},\ \sigma_n(\alpha) \geq 1)$$

with the natural inner product, and then construct the weighted orthogonal sum $\bigoplus_{n=1;\rho}^{\infty} G_{n,\delta_n}$ ($\rho = (\rho_n)_1^\infty$, $\rho_n \geq 1$, is the weight). Then

$$D = \operatorname{pr}\lim \bigoplus_{n=1;\rho}^{\infty} G_{n,\sigma_n},$$

where the projective limit is over all weights σ_n and ρ. Obviously, the imbedding $D \to H_+$ is topological, and D is dense in H_+.

Thus, the chain (4.4) has been constructed. We show that conditions 1) and 2) in the statement of the theorem are valid.

V. To prove the inclusion $A_x \restriction D \in \mathcal{L}(D \to H_+)$ ($x \in X = R^\tau$), it suffices to see that $A_x \restriction D_n \in \mathcal{L}(D_n \to H_{+,n})$ ($n = 1, 2, \ldots; x \in R^\tau$). If, for example, $n = 1$ and $u \in D_1$, then $u = Q_1 p$, where $p \in P_c(\mathsf{R}^\infty)$. According to (4.6) and (4.7), for a particular $x \in R^\tau$,

$$\begin{aligned} A_x u &= \int_{R_\tau} \exp i(\lambda, x)_{l_2}\, dF(\lambda) \left(\int_{R_\tau} p(\mu) \|\delta_\mu\|^{-1}_{A_{-\tau}(\mathsf{R}^\infty)}\, dF(\mu) l_1 \right) \\ &= \int_{R_\tau} \exp i(\lambda, x)_{l_2} p(\lambda) \|\delta_\lambda\|^{-1}_{A_{-\tau}(\mathsf{R}^\infty)}\, dF(\lambda) l_1. \end{aligned} \tag{4.16}$$

By Remark 2 in §4.5 of Chapter 1, the functions $\exp(i(\cdot, x)_{l_2}) p(\cdot) \in A_\tau(\mathsf{R}^\infty)$ (λ and x have now switched roles). Therefore, $A_x u = Q_1(\exp(i(\cdot, x)_{l_2}) p(\cdot)) \in \mathfrak{R}(Q_1) = H_{+,1}$. The operator $D_1 \ni u \mapsto A_x u \ni H_{+,1}$ is obviously continuous, since convergence of u means coordinatewise convergence when u is expanded in the basis $(Q_1 e_\alpha)_{\alpha\in A}$, with the coordinates uniformly finitely nonzero. This shows that condition 1) holds.

Condition 2) amounts to weak continuity of the vector-valued function $R^\tau \ni x \mapsto A_x u \in H_+$ for each $u \in D$. It suffices to consider $u \in D_n$; then $A_x u \in H_{+,n}$ ($n = 1, 2, \ldots$). For $n = 1$, $A_x u = Q_1(\exp(i(\cdot, x)_{l_2}) p(\cdot))$ by (4.16). But by the same Remark 2 in §4.5 of Chapter 1, the vector-valued function $R^\tau \ni x \mapsto \exp(i(\cdot, x)_{l_2}) p(\cdot) \in A_\tau(\mathsf{R}^\infty)$ is strongly continuous. Since $Q_1 \in \mathcal{L}(A_\tau(\mathsf{R}^\infty) \to H_0)$, it follows that $R^\tau \ni x \mapsto Q_1(\exp(i(\cdot, x)_{l_2}) p(\cdot)) = A_x u \in H_{+,1}$ is even strongly continuous.

We prove the last assertion of the theorem: that E is uniquely determined by the family $(A_x)_{x\in X}$. Let us choose an orthonormal basis in X and realize this space as the real space l_2. Then (4.5) can be rewritten in the form

$$A_x = \int_{l_2} \exp(i(\lambda, x)_{l_2})\, dE(\lambda) \qquad (x \in l_2). \tag{4.17}$$

Fix a $p = 1, 2, \ldots$; consider an $x^{(p)} \in \mathbb{R}^p$. Let $x = (x_1^{(p)}, \ldots, x_p^{(p)}, 0, 0, \ldots)$. Substituting this x in (4.17) and writing $A_x = A_{x^{(p)}}^{(p)}$, we obtain for the unitary representation $\mathbb{R}^p \ni x^{(p)} \mapsto A_{x^{(p)}}^{(p)}$ the formula

$$A_{x^{(p)}}^{(p)} = \int_{\mathbb{R}^p} \exp\left(i \sum_{n=1}^{p} \lambda_n x_n \right) dE^{(p)}(\lambda) \qquad (x^{(p)} \in \mathbb{R}^p), \tag{4.18}$$

where

$$\mathcal{B}(\mathbb{R}^p) \ni \delta \mapsto E^{(p)}(\delta) = E(\mathfrak{C}(x_1^{(p)}, \ldots, x_p^{(p)}; \delta) \cap l_2).$$

By the usual p-dimensional Stone's theorem, the resolution of the identity $E^{(p)}$ in (4.18) is uniquely determined by $(A_{x^{(p)}}^{(p)})_{x^{(p)} \in \mathbb{R}^p}$, hence uniquely determined by $(A_x)_{x \in l_2}$. But it is clear that the $E^{(p)}$ $(p = 1, 2, \ldots)$ uniquely determine E. ■

It remains for us to prove the following general lemma.

LEMMA 4.1. *Let $\mathcal{B}(\mathbb{R}^p) \ni \delta \mapsto \omega(\delta)$ be a complex-valued finite measure such that for some $\varepsilon_1, \ldots, \varepsilon_p > 0$*

$$\int_{\mathbb{R}^p} h_{\alpha_1}(x_1) \cdots h_{\alpha_p}(x_p) \exp\left(- \sum_{k=1}^{p} x_k^2 \varepsilon_k \right) d\omega(x_1, \ldots, x_p) = 0$$
$$(\alpha_1, \ldots, \alpha_p = 0, 1, \ldots).$$

Then $\omega = 0$.

PROOF. Assume that the lemma has been proved for $p = 1$ and $p = n - 1$. We show that it is then valid also for $p = n$ $(n = 2, 3, \ldots)$. Fix a $\delta_0 \in \mathcal{B}(\mathbb{R}^1)$ and consider the finite complex-valued measure

$$\mathcal{B}(\mathbb{R}^{n-1}) \ni \delta \mapsto \int_{\delta_0 \times \delta} d\omega(x_1, x_2, \ldots, x_n) = \omega_{\delta_0}(\delta);$$

then fix $\alpha_2, \ldots, \alpha_n$ and consider the analogous measure

$$\begin{aligned} \mathcal{B}(\mathbb{R}^1) \ni \delta' \mapsto & \int_{\mathbb{R}^{n-1}} h_{\alpha_2}(x_2) \cdots h_{\alpha_n}(x_n) \\ & \qquad \times \exp\left(- \sum_{k=2}^{n} x_k^2 \varepsilon_k \right) d\omega_{\delta'}(x_2, \ldots, x_n) \\ = & \int_{\delta' \times \mathbb{R}^{n-1}} h_{\alpha_2}(x_2) \cdots h_{\alpha_n}(x_n) \\ & \qquad \times \exp\left(- \sum_{k=2}^{n} x_k^2 \varepsilon_k \right) d\omega(x_1, x_2, \ldots, x_n) \\ = & \ \omega_{\alpha_2, \ldots, \alpha_n}(\delta'). \end{aligned} \tag{4.19}$$

We have that

$$0 = \int_{\mathbf{R}^n} h_{\alpha_1}(x_1) \cdots h_{\alpha_n}(x_n) \exp\left(-\sum_{k=1}^{n} x_k^2 \varepsilon_k \right) d\omega(x_1, \ldots, x_n)$$
$$= \int_{\mathbf{R}^1} h_{\alpha_1}(x_1) \exp(-x_1^2 \varepsilon_1) \, d\omega_{\alpha_2, \ldots, \alpha_n}(x_1) \qquad (\alpha_1 = 0, 1, \ldots),$$

which by assumption implies that $\omega_{\alpha_2,\ldots,\alpha_n} = 0$, i.e., $\omega_{\alpha_2,\ldots,\alpha_n}(\delta') = 0$ ($\delta' \in \mathcal{B}(\mathbf{R}^1)$). Fix δ'; then from the last equality for $\alpha_2, \ldots, \alpha_n = 0, 1, \ldots$, from (4.19), and from our assumption we conclude that $\omega_{\delta'} = 0$, i.e., $\omega_{\delta'}(\delta) = 0$ ($\delta \in \mathcal{B}(\mathbf{R}^{n-1})$. Since δ' is arbitrary, this gives us that $\omega = 0$.

It is thus clear that it suffices to establish the lemma for $p = 1$. We have that

$$0 = \int_{\mathbf{R}^1} h_{\alpha_1}(x_1) \exp(-x_1^2 \varepsilon_1) \, d\omega(x_1) = \int_{\mathbf{R}^1} h_{\alpha_1}(x_1) \exp(-x_1^2) \, d\omega_1(x_1),$$
$$\mathcal{B}(\mathbf{R}^1) \ni \delta \mapsto \omega_1(\delta) = \int_\delta \exp(x_1^2(1 - \varepsilon_1)) \, d\omega(x_1) \qquad (\alpha_1 = 0, 1, \ldots).$$

For $0 < \varepsilon_1 < 1$ the function $\omega_1(x_1) = \int_0^{x_1} \exp(t^2(1 - \varepsilon_1)) \, d\omega(t)$ is unbounded in general, but it satisfies the estimate $|\omega_1(x_1)| \leq c \exp(x_1^2(1 - \varepsilon_1))$ ($x_1 \in \mathbf{R}^1$). Therefore, by (4.38) and (4.40) in Chapter 1, integration by parts gives us that

$$0 = \int_{\mathbf{R}^1} h_{\alpha_1}(x_1) \exp(-x_1^2) \, d\omega_1(x_1) = -\int_{\mathbf{R}^1} \omega_1(x_1) \, d(h_{\alpha_1}(x_1) \exp(-x_1^2))$$
$$= \sqrt{2(\alpha_1 + 1)} \int_{\mathbf{R}^1} \omega_1(x_1) h_{\alpha_1 + 1}(x_1) \exp(-x_1^2) \, dx_1 \qquad (\alpha_1 = 0, 1, \ldots).$$

Hence, $\omega_1(x_1) = c$ ($x_1 \in \mathbf{R}^1$), i.e., $\omega_1(\delta) = 0$ ($\delta \in \mathcal{B}(\mathbf{R}^1)$), and so $\omega = 0$. The case when $\varepsilon_1 \geq 1$ is treated similarly. ■

REMARK 1. Let $X = \mathbf{R}^p$ be finite-dimensional. Then a rigging (4.4) with the required properties 1) and 2) can be constructed from a given unitary representation $\mathbf{R}^p \ni x \mapsto A_x \in \mathcal{L}(H_0, H_0)$ by means of constructions close to those described below in §5.5 (with use of the space $\mathcal{D}(\mathbf{R}^p)$). Thus, Theorem 4.2 passes into the usual Stone's theorem in this case: the requirements connected with the rigging are automatically satisfied. An analogous situation holds if X is a locally compact separable abelian group with a countable neighborhood base.

REMARK 2. A treatment analogous to that in Theorem 4.2 applies to representations of a real separable Hilbert space X by selfadjoint unbounded (in general) operators acting in H_0 (an infinite-dimensional rigging of the Sz.-Nagy–Hille theorem for groups). The formula of type (4.5) now has the form

(if we assume that $A_0 = 1$)

$$A_x = \int_X \exp((\lambda, x)_X)\, dE(\lambda),$$
$$\mathfrak{D}(A_x) = \left\{ f \in H_0 \mid \int_X \exp(2(\lambda, x)_X)\, d(E(\lambda)f, f)_{H_0} < \infty \right\}$$
$$(x \in X). \quad (4.20)$$

The analogue of Remark 1 is valid in the case of finite-dimensional X. It is also possible to consider representations of X by normal operators.

4.3. *Representations of a semigroup.* Let X be an abelian semigroup, i.e., a set with an associative and commutative operation $X \times X \ni (x, y) \mapsto x + y \in X$; the semigroup may or may not have a zero (i.e., an element $0 \in X$ such that $x + 0 = x$ for $x \in X$). If in addition X is a topological space and the mapping $X \times X \to X$ just written is continuous, then the semigroup is called a topological semigroup. The definitions of a generalized character, a character, etc., and a representation of X given above in §4.1 are preserved without changes for semigroups. Theorem 4.1 carries over to semigroups without any changes in the formulation and proof.

Matters are more complicated in the case of Theorem 4.2 and formula (4.20). For semigroups it is common to consider representations by selfadjoint or normal operators. Let X be a semigroup coinciding with the open positive octant of some real separable Hilbert space Y. In other words, a particular orthonormal basis $(e_j)_1^\infty$ is considered in Y, and

$$X = \left\{ x = \sum_{j=1}^\infty x_j e_j \mid \sum_{j=1}^\infty x_j^2 < \infty,\ x_j > 0;\ j = 1, 2, \ldots \right\}$$

with the addition and the topology induced from Y. For a representation of X by selfadjoint operators we can repeat the scheme of proof for Theorem 4.2 to Remark 2, bearing in mind two facts: 1) the real continuous characters on X have the form of the functions $\chi(x) = \chi_\lambda(x) = e^{(\lambda, x)_Y}$ $(\lambda \in Y,\ x \in X)$ and 0; and 2) it is more complicated to establish the converse assertion of the theorem, i.e., the construction of a rigging from a formula of type (4.20). We do not formulate the result itself here.

As in §4.2, we can show that local compactness, separability, and the existence of a countable neighborhood base in X automatically imply the existence of the required quasinuclear rigging of H_0 connected in the standard way with the family $A = (A_x)_{x \in X}$ realizing the representation. Representations by normal operators can also be considered in an analogous way for semigroups.

4.4. *Representation of a linear space.* Let X be a complex linear space (without a topology), and X_a^* the space of all linear functionals $X \ni x \mapsto l(x) \in \mathbb{C}^1$ on X; if X is a linear topological space, then X^* is the dual space of continuous linear functionals on it. Let $\Phi' \supseteq H_0 \supseteq \Phi$ be a nuclear rigging of a separable Hilbert space H_0. An operator-valued function $X \ni x \mapsto A_x$ whose values are commuting normal operators acting in H_0 and connected in the standard way with this rigging in the sense that $\Phi \subseteq \mathfrak{D}(A_x)$ and $A_x \upharpoonright \Phi \in \mathcal{L}(\Phi, \Phi)$ $(x \in X)$ is called a nuclear representation of X if $A_{\alpha x+\beta y}\varphi = \alpha A_x\varphi + \beta A_y\varphi$ $(x, y \in X;\ \alpha, \beta \in \mathbb{C}^1;\ \varphi \in \Phi)$. By definition, such a representation is a nuclear continuous representation if X is a linear topological space and the vector-valued function $X \ni x \mapsto A_x\varphi \in \Phi$ is weakly continuous for each $\varphi \in \Phi$.

THEOREM 4.3. *Suppose that X is a linear space and $X \ni x \mapsto A_x$ is a nuclear representation of it by commuting normal operators. Then the following spectral integral representation is valid:*

$$A_x = \int_{X_a^*} l(x)\, dE(l),$$

$$\mathfrak{D}(A_x) = \left\{ f \in H_0 \mid \int_{X_a^*} |l(x)|^2\, d(E(l)f, f)_{H_0} < \infty \right\} \qquad (x \in X), \quad (4.21)$$

where E is some RI on the σ-algebra $\mathcal{C}_\sigma(X_a^*) = \{\alpha \cap X_a^* \mid \alpha \in \mathcal{C}_\sigma(\mathbb{C}^X)\}$. If X is a linear topological space and the representation is nuclear and continuous, then the X_a^* above is replaced by X^*.

PROOF. In essence, the proof repeats the scheme of proof for Theorem 4.1. Thus, consider the case of a nontopological X. As before, let $g(A)$ be the generalized spectrum of the commuting family $A = (A_x)_{x\in X}$ (with Remark 3 in §2.10 taken into account). The inclusion $g(A) \subseteq X_a^*$ now holds instead of (4.2). Indeed, let $\lambda(\cdot) \in g(A)$ and let $\varphi \in \Phi'$ be a corresponding generalized joint eigenvector. By the second equality in (2.19),

$$\begin{aligned}\overline{\lambda(\alpha x + \beta y)}(\varphi, u)_{H_0} &= (\varphi, A_{\alpha x+\beta y}u)_{H_0} \\ &= (\varphi, (\alpha A_x + \beta A_y)u)_{H_0} \\ &= \overline{(\alpha\lambda(x) + \beta\lambda(y))}(\varphi, u)_{H_0} \\ &\qquad (x, y \in X;\ \alpha, \beta \in \mathbb{C}^1;\ u \in \Phi), \quad (4.22)\end{aligned}$$

from which we get $\lambda(\alpha x + \beta y) = \alpha\lambda(x) + \beta\lambda(y)$ $(x, y \in X;\ \alpha, \beta \in \mathbb{C}^1)$, i.e., $\lambda(\cdot) \in X_a^*$ and $g(A) \subseteq X_a^*$. We then use Theorem 2.7 (more precisely, Remark 1 after it and Remark 2 in §2.10): it is possible to write (2.7) with π replaced by $\tau = X_a^* \supseteq g(A) \supseteq \pi$. As a result we arrive at (4.21).

In the case of a nuclear continuous representation, we have that $g(A) \subseteq X^*$: use also §2.9, which implies that a $\lambda(\cdot) \in g(A)$ is continuous. ■

REMARK 1. Representations of a real linear space X by selfadjoint operators could be treated in a way analogous to that in Theorem 4.3. Or for a complex X with an involution $X \ni x \mapsto x^* \in X$ we could consider *-representations of it by commuting normal operators (i.e., we could require in addition that $A_{x^*} = A_x^*$ ($x \in X$); in this case the integration in (4.21) is over those $l \in X_a^*$ such that $l(x^*) = \overline{l(x)}$ ($x \in X$), and the standard connection between the rigging and A must now be understood in the full scope of §2.5).

We show that an analogue of Theorem 4.2 is valid.

THEOREM 4.4. *Consider the chain* (4.4). *Suppose that X is a real Hilbert space, and $(A_x)_{x\in X}$ is a family of commuting selfadjoint operators acting in H_0 and with domains $\mathfrak{D}(A_x) \supseteq D$ such that $A_{\alpha x+\beta y}u = \alpha A_x u + \beta A_y u$ ($x, y \in X$; $\alpha, \beta \in \mathbf{R}^1$; $u \in D$). If conditions* 1) *and* 2) *in Theorem* 4.2 *hold for A_x, then this family admits the representation*

$$A_x = \int_X (\lambda, x)_X \, dE(\lambda),$$

$$\mathfrak{D}(A_x) = \left\{ f \in H_0 \mid \int_X (\lambda, x)_X^2 \, d(E(\lambda)f, f)_{H_0} < \infty \right\} \qquad (x \in X), \quad (4.23)$$

where E is some RI on $\mathcal{B}(X)$. Conversely, the existence of the representation (4.23) *for some family $(A_x)_{x\in X}$ implies the existence of the chain* (4.4) *and the remaining requirements of the theorem. The resolution of the identity E in* (4.23) *is uniquely determined by $(A_x)_{x\in X}$.*

PROOF. The validity of (4.23) follows from the modification of Theorem 4.3 applied to the real Hilbert space X. The fact that $\mathcal{C}_\sigma(X) = \mathcal{B}(X)$ here was explained in the proof of Theorem 4.2. The proof of the converse assertion of the theorem mainly repeats the corresponding part of the proof of Theorem 4.2. Namely, we rewrite the representation (4.23) in the form $A_x = \int_{R_\tau} (\lambda, x)_{l_2} \, dF(\lambda)$ ($x \in R^\tau$) analogous to (4.6), and then we repeat steps I–IV. Step V changes in that the strong continuity of the vector-valued function $R^\tau \ni x \mapsto (\cdot, x)_{l_2} p(\cdot) \in A_\tau(\mathbf{R}^\infty)$ must be used. The proof of this fact is completely analogous to that of Remark 2 in §4.5 of Chapter 1. Thus, the analogue of Lemma 4.6 in Chapter 1 with $e^{i(\lambda,x)_{l_2}}$ replaced by $(\lambda, x)_{l_2}$ is used. Its proof is similar to the derivation of the lemma, with the formula

$$\int_{\mathbf{R}^1} t h_j(t) \, dg(t) = \sqrt{\frac{1}{2}} \delta_{j,1} \qquad (j = 0, 1, \ldots)$$

instead of (4.42) of Chapter 1 (see (4.40) in Chapter 1). The corresponding computations will not be given. The standard procedure establishes that E is unique. ∎

We remark that it is not possible to obtain Theorem 4.2 from Theorem 4.4 (nor conversely) by taking the corresponding functions of operators, because, for example, satisfaction of conditions 1) and 2) for A_x is not enough to imply that they hold for e^{iA_x} ($x \in X$).

4.5. ***Generalized quantum processes.*** By analogy with generalized random processes we define a generalized selfadjoint quantum process to be a family $(A_x)_{x\in X}$ of commuting selfadjoint operators acting in H_0 that has a cyclic vector Ω, where the indexing set X is a real linear topological space, and

$$A_{\alpha x+\beta y}u = \alpha A_x u + \beta A_y u \qquad (x, y \in X;\ \alpha, \beta \in \mathbb{R}^1).$$

Here u is an arbitrary element of a linear set $D \subset \mathfrak{D}(A_x)$ $(x \in X)$ dense in H_0.

We explain that, as a rule, an ordinary selfadjoint quantum process $(A_t)_{t\in\mathbb{R}^N}$ is imbedded in a generalized quantum process in the standard way: we consider the real linear topological space X of test functions $\mathbb{R}^N \ni t \mapsto x(t) \in \mathbb{R}^1$ and form the operators

$$A_x = \int_{\mathbb{R}^N} A_t x(t)\, dt \qquad (x \in X).$$

It is also clear how to define a generalized process corresponding to normal operators A_t.

Theorems 4.3 and 4.4 and the modifications of them mentioned for this or that X lead to spectral representations of such processes close to the well-known representations for random processes in the nongeneralized (Gikhman and Skorokhod [2], Chapter 4, §5) and the generalized stationary cases (Gel′fand and Vilenkin [1], Chapter 3, §3.4). We present the result that is simplest to formulate and follows from the corresponding modification of Theorem 4.3. Let $(A_x)_{x\in X}$ be a generalized selfadjoint quantum process. Assume that H_0 is separable and D is a nuclear space such that $A_x \restriction D \in \mathcal{L}(D \to D)$ $(x \in X)$ and the vector-valued function $X \ni x \mapsto A_x u \in D$ is weakly continuous for each $u \in D$. Then the following representation is valid:

$$A_x = \int_{X^*} \lambda(x)\, dE(\lambda) \qquad (x \in X),$$

where E is a resolution of the identity on $\mathcal{C}_\sigma(X^*)$.

4.6. *Representations of algebras.* We can also consider representations of a commutative algebra X over the field of complex numbers. A functional $\omega \in X_a^*$ is said to be multiplicative if $\omega(xy) = \omega(x)\omega(y)$ $(x, y \in X)$; let M_a be the set of such functionals. If X is a topological algebra, then M denotes the collection of all continuous multiplicative functionals. Suppose that the previous H_0 is given along with a nuclear rigging of it; an operator-valued function $X \ni x \mapsto A_x$ whose values are normal operators acting in H_0 and connected in the standard way with the rigging in the sense indicated at the beginning of §4.4 is called a nuclear representation of X if $A_{\alpha x+\beta y}\varphi = \alpha A_x\varphi + \beta A_y\varphi$ and $A_{xy} = A_x A_y$ $(x, y \in X;\ \alpha, \beta \in \mathbb{C}^1;\ \varphi \in \Phi)$. In the case of a topological algebra X the concept of a nuclear continuous representation is introduced as before.

As in the case of Theorems 4.1 and 4.3 we can prove the spectral integral representation

$$A_x = \int_{M_a} \omega(x)\, dE(\omega),$$

$$\mathfrak{D}(A_x) = \left\{ f \in H_0 \mid \int_{M_a} |\omega(x)|^2 \, d(E(\omega)f, f)_{H_0} < \infty \right\}$$

$$(x \in X), \quad (4.24)$$

where E is an RI on the σ-algebra $\mathcal{C}_\sigma(M_a) = \{\alpha \cap M_a \mid \alpha \in \mathcal{C}_\sigma(\mathbf{C}^X)\}$; the M_a above is replaced by M in the case of a nuclear continuous representation. (The inclusion $g(A) \subseteq M_a$, M is established on the basis of computations like (4.3) and (4.22), and then Theorem 2.7 is applied.)

Clearly, it is now possible to consider nuclear *-representations if X is a *-algebra, and so on.

4.7. ***Representations of more general structures.*** It is obvious from subsections 1–6 that the character of the relations between the operators A_x in the family $A = (A_x)_{x \in X}$ of commuting normal operators can be very general: it is necessary only that the corresponding relation, when applied to a generalized joint eigenvector, give some equation for $\lambda(\cdot)$. (Roughly speaking, the coefficients in the relation must, for example, be scalars.) Here we confine ourselves to two examples.

1) Let $X = \mathbf{R}^1$, and let $A = (A_x)_{x \in \mathbf{R}^1}$ be a family of commuting normal operators acting in H_0, connected with the nuclear rigging $\Phi' \supseteq H_0 \supseteq \Phi$ in the standard way in the sense of §4.4 and satisfying the relations $A_x A_y = \frac{1}{2}(A_{x+y} + A_{x-y})$ $(x, y \in \mathbf{R}^1)$. As in the cases of (4.3) and (4.22), we get the functional equation

$$\lambda(x)\lambda(y) = \tfrac{1}{2}(\lambda(x+y) + \lambda(x-y)) \qquad (x, y \in \mathbf{R}^1);$$

for an eigenvalue $\lambda(\cdot)$; the spectral integral for A_x has the form (4.1), where the integration is over all solutions $\mathbf{R}^1 \ni x \mapsto \chi(x) \in \mathbf{C}^1$ of this functional equation.

If A_x depends continuously on x in the sense of §4.4, then we must consider only continuous solutions of the functional equation. Suppose also that the operators A_x are selfadjoint and $A_0 = 1$; setting $x = 0$ in the equation of the relation, we get that $A_{-x} = A_x$ $(x \in \mathbf{R}^1)$. Only continuous real-valued (even) solutions χ of the written equation such that $\chi(0) = 1$ will appear in the spectral representation. They have the form $\chi(x) = \chi_\lambda(x) = \cos\sqrt{\lambda}x$ $(x \in \mathbf{R}^1)$, where $\lambda \in \mathbf{R}^1$. Transporting the RI from the set of such solutions to $\mathbf{R}^1$, we get the spectral integral representation

$$A_x = \int_{\mathbf{R}^1} \cos\sqrt{\lambda}x \, dE(\lambda),$$

$$\mathfrak{D}(A_x) = \left\{ f \in H_0 \mid \int_{\mathbf{R}^1} (\cos\sqrt{\lambda}x)^2 \, d(E(\lambda)f, f)_{H_0} < \infty \right\} \qquad (x \in X)\ (4.25)$$

We can also consider a real separable Hilbert space as X and obtain a theorem of type 4.2 generalizing (4.25).

The relation here between the operators A_x is an example of a representation by operators in a so-called commutative hypercomplex system with continuous base. References to corresponding papers are given in the Comments on the Literature.

2) We consider a relation of an entirely different character. Let $X = G$, where $G \subseteq \mathbb{R}^N$ is a domain, let $A = (A_x)_{x \in G}$ be a family of bounded (for simplicity) normal commuting operators acting in H_0 and connected with the nuclear rigging $\Phi' \supseteq H_0 \supseteq \Phi$ in the standard way in the sense of §4.4, and let the function $G \ni x \mapsto A_x \in \mathcal{L}(H_0, H_0)$ be r times strongly continuously differentiable. The relation has the form of an rth-order differential equation:

$$\sum_{|\alpha| \leq r} a_\alpha(x)\, D^\alpha A_x = f(x)1 \qquad (x \in G), \tag{4.26}$$

where α is a multi-index, D^α is the corresponding mixed derivative, and $G \ni x \mapsto a_\alpha(x)$, $f(x) \in \mathbb{C}^1$, are complex-valued coefficients. The spectral representation for A appears as follows. Denote by $Q \subset \mathbb{C}^X$ the collection of all complex-valued r times continuously differentiable solutions $G \ni x \mapsto \lambda(x) \in \mathbb{C}^1$ of the equation

$$\sum_{|\alpha| \leq r} a_\alpha(x)(D^\alpha \lambda)(x) = f(x).$$

Then there exists an RI E on $\mathcal{C}_\sigma(Q) = \{\alpha \cap Q \mid \alpha \in \mathcal{C}_\sigma(\mathbb{C}^G)\}$ such that

$$A_x = \int_Q \lambda(x)\, dE(\lambda(\cdot)) \qquad (x \in G). \tag{4.27}$$

The proof of (4.27) proceeds according to the scheme already used: Theorem 2.7, Remark 1 after it, and Remarks 2 and 3 in §2.10 are used. Here $Q \supseteq g(A)$, since each eigenvalue $\lambda(\cdot) \in g(A)$ is a sufficiently smooth function (see §2.9) satisfying the required scalar differential equation; this follows from the second equality in (2.19) on the basis of (4.26) (cf. (4.3) and (4.22)). Thus, we can set $\tau = Q$, which leads to (4.27).

It is clear that if the functions $G \ni x \mapsto A_x \in \mathcal{L}(H_0, H_0)$ satisfy certain boundary-type conditions, then scalar solutions satisfying these boundary conditions are in Q. It is also clear that we can consider functional equations, differential-functional equations, and so on.

§5. Supplementary results on expansions in generalized eigenvectors

This section supplements the results in §2. Here we are first concerned with expansions in generalized eigenvectors of a single selfadjoint or normal operator. Then we show that the existence of "nice" spectral representations implies the existence of a rigging. We also present a comparison with the nuclear spectral theorem.

5.1. *A converse theorem.* We show that if a rigging (2.1) is such that (2.13) holds for any selfadjoint operator A defined in H_0, then the imbedding $H_+ \to H_0$ must be quasinuclear. (Moreover, this imbedding is quasinuclear if such a formula holds for any family of commuting selfadjoint operators.) We require even less of the function $R^1 \ni \lambda \mapsto P(\lambda)$ than was established by Theorem 2.2. Namely, the following result is valid.

THEOREM 5.1. *Suppose that the chain* (2.1) *is such that formula* (2.13) *is valid for the resolution E of the identity of any selfadjoint operator A defined in H_0, where $\mathcal{B}(\mathbf{R}^1) \ni \alpha \mapsto \rho(\alpha)$ is a finite measure and $\mathbf{R}^1 \ni \lambda \mapsto P(\lambda) \in \mathcal{L}(H_+ \to H_-)$ is an operator-valued function defined ρ-almost everywhere and such that $\|P(\lambda)\| \leq c < \infty$ for ρ-almost all $\lambda \in \mathbf{R}^1$. Then the imbedding $H_+ \to H_0$ is quasinuclear.*

LEMMA 5.1. *Let H be a Hilbert space, $(e_j)_1^\infty$ an orthonormal basis in it, and $A \in \mathcal{L}(H \to H)$. If the matrix of this operator has the form $(A_{jk})_{j,k=1}^\infty$, with $A_{jk} = (Ae_k, e_j)_H = \alpha_j \bar{\alpha}_k$, where $\alpha = (\alpha_j)_1^\infty \in l_2$, then A is a Hilbert-Schmidt operator, and*

$$|A| = \|A\| = \sum_{j=1}^{\infty} |\alpha_j|^2. \tag{5.1}$$

PROOF. Obviously,

$$|A|^2 = \sum_{j,k=1}^{\infty} |A_{jk}|^2 = \left(\sum_{j=1}^{\infty} |\alpha_j|^2 \right)^2 = \|\alpha\|_{l_2}^4 < \infty,$$

and $\|A\| \leq |A|$ always holds. Therefore, to prove (5.1) it suffices to verify the inequality $\|A\| \geq \|\alpha\|_{l_2}^2$. It follows from the relation $Af = \|\alpha\|_{l_2}^2 f$, where $f = \sum_1^\infty \alpha_j e_j \in H$. ■.

PROOF OF THE THEOREM. Consider the operators O, O^+, J, and $\mathbf{J}$ connected with the chain (2.1), and an arbitrary resolution $\mathcal{B}(\mathbf{R}^1) \ni \delta \mapsto E(\delta)$ of the identity in H_0. Let $C = OJ\colon H_0 \to H_0$. Obviously, $C^* = \mathbf{J}O^+\colon H_0 \to H_0$. For any disjoint $\delta_l \in \mathcal{B}(\mathbf{R}^1)$ $(l = 1, 2, \ldots)$ such that $\bigcup_1^\infty \delta_l = \mathbf{R}^1$ we get with the help of (2.13) and the conditions of the theorem that

$$\begin{aligned} \sum_{l=1}^{\infty} \|C^* E(\delta_l) C\| &= \sum_{l=1}^{\infty} \mathbf{J}O^+ E(\delta_l) OJ\| \leq \sum_{l=1}^{\infty} \|O^+ E(\delta_l) O\| \\ &= \sum_{l=1}^{\infty} \left\| \int_{\delta_l} P(\lambda)\, d\rho(\lambda) \right\| \\ &\leq c \sum_{l=1}^{\infty} \rho(\delta_l) = c\rho(\mathbf{R}^1) < \infty. \end{aligned} \tag{5.2}$$

We now construct the following one-dimensional resolution of the identity in H_0. Let $(e_l)_1^\infty$ be an orthonormal basis in H_0, P_l the projection on the one-dimensional subspace spanned by e_l and $(\lambda_l)_1^\infty$ a particular sequence of real numbers monotonically increasing to $+\infty$. Let $\mathcal{B}(\mathbb{R}^1) \ni \delta \mapsto E(\delta) = \sum_{\lambda_l \in \delta} P_l$; the set function constructed in this way is the required resolution of the identity. We apply to it the estimate (5.2), where $\delta_1 = (-\infty, \lambda_1]$, $\delta_2 = (\lambda_1, \lambda_2]$, $\delta_3 = (\lambda_2, \lambda_3], \dots$. As a result,

$$\sum_{l=1}^{\infty} \|C^* P_l C\| = \sum_{l=1}^{\infty} \|C^* E(\delta_l) C\| < \infty. \tag{5.3}$$

But it is easy to compute that the matrix $(A_{jk}^{(l)})_{j,k=1}^\infty$ of the operator $C^* P_l C$ has the form $A_{jk}^{(l)} = \bar{C}_{lj} C_{lk}$, where $(C_{jk})_{j,k=1}^\infty$ is the matrix of C. Therefore, Lemma 5.1 is applicable, and on the basis of it $\|C^* P_l C\| = \sum_{j=1}^\infty |C_{lj}|^2$. Thus, (5.3) leads to the condition $\sum_{l,j=1}^\infty |C_{lj}|^2 < \infty$, i.e., $C = OJ$ is a Hilbert-Schmidt operator on H_0. But since J is an isometry between H_0 and H_+, this implies that O is a Hilbert-Schmidt operator from H_+ to H_0. ■

5.2. *The case of an imbedding that is not quasinuclear.* We have shown that if the chain (2.1) is suitable for constructing expansions in generalized eigenvectors of any selfadjoint operator A acting in H_0, then the imbedding $H_+ \to H_0$ must be quasinuclear. However, for a particular A the choice of the chain (2.1) can turn out to be more free. We give some results.

Thus, let A be a selfadjoint operator acting in H, and E its resolution of the identity. If we can construct the chain (2.1) in such a way that the operator-valued measure $\mathcal{B}(\mathbb{R}^1) \ni \delta \mapsto \theta(\delta) = O^+ E(\delta) O$ has a σ-finite trace $\rho(\delta)$, then according to Remark 1 at the end of §2.1, it can be differentiated with respect to this trace and it is possible to repeat all the arguments in §§2.1–2.3 (in the case of a single operator). The formulations of Theorems 2.2 and 2.6 do not change, except that the spectral measure $\mathcal{B}(\mathbb{R}^1) \ni \delta \mapsto \rho(\delta) = \operatorname{Tr}(O^+ E(\delta) O)$ is not finite in general.

THEOREM 5.2. *Suppose that the operator A and the chain* (2.1) *are such that there exists a bounded continuous nonzero complex-valued function $\alpha(\lambda)$ on the spectrum $s(A)$ of A such that the operator $\alpha(A)O$: $H_+ \to H_0$ is quasinuclear. Then the operator-valued measure $\mathcal{B}(\mathbb{R}^1) \ni \delta \mapsto O^+ E(\delta) O$ has a σ-finite trace, and consequently, the chain* (2.1) *is suitable for constructing expansions in generalized eigenvectors of A.*

We first prove the following lemma.

LEMMA 5.2. *If there exists a bounded continuous positive function $\beta(\lambda)$ on $s(A)$ such that $\operatorname{Tr}(\mathbf{J} O^+ \beta(A) O J) < \infty$, then the measure $\mathcal{B}(\mathbb{R}^1) \ni \delta \mapsto O^+ E(\delta) O$ has a σ-finite trace.*

PROOF. Let $s_n(A) = (s(A)) \cap [-n, n]$ $(n = 1, 2, \ldots)$. Then for each n there exists an $\varepsilon_n > 0$ such that $\alpha(\lambda) \geq \varepsilon_n$ $(\lambda \in s_n(A))$. Therefore,

$$\begin{aligned} 0 &\leq \varepsilon_n E([-n,n]) = \varepsilon_n E(s_n(A)) \\ &\leq \int_{s_n(A)} \beta(\lambda)\, dE(\lambda) \leq \int_{s(A)} \beta(\lambda)\, dE(\lambda) = \beta(A). \end{aligned}$$

Thus,

$$0 \leq \varepsilon_n C^* E([-n,n])C \leq C^* \beta(A) C$$

for any $C \in \mathcal{L}(H_0 \to H_0)$. Setting $C = OJ$ and considering that $C^* = \mathbf{J}O^+$, we get that

$$0 \leq \mathbf{J}O^+ E([-n,n])O\mathbf{J} \leq \varepsilon_n^{-1} \mathbf{J}O^+ \beta(A) OJ$$

and hence also

$$\operatorname{Tr}(\mathbf{J}O^+ E([-n,n])OJ) \leq \varepsilon_n^{-1}(\mathbf{J}O^+\beta(A)OJ) < \infty \qquad (n = 1, 2, \ldots). \quad (5.4)$$

On the other hand, if $C \in \mathcal{L}(H_+ \to H_-)$ is nonnegative, then $\mathbf{J}CJ \in \mathcal{L}(H_0 \to H_0)$ is clearly also nonnegative, and

$$\operatorname{Tr}(C) = \operatorname{Tr}(\mathbf{J}CJ). \quad (5.5)$$

Indeed, if $(e_j)_1^\infty$ is an orthonormal basis in H_+, then $(J^{-1}e_i)_1^\infty$ is an orthonormal basis in H_0, and we have that

$$\operatorname{Tr}(C) = \sum_{j=1}^{\infty} (Ce_j, e_j)_{H_0} = \sum_{j=1}^{\infty} (\mathbf{J}CJJ^{-1}e_i, J^{-1}e_j)_{H_0} = \operatorname{Tr}(\mathbf{J}CJ).$$

With the help of (5.5) we now get from (5.4) that

$$\operatorname{Tr}(O^+E([-n,n])O) = \operatorname{Tr}(\mathbf{J}O^+E([-n,n])OJ) < \infty \qquad (n = 1, 2, \ldots). \quad \blacksquare$$

PROOF OF THE THEOREM. We use the obvious relation: $\operatorname{Tr}(C^*C) = |C|^2 \leq \infty$ for any $C \in \mathcal{L}(H_0 \to H_0)$. Applying it to $C = \alpha(A)OJ \in \mathcal{L}(H_0 \to H_0)$, we get the general formula

$$\begin{aligned} \operatorname{Tr}(\mathbf{J}O^+(|\alpha|^2(A))OJ) &= \operatorname{Tr}(\mathbf{J}O^+(\alpha(A))^*\alpha(A)OJ) \\ &= |\alpha(A)OJ|^2 = |\alpha(A)O|^2 < \infty. \end{aligned} \quad (5.6)$$

It remains to use Lemma 5.2 with $\beta(\lambda) = |\alpha(\lambda)|^2$. $\blacksquare$.

Let us proceed to the question of verifying the condition $|\alpha(B)O| < \infty$ for a perturbation B of A which itself satisfies such a condition. We establish two lemmas, the first of which is a simple observation that is, however, useful for the verification, while the second generalizes to unbounded operators a fact well known in the finite-dimensional case.

LEMMA 5.3. *Suppose that the conditions of Theorem* 5.2 *are satisfied and, moreover,* B *is a selfadjoint operator acting in* H_0 *and* $s(B)$ *is its spectrum. Assume that there is a nonnegative bounded function* $\gamma \in C(\mathbb{R}^1)$ *that is positive on* $s(B)$, *does not exceed* $|\alpha(\lambda)|^2$ *on* $s(A)$, *and is such that* $\gamma(B) \leq \gamma(A)$ *or, more generally,*

$$\mathrm{Tr}(\mathbf{J}O^+\gamma(B)OJ) \leq \mathrm{Tr}(\mathbf{J}O^+\gamma(A)OJ).$$

Then $|\gamma^{1/2}(B)O| < \infty$.

PROOF. According to (5.6),

$$\begin{aligned}\left|\gamma^{1/2}(B)O\right|^2 &= \mathrm{Tr}(\mathbf{J}O^+\gamma(B)OJ)\\ &\leq \mathrm{Tr}(\mathbf{J}O^+\gamma(A)OJ) \leq \mathrm{Tr}(\mathbf{J}O^+(|\alpha|^2(A))OJ)\\ &\leq \left|\alpha(A)O\right|^2 < \infty. \quad \blacksquare\end{aligned}$$

Recall that a linear set $F \subseteq \mathfrak{D}(A)$ is called a core of a closed operator A if $(A \restriction F)^{\sim} = A$.

LEMMA 5.4. *Let* H *be a Hilbert space, and let* A *and* B *be nonnegative selfadjoint operators such that the inverses* A^{-1} *and* B^{-1} *exist and are bounded. If* $A \leq B$ *on some linear set* F *which is a core for both operators* (*i.e.*, $(Af, f)_H \leq (Bf, f)_H$, $f \in F$), *then* $B^{-1} \leq A^{-1}$.

PROOF. The nonnegative operator $C = B^{-1/2}AB^{-1/2}$ is defined on the domain $\mathfrak{D}(C)$ consisting of the vectors of the form $f = B^{1/2}h$ $(h \in F)$. Since F is a core of B and B^{-1} exists, $\mathfrak{D}(C)$ is dense in H. We have that

$$\begin{aligned}(Cf, f)_H = (B^{-1/2}AB^{-1/2}f, f)_H &= (Ah, h)_H\\ &\leq (Bh, h)_H = \|f\|_H^2 \qquad (f \in \mathfrak{D}(C)),\end{aligned} \tag{5.7}$$

and so by the Cauchy-Schwarz-Bunyakovskiĭ inequality, the quasi-inner product $(f, g) = (Cf, g)_H$ $(f, g \in \mathfrak{D}(C))$ satisfies

$$|(f, g)|^2 \leq (f, f)(g, g) \leq \|f\|_H^2 \|g\|_H^2 \qquad (f, g \in \mathfrak{D}(C)).$$

The last estimate shows that the nonnegative bilinear form $\mathfrak{D}(C) \ni f, g \mapsto B(f, g) = (f, g)$ can be extended by continuity to such a form on the closure of $\mathfrak{D}(C)$ in H, i.e., to H. We keep the previous notation (f, g) $(f, g \in H)$ for it. Note that $(f, g) = (Cf, g)_H$ for $f \in \mathfrak{D}(C)$ and $g \in H$. By (5.7), $(f, f) \leq \|f\|_H^2$ $(f \in H)$.

Now define the operator $D = B^{1/2}A^{-1}B^{1/2}$ on the set $\mathfrak{D}(D)$ consisting of the vectors of the form $u = B^{-1/2}Av$ $(v \in F)$. It is clear that the range satisfies $\mathfrak{R}(D) = \mathfrak{D}(C)$, and $CDu = u$ $(u \in \mathfrak{D}(D))$. Applying the Cauchy-Schwarz-Bunyakovskiĭ inequality to (f, g) $(f, g \in H)$ and using the estimate $(f, f) \leq \|f\|_H^2$, we get for a vector $u \in \mathfrak{D}(D)$ that

$$\begin{aligned}(u, u)_H^2 = (CDu, u)_H^2 &= (Du, u)^2 \leq (Du, Du)(u, u)\\ &= (CDu, Du)_H(u, u) \leq (u, Du)_H(u, u)_H,\end{aligned}$$

and so $(u,u)_H \leq (u, Du)_H$, i.e.,

$$(u,u)_H \leq (u, B^{1/2}A^{-1}B^{1/2}u)_H = (B^{1/2}u, A^{-1}B^{1/2}u)_H \qquad (u \in \mathfrak{D}(D)).$$

Let $B^{1/2}u = w$. Then the last inequality can be rewritten in the form

$$(w, B^{-1}w)_H \leq (w, A^{-1}w)_H. \tag{5.8}$$

But $u = B^{-1/2}Av$ $(v \in F)$, so $w = Av$, where v runs through F. Since $(A \upharpoonright F)^{\sim} = A$ and A^{-1} exists, the vectors w run through a dense subset of H. The operators A^{-1} and B^{-1} are continuous, so (5.8) means that $B^{-1} \leq A^{-1}$. ■

Combination of the last two lemmas leads to the following theorem.

THEOREM 5.3. *Suppose that A is a selfadjoint operator which is semi-bounded below by a number a and that the chain* (2.1) *is such that $|\alpha(A)O| < \infty$ for the function $\alpha(\lambda) = (\lambda - z)^{-1/2}$ $(\lambda \in [a, \infty)$, and $z < a$ is fixed). Then the estimate $|\alpha(B)O| < \infty$ is valid for any selfadjoint operator B having a core common with A and such that $B \geq A$ on this core. Thus, the chain* (2.1) *is also suitable for constructing expansions in generalized eigenvectors of B.*

PROOF. Let F be a common core of A and B. Then it is a common core also of the operators $A - z1$ and $B - z1$, and $A - z1 \leq B - z1$ on F. According to Lemma 5.4, this implies that $(B - z1)^{-1} \leq (A - z1)^{-1}$, i.e., the condition $\gamma(B) \leq \gamma(A)$ is satisfied with the function $\gamma(\lambda)$ equal to $(\lambda - z)^{-1}$ for $\lambda \geq a$ and to $(a - z)^{-1}$ for $\lambda < a$ (by the assumption, $s(A) \subseteq [a, \infty)$, and $s(B) \subseteq [a, \infty)$ due to the inequality $B \geq A$). This function is positive on $s(B)$, and $\gamma(\lambda) = |\alpha(\lambda)|^2$ on $s(A)$. The theorem now follows from Lemma 5.3. ■

5.3. *Expansion in eigenfunctions of Carleman operators.* We apply the results in §5.2 to more concrete situations. Let $H_0 = L_2(R, d\mu(x))$, where R is a measure space with measure μ defined on a σ-algebra of subsets of R; $\mu(R) \leq \infty$. A selfadjoint operator A acting in H_0 is called a Carleman operator if there exists a bounded continuous nonzero complex-valued function $\alpha(\lambda)$ on its spectrum such that $\alpha(A)$ is a Carleman integral operator. The latter means that there exists a $(\mu \times \mu)$-measurable kernel $K(x,y)$, defined for $(\mu \times \mu)$-almost all $(x, y) \in R \times R$, such that the representation

$$(\alpha(A)f)(x) = \int_R K(x,y) f(y)\, d\mu(u)$$

is valid for some set of functions f dense in $L_2(R, d\mu(x))$ and such that

$$\int_R |K(x,y)|^2\, d\mu(x) < \infty \tag{5.9}$$

for μ-almost every $y \in R$.

Examples of Carleman operators (elliptic operators with smooth coefficients and others) are given in Chapter 6 of the author's book [4] and below.

THEOREM 5.4. *Assume A is a selfadjoint Carleman operator acting in $L_2(R, d\mu(x))$. Then there exists a μ-measurable weight $p(x) \geq 1$ $(x \in R)$ such that the chain (see* (3.2) *in Chapter* 1)

$$L_2(R, p^{-1}(x)\, d\mu(x)) \supseteq L_2(R, d\mu(x)) \supseteq L_2(R, p(x)\, d\mu(x)). \tag{5.10}$$

is suitable for the construction of expansions in generalized eigenfunctions of A.

PROOF. Consider an arbitrary chain of the form (5.10). According to (3.3) in Chapter 1, $(If)(x) = p^{-1}(x)f(x)$ $(f \in L_2(R, d\mu(x)))$ for it; therefore,

$$(Jf)(x) = p^{-1/2}(x)f(x) \qquad (f \in L_2(R, d\mu(x))). \tag{5.11}$$

Consequently, the kernel $K_1(x, y) = K(x, y)p^{-1/2}(y)$ corresponds to the operator $\alpha(A)OJ$ acting continuously in $L_2(R, d\mu(x))$, and

$$\begin{aligned} |\alpha(A)O|^2 &= |\alpha(A)OJ|^2 \\ &= \int_R \int_R |K(x, y)|^2 p^{-1}(y)\, d\mu(x)\, d\mu(y) < \infty. \end{aligned} \tag{5.12}$$

According to (5.9), $k(y) = \int_R |K(x, y)|^2\, d\mu(x)$ is measurable and finite almost everywhere. Therefore, there is a measurable weight $p(y) \geq 1$ such that $\int_R k(y)p^{-1}(y)\, d\mu(y) < \infty$. The last condition, as follows from (5.12), means that $|\alpha(A)O| < \infty$. We see that the condition of Theorem 5.2 holds with the chain (5.10) and the weight $p(x)$ chosen in this way. ■

We emphasize that the weight $p(x)$ in the theorem is an arbitrary measurable weight $p(x) \geq 1$ $(x \in R)$ for which the integral (5.12) converges.

Thus, for Carleman operators the fact that an eigenfunction is generalized means only that it belongs not to $L_2(R, d\mu(x))$ but to $L_2(R, p^{-1}(x)\, d\mu(x))$ with the indicated weight $p(x)$. Expansions in individual eigenfunctions are carried out according to the general scheme in §2.11. We give four simple assertions relating to a Carleman operator A.

I. *The operator $P(\lambda)$: $L_2(R, p(x)\, d\mu(x)) \to L_2(R, p^{-1}(x)\, d\mu(x))$ is an integral operator, i.e.,*

$$(P(\lambda)u)(x) = \int_R P(x, y; \lambda)u(y)\, d\mu(y) \qquad (u \in L_2(R, p(x)\, d\mu(x))), \tag{5.13}$$

and the kernel $P(x, y; \lambda)$ (the spectral kernel of A) satisfies for ρ-almost all $\lambda \in \mathbb{R}^1$ the estimate

$$\int_R \int_R |P(x, y; \lambda)|^2 p^{-1}(x)p^{-1}(y)\, d\mu(x)\, d\mu(y) \leq 1 \tag{5.14}$$

and is jointly $(\mu \times \mu \times \rho)$-measurable.

Indeed, $|P(\lambda)| \leq \operatorname{Tr}(P(\lambda)) = 1$ for ρ-almost all λ. Consider the operator $\mathbf{J}P(\lambda)J$: $L_2(R, d\mu(x)) \to L_2(R, d\mu(x))$; $|\mathbf{J}P(\lambda)J| = |P(\lambda)| \leq 1$ holds ρ-almost everywhere. Therefore, for a corresponding particular λ, this operator is an integral operator acting in $L_2(R, d\mu(x))$ whose kernel $K(x,y)$ satisfies the estimate

$$\int_R \int_R |K(x,y)|^2 \, d\mu(x) \, d\mu(y) \le 1.$$

By (5.11), $P(x,y;\lambda) = K(x,y)p^{1/2}(x)p^{1/2}(y)$ is the kernel of $P(\lambda)$, which implies (5.13) and (5.14). The measurability of P follows from Orochko's paper [8]. ■

II. *Let* $\mathbf{R}^1 \ni \lambda \mapsto \gamma(\lambda) \in \mathbf{C}^1$ *be a Borel-measurable function such that* $|\gamma(\lambda)| \le |\alpha(\lambda)|^2$ $(\lambda \in s(A))$. *Then* $\int_{\mathbf{R}^1} |\gamma(\lambda)| \, d\rho(\lambda) < \infty$.

Indeed, for $\delta \in \mathcal{B}(\mathbf{R}^1)$,

$$\begin{aligned} \rho(\delta) &= \operatorname{Tr}(O^+ E(\delta) O) = \operatorname{Tr}(\mathbf{J} O^+ E(\delta) O J) \\ &= \sum_{j=1}^{\infty} (\mathbf{J} O^+ E(\delta) O J e_j, e_j)_{H_0} = \lim_{n\to\infty} \rho_n(\delta), \end{aligned}$$

$$\rho_n(\delta) = \sum_{j=1}^{n} (\mathbf{J} O^+ E(\delta) O J e_j, e_j)_{H_0},$$

where $(e_j)_1^\infty$ is an orthonormal basis in $H_0 = L_2(R, d\mu(x))$. It is easy to prove the following limit relation with the help of Helly's theorems:

$$\begin{aligned} \int_{\mathbf{R}^1} |\gamma(\lambda)| \, d\rho(\lambda) &= \lim_{n\to\infty} \int_{\mathbf{R}^1} |\gamma(\lambda)| \, d\rho_n(\lambda) \\ &= \lim_{n\to\infty} \int_{\mathbf{R}^1} \sum_{j=1}^{n} |\gamma(\lambda)| \, d(\mathbf{J} O^+ E(\lambda) O J e_j, e_j)_{H_0} \\ &= \lim_{n\to\infty} \sum_{j=1}^{n} (\mathbf{J} O^+ (|\gamma|(A)) O J e_j, e_j)_{H_0} \\ &\le \lim_{n\to\infty} \sum_{j=1}^{n} (\mathbf{J} O^+ (|\alpha|^2(A)) O J e_j, e_j)_{H_0} \\ &= \operatorname{Tr}(\mathbf{J} O^+ (|\alpha|^2(A)) O J) = |\alpha(A) O|^2 < \infty. \quad ■ \end{aligned}$$

III. *For every function* $\gamma(\lambda)$ *in assertion* II *the operator*

$$O^+\gamma(A)O \colon L_2(R, p(x)\, d\mu(x)) \to L_2(R, p^{-1}(x)\, d\mu(x))$$

is an integral operator, and its kernel $K(x,y)$ *is in*

$$L_2(R \times R, p^{-1}(x) p^{-1}(y) \, d\mu(x) \times d\mu(y)).$$

Indeed, by (2.13),

$$O^+\gamma(A)O = \int_{\mathbf{R}^1} \gamma(\lambda) \, d(O^+ E(\lambda) O) = \int_{\mathbf{R}^1} \gamma(\lambda) P(\lambda) \, d\rho(\lambda).$$

But by assertion I, $P(\lambda)$ is an integral operator, so, formally, $O^+\gamma(A)O$ is also an integral operator, with kernel

$$K(x,y) = \int_{\mathbf{R}^1} \gamma(\lambda)P(x,y;\lambda)\,d\rho(\lambda) \qquad (x,y \in R). \tag{5.15}$$

To prove the assertion it suffices to establish that the function (5.15) is in

$$L_2(R \times R, p^{-1}(x)p^{-1}(y)\,d\mu(x) \times d\mu(y)).$$

By the Cauchy-Schwarz-Bunyakovskiĭ inequality and (5.14),

$$\begin{aligned}
&\int_R \int_R \left| \int_{\mathbf{R}^1} \gamma(\lambda)P(x,y;\lambda)\,d\rho(\lambda) \right|^2 p^{-1}(x)p^{-1}(y)\,d\mu(x)\,d\mu(y) \\
&\quad \leq \int_{\mathbf{R}^1} |\gamma(\lambda)|\,d\rho(\lambda) \int_R \int_R \left(\int_{\mathbf{R}^1} |\gamma(\lambda)|\,|P(x,y;\lambda)|^2\,d\rho(\lambda) \right) \\
&\qquad\qquad \times p^{-1}(x)p^{-1}(y)\,d\mu(x)\,d\mu(y) \\
&\quad \leq \left(\int_{\mathbf{R}^1} |\gamma(\lambda)|\,d\rho(\lambda) \right)^2 < \infty. \quad \blacksquare
\end{aligned}$$

IV. *If it is possible to choose the rigging* (5.10) *such that*

$$|\alpha(A)O|^2 = \mathrm{Tr}(\mathbf{J}O^+(|\alpha|^2(A))OJ) < \infty$$

where A is a selfadjoint operator acting in $L_2(R, d\mu(x))$ and $\alpha(\lambda)$ is some function as in the definition of the Carleman property, then A is a Carleman operator.

Indeed, it is obvious that $|\alpha(A)OJ| < \infty$. Therefore, $\alpha(A)OJ$, which acts in $L_2(R, d\mu(x))$, is an integral operator whose kernel $K_1(x,y)$ is square-integrable with respect to x and y. But then $K(x,y) = K_1(x,y)p^{1/2}(y)$ is the required kernel for $\alpha(A)$. $\blacksquare$

We proceed to an investigation, based on Theorem 5.3 and Lemma 5.3, of the Carleman property for a perturbation of a Carleman operator.

THEOREM 5.5. *Let A be a selfadjoint operator acting in the space $L_2(\mathbf{R}^N)$ with respect to Lebesgue measure and semibounded below by a number a. Assume that the resolvent $R_z(A)$ is an integral operator for some $z < a$, and its kernel is locally bounded. Consider a selfadjoint operator B acting in the same space, having a core common with A, and such that $B \geq A$ on this core. Then B is a Carleman operator.*

PROOF. It suffices to show that by choosing a chain of the form (5.10) we can ensure that $|\alpha(A)O| < \infty$, where $\alpha(\lambda) = (\lambda - z)^{-1/2}$ ($\lambda \in [a, \infty)$). Then the same inequality is valid also when A is replaced by B, by Theorem 5.3. Therefore, according to IV, B is a Carleman operator.

From (5.6) it follows that the weight $p(x) \geq 1$ ($x \in \mathbf{R}^N$) must be chosen so that $\mathrm{Tr}(\mathbf{J}O^+COJ) < \infty$, where $C = R_z(A) > 0$.

We introduce the averaging operator $(S_\varepsilon f)(x) = (\omega_\varepsilon * f)(x)$ acting in $L_2(\mathbf{R}^N)$. Here $*$ denotes the convolution centered at 0 and

$$\omega_\varepsilon(x) = (m(K_\varepsilon))^{-1}\kappa_{K_\varepsilon}(x).$$

(K_ε is the open ball of radius ε about 0, $m(K_\varepsilon)$ is the Lebesgue measure for it, and κ_{K_ε} is the characteristic function of K_ε.) It is not hard to see that S_ε is a bounded selfadjoint operator with nonnegative symmetric kernel $S_\varepsilon(x, y)$, $S_\varepsilon \to 1$ in the strong sense as $\varepsilon \to 0$, and

$$\sup_{x\in G_1;\ 0<\varepsilon\le\delta} |(S_\varepsilon f)(x)| \le \sup_{x\in G_1} |f(x)|,$$

for each pair of domains $G_1 \subseteq G_2 \subseteq \mathbf{R}^N$, where $\delta = \rho(G_1, \mathbf{R}^V\backslash G_2) > 0$. The operator $\mathbf{J}O^+S_\varepsilon CS_\varepsilon O\mathbf{J}$ acts continuously in $L_2(\mathbf{R}^N)$ and has the kernel

$$K_\varepsilon(x,y) = \int_{\mathbf{R}^N}\int_{\mathbf{R}^N} S_\varepsilon(x,s)C(s,t)S_\varepsilon(t,y)\,ds\,dt\,p^{-1/2}(x)p^{-1/2}(y) \qquad (x, y \in \mathbf{R}^N). \tag{5.16}$$

where $C(x, y)$ is the kernel of C. Assume that $p \in C(\mathbf{R}^N)$; then the kernel (5.16) depends continuously on (x, y) and is positive-definite. Therefore,

$$\operatorname{Tr}(\mathbf{J}O^+S_\varepsilon CS_\varepsilon O J) = \int_{\mathbf{R}^N} K_\varepsilon(x,x)\,dx \le \infty \qquad (\varepsilon > 0). \tag{5.17}$$

We estimate the integral (5.17). Let $G_1 \subset G_2 \subset \cdots$ be a sequence of balls $G_n = \{x \in \mathbf{R}^N \mid |x| < n\}$. Then we write $c_n = \sup_{x,y\in G_n} |C(x,y)|$ $(n = 1, 2, \ldots)$ and consider the kernel $K_{1,\varepsilon}(x,y) = K_\varepsilon(x,y)p^{1/2}(x)p^{1/2}(y)$ $(x, y \in \mathbf{R}^N)$. According to (5.16), it can be understood as the result of applying the tensor product $S_\varepsilon \otimes S_\varepsilon$ of the two averaging operators to $C(s, t)$; therefore,

$$\sup_{x,y\in G_n;\ 0<\varepsilon\le 1} |K_{1,\varepsilon}(x,y)| \le c_{n+1} \qquad (n = 1, 2, \ldots)$$

in view of the properties of these operators. From this we get $(G_0 = \varnothing)$

$$\begin{aligned}\int_{\mathbf{R}^N} K_\varepsilon(x,x)\,dx &= \int_{\mathbf{R}^N} K_{1,\varepsilon}(x,x)p^{-1}(x)\,dx \\ &= \sum_{n=0}^{\infty}\int_{G_{n+1}\backslash G_n} K_{1,\varepsilon}(x,x)p^{-1}(x)\,dx \\ &\le \sum_{n=0}^{\infty} c_{n+2}\int_{G_{n+1}\backslash G_n} p^{-1}(x)\,dx \qquad (0 < \varepsilon \le 1).\end{aligned} \tag{5.18}$$

Choose a function $p(x)$ that increases so rapidly as $|x| \to \infty$ that the series on the right-hand side of (5.28) converges. Then

$$\operatorname{Tr}(\mathbf{J}O^+S_\varepsilon CS_\varepsilon O J) \le c < \infty \qquad (0 < \varepsilon \le 1). \tag{5.19}$$

It is not hard to see that we can pass to the limit as $\varepsilon \to 0$ in (5.19) and obtain the required estimate $\mathrm{Tr}(\mathbf{J}O^+COJ) \le c$. Indeed, if $(e_j)_1^\infty$ is an orthonormal basis in $L_2(\mathsf{R}^N)$, then the left-hand side of (5.19) is equal to

$$\sum_{j=1}^{\infty}(CS_\varepsilon OJe_j, S_\varepsilon OJe_j)_{L_2(\mathsf{R}^N)}.$$

For each term,

$$(CS_\varepsilon OJe_j, S_\varepsilon OJe_j)_{L_2(\mathsf{R}^N)} \to (COJe_j, OJe_j)_{L_2(\mathsf{R}^N)}$$

as $\varepsilon \to 0$; therefore, since the terms are nonnegative, (5.19) gives us that

$$\mathrm{Tr}(\mathbf{J}O^+COJ) = \sum_{j=1}^{\infty}(COJe_j, OJe_j)_{L_2(\mathsf{R}^N)} \le c < \infty. \quad \blacksquare$$

We emphasize that a chain of the form (5.10), where $p(x) \ge 1$ $(x \in \mathsf{R}^V)$ is continuous and is chosen so that the series (5.18) converges, is suitable for any operator B satisfying the conditions of the theorem.

EXAMPLE. Consider in $L_2(\mathsf{R}^N)$ the differential expression

$$\begin{aligned}(\mathcal{M}u)(x) &= \sum_{|\alpha|\le r} a_\alpha(x)(D^\alpha u)(x) + q(x)u(x)\\ &= (\mathcal{L}u)(x) + q(x)u(x). \qquad (5.20)\end{aligned}$$

Here $\mathcal{L}$ is an elliptic, formally selfadjoint expression of order $r > N$ with coefficients $a_\alpha \in C^\infty(\mathsf{R}^N)$, $q \in L_{2,\mathrm{loc}}(\mathsf{R}^V)$ is nonnegative almost everywhere, and $\mathcal{L}$ is assumed to be semibounded below on $C_0^\infty(\mathsf{R}^N)$ by some a (i.e., $(\mathcal{L}u, u)_{L_2(\mathsf{R}^N)} \ge a\|u\|^2_{L_2(\mathsf{R}^N)}$, $u \in C_0^\infty(\mathsf{R}^N)$). We introduce the Hermitian operators $C_0^\infty(\mathsf{R}^N) \ni u \mapsto \mathcal{L}u$ and $C_0^\infty(\mathsf{R}^N) \ni u \mapsto \mathcal{M}u$ defined in $L_2(\mathsf{R}^N)$ and assume that they have selfadjoint extensions A and B acting in $L_2(\mathsf{R}^N)$ which are semibounded below and have a common core. It is claimed that A and B are Carleman operators with common $\alpha(\lambda) = (\lambda - z)^{-1/2}$ ($z < a$ is fixed, and $\lambda \in [a, \infty)$) and common weight $p(x)$. This follows from the foregoing, since it is known (Berezanskiĭ [4], Theorem 5.1 in Chapter 3) that for $r > N$ the resolvent $R_z(A)$ of the A introduced here is an integral operator whose kernel is locally bounded.

The scheme we have presented does not work in the case of differential expressions (5.20) whose order is not high enough (at most N), because to prove the Carleman property here it is necessary to take a function $\alpha(\lambda)$ of the type $(\lambda^l - z)^{-1}$, where the integer $l > 0$ is sufficiently large, and the corresponding analogue of Lemma 5.4 is false. We present a somewhat different approach.

Let A be a selfadjoint Carleman operator which acts in $L_2(R, d\mu(x))$ and is semibounded below, and let $\alpha(\lambda)$ be the corresponding function. Assume that for any $t \in (0,1]$ there exists a $c_t > 0$ such that

$$|\alpha(\lambda)|^2 \geq c_t e^{-t\lambda} \qquad (\lambda \in s(A)). \tag{5.21}$$

According to assertion III above, each operator

$$O^+ e^{-tA} O \colon L_2(R, p(x)\, d\mu(x)) \to L_2(R, p^{-1}(x)\, d\mu(x))$$

is a integral operator; let

$$U(x,y;t) \in L_2(R \times R, p^{-1}(x) p^{-1}(y)\, d\mu(x) \times d\mu(y)) \qquad (t \in (0,1])$$

be the corresponding kernel. The operator A is said to satisfy the semigroup positivity condition if $U(x,y;t) \geq 0$ for $(\mu \times \mu)$-almost all $(x,y) \in R \times R$ for each $t \in (0,1]$.

THEOREM 5.6. *Suppose A is a Carleman operator acting in $L_2(R, d\mu(x))$ that is semibounded below and satisfies the semigroup positivity condition, and $R \ni x \mapsto v(x) \geq 0$ is a measurable function such that A has a core F for which the closure V of the operator $F \ni f(x) \mapsto v(x)f(x)$ is selfadjoint. Denote by B the closure of the operator $F \ni f \mapsto Af + Vf$, and assume that B is selfadjoint.*

Then B is a Carleman operator and the chain (5.10) *constructed from A is suitable for constructing an expansion in its generalized eigenfunctions.*

PROOF. We use a known multiplicative formula: suppose that selfadjoint operators A and V are semibounded below and act in the Hilbert space H. Denote by B the closure of the operator $\mathfrak{D}(A) \cap \mathfrak{D}(V) \ni f \mapsto Af + Vf$, and assume that B is selfadjoint. Then for each $t > 0$ and $f \in H$

$$e^{-tB} f = \lim_{n \to \infty} (e^{-(t/n)A} e^{-(t/n)V})^n f, \tag{5.22}$$

in the sense of strong convergence. Considering the estimate (5.21) and applying Lemma 5.3, where $\gamma(\lambda)$ equals $c_1 e^{-\lambda}$ for $\lambda \geq a$ and $c_1 e^{-a}$ for $\lambda < a$, we get that it suffices to prove the inequality

$$\mathrm{Tr}(\mathbf{J} O^+ e^{-B} O J) \leq \mathrm{Tr}(\mathbf{J} O^+ e^{-A} O J)$$

or, what is the same,

$$\left| e^{-B/2} O J \right| \leq \left| e^{-A/2} O J \right|$$

(see (5.6)).

By (5.22), for each $f \in L_2(R, d\mu(x))$,

$$(e^{-B/2} O J f)(x) = \lim_{n \to \infty} ((e^{-A/2n} e^{-V/2n})^n O J f)(x). \tag{5.23}$$

Let $e_n(x) = e^{-v(x)/2n} \in [0,1]$ $(x \in R; n = 1, 2, \ldots)$. Since the functions $U(x,y;t)$, $e_n(x) \leq 1$, and $p(x)$ are nonnegative, Fubini's theorem gives us for the kernel $K_n(x,y)$ of the operators after the limit sign on the right-hand side in (5.23) that for $(\mu\times\mu)$-almost all $(x,y)\in R\times R$

$$\begin{aligned}
0 &\le K_n(x,y)\\
&= \int_R U(x,z_1;(2n)^{-1})e_n(z_1)\left(\int_R U(z_1,z_2;(2n)^{-1})e_n(z_2)\right.\\
&\qquad\left.\times\left(\cdots\left(\int_R U(z_n,y;(2n)^{-1})\,dz_n\right)\cdots\right)dz_2\right)dz_1 e_n(y)p^{-1/2}(y)\\
&= \int_R\cdots\int_R U(x,z_1;(2n)^{-1})e_n(z_1)U(z_1,z_2;(2n)^{-1})e_n(z_2)\\
&\qquad\cdots U(z_n,y;(2n)^{-1})\,dz_1\,dz_2\cdots dz_n e_n(y)p^{-1/2}(y)\\
&\le \int_R\cdots\int_R U(x,z_1;(2n)^{-1})U(z_1,z_2;(2n)^{-1})\cdots U(z_n,y;(2n)^{-1})\\
&\qquad\times dz_1\,dz_2\cdots dz_n p^{-1/2}(y).
\end{aligned}$$

The last integral is the kernel of the operator

$$(e^{-A/2n})^n OJ = e^{-A/2}OJ.$$

Denote it by $L(x,y)$. Thus, $0\le K_n(x,y)\le L(x,y)$ for $(\mu\times\mu)$-almost all $(x,y)\in R\times R$. From this we conclude that

$$\begin{aligned}
\left|(e^{-A/2n}e^{-V/2n})^n OJ\right|^2 &= \int_R\int_R K_n^2(x,y)\,d\mu(x)\,d\mu(y)\\
&\le \int_R\int_R L^2(x,y)\,d\mu(x)\,d\mu(y) = \left|e^{-A/2}OJ\right|^2.
\end{aligned}$$

This inequality and (5.22) imply that

$$\left|e^{-B/2}OJ\right|^2 \le \left|e^{-A/2}OJ\right|^2. \quad \blacksquare$$

EXAMPLE. Consider in $L_2(\mathsf{R}^N)$ the Schrödinger operator B with singular potential semibounded below and given by the closure of the operator $C_0^\infty(\mathsf{R}^N)\ni u(x)\mapsto -(\Delta u)(x)+q(x)u(x)$, where $q\in L_{2,\mathrm{loc}}(\mathsf{R}^N)$ and $q(x)\ge a$ $(x\in\mathsf{R}^N)$. It is known that B is selfadjoint (Kato [1]; see also Theorem 3.1 or Lemma 3.4 in §3.4 of Chapter 3). With the help of Theorem 5.6 it is easy to show that B is a Carleman operator.

Indeed, define A to be the same as B except with the differential expression $-\Delta u + au$. It is known that A is semibounded below by the number a, is selfadjoint, and is a Carleman operator, with $\alpha(\lambda) = (\lambda^l - z)^{-1}$, where l is an integer $> N/4$ and $z < a$ (Berezanskiĭ [4], Chapter 6, Lemma 2.1). Moreover, A satisfies the semigroup positivity condition: (5.21) is obvious, and the required properties of the kernel $U(x,y;t)$ (the fundamental solution of the heat conduction equation, shifted by a) follow from the well-known representation of this kernel. (Recall that if $a = 0$, then

$$U(x, y; t) = (2\sqrt{\pi t})^{-N} \exp\left(-\frac{|x-y|^2}{4t}\right)$$

for $x, y \in \mathbb{R}^N$ and $t \in (0, 1]$). Here $F = C_0^\infty(\mathbb{R}^N)$ and $v(x) = q(x) - a \geq 0$ ($x \in \mathbb{R}^N$).

In the case of Carleman operators acting in $L_2(R, d\mu(x))$, where R is a locally compact separable space and μ is a Borel measure which is positive on open sets and finite on compact sets, we can draw some useful conclusions from the a priori assumed continuity of the spectral kernel $P(x, y; \lambda)$ with respect to $(x, y) \in R \times R$ (for example, we can prove that the individual eigenfunctions $\varphi_\alpha(x; \lambda)$ are continuous). We shall not dwell on these results (Berezanskiĭ [4], §4.3 in Chapter 5). We also shall not touch on growth estimates at ∞ for generalized eigenfunctions of Carleman operators and, in particular, differential operators. On this topic see Berezanskiĭ's book [4] (§5 in Chapter 5), the recent survey of Simon [5], and the Comments on the Literature.

5.4. *Expansion in generalized eigenfunctions of operators acting in L_2-spaces.* For the construction of an expansion in the case of a family $(A_x)_{x \in X}$ of general commuting selfadjoint operators acting in the space $H_0 = L_2(G)$, where $G \subseteq \mathbb{R}^N$, it is necessary to construct a rigging (2.1) of this space with a quasinuclear imbedding $H_+ \to H_0$. It is frequently convenient to take H_+ to be the Sobolev space $W^l(G)_2$ ($l > N/2$) in the case when G is bounded (see Theorem 3.2 in Chapter 1), or the spaces $W_2^{(l,q_1)}(G)$ and $W_2^l(G, q_2(x)\,dx)$ ($l > N/2$) in the case when G is unbounded and regular. (The conditions on the weights q_1 and q_2 are specified by Theorems 3.5 and 3.6 in Chapter 1.) To obtain explicit formulas it is sometimes useful to take H_+ to be the positive space constructed according to §1.8 in Chapter 1 from the closure D of the operator

$$C_0^\infty(G) \ni u(x) \mapsto \delta^{-1}(x)(\mathcal{D}u)(x)$$

acting in $L_2(G)$, where $\mathcal{D} = \partial^N/(\partial x_1 \cdots \partial x_N)$, and δ is chosen according to §3.11 in Chapter 1. (The quasinuclearity of the imbedding $H_+ \to H_0$ follows from (3.53) in Chapter 1.)

For operators acting in $H_0 = L_2(\mathbb{R}^\infty, d\theta(x))$, where $d\theta(x)$ is a product measure, the rigging (2.1) can be constructed according to (4.3) and (4.14) in Chapter 1 with the rules indicated there for choosing the positive spaces whose product is taken and the weight.

The first of the formulas in (2.14) can be rewritten in the form

$$(E(\alpha)u, v)_{H_0} = \int_\alpha (Q(\lambda(\cdot))J^{-1}u, J^{-1}v)_{H_0}\, d\rho(\lambda(\cdot))$$

$$(\alpha \in \mathcal{C}_\sigma(\mathbb{R}^X);\ u, v \in H_+), \quad (5.24)$$

where $Q(\lambda(\cdot)) = \mathbf{J}P(\lambda(\cdot))J\colon H_0 \to H_0$ is an operator acting in H_0 with trace 1 for ρ-almost all $\lambda(\cdot)$. The operator $Q(\lambda(\cdot))$ is a fortiori quasinuclear; therefore, for example, in the case $H_0 = L_2(G)$ the integrand in (5.24) takes the form

$$(Q(\lambda(\cdot))J^{-1}u, J^{-1}v)_{L_2(G)} = \int_G \int_G Q(x,y;\lambda(\cdot))(J^{-1}u)(y)\overline{(J^{-1}v)(x)}\,dx\,dy \quad (u, v \in H_+), \quad (5.25)$$

where $Q(\cdot,\cdot;\lambda(\cdot)) \in L_2(G \times G)$ for ρ-almost all $\lambda(\cdot)$. This formula is reminiscent of the corresponding equality for Carleman operators. (If we transfer the J^{-1} in it, then

$$(\mathbf{J}^{-1} \otimes \mathbf{J}^{-1})Q(x,y;\lambda(\cdot)) = P(x,y;\lambda(\cdot));$$

it is true that the last kernel is a generalized one.)

In particular, if H_+ is generated by the operator D, then with the help of arguments like those in §3.11 of Chapter 1 we obtain the more explicit representation

$$(Q(\lambda(\cdot))J^{-1}u, J^{-1}v)_{L_2(G)} = \int_G \int_G Q_1(x,y;\lambda(\cdot))(\mathcal{D}u)(y)\overline{(\mathcal{D}v)(x)}\,dx\,dy \quad (u, v \in C_0^\infty(G)). \quad (5.26)$$

Here

$$Q_1(x,y;\lambda(\cdot)) = ((U^{-1} \otimes U^{-1})Q(x,y;\lambda(\cdot)))\delta^{-1}(x)\delta^{-1}(y) \in L_{2,\mathrm{loc}}(G \times G),$$

with the unitary operator U determined by the equality $\sqrt{D^*D} = UD$. From (5.26) and (5.24)

$$(E(\alpha)u, v)_{L_2(G)} = \int_\alpha \left(\int_G \int_G Q_1(x,y;\lambda(\cdot))(\mathcal{D}u)(y)\overline{(\mathcal{D}v)(x)}\,dx\,dy\right) d\rho(\lambda(\cdot)) \quad (\alpha \in \mathcal{C}_\sigma(\mathsf{R}^X);\ u, v \in C_0^\infty(G)). \quad (5.27)$$

The kernel $Q_1(x,y;\lambda(\cdot))$ in (5.27) is not uniquely determined. (5.27) does not change when an arbitrary term $K(x,y)$ for which $(\mathcal{D}_x\mathcal{D}_yK)(x,y) = 0$ is added to this kernel. One of the kernels $Q_1(x,y;\lambda(\cdot))$ satisfying (5.27) has the following simple form:

$$Q_1(x,y;\lambda(\cdot)) = ((d/d\rho)(E\omega(y,\cdot),\omega(x,\cdot))_{L_2(G)})(\lambda(\cdot)) \qquad (x, y \in G), \quad (5.28)$$

where the $\omega(x,\xi)$ are defined in §3.11 of Chapter 1. To prove (5.28) we write (according to Theorem 3.10 in Chapter 1) the representation

$$(E(\alpha)u, v)_{L_2(G)} = \int_G \int_G (E(\alpha)\omega(y,\cdot),\omega(x,\cdot))_{L_2(G)}(\mathcal{D}u)(y)\overline{(\mathcal{D}v)(x)}\,dx\,dy \quad (\alpha \in \mathcal{C}_\sigma(\mathsf{R}^X);\ u, v \in C_0^\infty(G)). \quad (5.29)$$

For each particular pair $x, y \in G$ the measure

$$\mathcal{C}_\sigma(\mathsf{R}^X) \ni \alpha \mapsto (E(\alpha)\omega(y,\cdot),\omega(x,\cdot))_{L_2(G)}$$

obviously is absolutely continuous with respect to ρ. Using the usual Radon-Nikodým theorem and representing the integrals in (5.29), we arrive at (5.27) with the kernel (5.28). The other formulas in §§2 and 3 can be written similarly.

5.5. *Existence of a rigging.* It was shown in §2 that if a family $A = (A_x)_{x\in X}$ of commuting normal operators admits a quasinuclear rigging, then, in particular, the representation (2.24) is valid. We now give a natural construction which shows that the existence of a representation of the type (2.24) with π replaced by $\tau \supseteq \pi$ together with the presence of certain spaces of test functions on τ implies the existence of a quasinuclear rigging connected in the standard way with A. This construction is, in essence, an outgrowth of the method for proving necessity in the generalization of Stone's theorem (Theorem 4.2) and Theorem 4.4.

Accordingly, let $\tau \subset \mathbb{C}^X$ be a set, let $\mathcal{C}_\sigma(\tau) = \{\alpha' = \alpha \cap \tau \mid \alpha \in \mathcal{C}_\sigma(\mathbb{C}^X)\}$ be the previous σ-algebra of subsets of it, and let $\mathcal{C}_\sigma(\tau) \ni \alpha' \mapsto E(\alpha')$ be a resolution of the identity in a space H_0. The function $\tau \ni \lambda(\cdot) \mapsto \lambda(x) \in \mathbb{C}^1$ ($x \in X$ fixed) is $\mathcal{C}_\sigma(\tau)$-measurable; hence as in §1.11 we can define a family $A = (A_x)_{x\in X}$ of operators by the formula

$$A_x = \int_\tau \lambda(x)\, dE(\lambda(\cdot)),$$

$$\mathfrak{D}(A_x) = \left\{ f \in H_0 \mid \int_\tau |\lambda(x)|^2\, d(E(\lambda(\cdot))f, f)_{H_0} < \infty \right\} \qquad (x \in X). \quad (5.30)$$

It is easy to prove at once that the operators A_x are normal and commuting, and A_x^* can be represented by the same integral (5.30) with $\lambda(x)$ replaced by $\overline{\lambda(x)}$. (This follows from the preceding: extend E to an RI on $\mathcal{C}_\sigma(\mathbb{C}^X)$ by the procedure in §1.8; then use Remark 1 in §1.3 and §1.11.)

Fix $\tau \subseteq \mathbb{C}^X$, and consider a chain

$$\mathcal{H}(\tau) \supseteq D(\tau) \quad (5.31)$$

where $\mathcal{H}(\tau)$ is a Hilbert space consisting of $\mathcal{C}_\sigma(\tau)$-measurable functions $\tau \ni \lambda(\cdot) \mapsto a(\lambda(\cdot)) \in \mathbb{C}^1$, $D(\tau)$ is a linear topological space, and the imbedding (5.31) is topological. Assume that the following conditions hold:

a) $|a(\lambda(\cdot))| \le c_{\lambda(\cdot)} \|a\|_{\mathcal{H}(\tau)}$ ($a \in \mathcal{H}(\tau)$) for each $\lambda(\cdot) \in \tau$;

b) the multiplication operators

$$D(\tau) \ni a(\lambda(\cdot)) \mapsto \lambda(x) a(\lambda(\cdot)), \qquad \overline{\lambda(x)} a(\lambda(\cdot)) \in \mathcal{H}(\tau),$$

are defined and continuous for each $x \in X$; and

c) with $\varepsilon(\lambda(\cdot))$ equal to 1 divided by the norm of the functional $\mathcal{H}(\tau) \ni a \mapsto a(\lambda(\cdot)) \in \mathbb{C}^1$, it is required that a finite charge $\mathcal{C}_\sigma(\tau) \ni \alpha \mapsto \mu(\alpha) \in \mathbb{C}^1$ equals zero if $\int_\tau a(\lambda(\cdot))\varepsilon(\lambda(\cdot))\, d\mu(\lambda(\cdot)) = 0$ ($\alpha \in \mathcal{H}(\tau)$).

A chain (5.31) for which conditions a)–c) hold is said to be admissible. The functional in c) is the "delta-function $\delta_{\lambda(\cdot)}$ concentrated at the point $\lambda(\cdot)$", whose existence follows from a), and the integrand in c) is bounded and is easily seen to be $\mathcal{C}_\sigma(\tau)$-measurable.

THEOREM 5.7. *Suppose that an admissible chain* (5.31) *exists for the set* $\tau \subseteq \mathbf{C}^X$. *Then it is possible to construct a quasinuclear rigging* (2.1) *connected in the standard way with the family of operators* (5.30), *and* D *is a separable projective limit of Hilbert spaces.*

PROOF. I. We determine a formula for computing $\varepsilon(\lambda(\cdot))$. Let $(e_j)_1^\infty$ be an orthonormal basis in $\mathcal{H}(\tau)$; with $\delta_{\lambda(\cdot)}$ understood as a vector in $\mathcal{H}(\tau)$, we get that

$$\delta_{\lambda(\cdot)} = \sum_{j=1}^{\infty} (\delta_{\lambda(\cdot)}, e_j)_{\mathcal{H}(\tau)} e_j = \sum_{j=1}^{\infty} \overline{e_j(\lambda(\cdot))} e_j,$$

and so

$$\varepsilon(\lambda(\cdot)) = \|\delta_{\lambda(\cdot)}\|^{-1}_{\mathcal{H}(\tau)} = \left(\sum_{j=1}^{\infty} |e_j(\lambda(\cdot))|^2 \right)^{-1/2} \qquad (\lambda(\cdot) \in \tau).$$

(This formula implies, in particular, that $\varepsilon(\cdot)$ is measurable.)

II. Fix a vector $l_1 \in H_0$ with $\|l_1\|_{H_0} = 1$, and consider the linear mapping

$$\mathcal{H}(\tau) \ni a \mapsto Q_1 a = \int_\tau a(\lambda(\cdot)) \varepsilon(\lambda(\cdot))\, dE(\lambda(\cdot)) l_1 \in H_0. \tag{5.32}$$

It is a Hilbert-Schmidt operator: by I,

$$|Q_1|^2 = \sum_{j=1}^{\infty} \|Q_1 e_j\|^2_{H_0} = \sum_{j=1}^{\infty} \int_\tau |e_j(\lambda(\cdot))|^2 \varepsilon^2(\lambda(\cdot))\, d(E(\lambda(\cdot)) l_1, l_1)_{H_0} = 1.$$

Consider the kernel of the operators Q_1, i.e., the subspace $\{a \in \mathcal{H}(\tau) | Q_1 a = 0\}$. Let F be its orthogonal complement, and let $R = Q_1 \restriction F$. Choosing an orthonormal basis $(e_j)_1^\infty$ whose vectors are contained in this kernel and in F, we find that $|R| = |Q_1| = 1$, and thus $\|R\| \leq 1$. The closed inverse operator R^{-1} is defined on the range $\mathfrak{R}(R) = \mathfrak{R}(Q_1) \subseteq H_0$, and $\|R^{-1} f\|_{\mathcal{H}(\tau)} \geq \|f\|_{H_0}$ $(f \in \mathfrak{R}(R))$.

We turn $\mathfrak{R}(R) \subseteq H_0$ into a Hilbert space $H_{+,1}$ by setting

$$(u, v)_{H_{+,1}} = \kappa_1 (R^{-1} u, R^{-1} v)_{\mathcal{H}(\tau)} \qquad (u, v \in \mathfrak{R}(R)),$$

where $\kappa_1 \geq 1$ is a fixed number. Obviously, $H_{+,1}$ is complete, and $\|\cdot\|_{H_{+,1}} \geq \|\cdot\|_{H_0}$. In other words, if P is the projection from $\mathcal{H}(\tau)$ onto F, then

$$(Q_1 a, Q_1 b)_{H_{+,1}} = \kappa_1 (Pa, Pb)_{\mathcal{H}(\tau)} \qquad (a, b \in \mathcal{H}(\tau)). \tag{5.33}$$

The imbedding $O_1 \colon H_{+,1} \to H_0$ is quasinuclear, and $|O_1| = \kappa_1^{-1/2}$. Indeed, let $(e_j)_1^\infty$ be an orthonormal basis in F; then $(\kappa_1^{-1/2} R e_j)_1^\infty$ is an orthonormal basis in $H_{+,1}$. For it we have that

$$|O_1|^2 = \sum_{j=1}^{\infty} \|\kappa_1^{-1/2} Re_j\|_{H_0}^2 = \kappa_1^{-1} |R|^2 = \kappa_1^{-1}.$$

III. Use the orthogonalization procedure to construct an orthonormal basis $(h_j)_1^\infty$ in $\mathcal{H}(\tau)$ consisting of vectors in $D(\tau)$, and fix it. Let $D_0(\tau)$ be the collection of all vectors in $\mathcal{H}(\tau)$ having a finitely nonzero sequence of coordinates in this basis; $D_0(\tau) \subseteq D(\tau)$ and is dense in $\mathcal{H}(\tau)$.

Define $D_1 = Q_1 D_0(\tau) \subseteq \mathfrak{R}(Q_1) = H_{+,1}$. From (5.33) and the denseness of $D_0(\tau)$ in $\mathcal{H}(\tau)$ it follows that D_1 is dense in $H_{+,1}$. Each vector in D_1 has the form $u = \sum_1^\infty \alpha_j (Q_1 h_j)$, where $(\alpha_j)_1^\infty$, $\alpha_j \in \mathbb{C}^1$, is an arbitrary finitely nonzero sequence; $\sum_1^\infty \alpha_j h_j = a$ is an arbitrary vector in $D_0(\tau)$. The vectors $Q_1 h_1, Q_1 h_2, \ldots$ are linearly dependent in general. Sifting through this sequence from left to right, we construct a subsequence $(Q_1 f_{1,k})_1^\infty$, $f_{1,k} = h_{j_k}$, of it that now consists of linearly independent vectors. Then the vectors in D_1 have the form $u = \sum_1^\infty \xi_k (Q_1 f_{1,k})$, where $(\xi_k)_1^\infty$, $\xi_k \in \mathbb{C}^1$, is an arbitrary finitely nonzero sequence uniquely determined by u (the sequence of coordinates of u).

We introduce a convergence in D_1: $D_1 \ni u^{(n)} \to u \in D_1$ as $n \to \infty$ if the corresponding coordinates $\xi_k^{(n)}$ are uniformly finitely nonzero and $\xi_k^{(n)} \to \xi_k$ as $n \to \infty$ for each $k = 1, 2, \ldots$. It is clear that this convergence corresponds to the following topology. Consider the Hilbert space

$$G_{1,\sigma} = \left\{ \sum_{k=1}^{\infty} \xi_k (Q_1 f_{1,k}) \,\middle|\, \sum_{k=1}^{\infty} |\xi_k|^2 \sigma_k < \infty \right\},$$

$\sigma = (\sigma_k)_1^\infty$, $\sigma_k \geq 1$ a weight, with the corresponding inner product; then $D_1 = \operatorname{pr\,lim} G_{1,\sigma}$, where the projective limit is taken over all weights σ.

IV. The imbedding $D_1 \subseteq H_{+,1}$ is topological. Thus, suppose that $D_1 \ni u^{(n)} \to 0$, $u^{(n)} = \sum_{k=1}^\infty \xi_k^{(n)} (Q_1 f_{1,k})$ in D_1 as $n \to \infty$. This means that the $\xi_k^{(n)}$ are uniformly finitely nonzero and that $\xi_k^{(n)} \to 0$ for each $k = 1, 2, \ldots$. Therefore, $D_0(\tau) \ni \sum_{k=1}^\infty \xi_k^{(n)} f_{1,k} = a^{(n)} \to 0$ in $D(\tau)$ as $n \to \infty$, and, since the imbedding $D(\tau) \subseteq \mathcal{H}(\tau)$ is topological, it follows that $a^{(n)} \to 0$ in $\mathcal{H}(\tau)$. Consequently, $Pa^{(n)} \to 0$ in $\mathcal{H}(\tau)$, and $\|Q_1 a^{(n)}\|_{H_{+,1}} = \kappa_1^{1/2} \|Pa^{(n)}\|_{\mathcal{H}(\tau)} \to 0$ by (5.33), i.e., $u^{(n)} = Q_1 a^{(n)} \to 0$ in $H_{+,1}$.

V. Let us show that $D_1 \subseteq \mathfrak{D}(A_x)$ $(x \in X)$. It follows from (5.30) that the vector $f = \int_\tau F(\lambda(\cdot))\, dE(\lambda(\cdot)) l$, where l is in H_0 and $\tau \ni \lambda(\cdot) \mapsto F(\lambda(\cdot)) \in \mathbb{C}^1$ is $\mathcal{C}_\sigma(\tau)$-measurable, is in $\mathfrak{D}(A_x)$ if and only if

$$\int_\tau |\lambda(x) F(\lambda(\cdot))|^2 \, d(E(\lambda(\cdot))l, l)_{H_0} < \infty;$$

$$A_x f = \int_\tau \lambda(x) F(\lambda(\cdot))\, dE(\lambda(\cdot)) l. \tag{5.34}$$

For $a \in D_0(\tau) \subseteq D(\tau)$ condition b) gives us that $\lambda(x)a(\lambda(\cdot)) \in \mathcal{H}(\tau)$, and consequently, the function $\tau \ni \lambda(\cdot) \mapsto \lambda(x)a(\lambda(\cdot))\varepsilon(\lambda(\cdot)) \in \mathbb{C}^1$ is bounded. Therefore, condition (5.34) is satisfied for the vector $u = Q_1 a = f$ given by (5.32). (The vectors u run through the whole of D_1.)

VI. We show that the operator $A_x \restriction D_1$ acts continuously from D_1 to $H_{+,1}$ ($x \in X$). Suppose that $D_1 \ni u^{(n)} \to 0$ in D_1 as $n \to \infty$, where

$$u^{(n)} = \sum_{k=1}^{\infty} \xi_k^{(n)}(Q_1 f_{1,k}) = Q_1 a^{(n)} \quad \text{with} \quad a^{(n)} = \sum_{k=1}^{\infty} \xi_k^{(n)} f_{1,k} \in D_0(\tau).$$

Using formula (5.32) for $u^{(n)}$ and then acting with the operator A_x, we get that $A_x u^{(n)} = Q_1(\lambda(x)a^{(n)}(\lambda(\cdot)))$ by (5.34). Therefore,

$$\|A_x u^{(n)}\|_{H_{+,1}} = \kappa_1^{1/2} \|P(\lambda(x)a^{(n)}(\lambda(\cdot)))\|_{\mathcal{H}(\tau)},$$

by (5.33). But, just as in IV, $a^{(n)} \to 0$ in $D(\tau)$ as $n \to \infty$; hence

$$\lambda(x)a^{(n)}(\lambda(\cdot)) \to 0$$

in $\mathcal{H}(\tau)$ due to b), and consequently $P(\lambda(x)a^{(n)}(\lambda(\cdot))) \to 0$ in $\mathcal{H}(\tau)$. Thus, $A_x u^{(n)} \to 0$ in $H_{+,1}$, which is what was to be proved.

Similarly, the operator $A_x^* \restriction D_1$ acts continuously from D_1 to $H_{+,1}$ ($x \in X$).

VII. Summarizing II–VI, we conclude that we have constructed a chain

$$H_0 \supseteq H_{+,1} \supseteq D_1, \qquad \|u\|_{H_0} \le \|u\|_{H_{+,1}} \quad (u \in H_{+,1}), \tag{5.35}$$

where the imbedding $O_1\colon H_{+,1} \to H_0$ is quasinuclear, $|O_1| = \kappa_1^{-1/2}$, the imbedding $D_1 \subseteq H_{+,1}$ is topological, and D_1 is a separable projective limit of Hilbert spaces. For each $x \in X$ we have that $D_1 \subseteq \mathfrak{D}(A_x)$ and the restrictions $A_x \restriction D_1$ and $A_x^* \restriction D_1$ act continuously from D_1 to $H_{+,1}$.

VIII. Assume that $H_{+,1}$ is not dense in H_0, and let $l_2 \in H_0$ be a unit vector orthogonal to $H_{+,1}$ in H_0. Repeating the arguments in II–VI with l_1 replaced by l_2 and fixing another $\kappa_2 \in [1, \infty)$ in place of κ_1, we get a chain with analogous properties in place of (5.35):

$$H_0 \supseteq H_{+,2} \supseteq D_2. \tag{5.36}$$

With the help of condition c) it is not hard to prove that the linear sets $H_{+,1}$ and $H_{+,2}$ are orthogonal in H_0. Indeed, since $l_2 \perp H_{+,1}$, it follows from (5.32) that for

$$0 = (Q_1 a, l_2)_{H_0} = \int_\tau a(\lambda(\cdot))\varepsilon(\lambda(\cdot))\, d(E(\lambda(\cdot))l_1, l_2)_{H_0} \qquad (a \in \mathcal{H}(\tau)).$$

From this we see by c) that the finite charge $\mathcal{C}_\sigma(\tau) \ni \alpha \mapsto (E(\alpha)l_1, l_2)_{H_0} \in \mathbb{C}^1$ is equal to 0. Therefore, denoting by Q_2 the operator analogous to Q_1 but constructed from l_2, we get from (5.32) that for $(a, b \in \mathcal{H}(\tau))$

$$(Q_1 a, Q_2 b)_{H_0} = \int_\tau a(\lambda(\cdot))\overline{b(\lambda(\cdot))}\varepsilon^2(\lambda(\cdot))\, d(E(\lambda(\cdot))l_1, l_2)_{H_0} = 0,$$

i.e., $H_{+,1} \perp H_{+,2}$.

IX. It follows from (5.35) and (5.36) that

$$H_0 \supseteq H_{+,1} \oplus H_{+,2} \supseteq D_1 \oplus D_2,$$

where $\oplus$ denotes the orthogonal sum in H_0. If $H_{+,1} \oplus H_{+,2}$ is dense in H_0, then the construction is finished. If not, then choose a vector $l_3 \in H_0$, $\|l_3\|_{H_0} = 1$, orthogonal to $H_{+,1} \oplus H_{+,2}$ and a number $\kappa_3 \in [1, \infty)$, and construct a chain analogous to (5.35) and (5.36), and so on. The result is a chain

$$H_0 \supseteq \bigoplus_{m=1}^{\infty} H_{+,m} \supseteq \bigoplus_{m=1}^{\infty} D_m, \tag{5.37}$$

where $\bigoplus_1^\infty H_{+,m}$ is now dense in H_0. (There can be finitely many terms, but for definiteness we consider the case of infinitely many.) The numbers κ_m are chosen so that $\sum_1^\infty \kappa_m^{-1} < \infty$.

We turn $H_+ = \bigoplus_1^\infty H_{+,m}$ into a Hilbert space by defining $(u, v)_{H_+} = \sum_1^\infty (u_m, v_m)_{H_{+,m}}$ for $H_+ \ni u = (u_m)_1^\infty$ $(u_m \in H_{+,m})$ and an analogous $v \in H_+$. Next, denote by D the collection of all finitely nonzero sequences $u = (u_m)_1^\infty$ $(u_m \in D_m)$ with the following convergence: $D \ni u^{(n)} \to u \in D$ as $n \to \infty$ if the corresponding coordinates $u_m^{(n)}$ are uniformly finitely nonzero and $u_m^{(n)} \to u_m$ in D_m for each $m = 1, 2, \ldots$. Since each D_m is a separable projective limit of Hilbert spaces (see III), it follows that D is the same kind of space: the corresponding Hilbert space is

$$G_\sigma = \left\{ \sum_{m,k=1}^{\infty} \xi_{m,k}(Q_m f_{m,k}) \,\middle|\, \sum_{m,k=1}^{\infty} |\xi_{m,k}|^2 \sigma_{m,k} < \infty \right\},$$

where Q_m and $f_{m,k}$ are the operator Q_1 and the vectors $f_{1,k}$ associated with the mth chain of form (5.35), and $\sigma = (\sigma_{m,k})_{m,k=1}^\infty$, $\sigma_{m,k} \geq 1$, is a weight. Now $D = \operatorname{pr\,lim} G_\sigma$, with the projective limit over all weights σ.

It is clear that $\|\cdot\|_{H_0} \leq \|\cdot\|_{H_+}$, and D is topologically imbedded in H_+ by virtue of IV.

Thus, we have constructed the chain

$$H_- \supseteq H_0 \supseteq H_+ \supseteq D \tag{5.38}$$

corresponding to (5.37). The imbedding $O\colon H_+ \to H_0$ is quasinuclear because $O = \bigoplus_1^\infty O_m$, where $O_m\colon H_{+,m} \to H_0$, and hence

$$|O|^2 = \sum_{m=1}^{\infty} |O_m|^2 = \sum_{m=1}^{\infty} \kappa_m^{-1} < \infty.$$

X. Let us show that the family of operators (5.30) is connected in the standard way with (5.38). Indeed, a $u \in D$ has the form of a finitely nonzero sequence $u = (u_m)_1^\infty$, where $u_m \in D_m$ and $u_m = 0$ for all large m. By

V, $D \subseteq \mathfrak{D}(A_x)$, and $A_x u = (A_x u_m)_1^\infty$ $(x \in X)$. Further, if $(u_m^{(n)})_{m=1}^\infty = u^{(n)} \to 0$ in D as $n \to \infty$, then in view of the nature of this convergence the sequence $A_x u^{(n)} = (A_x u_m^{(n)})_1^\infty$ is finitely nonzero uniformly with respect to n, and $A_x u_m^{(n)} \to 0$ in $H_{+,m}$ for each $m = 1, 2, \cdots$ (see VI). This implies that $A_x u^{(n)} \to 0$ in H_+. ■

REMARK 1. Instead of the operators (5.30) we could consider functions of them. For example, the operators A_x given by (5.30) with $\tau \ni \lambda(\cdot) \to \lambda(x) \in \mathbb{C}^1$ replaced by some function $\tau \ni \lambda(\cdot) \mapsto F_x(\lambda(\cdot)) \in \mathbb{C}^1$ is $\mathcal{C}_\sigma(\tau)$-measurable $(x \in X)$. It is easy to see that Theorem 5.7 is preserved also in this case, provided that $\lambda(x)$ is replaced by $F_x(\lambda(\cdot))$ in condition b) for admissibility of the chain (5.31).

REMARK 2. Let X be a topological space, and $(A_x)_{x \in X}$ a family of operators (5.30) given by functions $F_x(\lambda(\cdot))$. The requirement (2.49) ensuring continuity of the eigenvalues is satisfied under a natural additional condition on these functions. Namely, it is easy to see from VI that this requirement holds if for each $a \in D(\tau)$ the vector-valued function $X \ni x \mapsto \overline{F_x(\lambda(\cdot))} a(\lambda(\cdot)) \in \mathcal{H}(\tau)$ is weakly continuous. (It suffices to consider the chain (5.35). For each $u \in D_1$ there is an $a \in D_0(\tau) \subseteq D(\tau)$ such that $u = Q_1 a$; therefore, $A_x^* u = Q_1(\overline{F_x(\lambda(\cdot))} a(\lambda(\cdot)))$ and, by (5.33),

$$\begin{aligned}(A_x^* u, Q_1 b)_{H+,1} &= \kappa_1(P(\overline{F_x(\lambda(\cdot))} a(\lambda(\cdot))), Pb)_{\mathcal{H}(\tau)} \\ &= \kappa_1(\overline{F_x(\lambda(\cdot))} a(\lambda(\cdot)), Pb)_{\mathcal{H}(\tau)}.\end{aligned}$$

This gives us the assertion, since $Q_1 b$ varies over the whole of $H_{+,1}$ as b varies over $\mathcal{H}(\tau)$.)

REMARK 3. The results in this subsection are also valid for operators of the type (5.30) constructed from an RI E on an abstract space τ.

We present an example of an application of Theorem 5.7.

THEOREM 5.8. *For an arbitrary at most countable family $(A_x)_{x \in X}$ $(X = \{1, \ldots, p\},\ p \leq \infty)$ of commuting normal operators, there always exists a quasinuclear rigging connected with it in the standard way.*

PROOF. Suppose first that $p < \infty$. According to Theorem 5.7 and §1.4, it suffices to construct an admissible chain (5.31) in the case when $\tau = \mathbb{C}^p, \mathcal{C}_\sigma(\tau) = \mathcal{B}(\mathbb{C}^p)$ and $\lambda(\cdot) = \lambda \in \mathbb{R}^{2p}$. We identify $\mathbb{C}^p$ with $\mathbb{R}^{2p}$ and let $\mathcal{H}(\tau) = W_2^l(\mathbb{R}^{2p}, (1 + |\lambda|^2)^l \, d\lambda)$ $(l > p)$ (the Sobolev space with weight $(1 + |\lambda|^2)^l$), and $D(\tau) = \mathcal{D}(\mathbb{R}^{2p})$. Conditions a) and b) are obviously satisfied on the basis of imbedding theorems (see also Theorems 3.4 and 3.6 in Chapter 1). Condition c) is also satisfied, because $a(\lambda)\varepsilon(\lambda)$ can be taken to be any function in $C_0^\infty(\mathbb{R}^{2p})$. The assertion is proved in the case of finitely many operators.

Suppose that $p = \infty$. Without loss of generality it can be assumed that the operators $(A_x)_{x \in X}$, $X = \{1, 2, \ldots\}$, are commuting selfadjoint operators. Here $\tau = \mathbb{R}^\infty$, $\mathcal{C}_\sigma(\tau) = \mathcal{B}(\mathbb{R}^\infty)$ and $\lambda(\cdot) = \lambda = (\lambda_n)_1^\infty$. We make use of the following fact: if $\mathcal{B}(\mathbb{R}^\infty) \ni \alpha \mapsto E(\alpha)$ is some resolution of the identity, then there exists a weight $\sigma = (\sigma_n)_1^\infty$ ($\sigma_n > 0$) such that $l_{2,\sigma} = \{\lambda \in \mathbb{R}^\infty | \sum_1^\infty \lambda_n^2 \sigma_n < \infty\} \in \mathcal{B}(\mathbb{R}^\infty)$ is of full E-measure. (This assertion is known in the case when E is replaced by a nonnegative finite measure ρ (Kosyak and Samoĭlenko [1], and Samoĭlenko [2], Chapter 1, §1, Lemma 3); taking ρ to be a spectral measure of E, we get the stated result.) Due to this, $\tau = \mathbb{R}^\infty$ can be replaced by $l_{2,\sigma}$ in the integrals (5.30), and it can be assumed that $\sum_1^\infty \sigma_n < \infty$, $\sigma_n < 1$. Thus, according to Theorem 5.7, it suffices to construct an admissible chain (5.31) in the case when $\tau = l_{2,\sigma}$ and $\mathcal{C}_\sigma(\tau) = \mathcal{B}(l_{2,\sigma})$. This chain can be constructed with the help of the spaces $A_t(\mathbb{R}^\infty)$ of test functions of infinitely many variables introduced in §4.5 of Chapter 1, i.e., $\mathcal{H}(\tau) = (A_t(\mathbb{R}^\infty)) \restriction l_{2,\sigma}$ ($t = (\sigma_n^{-1} - 1)_1^\infty$) and $D(\tau) = (P_c(\mathbb{R}^\infty)) \restriction l_{2,\sigma}$, where $P_c(\mathbb{R}^\infty)$ is the linear space of cylindrical polynomials of a point $\lambda \in \mathbb{R}^\infty$. The required properties of the chain (5.31) follow from §4.5 in Chapter 1 and §4.2 in Chapter 2. ■

REMARK 4. Theorem 5.8 (more precisely, Remarks 1 and 2 after it) applies in a natural way to the proof of necessity in Theorems 4.2 and 4.4. (In fact, this proof is essentially a repetition of the corresponding arguments in §4.)

5.6. ***Connection with the theory of commutative normed algebras and the nuclear spectral theorem.*** We have already dwelt in §3 on the connections between the theory developed in §§1 and 2 and commutative normed algebras. Here we look at them from some different points of view and clear up the interrelation between the spectral projection theorem in §2 and the nuclear spectral theorem established by Maurin [1]–[4].

Let $A = (A_x)_{x \in X}$ be a family of bounded commuting normal operators; without loss of generality it can be assumed that A contains I. Take the algebra generated by A and close it with respect to operator norm convergence; this gives a commutative C^*-algebra $\mathcal{A}$ of operators with identity. Let M be its compact space of maximal ideals ω. By the Gel′fand-Naĭmark theorem, $\mathcal{A}$ is isometrically isomorphic to the algebra $C(M)$ of all complex-valued continuous functions on M with the usual algebraic operations and the uniform norm.

This theorem can be used to prove that there is an RI $\mathcal{B}(M) \ni \alpha \mapsto F(\alpha)$ on the Borel subsets of M which gives a representation of each operator in $\mathcal{A}$ as a spectral integral

$$a = \int_M a(\omega)\, dF(\omega) \qquad (a \in \mathcal{A}). \tag{5.39}$$

Here $a(\omega)$ is the value of a on a maximal ideal ω, i.e., $a(\omega)$ is the image of a under the isomorphism $\mathcal{A} \leftrightarrow C(M)$ (Naĭmark [1], Chapter 4, §17.4). We are interested in a maximal ideal ω as a continuous multiplicative linear functional ω on $\mathcal{A}$: $\omega(a) = a(\omega)$ ($a \in \mathcal{A}$). Recall that the topology in M is the relative topology induced by the Tychonoff topology in the space $\mathbb{C}^{\mathcal{A}} \supseteq M$.

For $\omega \in M$, let $\lambda_\omega(x) = \omega(A_x)$ ($x \in X$); $\lambda_\omega = \omega \restriction A$. The mapping $\mathbb{C}^{\mathcal{A}} \supseteq M \ni \omega \mapsto \lambda_\omega(\cdot) \in \mathbb{C}^X$ is one-to-one: if $\omega_1(A_x) = \omega_2(A_x)$ ($x \in X$), then by linearity, multiplicativity, and continuity, this equality extends to the whole of $\mathcal{A}$, i.e., $\omega_1 = \omega_2$. Let $\mu \subseteq \mathbb{C}^X$ be the compact set which is the image of M under this mapping, topologized by the image of the topology, i.e., in essence μ is the same compact maximal ideal space. We introduce the RI $\mathcal{B}(\mu) \ni \alpha \mapsto G(\alpha)$ which is the image of the RI F. For $a = A_x$ the representation (5.39) takes the form

$$A_x = \int_\mu \lambda(x)\, dG(\lambda(\cdot)) \qquad (x \in X). \tag{5.40}$$

The intersections with M of the cylindrical sets $\mathfrak{C}_b(\mathbb{C}^{\mathcal{A}}) = \mathfrak{C}(a_1, \dots, a_p; u_1 \times \cdots \times u_p)$ ($a_n \in \mathcal{A}$, u_n open in C^1) in $\mathbb{C}^{\mathcal{A}}$ serve as basic neighborhoods in M, and their images give a neighborhood base $\Sigma(\mu)$ in μ. If $a_n = A_{x_n}$ $(n = 1, \dots, p)$, then such an image obviously coincides with a basic neighborhood in the space μ, topologized by the relative topology from $\mathbb{C}^X$; let $\Sigma'(\mu)$ be the neighborhood base for μ topologized in this way. Thus,

$$\Sigma'(\mu) \subseteq \Sigma(\mu). \tag{5.41}$$

Consider the σ-algebra $\mathcal{C}_\sigma(\mu) = \{\alpha' = \alpha \cap \mu | \alpha \in \mathcal{C}_\sigma(\mathbb{C}^X)\}$; it coincides with the σ-algebra generated by the base $\Sigma'(\mu)$, and is in $\mathcal{B}(\mu)$ by (5.41). The function $\mathbb{C}^X \supseteq \mu \ni \lambda(\cdot) \mapsto \lambda(x) \in \mathbb{C}^1$ is $\mathcal{C}_\sigma(\mu)$-measurable, and thus in (5.40) the measure G can be replaced by its restriction $E' = G \upharpoonright \mathcal{C}_\sigma(\mu)$, which is an RI as before:

$$A_x = \int_\mu \lambda(x)\, dE'(\lambda(\cdot)) \qquad (x \in X). \tag{5.42}$$

It is not hard to see that E' is the modification of the joint RI E of the family A with respect to μ. Indeed, from the RI E' on $\mathcal{C}_\sigma(\mu)$ we construct the RI E'' on $\mathcal{C}_\sigma(\mathbb{C}^X)$ according to §1.8, namely, $\mathcal{C}_\sigma(\mathbb{C}^X) \ni \alpha \mapsto E''(\alpha) = E'(\alpha \cap \mu)$. The integral (5.42) is written in the form (1.25) with E replaced by E''. But $E'' = E$: according to Remark 1 in §1.3, $E'' = \times_{x \in X} E''_x$, where $\mathcal{B}(\mathbb{C}^1) \ni \alpha \mapsto E''_x(\alpha)$ is a certain RI. For E'' it now follows from (1.25) that $A_x = \int_{\mathbb{C}^1} \lambda\, dE''_x(\lambda)$, and so $E''_x = E_x$ $(x \in X)$, and $E'' = E$. Thus, $\mathcal{C}_\sigma(\mathbb{C}^X) \ni \alpha \mapsto E(\alpha) = E''(\alpha \cap \mu)$, from which it follows that μ is of full outer E-measure, and the modification of E with respect to μ leads to E'. ■

The inclusion (5.41) is strict in general. However, the topologies determined by $\Sigma'(\mu)$ and $\Sigma(\mu)$ coincide: the topologization of $\mu \subseteq \mathbb{C}^X$ as the maximal ideal space coincides with the relative topology induced by the space $\mathbb{C}^X$. Indeed, it suffices to show that for any point $\omega_0 \in M$ and any neighborhood $\mathfrak{C}_b(\mathbb{C}^{\mathcal{A}}) \cap M$ of it in M there is a neighborhood of the point $\lambda_{\omega_0}(\cdot) \in \mu$ in μ of the form $\mathfrak{C}_b \cap \mu$ ($\mathfrak{C}_b = \mathfrak{C}_b(x_1, \dots, x_q; v_1 \times \cdots \times v_q)$, $x_k \in X$, v_k open in $\mathbb{C}^1$) such that if $\lambda_\omega(\cdot) \in \mathfrak{C}_b \cap \mu$, then $\omega \in \mathfrak{C}_b(\mathbb{C}^{\mathcal{A}}) \cap M$. Let $u_j = \{z \in \mathbb{C}^1 | |z - \omega_0(a_j)| < \varepsilon\}$ $(j = 1, \dots, p;\ \varepsilon > 0)$, and let $\mathcal{A}'$ be the algebra generated by the family $(A_x)_{x \in X}$. For each a_j we find a $b_j \in \mathcal{A}'$ such that $\|a_j - b_j\| < \varepsilon$ $(j = 1, \dots, p)$, and let $A_{x_1}, \dots, A_{x_q}$ be operators generating an algebra containing $b_1, \dots, b_p$. Thus,

$$b_j = \sum_{|\alpha| \leq n_j} c_{\alpha,j} A_{x_1}^{\alpha_1} \cdots A_{x_q}^{\alpha_q},$$

$$\omega(b_j) = \sum_{|\alpha| \leq n_j} c_{\alpha,j} \lambda_\omega^{\alpha_1}(x_1) \cdots \lambda_\omega^{\alpha q}(x_q) = P_j(\lambda_\omega(x_1), \dots, \lambda_\omega(x_q))$$

$$(\alpha = (\alpha_1, \dots, \alpha_q),\ |\alpha| = \alpha_1 + \cdots + \alpha_q;\ j = 1, \dots, p;\ \omega \in M), \quad (5.43)$$

where the $c_{\alpha,j} \in \mathbb{C}^1$ are coefficients. The polynomial $P_j(z_1, \dots, z_q)$ ($z_k \in \mathbb{C}^1$) takes the value $\omega_0(b_j)$ at the point $(z_1^0, \dots, z_q^0) = (\lambda_{\omega_0}(x_1), \dots, \lambda_{\omega_0}(x_q))$. Since it is continuous, there is a $\delta > 0$ such that $|\omega_0(b_j) - P_j(z_1, \dots, z_q)| < \varepsilon/3$ for all $j = 1, \dots, p$ if $|z_k - z_k^0| < \delta$ ($k = 1, \dots, q$). The neighborhood $\mathfrak{C}_\mathrm{b}$ with the x_k chosen above and with $v_k = \{z \in \mathbb{C}^1 \mid |z - z_k^0| < \delta\}$ is the required one: if $\lambda_\omega(\cdot) \in \mathfrak{C}_\mathrm{b} \cap \mu$, then $|\omega_0(b_j) - \omega(b_j)| < \varepsilon/3$ ($j = 1, \dots, p$) by the second equality in (5.43), and, since the norm of a nonzero multiplicative functional is equal to 1, we have that

$$\begin{aligned} |\omega(a_j) - \omega_0(a_j)| &\leq |\omega(a_j) - \omega(b_j)| + |\omega(b_j) - \omega_0(b_j)| + |w_0(b_j) - w_0(a_j)| \\ &\leq 2\|a_j - b_j\| + |\omega(b_j) - \omega_0(b_j)| < \varepsilon \qquad (j = 1, \dots, p), \end{aligned}$$

i.e., $\omega \in \mathfrak{C}_\mathrm{b}(\mathbb{C}^{\mathcal{A}}) \cap M$. ■

The support of the joint RI E of the family of operators under consideration coincides with the maximal ideal space of the algebra $\mathcal{A}$; more precisely, $\operatorname{supp} E = \mu$. Indeed, if $\varphi \in C_\sigma(\mathbb{C}^X)$ is of full E-measure and is closed in $\mathbb{C}^X$, then $\varphi \supseteq \mu$. Indeed, assume the contrary: there is a point $\lambda_0(\cdot) \in (\mathbb{C}^X \backslash \varphi) \cap \mu$. The last set is open in μ in the relative topology; let $\mathfrak{C}_\mathrm{b} \cap \mu$ be a basic neighborhood of $\lambda_0(\cdot)$ contained in $(\mathbb{C}^X \backslash \varphi) \cap \mu$. Since $E'((\mathbb{C}^X \backslash \varphi) \cap \mu) = E(\mathbb{C}^X \backslash \varphi) = 0$, it follows that $G(\mathfrak{C}_\mathrm{b} \cap \mu) = E'(\mathfrak{C}_\mathrm{b} \cap \mu) = 0$. Let o be the inverse image of $\mathfrak{C}_\mathrm{b} \cap \mu$ under the mapping $M \to \mu$. The set o is open in the topology of M, and $F(o) = 0$—this contradicts (5.39) and the isomorphism $\mathcal{A} \leftrightarrow C(M)$. Accordingly, $\varphi \supseteq \mu$, and thus $\operatorname{supp} E \supseteq \mu$. On the other hand, as proved above, μ is a compact set in the relative topology of $\mathbb{C}^X$; hence, μ is closed in $\mathbb{C}^X$. Assume that $\operatorname{supp} E \neq \mu$. Then there is a basic neighborhood $\mathfrak{C}_\mathrm{b} \subseteq \mathbb{C}^X \backslash \mu$ having nonempty intersection with $\operatorname{supp} E$. Therefore, $E(\mathfrak{C}_\mathrm{b}) \neq 0$, and the set $C_\sigma(\mathbb{C}^X) \ni \mathbb{C}^X \backslash \mathfrak{C}_\mathrm{b} \supseteq \mu$ is not of full E-measure. Consequently, μ is not of full outer E-measure, which is absurd. ■

Thus, in the case of a family of bounded operators the spectral representation (5.39) is valid in the "algebra" approach, and it implies (5.42), i.e., (1.25). For subsequently constructing a theory of expansions we can use two schemes: 1) differentiate the RI in (1.25) or (5.42), which is the path taken above in §2 (the differentiation of F in (5.39) is less rational, since the measure here is defined on the special subset M of $\mathbb{C}^{\mathcal{A}}$); or 2) extend (5.39) to the full spectral theorem in the von Neumann form and use Fubini's theorem.

We outline scheme 2). For simplicity assume that the family A has a cyclic vector (of norm 1) $\Omega \in H_0$, i.e., the set $\{a\Omega | a \in \mathcal{A}\}$ is dense in H_0. It is well known that in this case the isomorphism $\mathcal{A} \leftrightarrow C(M)$ gives rise to an isometry $H_0 \ni f \mapsto \hat{f} = Gf \in L_2(M, d\sigma(\omega))$, where $\mathcal{B}(M) \ni \alpha \mapsto \sigma(\alpha) \in [0, \infty)$ is a finite measure, and $(af)\hat{}(\omega) = \omega(a)\hat{f}(\omega)$ $(f \in H_0,\ a \in \mathcal{A})$ for almost all ω with respect to σ (see Naĭmark [1], §§40 and 41 in Chapter 8; also §3.3). Thus, we have a "Parseval's equality" of the type (2.56) (with $N_{\lambda(\cdot)} = 1$):

$$(f, g)_{H_0} = \int_M \hat{f}(\omega)\overline{\hat{g}(\omega)}\, d\sigma(\omega) \qquad (f, g \in H_0). \tag{5.44}$$

In large part the theory of expansions in generalized eigenvectors reduces to obtaining representations of type (2.53) for $\hat{f}(\omega)$: $\hat{f}(\omega) = (f, \varphi(\omega))_{H_0}$ for σ-almost all $\omega \in M$ for f in some dense linear subset of H_0, where $\varphi(\omega)$ is a generalized eigenvector in some sense. This can be obtained in the following simple form.

Assume that we have a quasinuclear chain (2.1). Let $(e_j)_1^\infty$ be an orthonormal basis in H_+. Since O is quasinuclear, the operator $GO\colon H_+ \to L_2(M, d\sigma(\omega))$ is quasinuclear; therefore, by Fubini's theorem,

$$\infty > \sum_{j=1}^\infty \|GOe_j\|^2_{L_2(M, d\sigma(\omega))} = \sum_{j=1}^\infty \int_M |(GOe_j)(\omega)|^2\, d\sigma(\omega)$$
$$= \int_M \left(\sum_{j=1}^\infty |(GOe_j)(\omega)|^2 \right) d\sigma(\omega).$$

From this we conclude that $\sum_1^\infty |(GOe_j)(\omega)|^2 < \infty$ for σ-almost all $\omega \in M$; therefore, the following vector is defined for such ω:

$$w(\omega) = \sum_{j=1}^\infty \overline{(GOe_j)(\omega)} e_j \in H_+;$$

$$(u, w(\omega))_{H_+} = \sum_{j=1}^\infty (u, e_j)_{H_+} (GOe_j)(\omega) \qquad (u \in H_+). \tag{5.45}$$

But $u = \sum_1^\infty (u, e_j)_{H_+} e_j$; applying the operator GO to this equality and using the second of the relations (5.45), we get that

$$\hat{u}(\omega) = (GOu)(\omega) = (u, w(\omega))_{H_+} = (u, I^{-1}w(\omega))_{H_0} \qquad (u \in H_+).$$

Accordingly, if we set $\varphi(\omega) = I^{-1}w(\omega) \in H_-$, then for σ-almost all $\omega \in M$

$$\hat{u}(\omega) = (u, \varphi(\omega))_{H_0}, \qquad \varphi(\omega) \in H_-. \tag{5.46}$$

Let the operator $a \in \mathcal{A}$ be such that $a^* \restriction D$ acts from D to H_+ (possibly not in a continuous way; if the family A is connected with (2.1) in the standard way, then this property is automatically satisfied for $a = A_x \in A$). The vector

$\varphi(\omega)$ is a generalized eigenvector for a with eigenvalue $\omega(a)$ in the sense that $(\varphi(\omega), a^*u)_{H_0} = \omega(a)(\varphi(\omega), u)_{H_0}$ $(u \in D)$: according to (5.46),

$$(\varphi(\omega), a^*u)_{H_0} = (a^*u)\hat{}(\omega) = \overline{\omega(a^*)\,\hat{u}(\omega)} = \omega(a)(\varphi(\omega), u)_{H_0}.$$

We introduce the operator $P(\omega)$: $H_+ \to H_-$, with $P(\omega)u = (u, \varphi(\omega))_{H_0}\varphi(\omega)$ $(u \in H_+)$. Then $(P(\omega)u, v)_{H_0} = (u, \varphi(\omega))_{H_0}(v, \varphi(\omega))_{H_0}\cdot$, and (5.44) can be rewritten in a form analogous to (2.50) (with $\alpha = \tau$):

$$(u, v)_{H_0} = \int_M (P(\omega)u, v)_{H_0}\, d\sigma(\omega) \qquad (u, v \in H_+). \tag{5.47}$$

Above, it is assumed that there is a cyclic vector Ω. In the general case one should proceed as follows (derivation of the spectral theorem in the von Neumann form): the separable space H_0 is representable in the form of an orthogonal sum of Hilbert spaces $H_0^{(\gamma)}$: $H_0 = \bigoplus_1^\infty H_0^{(\gamma)}$ (the sum can be finite), where each $H_0^{(\gamma)}$ contains a cyclic vector Ω_γ. To prove this, fix a unit vector Ω_1 and let $H_0^{(1)}$ be the closure of $\{a\Omega_1 | a \in \mathcal{A}\}$. Then take a unit vector $\Omega_2 \perp H_0^{(1)}$ and construct $H_0^{(2)}$ in a similar way; $H_0^{(2)} \perp H_0^{(1)}$, since $(b\Omega_2, a\Omega_1)_{H_0} = (\Omega_2, b^*a\Omega_1)_{H_0} = 0$ ($b^*a \in \mathcal{A}$: $\mathcal{A}$ is a C^*-algebra), and so on. Each $H_0^{(\gamma)}$ is isometric to $L_2(M, d\sigma_\gamma(\omega))$ with some measure on $\mathcal{B}(M)$, and as a result we get the isometry

$$H_0 \ni f \mapsto \hat{f} = Gf \in \bigoplus_{\gamma=1}^{\infty} L_2(M, d\sigma_\gamma(\omega)); \qquad \hat{f}(\omega) = (\hat{f}_1(\omega), \hat{f}_2(\omega), \ldots).$$

It is easy to see that the measure σ_γ can be chosen to be absolutely continuous with respect to some single measure σ on $\mathcal{B}(M)$. By introducing in $\hat{f}$ the corresponding factors connected with the derivatives $(d\sigma_\gamma/d\sigma)(\omega)$, it is possible to pass to the orthogonal sum of the spaces $L_2(M, d\sigma(\omega))$ and write a Parseval's equality of type (5.44) for the vector-valued functions $\hat{f}(\omega)$. Then repeat (with minor additions) the arguments leading to (5.46) and (5.47).

Let us modernize the last constructions by approximating them to scheme 1). As before, suppose first that there is a cyclic vector Ω. The mapping $\mathbb{C}^{\mathcal{A}} \supseteq M \ni \omega \mapsto \lambda_\omega(\cdot) \in \mu \subseteq \mathbb{C}^X$ $(\lambda_\omega(x) = \omega(A_x),\ x \in X)$ carries $\mathcal{B}(M)$ into $\mathcal{B}(\mu)$, the measure σ is carried into some measure $\mathcal{B}(\mu) \ni \alpha \mapsto \rho(\alpha) \in [0, \infty)$, and $L_2(M, d\sigma(\omega))$ is carried isometrically into $L_2(\mu, d\rho(\lambda(\cdot)))$. Using this symmetry, we can assume that $H_0 \ni f \mapsto Gf = \tilde{f}(\lambda(\cdot)) \in L_2(\mu, d\rho(\lambda(\cdot)))$ $(\tilde{f}(\lambda(\cdot)) = \hat{f}(\omega) \restriction \mu,\ \lambda(\cdot) \leftrightarrow \omega)$, and (5.44), (5.46), and (5.47) take the following forms:

$$(f, g)_{H_0} = \int_\mu \tilde{f}(\lambda(\cdot))\overline{\tilde{g}(\lambda(\cdot))}\, d\rho(\lambda(\cdot)) \qquad (f, g \in H_0);$$

$$\tilde{u}(\lambda(\cdot)) = (u, \varphi(\lambda(\cdot)))_{H_0} \qquad (u \in H_+, \varphi(\lambda(\cdot)) \in H_-);$$

$$(u, v)_{H_0} = \int_\mu (P(\lambda(\cdot))u, v)_{H_0}\, d\rho(\lambda(\cdot)) \qquad (u, v \in H_+). \tag{5.48}$$

The vector $\varphi(\lambda(\cdot)) = \varphi(\omega)$ $(\lambda(\cdot) \leftrightarrow \omega)$ is a generalized eigenvector for A_x with eigenvalue $\omega(A_x) = \lambda(x)$ $(x \in X)$. Therefore, the formulas (5.48) are similar to (2.54) and (2.50) (with $\alpha = \tau = \pi$). In particular, the integrals are taken as before over the set of eigenvalues. An analogous situation holds also in the general case when there is no cyclic vector.

The impression may have been created that the formulas (5.48) (and the corresponding more general formulas, when there is no cyclic vector) are the same as in the main scheme 1) and that this scheme can thereby be avoided in the case of bounded operators A_x $(x \in X)$ by considering the C^*-algebra generated by these operators. However, this is not so when X is more than countable. Indeed, when scheme 2) is used, the RI G and the measure ρ are defined on the σ-algebra $\mathcal{B}(\mu)$ of Borel sets, and the generalized projection $\mu \ni \lambda(\cdot) \mapsto P(\lambda(\cdot))$ is weakly measurable with respect to this σ-algebra. In the case of scheme 1), E, ρ, and $P(\lambda(\cdot))$ are connected with the σ-algebra $\mathcal{C}_\sigma(\mu) = \{\alpha' = \alpha \cap \mu | \alpha \in \mathcal{C}_\sigma(\mathbb{C}^X)\} \subseteq \mathcal{B}(\mu)$, which is smaller than $\mathcal{B}(\mu)$. (For instance, the singleton set $\{\lambda(\cdot)\} \in \mathcal{B}(\mu)$ is not in $\mathcal{C}_\sigma(\mu)$ for sufficiently massive μ: it is not in $\mathcal{C}_\sigma(\mathbb{C}^X)$ when X is more than countable; see §1.4.) The information obtained about the function $\mu \ni \lambda(\cdot) \mapsto P(\lambda(\cdot))$ is thus more sparse in scheme 2) than in 1): the supply of Lebesgue sets of this function, i.e., of sets $\{\lambda(\cdot) \in \mu | (P(\lambda(\cdot))u, v)_{H_0} \in \delta \in \mathcal{B}(\mathbb{C}^1);\ u, v \in H_+\}$, is automatically smaller in reality than for $\mathcal{B}(\mu)$. (Of course, we could first restrict G and ρ to $\mathcal{C}_\sigma(\mu)$, and then perform the differentiation, but this means repeating 1)). ■

Let us dwell briefly on scheme 2) in the general case when some of the operators A_x can be unbounded.

Below we use known facts presented, for example, in §3.3. Take the algebra generated by the family $(E_x(\alpha))_{x \in X, \alpha \in \mathcal{B}(\mathbb{C}^1)}$ of projections, and close it with respect to operator norm convergence. Let M be the maximal ideal space of the C^*-algebra $\mathcal{A}$ with identity obtained in this way. Assume that $\mathcal{A}$ has a unit cyclic vector $\Omega \in H_0$: the set $\{a\Omega | a \in \mathcal{A}\}$ is dense in H_0. As before, the isomorphism $\mathcal{A} \leftrightarrow C(M)$ gives rise to an isometry $H_0 \ni f \mapsto \hat{f} = Gf \in L_2(M, d\sigma(\omega))$, and the relations (5.44)–(5.47) are preserved here.

However, if the operator A_x is unbounded, then it is not in $\mathcal{A}$, and for it we can no longer integrate $\varphi(\omega)$ directly as a generalized eigenvector. At the same time, it is known that the correspondence between the $a \in \mathcal{A}$ and the operators of multiplication by the functions $a(\omega) = \omega(a)$ in the space $L_2(M, d\sigma(\omega))$ can be extended to the operators A_x $(x \in X)$, though the functions $A_x(\omega)$ corresponding to such operators are not continuous with respect to ω, but are measurable and σ-almost everywhere finite. As before, we conclude from this that $\varphi(\omega)$ is a generalized eigenvector for A_x corresponding to the eigenvalue $A_x(\omega)$. If A does not have a cyclic vector Ω, then we repeat the above arguments with the decomposition $H_0 = \bigoplus_1^\infty H_0^{(\gamma)}$ and easily obtain a Parseval's equality in the form (5.47).

It is no longer possible to drop from (5.47) to (5.48) and pass to integration over the spectrum: the function $X \ni x \mapsto A_x(\omega) \in \mathbb{C}^1$ does not extend by linearity, multiplicativity, and continuity to the whole algebra $\mathfrak{A}$; what is more, for a particular $\omega \in M$ it is not well defined in general, since $A_k(\omega)$ is defined only for σ-almost all ω for each $x \in X$. In this way the representation (5.47) differs from the representation in the spectral projection Theorem 2.7.

§6. Selfadjointness of operators and uniqueness of the solution of the Cauchy problem for evolution equations

For proving that operators are selfadjoint we present a simple and often effective method that reduces the question to a study of a corresponding Cauchy problem. In particular, it leads easily to selfadjointness criteria in terms of quasi-analytic vectors. We also give some devices associated with this approach for verifying that selfadjoint operators commute.

6.1. ***General theorems on the connection between selfadjointness and uniqueness.*** Let H be a Hilbert space in which an operator B with dense domain $\mathfrak{D}(B)$ acts, let $I \subseteq \mathbb{R}^1$ be a finite or infinite closed, open, or half-open interval, and let $r = 1, 2, \ldots$. By a strong solution of the equation

$$(d^r u/dt^r)(t) + Bu(t) = 0 \qquad (t \in I) \tag{6.1}$$

on I we mean an r times strongly continuously differentiable vector-valued function $I \ni t \mapsto u(t) \in H$ (i.e., a function having r strong derivatives on I, the last of which is continuous) such that $u(t) \in \mathfrak{D}(B)$ and (6.1) is satisfied for each $t \in I$.

An r times strongly continuously differentiable vector-valued function $I \ni t \mapsto u(t) \in H$ is a strong solution of the equation

$$(d^r u/dt^r)(t) + B^* u(t) = 0 \qquad (t \in I) \tag{6.2}$$

if and only if the following "weak" equality holds:

$$((d^r u/dt^r)(t), f)_H + (u(t), Bf)_H = 0 \qquad (f \in \mathfrak{D}(B);\ t \in I). \tag{6.3}$$

This assertion follows at once from the definition of an adjoint operator, since (6.3) implies that $u(t) \in \mathfrak{D}(B^*)$ for each $t \in I$ because of the inclusion $-u^{(r)}(t) \in H$.

We say that uniqueness holds for strong solutions of the Cauchy problem for equation (6.1) on $I = [0, b)$ $(0 < b \leq \infty)$ if each strong solution $u(t)$ of this equation on $[0, b)$ such that $u(0) = \cdots = u^{(r-1)}(0) = 0$ vanishes for all $t \in (0, b)$. Of course, *if uniqueness holds on* $[0, b)$ *for some* $b > 0$, *then it holds also on* $[0, b)$. Indeed, suppose that $[0, \infty) \ni t \mapsto u(t)$ is a strong solution of (6.1) on $[0, \infty)$ such that $u(0) = \cdots = u^{(r-1)}(0) = 0$. From the assumed uniqueness it can be concluded that $u(t) = 0$ for $t \in (0, b)$; in particular, $u(t) = 0$ in a neighborhood of the point $c = b/2$, and so $u(c) = \cdots = u^{(r-1)}(c) = 0$.

The function $[0,\infty) \ni t \mapsto u_1(t) = u(t+c)$ is a strong solution of (6.1) on $[0,\infty)$ such that $u_1(0) = u(c) = 0, \dots, u_1^{(r-1)}(0) = u^{(r-1)}(c) = 0$, and hence $u_1(t) = 0$ for $t \in (0,b)$. Repeating the previous argument, we conclude that the function $[0,\infty) \ni t \mapsto u_2(t) = u_1(t+c) = u(t+2c)$ vanishes for $t \in (0,b)$. We then construct the function $u_3(t)$, and so on. As a result, we get that $u(t) = 0$ $(t \in (0,\infty))$.

In what follows, it will be assumed that r is 1 or 2, though generalization of some of the subsequent results to the case of general r is possible (and makes sense).

THEOREM 6.1. *Let A be an Hermitian operator acting in H. For A to be essentially selfadjoint it is necessary that uniqueness hold for strong solutions of the Cauchy problem for both the equations*

$$(du/dt)(t) \pm (iA)^* u(t) = 0 \qquad (t \in [0,b)) \tag{6.4}$$

on $[0,b)$ for any $b \in (0,\infty]$, and sufficient that such uniqueness hold for some such b.

PROOF. The proof will be carried out in several steps.

I. We establish sufficiency, assuming for the time being that A has equal deficiency numbers. Assume the opposite: suppose that the closure $\tilde{A}$ is not selfadjoint. Then A has two different selfadjoint extensions A_1 and A_2 defined in H. Let E_1 and E_2 be the corresponding resolutions of the identity. For any $g \in \mathfrak{D}(A) \subseteq \mathfrak{D}(A_1)$ the integral $\int_{\mathbf{R}^1} \lambda^2 \, d(E_1(\lambda)g, g)_H$ converges; therefore, the vector-valued function

$$[0,\infty) \ni t \mapsto u_1(t) = \int_{\mathbf{R}^1} e^{i\lambda t} \, dE_1(\lambda) g \tag{6.5}$$

is once strongly continuously differentiable, and

$$u_1'(t) = i \int_{\mathbf{R}^1} \lambda e^{i\lambda t} \, dE_1(\lambda) g.$$

It is easily seen to be a strong solution of (6.4) on $[0,\infty)$ with the $+$ sign. Indeed, it is necessary to verify the corresponding weak equality (6.3), which here looks like

$$((du_1/dt)(t), f)_H + (u_1(t), iAf)_H = 0 \qquad (f \in \mathfrak{D}(A);\ t \in [0,\infty)).$$

Since

$$\begin{aligned} d(E_1(\lambda)g, Af)_H &= d(E_1(\lambda)g, A_1 f)_H \\ &= d\left(\int_{-\infty}^{\lambda} \mu d(E_1(\mu)g, f)_H\right) = \lambda d(E_1(\lambda)g, f)_H, \end{aligned}$$

it follows that

$$\begin{aligned}((du_1/dt)(t), f)_H &+ (u_1(t), iAf)_H \\ &= i\int_{\mathbf{R}^1} \lambda e^{i\lambda t}\, d(E_1(\lambda)g, f)_H - i\int_{\mathbf{R}^1} e^{i\lambda t}\, d(E_1(\lambda)g, Af)_H \\ &= 0 \qquad (f \in \mathfrak{D}(A);\ t \in [0,\infty)),\end{aligned}$$

i.e., the required relation is satisfied. Similarly, the function $u_2(t)$ constructed from E_2 according to (6.5) is a strong solution of the same equation, and $u_1(0) = u_2(0) = g$. Thus, $u(t) = u_1(t) - u_2(t)$ is a strong solution of (6.4) with the + sign on $[0,\infty)$, and $u(0) = 0$. By the remark made earlier, uniqueness holds on $[0,\infty)$; therefore, $u(t) = 0$ for $t \geq 0$. Therefore,

$$\int_{\mathbf{R}^1} e^{i\lambda t}\, d((E_1(\lambda) - E_2(\lambda))g, h)_H = 0$$

$$(g \in \mathfrak{D}(A),\ h \in H,\ t \in [0,\infty)). \qquad (6.6)$$

Consider now equation (6.4) with the − sign. We can repeat the argument given, with $e^{i\lambda t}$ replaced by $e^{-i\lambda t}$ in (6.5). As a result we obtain (6.6), with the same substitution. Thus, if we introduce the function of bounded variation $\omega(\lambda) = ((E_1(\lambda) - E_2(\lambda))g, h)_H$ ($\lambda \in \mathbf{R}^1$) then $\int_{\mathbf{R}^1} e^{i\lambda t}\, d\omega(\lambda) = 0$ for all $t \in \mathbf{R}^1$, by (6.6) and this modification of it. From this, the uniqueness theorem for the Fourier-Stieltjes transform gives us that $d\omega(\lambda) = 0$, i.e.,

$$((E_1(\lambda) - E_2(\lambda))g, h)_H = 0 \qquad (\lambda \in \mathbf{R}^1).$$

Since $g \in \mathfrak{D}(A)$ here, while $\mathfrak{D}(A)$ is dense in H and $h \in H$, it follows that $E_1(\lambda) = E_2(\lambda)$ ($\lambda \in \mathbf{R}^1$), a contradiction.

II. We use the following simple lemma in the case when the deficiency numbers of A are different.

LEMMA 6.1. *Construct the space $\mathfrak{H} = H \oplus H$ of vectors $f = (f_1, f_2)$ ($f_1, f_2 \in H$) and in it an operator $\mathfrak{A}$ with dense domain $\mathfrak{D}(\mathfrak{A}) = \mathfrak{D}(A) \oplus \mathfrak{D}(A)$ by setting $\mathfrak{A}f = (Af_1, -Af_2)$ ($f \in \mathfrak{D}(\mathfrak{A})$). Consider the following equation in $\mathfrak{H}$-valued functions ($b \in (0,\infty]$):*

$$(du/dt)(t) + (i\mathfrak{A})^* u(t) = 0 \qquad (t \in [0,b)). \qquad (6.7)$$

If uniqueness holds for strong solutions for the Cauchy problem on $[0,b)$ for both the equations (6.4), *then the same uniqueness holds also for strong solutions of* (6.7), *and conversely.*

PROOF. Let $[0,b) \ni t \mapsto u(t) = (u_1(t), u_2(t)) \in \mathfrak{H}$ be a strong solution of the Cauchy problem for equation (6.7). Since $\mathfrak{A}^* f = (A^* f_1, -A^* f_2)$ ($f \in \mathfrak{D}(\mathfrak{A}^*) = \mathfrak{D}(A^*) \oplus \mathfrak{D}(A^*)$), it follows that $[0,b) \ni t \mapsto u_1(t) \in H$ and $[0,b) \ni t \mapsto u_2(t) \in H$ are strong solution of (6.4) with the + and − signs, respectively. This and the assumed uniqueness of the strong solutions of the

Cauchy problem on $[0, b)$ for (6.4) give us uniqueness for (6.7). The converse assertion follows just as simply. ■

III. We now prove the sufficiency in the theorem for an operator A with deficiency number (m, n). Construct the operator $\mathfrak{A}$ just as in Lemma 6.1. It is easy to check that the deficiency number of this operator is equal to $(m+n, m+n)$ (see, for example, Akhiezer and Glazman [1], Chapter 9, §111). By Lemma 6.1, uniqueness holds for (6.7) on $[0, b)$. Applying this lemma to the case when A is replaced by $-A$, we get that the same uniqueness holds for (6.7) with the $+$ sign replaced by $-$. The deficiency numbers of $\mathfrak{A}$ are both equal to $m+n$, so we can apply step I to $\mathfrak{A}$ and conclude that it is essentially selfadjoint. But then $m+n=0$; hence, $m=n=0$, i.e., A is also essentially selfadjoint.

IV. To prove the necessity we establish a general lemma which reflects the Holmgren principle in our situation. This lemma will also be used for other purposes in what follows.

LEMMA 6.2. *Consider equation* (6.2) *on* $[0, b)$ ($b \in (0, \infty]$). *Suppose that* Φ *is a dense subset of* H *such that for any* $T \in (0, b)$ *and* $\varphi_0, \ldots, \varphi_{r-1} \in \Phi$ *there exists a strong solution of the Cauchy problem*

$$(d^r\varphi/dt^r)(t) + (-1)^r B\varphi(t) = 0 \qquad (t \in [0, T]),$$
$$\varphi(T) = \varphi_0, \ldots, \varphi^{(r-1)}(T) = \varphi_{r-1}. \tag{6.8}$$

Then uniqueness holds for strong solutions of the Cauchy problem for (6.2) *on* $[0, b)$.

PROOF. We carry out the proof, for example, for the case $r = 2$. It is simple to verify the following integration by parts formula: if $[0, T] \ni t \mapsto \alpha(t), \beta(t) \in H$ are twice strongly continuously differentiable vector-valued functions, then

$$\int_0^T (\alpha''(t), \beta(t))_H \, dt$$
$$= \int_0^T (\alpha(t), \beta''(t))_H \, dt + [(\alpha'(t), \beta(t))_H - (\alpha(t), \beta'(t))_H]_0^T. \tag{6.9}$$

Let $u(t)$ be a strong solution of the Cauchy problem for (6.2) with $r = 2$ on $[0, b)$ such that $u(0) = u'(0) = 0$, and let $\varphi(t)$ be the strong solution in the statement of the lemma. By (6.9),

$$\int_0^T ((u''(t), \varphi(t))_H - (u(t), \varphi''(t))_H) \, dt = (u'(T), \varphi_0)_H - (u(T), \varphi_1)_H. \tag{6.10}$$

Since $\varphi(s) \in \mathfrak{D}(B)$ for each $s \in [0, T]$, equality (6.3) with $f = \varphi(s)$ permits us to write $(u''(t), \varphi(s))_H + (u(t), B\varphi(s))_H = 0$ ($t \in [0, b)$). Setting $t = s$ here and then replacing s by t, we get that

$$(u''(t), \varphi(t))_H = -(u(t), B\varphi(t))_H \qquad (t \in [0, T]).$$

By (6.8) with $r = 2$,

$$(u(t), \varphi''(t))_H = -(u(t), B\varphi(t))_H \qquad (t \in [0, T]).$$

We conclude from these two equalities that the expression on the left-hand side of (6.10) vanishes; therefore,

$$(u'(T), \varphi_0)_H - (u(T), \varphi_1)_H = 0 \qquad (\varphi_0, \varphi_1 \in \Phi).$$

The denseness of Φ in H now allows us to conclude that $u(T) = u'(T) = 0$. Since $T \in (0, b)$ is arbitrary, the assertion follows from this.

The arguments are similar in the case $r = 1$; it is only necessary to use the integration by parts formula

$$\int_0^T (\alpha'(t), \beta(t))_H \, dt = -\int_0^T (\alpha(t), \beta'(t))_H \, dt + [(\alpha(t), \beta(t))_H]_0^T, \qquad (6.11)$$

which is valid for a once continuously differentiable vector-valued function $[0, T] \ni t \mapsto \alpha(t),\ \beta(t) \in H$. In the case of general r formula (6.11) must be integrated r times. (Formula (6.10) is (6.11), integrated twice.) ■

V. We proceed to a proof of necessity. Suppose that $\tilde{A}$ is selfadjoint, and let E be its resolution of the identity. We apply Lemma 6.2, setting $r = 1$, $B = (iA)^* = -i\tilde{A}$, and $\Phi = \bigcup_1^\infty E((-n, n))H$. The strong solution of the Cauchy problem (6.8), which now has the form $\varphi'(t) + i\tilde{A}\varphi(t) = 0$ $(t \in [0, T])$, $\varphi(T) = \varphi_0$, exists and is equal to

$$\varphi(t) = \int_{\mathbf{R}^1} e^{-i\lambda(t-T)} \, dE(\lambda)\varphi_0 \qquad (t \in [0, T]). \qquad (6.12)$$

(Since $\varphi_0 \in \Phi$, the integral in (6.12) is really over a finite interval, and thus the function $[0, T] \ni t \mapsto \varphi(t)$ is once continuously differentiable; it is clearly a solution of the required problem.) Thus, by this lemma, uniqueness holds for (6.4) with the $+$ sign on $[0, b)$. The case of the $-$ sign is handled similarly, with $B = -(iA)^* = i\tilde{A}$.

THEOREM 6.2. *Suppose that A is an Hermitian operator acting in H. For A to be essentially selfadjoint it is necessary that uniqueness hold for strong solutions of the Cauchy problem for the equation*

$$(d^2u/dt^2)(t) + A^*u(t) = 0 \qquad (t \in [0, b)) \qquad (6.13)$$

on $[0, b)$ for any $b \in (0, \infty]$, and sufficient that A be semibounded below and that the uniqueness hold for strong solutions for some $b > 0$.

PROOF. We establish sufficiency. Assume that $\tilde{A}$ is not selfadjoint. Then A has two different selfadjoint extensions A_1 and A_2 acting in H and bounded below by a number $c > -\infty$. Let E_1 and E_2 be the corresponding resolutions of the identity. For any $g \in \mathfrak{D}(A) \subseteq \mathfrak{D}(A_1)$ the integral $\int_{\mathbf{R}^1} \lambda^2 \, d(E_1(\lambda)g, g)_H$ converges; therefore, the vector-valued function

$$[0, \infty) \ni t \mapsto u_1(t) = \int_c^\infty \cos\left(\sqrt{\lambda} t\right) \, dE_1(\lambda)g \qquad (6.14)$$

is twice strongly continuously differentiable. It is easy to see in a way similar to that in the proof of Theorem 6.1 that this function is a strong solution of (6.13) on $[0,\infty)$; to do this it must be verified that the corresponding weak equality of the form (6.3) holds. Moreover, $u_1(0) = g$ and $u_1'(0) = 0$. Similarly, replacing E_1 by E_2 in (6.14), we construct the function $u_2(t)$. The difference $u(t) = u_1(t) - u_2(t)$ is also a strong solution of (6.13) on $[0,\infty)$ such that $u(0) = u'(0) = 0$. By the assumed uniqueness of strong solutions of the Cauchy problem, $u(t) = 0$ for $t \geq 0$. Taking the inner product of this equality and an $h \in H$, we get that

$$\int_c^\infty \cos\left(\sqrt{\lambda}t\right)\, d((E_1(\lambda) - E_2(\lambda))g, h)_H = 0 \qquad (t \in [0,\infty)).$$

From this and the fact that a measure $d\omega(\lambda)$ is uniquely determined by its Fourier-Stieltjes cosine transform, we conclude that

$$\omega(\lambda) = ((E_1(\lambda) - E_2(\lambda))g, h)_H = 0 \qquad (\lambda \in \mathbb{R}^1).$$

Since $g \in \mathfrak{D}(A)$ and $h \in H$ are arbitrary, $E_1(\lambda) = E_2(\lambda)$ $(\lambda \in \mathbb{R}^1)$, which is absurd.

We establish the necessity. Let $\tilde{A}$ be selfadjoint, and let E be its resolution of the identity. We apply Lemma 6.2, setting $r = 2$, $B = A^* = \tilde{A}$, and $\Phi = \bigcup_1^\infty E((-n,n))H$. The strong solution of the Cauchy problem (6.8), which now takes the form $\varphi''(t) + \tilde{A}\varphi(t) = 0$ $(t \in [0,T])$ with $\varphi(T) = \varphi_0$ and $\varphi'(T) = \varphi_1$, exists and is equal to

$$\varphi(t) = \int_{\mathbb{R}^1} \cos\left(\sqrt{\lambda}(t-T)\right)\, dE(\lambda)\varphi_0 + \int_{\mathbb{R}^1} \frac{1}{\sqrt{\lambda}} \sin\left(\sqrt{\lambda}(t-T)\right)\, dE(\lambda)\varphi_1 \qquad (t \in [0,T]). \tag{6.15}$$

(As in (6.12), the integration in (6.15) is really over a finite interval.) Therefore, by this lemma, uniqueness holds for strong solutions of the Cauchy problem for (6.13) on $[0,b)$ for each $b \in (0,\infty]$. ■

Note that from the proof of necessity in Theorems 6.1 and 6.2 it is clear that essential selfadjointness of A implies uniqueness of strong solution of the Cauchy problem for the equation $(d^2u/dt^2)(t) + z\tilde{A}u(t) = 0$ $(r = 1, 2, \ldots;$ $z \in \mathbb{C}^1)$ on $[0,b)$ for any $b \in (0,\infty]$.

It is usually convenient to apply Theorem 6.2 in simple conjunction with Lemma 6.2. We formulate the corresponding result as a theorem.

THEOREM 6.3. *Let A be an Hermitian operator acting in H and semibounded below. Assume that there is a dense linear subset $\Phi \subseteq H$ such that for some $b > 0$ the Cauchy problem*

$$\begin{gathered}(d^2\varphi/dt^2)(t) + A\varphi(t) = 0 \qquad (t \in [0,T]),\\ \varphi(T) = \varphi_0, \qquad \varphi'(T) = \varphi_1\end{gathered} \tag{6.16}$$

has a strong solution for any $T \in (0, b)$ *and* $\varphi_0, \varphi_1 \in \Phi$. *Then* A *is essentially selfadjoint.*

PROOF. By Lemma 6.2, the conditions of the theorem imply that uniqueness holds for strong solutions of the Cauchy problem on $[0, b)$ for (6.13). But then $\tilde{A}$ is selfadjoint, according to Theorem 6.2. ■

Of course, such a theorem is valid also for Theorem 6.1. Note also that the set Φ in Lemma 6.2 and Theorem 6.3 can depend on T.

6.2. ***Some generalizations.*** We give three results extending the domain of application of Theorem 6.3 to operators A that are not semibounded, to operators such that the Cauchy problem (6.16) is poorly solvable, and to powers of A.

THEOREM 6.4. *Let* A *be an Hermitian operator acting in* H *and having equal deficiency indices. Assume that there exists a dense subset* Φ *of* H *such that for any* $\varphi_0, \varphi_1 \in \Phi$ *the Cauchy problem*

$$(d^2\varphi/dt^2)(t) + A\varphi(t) = 0 \qquad (t \in [0, \infty)),$$
$$\varphi(0) = \varphi_0, \qquad \varphi'(0) = \varphi_1 \tag{6.17}$$

has a strong solution. Moreover, assume that for each particular $\varphi_0 \in \Phi$ *the solution* $\varphi(t)$ *of the Cauchy problem (6.17) with* $\varphi_1 = 0$ *satisfies*

$$(\varphi(t), \varphi_0)_H \underset{t\to\infty}{=} O(e^{Ct^2}) \tag{6.18}$$

for some $C = C_{\varphi_0} \geq 0$. *Then* A *is essentially selfadjoint.*

PROOF. First of all observe that the conditions of the theorem imply the existence of a strong solution $\varphi(t)$ of the Cauchy problem (6.16) for any $T \in (0, \infty)$. Indeed, if $\psi(t)$ denotes a strong solution of the problem (6.17), then $\varphi(t) = \psi(T - t)$ $(t \in [0, T])$ is the required solution of (6.16).

Let $[0, \infty) \ni t \mapsto \varphi(t)$ be a strong solution of (6.17) such that $\varphi_1 = 0$. We have the representation

$$\varphi(t) = \int_{\mathbb{R}^1} \cos\left(\sqrt{\lambda} t\right) dE_1(\lambda)\varphi_0 \qquad (t \in [0, \infty)), \tag{6.19}$$

where E_1 is the resolution of the identity corresponding to any selfadjoint extension A_1 of A in H. Indeed, fix a $\Delta \in \mathcal{B}(\mathbb{R}^1)$ with compact closure. The vector-valued function $[0, \infty) \ni t \mapsto \chi_1(t) = E_1(\Delta)\varphi(t)$ is twice strongly continuously differentiable, because $\varphi(t)$ has this property and the operator $E_1(\Delta)$ is continuous, and $\chi_1''(t) = E_1(\Delta)\varphi''(t)$ $(t \in [0, \infty))$. Moreover,

$$\chi_1(t) = E_1(\Delta)\varphi(t) \in \mathfrak{D}(A_1) \subseteq \mathfrak{D}(A^*)$$

and

$$\begin{aligned} A^*\chi_1(t) &= A_1\chi_1(t) = A_1E_1(\Delta)\varphi(t) \\ &= E_1(\Delta)A_1\varphi(t) = E_1(\Delta)A\varphi(t) \end{aligned}$$

for each $t \in [0, \infty)$. Considering these relations, we can, by (6.17), write

$$\begin{aligned}\chi_1''(t) + A^*\chi_1(t) &= E_1(\Delta)\varphi''(t) + E_1(\Delta)A\varphi(t)\\ &= E_1(\Delta)(\varphi''(t) + A\varphi(t)) = 0 \qquad (t \in [0, \infty)),\end{aligned}$$

and $\chi_1(0) = E_1(\Delta)\varphi(0) = E_1(\Delta)\varphi_0$ and $\chi_1'(0) = E_1(\Delta)\varphi'(0) = E_1(\Delta)\varphi_1$. Thus, $\chi_1(t)$ is a strong solution of the Cauchy problem

$$\begin{aligned}\chi''(t) + A^*\chi(t) = 0 \qquad &(t \in [0, \infty)),\\ \chi(0) = E_1(\Delta)\varphi_0, \qquad &\chi'(0) = 0.\end{aligned}$$

The vector-valued function $[0, \infty) \ni t \mapsto \chi_2(t) = \int_\Delta \cos(\sqrt{\lambda}t)\, dE_1(\lambda)\varphi_0$ is obviously another such solution (see the proof of sufficiency in Theorems 6.2 and 6.1). On the basis of Lemma 6.2 these solutions must coincide, because the Cauchy problem (6.16), as already explained, has a strong solution. Accordingly,

$$E_1(\Delta)\varphi(t) = \int_\Delta \cos\left(\sqrt{\lambda}t\right) dE_1(\lambda)\varphi_0 \qquad (t \in [0, \infty)). \tag{6.20}$$

In (6.20) we set $\Delta = (-n, n)$ and take the limit as $n \to \infty$. For each t, $E_1((-n, n))\varphi(t) \to \varphi(t)$ in H as $n \to \infty$; therefore, the strong limit

$$\lim_{n\to\infty} \int_{-n}^{n} \cos\left(\sqrt{\lambda}t\right) dE_1(\lambda)\varphi_0$$

exists and is equal by definition to the integral on the right-hand side of (6.19). Thus, (6.19) is established.

Assume that $\tilde{A}$ is not selfadjoint. Then A has two different selfadjoint extensions A_1 and A_2 acting in H. Let E_1 and E_2 be the corresponding resolutions of the identity. By what has been proved, formula (6.19) is preserved for E_1 and for E_2. Therefore, by (6.18), we can write for each $t \in [0, \infty)$

$$\begin{aligned}\int_{\mathbb{R}^1} \cos\left(\sqrt{\lambda}t\right) d(E_1(\lambda)\varphi_0, \varphi_0)_H &= \int_{\mathbb{R}^1} \cos\left(\sqrt{\lambda}t\right) d(E_2(\lambda)\varphi_0, \varphi_0)_H\\ &= (\varphi(t), \varphi_0)_H \underset{t\to\infty}{=} O(e^{Ct^2}).\end{aligned}$$

From this, as is known (see below), it follows that the measures coincide:

$$(E_1(\Delta)\varphi_0, \varphi_0)_H = (E_2(\Delta)\varphi_0, \varphi_0)_H \qquad (\Delta \in \mathcal{B}(\mathbb{R}^1);\ \varphi_0 \in \Phi).$$

Since Φ is dense in H, we conclude from the last equality that $E_1 = E_2$, which is absurd. ■

Above we used the following result, which improves the theorem that a measure can be uniquely recovered from its Fourier-Stieltjes cosine transform: if $\mathbb{R}^1 \ni \lambda \mapsto \omega(\lambda) \in \mathbb{C}^1$ is a function of bounded variation such that

$$\int_{\mathbb{R}^1} \cos\left(\sqrt{\lambda}t\right) d\omega(\lambda) = 0 \qquad (t \in \mathbb{R}^1),$$

$$\int_{\mathbb{R}^1} \cos\left(\sqrt{\lambda}t\right) |d\omega(\lambda)| \underset{t\to\infty}{=} O(e^{Ct^2}) \tag{6.21}$$

for some $C \geq 0$, then $d\omega(\lambda) = 0$—see, for example, Berezanskiĭ [4], Chapter 8, Theorem 3.18 and its corollary; we must set

$$\omega(\lambda) = (E_1(\lambda)\varphi_0, \varphi_0)_H - (E_2(\lambda)\varphi_0, \varphi_0)_H.$$

This result can be improved (Vul [1], Chaus [1], or Theorem 2 in Chaus [2]) in the respect that the constant C in (6.21) can be replaced by a slowly increasing function $C(t)$, i.e., by a twice continuously differentiable function $(0, \infty) \ni t \mapsto C(t) > 0$ that satisfies the following conditions:

1) $C'(t) \geq 0$ $(t \in (0, \infty))$ and $\lim_{t\to\infty} C(t) = \infty$;
2) for any $\varepsilon > 0$ there exist t_ε, $c_\varepsilon > 0$ such that $C(t) \leq c_\varepsilon t^\varepsilon$ for $t \geq t_\varepsilon$;
3) the limit $\lim_{t\to\infty} tC''(t)(C'(t))^{-1}$ exists, finite or infinite;
4) the function $tC'(t)(C(t))^{-1}$ is nonincreasing for sufficiently large t; and
5) $\int_1^\infty t^{-1}(C(t))^{-1}\, dt = \infty$.

It is clear from the proof of Theorem 6.4 that the theorem is preserved if (6.18) is replaced by the condition

$$(\varphi(t), \varphi_0)_H \underset{t\to\infty}{=} O(e^{C(t)t^2}), \tag{6.22}$$

where $C(t)$ is a slowly increasing function.

In essence, this theorem generalizes Theorem 6.3 to operators A that are not semibounded. The following assertion generalizes it to a case when it is not clear whether there exists a strong solution of the Cauchy problem (6.16).

THEOREM 6.5. *Suppose that A is an Hermitian operator acting in H and semibounded below. Assume that there exist a sequence $(A_n)_1^\infty$ of operators acting in H with domains $\mathfrak{D}(A_n)$, a dense subset Φ of H, and a $b > 0$ having the following properties:*

1) for any $T \in (0, b)$ and $\varphi_0, \varphi_1 \in \Phi$ there exist sequences $(\varphi_{0,n})_1^\infty$ and $(\varphi_{1,n})_1^\infty$ of vectors in H such that $\varphi_{0,n} \to \varphi_0$, $\varphi_{1,n} \to \varphi_1$ in H as $n \to \infty$, and there exist strong solutions of the Cauchy problem

$$\begin{aligned} &(d^2\varphi_n/dt^2)(t) + A_n\varphi_n(t) = 0 \qquad && (t \in [0, T]) \\ &\varphi_n(T) = \varphi_{0,n} \qquad \varphi_n'(T) = \varphi_{1,n} \qquad && (n = 1, 2, \ldots), \end{aligned} \tag{6.23}$$

with $\varphi_n(t) \in \mathfrak{D}(A)$ $(t \in [0, T])$;

2) for any $T \in (0, b)$ and $\varphi_0, \varphi_1 \in \Phi$

$$\int_0^T (u(t), (A_n - A)\varphi_n(t))_H \, dt \underset{n\to\infty}{\to} 0, \tag{6.24}$$

where $u(t)$ is a strong solution of (6.13) such that $u(0) = u'(0) = 0$.

Then A is essentially selfadjoint.

PROOF. According to Theorem 6.2, it suffices to show that $u(t) = 0$ for $t \in (0, b)$. Let $T \in (0, b)$ and $\varphi_0, \varphi_1 \in \Phi$. We construct corresponding approximating sequences $(\varphi_{0,n})_1^\infty$ and $(\varphi_{1,n})_1^\infty$ and solutions $\varphi_n(t)$. By (6.9),

$$\int_0^T ((u''(t), \varphi_n(t))_H - (u(t), \varphi_n''(t))_H)\, dt = (u'(T), \varphi_{0,n})_H - (u(T), \varphi_{1,n})_H \quad (n = 1, 2, \ldots). \tag{6.25}$$

Since $\varphi_n(t) \in \mathfrak{D}(A)$, we get $(u''(t), \varphi_n(t))_H = -(u(t), A\varphi_n(t))_H$ $(t \in [0, T])$ with the help of (6.3) as in the proof of Lemma 6.2; by (6.23), we also have that

$$(u(t), \varphi_n''(t))_H = -(u(t), A_n\varphi_n(t))_H \qquad (t \in [0, T]).$$

These two relations give us that the left-hand side of (6.25) is equal to the left-hand side of (6.24); therefore,

$$(u'(T), \varphi_0)_H - (u(T), \varphi_1)_H = \lim_{n\to\infty} [(u'(T), \varphi_{0,n})_H - (u(T), \varphi_{1,n})_H] = 0.$$

Since $\varphi_0, \varphi_1 \in \Phi$ are arbitrary and Φ is dense in H, this leads to the conclusion that $u(T) = 0$. ■

REMARK 1. *Assume that the conditions of Theorem* 6.5 *are satisfied, and the operators* A_n $(n = 1, 2, \ldots)$ *are selfadjoint, positive, and invertible, and that there exists a sequence* $(m_n)_1^\infty$ *of positive integers* m_n *such that the following conditions hold:*

1) *the strong solutions are such that* $\varphi_n(t) \in \mathfrak{D}(A_n^{m_n+1})$ $(t \in [0, T]$; $n = 1, 2, \ldots)$;

2) *the norms* $\|A_n^{m_n}\varphi_n(T)\|_H$ *and* $\|A_n^{m_n-1/2}\varphi_n'(T)\|_H$ *are bounded with respect to* $n = 1, 2, \ldots$; *and*

3) $\mathfrak{R}(A_n^{-m_n}) \subseteq \mathfrak{D}(A_n - A)$ $(n = 1, 2, \ldots)$ *and* $\|(A_n - A)A_n^{-m_n}\| \to 0$ *as* $n \to \infty$.

Then condition (6.24) *holds, and consequently, A is essentially selfadjoint.*

Indeed, applying the operator $A_n^{m_n}$ to (6.23), we conclude that the function $\psi_n(t) = A_n^{m_n}\varphi_n(t)$ satisfies the same equation (6.23) along with the initial conditions $\psi_n(T) = A_n^{m_n}\varphi_n(T)$ and $\psi_n'(T) = A_n^{m_n}\varphi_n'(T)$. Since A_n is selfadjoint, a representation of the form (6.15) can be written for $\psi_n(t)$, and it implies the estimate

$$\|\psi_n(t)\|_H \le \|\psi_n(T)\|_H + \|A_n^{-1/2}\psi_n'(T)\|_H \qquad (t \in [0, T]).$$

(Use the upper bound 1 to estimate the moduli of the cosine and sine in the integrals in (6.15).) Therefore, by 2), $\|A_n^{m_n}\varphi_n(t)\|_H \le c$ $(t \in [0, T]; n = 1, 2, \ldots)$. The integrand in the integral in (6.24) can now be written in the form

$$(u(t), (A_n - A)A_n^{-m_n}A_n^{m_n}\varphi_n(t))_H,$$

and so, by the Cauchy-Schwarz-Bunyakovskiĭ inequality, its modulus has the upper estimate

$$c \max_{t\in[0,T]} \|u(t)\|_H \|(A_n - A)A_n^{-m_n}\| \underset{n\to\infty}{\to} 0. \quad ■$$

The next theorem shows that, roughly speaking, if Theorem 6.3 can be used to prove that A is essentially selfadjoint, then any power of A has automatically been proved to be essentially selfadjoint.

THEOREM 6.6. *Let $\mathfrak{D} \subseteq H$ be a dense subset of H, and let A be an Hermitian operator that is defined on it, semibounded below, and such that $A\mathfrak{D} \subseteq \mathfrak{D}$. Assume that for some $b > 0$ and for each $T \in (0, b)$ and each $\varphi_0, \varphi_1 \in \mathfrak{D}$ there exists a vector-valued function $[0,T] \ni t \mapsto \varphi(t) \in \mathfrak{D}$ such that 1) for each $n = 0, 1, \ldots$ the vector-valued function $[0,T] \ni t \mapsto A^n\varphi(t) \in H$ is twice strongly continuously differentiable, and 2) the equation and initial conditions*

$$(d^2\varphi/dt^2)(t) + A\varphi(t) = 0 \qquad (t \in [0,T]),$$
$$\varphi(T) = \varphi_0, \qquad \varphi'(T) = \varphi_1. \tag{6.26}$$

are satisfied.

Under these assumptions each of the operators $\mathfrak{D} \ni f \mapsto A^n f$ $(n = 1, 2, \ldots)$ defined in H is essentially selfadjoint.

PROOF. Let $A \geq c1$ $(c \in \mathbb{R}^1)$, and let $B = A + a1$, where $a = -c + 1$. Obviously, $B \geq 1$. We fix an $m = 0, 1, \ldots$ and introduce on $\mathfrak{D}$ the inner product

$$(f, g)_{\mathfrak{H}_m} = (B^m f, B^m g)_H \qquad (f, g \in \mathfrak{D}). \tag{6.27}$$

Since $\|Bf\|_H \geq \|f\|_H$, it follows that $\|B^m f\|_H \geq \|f\|_H$ $(f \in \mathfrak{D})$. Therefore, (6.27) does define an inner product in $\mathfrak{D}$. Let $\mathfrak{H}_m$ be the corresponding completion of $\mathfrak{D}$ $(\mathfrak{H}_0 = H)$. From A we construct an operator $\mathfrak{A}$ acting in $\mathfrak{H}_m$, taking $\mathfrak{D}$ as its domain $\mathfrak{D}(\mathfrak{A})$ and setting $\mathfrak{A}f = Af$ $(f \in \mathfrak{D}(\mathfrak{A}) = \mathfrak{D})$. This operator is Hermitian in $\mathfrak{H}_m$ and is semibounded:

$$(\mathfrak{A}f, g)_{\mathfrak{H}_m} = (B^m Af, B^m g)_H = (B^m f, B^m Ag)_H = (f, \mathfrak{A}g)_{\mathfrak{H}_m} \qquad (f, g \in \mathfrak{D}),$$

$$(\mathfrak{A}f, f)_{\mathfrak{H}_m} = (B^m Af, B^m f)_H = (AB^m f, B^m f)_H$$
$$\geq c\|B^m f\|_H^2 = c\|f\|_{\mathfrak{H}_m}^2 \qquad (f \in \mathfrak{D}).$$

It is also essentially selfadjoint. Indeed, for this to be true it suffices, by Theorem 6.3, that there be a dense subset Φ of $\mathfrak{H}_m$ such that for any $T \in (0, b)$ and $\varphi_0, \varphi_1 \in \Phi$ the Cauchy problem

$$(d^2\varphi/dt^2)(t) + \mathfrak{A}\varphi(t) = 0 \qquad (t \in [0,T]),$$
$$\varphi(T) = \varphi_0, \qquad \varphi'(T) = \varphi_1 \tag{6.28}$$

has a strong solution $[0,T] \ni t \mapsto \varphi(t) \in \mathfrak{H}_m$. It is easy to see that the function $\varphi(t)$ in the conditions of the theorem serves as such a solution. Moreover, $\Phi = \mathfrak{D}$. Indeed, it can be regarded as a vector-valued function $\varphi(t)$: $[0,T] \to \mathfrak{H}_m$, and it is twice strongly continuously differentiable. (The latter follows from the fact that $B^m\varphi(t) = A^m\varphi(t) + maA^{m-1}\varphi(t) + \cdots + a^m\varphi(t)$: $[0,T] \to H$

is twice strongly continuously differentiable.) For each $t \in [0, T]$ we have that $\varphi(t) \in \mathfrak{D} = \mathfrak{D}(\mathfrak{A})$, and (6.28) is satisfied because of (6.26).

Accordingly, $\mathfrak{A}$ is essentially selfadjoint, and therefore, for each $z \notin [c, \infty)$ the range of $\mathfrak{A} - z1$, i.e., $(A - z1)\mathfrak{D}$, is dense in $\mathfrak{H}_m$ $(m = 0, 1, \ldots)$.

Fix an $n = 1, 2, \ldots$ and a nonreal number z such that none of its nth roots $z_1, \ldots, z_n$ fall on $[c, \infty)$. To prove the theorem it suffices to establish that $(A^n - z1)\mathfrak{D}$ is dense in H. Obviously, the decomposition

$$(A^n - z1)f = \left(\prod_{j=1}^{n}(A - z_j 1)\right) f$$

is valid for $f \in \mathfrak{D}$; therefore, for any $f_0 \in H$ and $f_1, \ldots, f_n \in \mathfrak{D}$ we can use the triangle inequality to write the estimate

$$\begin{aligned} &\|f_0 - (A^n - z1)f_n\|_H \\ &\quad \leq \|f_0 - (A - z_1 1)f_1\|_H \\ &\quad\quad + \sum_{m=1}^{n-1} \left\| \left(\prod_{j=1}^{m}(A - z_j 1)\right)(f_m - (A - z_{m+1}1)f_{m+1})\right\|_H . \end{aligned} \tag{6.29}$$

For $\varsigma \in \mathbf{C}^1$ and $g \in \mathfrak{D}$ we have that

$$\|(A - \varsigma 1)g\|_H = \|(B - (\varsigma + a)1)g\|_H \leq (1 + |\varsigma + a|)\|Bg\|_H .$$

Iterating this estimate and using the fact that A and B commute along with (6.27), we can extend (6.29) as follows

$$\begin{aligned} &\|f_0 - (A^n - z1)f_n\|_H \\ &\quad \leq \|f_0 - (A - z_1 1)f_1\|_H + \sum_{m=1}^{n-1}\left(\prod_{j=1}^{m}(1 + |z_j + a|)\right) \\ &\quad\quad\quad \times \|B^m(f_m - A - z_{m+1}1)f_{m+1}\|_H \\ &\quad \leq c \sum_{m=0}^{n-1} \|f_m - (A - z_{m+1}1)f_{m+1}\|_{\mathfrak{H}_m}, \end{aligned} \tag{6.30}$$

where c is a constant depending on a and z_j.

Let $f_0 \in H$ and $\varepsilon > 0$ be given. Since $(A - z1)\mathfrak{D}$ is dense in $\mathfrak{H}_0$, there is an $f_1 \in \mathfrak{D}$ such that $\|f_0 - (A - z_1 1)f_1\|_{\mathfrak{H}_0} < \varepsilon$. Then, since $(A - z_2 1)\mathfrak{D}$ is dense in $\mathfrak{H}_1$, there is an $f_2 \in \mathfrak{D}$ such that $\|f_1 - (A - z_2 1)f_2\|_{\mathfrak{H}_1} < \varepsilon$, and so on; $f_n \in \mathfrak{D}$ is chosen so that $\|f_{n-1} - (A - z_n 1)f_n\|_{\mathfrak{H}_{n-1}} < \varepsilon$. As a result of (6.30) we have the estimate $\|f_0(A^n - z1)f_n\|_H < cn\varepsilon$, and this proves that $(A^n - z1)\mathfrak{D}$ is dense in H. ■

In applications of this theorem to differential operators, $\mathfrak{D}$ consists of infinitely differentiable functions; therefore, the condition $A\mathfrak{D} \subseteq \mathfrak{D}$ implies the requirement of infinite differentiability of the coefficients. If we are interested in the essential selfadjointness of a definite nth power of A, then the condition can be avoided. To do this it is necessary to weaken somewhat the assumptions of the theorem. (Its proof remains as before.) Suppose that A is an Hermitian operator defined in H with dense domain $\mathfrak{D}(A)$. Let

$$\mathfrak{D}(A^m) = \{f \in \mathfrak{D}(A) | Af \in \mathfrak{D}(A), \dots, A^{m-1}f \in \mathfrak{D}(A)\},$$

and define the operator A^m in the natural way on $\mathfrak{D}(A^m)$ $(m = 1, 2, \dots)$. Obviously, $\mathfrak{D}(A) \supseteq \mathfrak{D}(A^2) \supseteq \cdots$. Let $n = 1, 2, \dots$ be fixed, and let A be an Hermitian operator defined on $\mathfrak{D}(A)$ in H and semibounded below, and suppose that $\mathfrak{D}(A^n)$ is dense in H. Assume that there is a $b > 0$ such that for each $T \in (0, b)$ and each $\varphi_0, \varphi_1 \in \mathfrak{D}(A^{n-1})$ there exists a vector-valued function $[0, T] \ni t \mapsto \varphi(t) \in \mathfrak{D}(A^n)$ such that 1) the vector-valued function $[0, T] \ni t \mapsto A^m\varphi(t) \in H$ is twice strongly continuously differentiable for each $m = 0, \dots, n$, and 2) condition 2) in Theorem 6.6 holds. Then the statement of the theorem is preserved for A^n.

We note also that a theorem analogous to Theorem 6.6 is valid also for first-order evolution equations, i.e., in the framework where Theorem 6.1 is applicable.

6.3. ***Essential selfadjointness and quasi-analytic vectors.*** Let us first recall some facts from the theory of quasi-analytic functions (see, for example, Mandelbrojt [1], Chapter 4).

Let $[a, b] \subset \mathbb{R}^1$ be a finite close interval, and $(m_n)_1^\infty$ a particular sequence of positive numbers. The class $C\{m_n\}$ is defined to be the linear collection of all functions $f \in C^\infty([a, b])$ such that

$$|(D^n f)(t)| \leq K_f^n m_n \qquad (t \in [a, b];\ n = 1, 2, \dots, \tag{6.31}$$

where K_f is a constant depending on f.

It is known that the class of analytic functions $f(t)$ on $[a, b]$ is characterized by the estimates (6.31) with $m_n = n!$ The following fact obviously holds for this class $C\{n!\}$: if $f \in C\{n!\}$ is such that $(D^n f)(t_0) = 0$ at a particular point $t_0 \in [a, b]$ for all $n = 0, 1, \dots$, then $f(t) = 0$ for $t \in [a, b]$. We introduce the following definition with the purpose of generalizing this situation. The class $C\{m_n\}$ is said to be quasi-analytic if $f(t) = 0$ $(t \in [a, b])$ whenever $f \in C\{m_n\}$ and the equalities $(D^n f)(t_0) = 0$ $(n = 0, 1, \dots)$ hold at some particular point $t_0 \in [a, b]$. We have the following theorem of Denjoy and Carleman: the class $C\{m_n\}$ is quasi-analytic if and only if

$$\sum_{n=1}^{\infty} \left(\inf_{k=n,n+1,\dots} m_k^{1/k} \right)^{-1} = \infty. \tag{6.32}$$

Let H be a Hilbert space, and A an Hermitian operator acting in it. A vector $\varphi \in H$ is said to be quasi-analytic (with respect to A) if $\varphi \in \bigcap_1^\infty \mathfrak{D}(A^n)$ and the class $C\{\|A^n\varphi\|_H\}$ is quasi-analytic.

LEMMA 6.3. *A vector* $\varphi \in \bigcap_1^\infty \mathfrak{D}(A^n)$ *is quasi-analytic if and only if*

$$\sum_{n=1}^{\infty} \|A^n\varphi\|_H^{-1/n} = \infty. \tag{6.33}$$

PROOF. It is clear that $C\{\|A^n\varphi\|_H\} = C\{\|A^n(\lambda\varphi)\|_H\}$, where $\lambda > 0$ is fixed. This implies that it suffices to verify the lemma for a vector φ such that $\|\varphi\|_H = 1$. For such a vector the sequence

$$(\|A^n\varphi\|_H^{1/n})_{n=1}^\infty \tag{6.34}$$

is nondecreasing. Indeed,

$$\|A\varphi\|_H^2 = (A\varphi, A\varphi)_H = (A^2\varphi, \varphi)_H \le \|A^2\varphi\|_H\|\varphi\|_H,$$

i.e., $\|A\varphi\|_H \le \|A^2\varphi\|_H^{\frac{1}{2}}$. Assume that the inequality

$$\|A^n\varphi\|_H^{1/n} \le \|A^{n+1}\varphi\|_H^{1/(n+1)}$$

has already been proved. We prove that also

$$\|A^{n+1}\varphi\|_H^{1/(n+1)} \le \|A^{n+2}\varphi\|_H^{1/(n+2)} \qquad (n = 1, 2, \ldots).$$

With the assumed inequality taken into account,

$$\begin{aligned}\|A^{n+1}\varphi\|_H^2 &= (A^{n+1}\varphi, A^{n+1}\varphi)_H = (A^{n+2}\varphi, A^n\varphi)_H \\ &\le \|A^{n+2}\varphi\|_H\|A^n\varphi\|_H \le \|A^{n+2}\varphi\|_H\|A^{n+1}\varphi\|_H^{n/(n+1)}.\end{aligned}$$

Hence, $\|A^{n+1}\varphi\|_H^{(n+2)/(n+1)} \le \|A^{n+2}\varphi\|_H$, as required. Accordingly, the sequence (6.34) is nondecreasing.

Let us apply the Denjoy-Carleman criterion to the class $C\{\|A^n\varphi\|_H\}$ ($\|\varphi\|_H = 1$). Since the sequence (6.34) is nondecreasing,

$$\inf_{k=n,n+1,\ldots} \|A^k\varphi\|_H^{1/k} = \|A^n\varphi\|_H^{1/n}.$$

Therefore, the condition (6.32) for quasi-analyticity of this class, i.e., quasi-analyticity of the vector φ, can be rewritten in the form (6.33). ■

THEOREM 6.7. *Suppose that* A *is a closed Hermitian operator acting in* H. *It is selfadjoint if and only if* H *contains a total set consisting of quasi-analytic vectors.*

PROOF. The assertion is trivial in one direction: if A is selfadjoint, then it suffices to prove that each vector of the form $\varphi = E((a, b))f$ is quasi-analytic, where E is the resolution of the identity corresponding to A, $a, b \in \mathbb{R}^1$ $(a < b)$ and $f \in H$. Obviously, $\varphi \in \bigcap_1^\infty \mathfrak{D}(A^n)$. Further,

$$\|A^n\varphi\|_H^2 = \int_a^b \lambda^{2n} d(E(\lambda)f, f)_H \le c^{2n}\|f\|_H^2$$
$$(c = \max(|a|, |b|);\ n = 1, 2, \ldots);$$

therefore, the series (6.33) diverges, and φ is quasi-analytic according to Lemma 6.3.

Assume now that A has a total set M of quasi-analytic vectors φ. Since A is closed, it suffices to prove that it is essentially selfadjoint, or according to Theroem 6.1, that uniqueness holds for the strong solutions of the Cauchy problem for (6.4) when $b = \infty$. Let $u(t)$ be a strong solution of the problem

$$(du/dt)(t) - (\varsigma A)^* u(t) = 0 \qquad (t \in [0, \infty),\ u(0) = 0), \tag{6.35}$$

where $\varsigma = \pm i$. It suffices to establish that $u(t) = 0$ for $t \in [0, T]$, where $T > 0$ is arbitrary.

The "weak" equality (6.3) for (6.35) with a quasi-analytic vector $f = \varphi \in \bigcap_1^\infty \mathfrak{D}(A^n)$ gives us that

$$(d/dt)(u(t), \varphi)_H = ((du/dt)(t), \varphi)_H = (u(t), (\varsigma A)\varphi)_H \qquad (t \in [0, T]).$$

But $(\varsigma A)\varphi \in \bigcap_1^\infty \mathfrak{D}(A^n)$; therefore,

$$(d/dt)(u(t), (\varsigma A)\varphi)_H = (u(t), (\varsigma A)^2 \varphi)_H \qquad (t \in [0, T])$$

and so on. This implies that $(u(t), \varphi)_H \in C^\infty([0, T])$ and

$$D^n (u(t), \varphi)_H = D^{n-1}(u(t), (\varsigma A)\varphi)_H = \cdots = (u(t), (\varsigma A)^n \varphi)_H$$
$$(t \in [0, T];\ n = 0, 1, \ldots). \tag{6.36}$$

Since the values of $u(t)$ on $[0, T]$ are bounded, it can be concluded from (6.36) that

$$|D^n (u(t), \varphi)_H] \le c \|(\varsigma A)^n \varphi\|_H = c\|A^n \varphi\|_H \qquad (t \in [0, T]; n = 0, 1, \ldots),$$

i.e., the scalar function $[0, T] \ni t \mapsto f(t) = (u(t), \varphi)_H$ belongs to the class $C\{\|A^n \varphi\|_H\}$. (6.36) and the equality $u(0) = 0$ give that $(D^n f)(0) = 0$ ($n = 0, 1, \ldots$); therefore, the equality $(u(t), \varphi)_H = f(t) = 0$ ($t \in [0, T]$) holds because $C\{\|A^n \varphi\|_H\}$ is quasi-analytic. Since the set M of vectors φ is total, $u(t) = 0$ ($t \in [0, T]$).

The following definition is also useful. As before, suppose that A is an Hermitian operator acting in a Hilbert space H. A vector $\varphi \in H$ is said to be analytic (with respect to A) if $\varphi \in \bigcap_1^\infty \mathfrak{D}(A^n)$ and the power series

$$\sum_{n=0}^{\infty} \frac{\|A^n \varphi\|_H}{n!} z^n \tag{6.37}$$

has nonzero radius of convergence, and is said to be entire if this radius is equal to infinity. Of course, every analytic vector is quasi-analytic, but not conversely. We remark that in proving the first part of Theorem 6.7 we established the following stronger fact: *if A is selfadjoint, then it has a total set of entire vectors.* (Each vector $\varphi = E((a, b))f$ considered in the proof satisfies the estimate $\|A^n \varphi\|_H \le c^n \|f\|_H$ ($n = 1, 2, \ldots$) and is thus entire.)

We establish another theorem, which refines Theorem 6.7 for operators that are semibounded below.

Let A be an Hermitian operator acting in H. A vector $\varphi \in H$ is called a Stieltjes vector (with respect to A) if $\varphi \in \bigcap_1^\infty \mathfrak{D}(A^n)$ and $C\{\|A^n\varphi\|_H^{1/2}\}$ is quasi-analytic, or, in other words,

$$\sum_{n=1}^{\infty} \|A^n\varphi\|_H^{1/2n} = \infty. \tag{6.38}$$

(The fact that the Stieltjes condition is equivalent to (6.38) follows from the fact that for $\|\varphi\|_H = 1$ the sequence $(\|A^n\varphi\|_H^{1/2n})_1^\infty$ is nondecreasing along with (6.34); then use the Denjoy-Carleman criterion.)

It is clear that each quasi-analytic vector is a Stieltjes vector, but not conversely. Thus, if we denote the sets of all entire, analytic, quasi-analytic and Stieltjes vectors with respect to A by $\mathcal{E}(A)$, $\mathcal{A}(A)$, $\mathcal{O}(A)$, and $\mathfrak{S}(A)$, respectively, then we get the inclusions

$$\mathcal{E}(A) \subseteq \mathcal{A}(A) \subseteq \mathcal{O}(A) \subseteq \mathfrak{S}(A). \tag{6.39}$$

THEOREM 6.8. *Let A be a closed Hermitian operator which is semibounded below. If H contains a total set consisting of Stieltjes vectors, then A is selfadjoint.*

The converse assertion is obvious in view of what has already been proved and (6.39).

PROOF. According to Theorem 6.2, it suffices to prove the uniqueness of strong solutions of the Cauchy problem for (6.13) when $b = \infty$. Let $u(t)$ be a strong solution of this problem such that $u(0) = u'(0) = 0$. We show that $u(t) = 0$ ($t \in [0,T]$) for any $T > 0$. Let M be the total set of Stieltjes vectors φ in the hypotheses of the theorem. Define $[0,T] \ni t \mapsto f(t) = (u(t),\varphi)_H$, where $\varphi \in M$. The relation (6.3), written for (6.13), gives us that $f \in C^2([0,T])$ and $(d^2f/dt^2)(t) = -(u(t), A\varphi)_H$ ($t \in [0,T]$). Since $A\varphi \in \bigcap_1^\infty \mathfrak{D}(A^n)$, we conclude in the same way that $(u(t),A\varphi)_H \in C^2([0,T])$ and

$$(d^2/dt^2)(u(t),A\varphi)_H = -(u(t),A^2\varphi)_H \qquad (t \in [0,T]),$$

and so on. As a result, $f \in C^\infty([0,T])$, and

$$\begin{aligned}(D^{2k}f)(t) &= D^{2k}(u(t),\varphi)_H = -D^{2(k-1)}(u(t),A\varphi)_H \\ &= (-1)^k(u(t),A^k\varphi)_H \qquad (t \in [0,T];\ k = 0,1,\ldots).\end{aligned} \tag{6.40}$$

An analogous equality can be written also for the odd derivatives. Indeed, differentiating (6.40), we get

$$(D^{2k+1}f)(t) = (-1)^k(u'(t),A^k\varphi)_H \qquad (t \in [0,H];\ k = 0,1,\ldots). \tag{6.41}$$

The values of $u(t)$ and $u'(t)$ are bounded on $[0,T]$; therefore, (6.40), (6.41), the fact that A is Hermitian, and the Cauchy-Schwarz-Bunyakovskiĭ inequality give us that

$$|(D^{2k}f)(t)|, |(D^{2k+1}f)(t)| \le c\|A^k\varphi\|_H \le c\|\varphi\|_H^{1/2}\|A^{2k}\varphi\|_H^{1/2}$$
$$(t \in [0,T];\ k = 0,1,\ldots),$$

i.e., $f \in C\{m_n\}$, where $(m_n)_0^\infty$ is the sequence of numbers

$$(\|\varphi\|_H^{1/2}, \|\varphi\|_H^{1/2}, \|A\varphi\|_H^{1/2}, \|A\varphi\|_H^{1/2}, \|A^2\varphi\|_H^{1/2}, \|A^2\varphi\|_H^{1/2}, \ldots).$$

The class $C\{m_n\}$ does not change if we normalize φ. But then, as already noted, the sequence $(\|A^n\varphi\|_H^{1/2n})_1^\infty$ is nondecreasing. This implies that the Denjoy-Carleman condition (6.32) can be written in the form (6.38) for the sequence under consideration. Thus, the class $C\{m_n\}$ is quasi-analytic.

On the other hand, (6.40), (6.41) and the conditions $u(0) = u'(0) = 0$ give us that $(D^n f)(0) = 0$ $(n = 1,2,\ldots)$. Therefore, $(u(t),\varphi)_H = f(t) = 0$ $(t \in [0,T])$. The fact that M is total leads to the conclusion that $u(t) = 0$ $(t \in [0,T])$.

6.4. *Verification of commutativity for selfadjoint operators.* Suppose that A_1 and A_2 are essentially selfadjoint unbounded (in general) operators acting in a Hilbert space H. Recall that they are said to commute if $E_1(\delta_1)E_2(\delta_2) = E_2(\delta_2)E_1(\delta_1)$ for any $\delta_1\delta_2 \in \mathcal{B}(\mathbb{R}^1)$, where E_1 and E_2 are the resolutions of the identity for the respective operators A_1 and A_2. Verification of such commutativity often causes serious difficulties. We present some facts which facilitate this verification. First of all, we establish the following known lemma.

LEMMA 6.4. *Let A_1 and A_2 be two selfadjoint operators. They commute if and only if their resolvents $R_{z_1}(A_1)$ and $R_{z_2}(A_2)$ commute for some particular z_1 and z_2.*

PROOF. If A_1 and A_2 commute, then the fact that their resolvents commute follows from the representation

$$R_{z_j}(A_j) = \int_{\mathbb{R}_1} (\lambda - z_j)^{-1}\, dE_j(\lambda) \qquad (j = 1,2).$$

Conversely, suppose that $R_{z_1}(A_1)$ and $R_{z_2}(A_2)$ commute. We first prove that $R_{\varsigma_1}(A_1)$ and $R_{\varsigma_2}(A_2)$ commute for any nonreal ς_1 and ς_2. Indeed, let B be an operator acting in H, and let $z, \varsigma \in \mathbb{C}^1$ be two regular points of it. Then the operator $(1 - (z-\varsigma)R_\varsigma(B))^{-1}$ exists and is equal to $1 + (z-\varsigma)R_z(B)$; this can be seen by a simple check with the use of the Hilbert identity. With the help of this identity and the fact established here we have that

$$R_{\varsigma_1}(A_1) - R_{z_1}(A_1) = (\varsigma_1 - z_1)R_{\varsigma_1}(A_1)R_{z_1}(A_1),$$
$$R_{\varsigma_1}(A_1)(1 - (\varsigma_1 - z_1)R_{z_1}(A_1)) = R_{z_1}(A_1),$$
$$R_{\varsigma_1}(A_1) = (1 - (\varsigma_1 - z_1)R_{z_1}(A_1))^{-1}R_{z_1}(A_1).$$

From the last equality and the fact that $R_{z_1}(A_1)$ and $R_{z_2}(A_2)$ commute it follows that $R_{\varsigma_1}(A_1)$ and $R_{z_2}(A_2)$ also commute. From this it follows similarly that $R_{\varsigma_1}(A_1)$ and $R_{\varsigma_2}(A_2)$ commute.

We now use a well-known formula: for finite open intervals δ_1 and δ_2

$$\frac{1}{2}(E_j(\tilde{\alpha}_j) + E_j(\alpha_j)) = \lim_{\varepsilon_j \to +0} \frac{1}{2\pi i} \int_{\delta_j + i\varepsilon_j} (R_{\varsigma_j}(A_j) - R_{\bar{\varsigma}_j}(A_j))\, d\varsigma_j \qquad (j = 1, 2). \quad (6.42)$$

in the sense of strong convergence. By what has been proved, the integrals in (6.42) commute for $j = 1$ and $j = 2$. But then their strong limits commute; hence $E_1(\delta_1)$ and $E_2(\delta_2)$ are easily seen also to commute. That is, A_1 and A_2 commute. ■

The following simple assertion, which gives a sufficient condition for commutativity in terms of quasi-analytic vectors, is useful.

THEOREM 6.9. *Let A_1 and A_2 be two Hermitian operators acting in H with domains $\mathfrak{D}(A_1)$ and $\mathfrak{D}(A_2)$, and let $\mathfrak{D} \subseteq \mathfrak{D}(A_1) \cap \mathfrak{D}(A_2)$ be a linear set. Assume that they commute on $\mathfrak{D}$, i.e., $A_1\mathfrak{D} \subseteq \mathfrak{D}(A_2)$, $A_2\mathfrak{D} \subseteq \mathfrak{D}(A_1)$, and $A_1A_2f = A_2A_1f$ $(f \in \mathfrak{D})$. Suppose in addition that A_1, A_2, and the restriction $A_1 \restriction ((A_2 - z1)\mathfrak{D})$ $(f \in \mathfrak{D})$ for some nonreal z have total sets of quasi-analytic vectors. Then the closures $\tilde{A}_1$ and $\tilde{A}_2$ are selfadjoint and commute.*

PROOF. According to Theorem 6.7 the operators $\tilde{A}_1$ and $\tilde{A}_2$ are selfadjoint. Let $R_\varsigma(\tilde{A}_1)$ and $R_\varsigma(\tilde{A}_2)$ be their resolvents. By Lemma 6.4, to prove that $\tilde{A}_1$ and $\tilde{A}_2$ commute it suffices to verify that $R_z(\tilde{A}_1)$ and $R_z(\tilde{A}_2)$ commute. Since A_1 and A_2 commute,

$$\begin{aligned} R_z(\tilde{A}_1)R_z(\tilde{A}_2)(A_1 - z1)(A_2 - z1)f \\ &= R_z(\tilde{A}_1)R_z(\tilde{A}_2)(A_2 - z1)(A_1 - z1)f \\ &= f = R_z(\tilde{A}_2)R_z(\tilde{A}_1)(A_1 - z1)(A_2 - z1)f \end{aligned}$$

for $f \in \mathfrak{D}$. Therefore, to see that $R_z(\tilde{A}_1)$ and $R_z(\tilde{A}_2)$ commute it suffices to see that $(A_1 - z1)(A_2 - z1)\mathfrak{D}$ is dense in H. But this set coincides with the range $\mathfrak{R}(A_1 \restriction ((A_2 - z1)\mathfrak{D}) - z1)$, and in view of the same Theorem 6.7, the operator $A_1 \restriction ((A_2 - z1)\mathfrak{D})$ is essentially selfadjoint; therefore, the indicated range is dense in H. ■

We dwell on the frequently arising situation when it must be verified that selfadjoint operators commute. Let H_1 and H_2 be separable Hilbert spaces. Form their tensor product $\mathcal{H} = H_1 \otimes H_2$, and consider in $\mathcal{H}$ a quasi-inner product $\langle \cdot, \cdot \rangle$ such that for some $c > 0$

$$\langle\langle f \rangle\rangle = \langle f, f \rangle^{1/2} \leq c\|f\|_{\mathcal{H}} \qquad (f \in \mathcal{H}). \quad (6.43)$$

With the help of $\langle \cdot, \cdot \rangle$ we form a Hilbert space $\hat{\mathcal{H}}$ from $\mathcal{H}$ in the standard way by means of identification and completion.

Let A_j be an operator defined in H_j with dense domain $\mathfrak{D}(A_j)$ ($j = 1,2$). From A_1 and A_2 we construct densely defined operators acting in $\mathcal{H}$ by setting $\mathcal{A}_1 = A_1 \otimes 1$, $\mathfrak{D}(\mathcal{A}_1) = \text{a.}(\mathfrak{D}(A_1) \otimes H_2)$ and $\mathcal{A}_2 = 1 \otimes A_2$, $\mathfrak{D}(A_2) = \text{a.}(H_1 \otimes \mathfrak{D}(A_2))$. It is clear that these operators commute on $\mathfrak{D} = \text{a.}(\mathfrak{D}(A_1) \otimes \mathfrak{D}(A_2)) \subseteq \mathfrak{D}(A_1) \cap \mathfrak{D}(A_2)$: we have $\mathcal{A}_1\mathfrak{D} \subseteq \mathfrak{D}(\mathcal{A}_2)$, $\mathcal{A}_2\mathfrak{D} \subseteq \mathfrak{D}(\mathcal{A}_1)$, and $\mathcal{A}_1\mathcal{A}_2 f = \mathcal{A}_2\mathcal{A}_1 f$ ($f \in \mathfrak{D}$).

Assume that for $j = 1$ and 2 the operator $\mathcal{A}_j$ is Hermitian with respect to $\langle\cdot,\cdot\rangle$: $\langle \mathcal{A}_j f, g\rangle = \langle f, \mathcal{A}_j g\rangle$ ($f, g \in \mathfrak{D}(\mathcal{A}_j)$). Then the mapping $\mathcal{H} \supseteq \mathfrak{D}(\mathcal{A}_j) \ni f \mapsto \mathcal{A}_j f \in \mathcal{H}$ gives rise in a natural way to an operator $\hat{\mathcal{A}}_j$ acting in $\hat{\mathcal{H}}$, because if $\langle f, f\rangle = 0$, then $\langle \mathcal{A}_j f, \mathcal{A}_j f\rangle = 0$ ($f \in \mathfrak{D}(\mathcal{A}_j)$). Let us see that this is so. For any $g \in \mathfrak{D}(\mathcal{A}_j)$ we have that

$$|\langle \mathcal{A}_j f, g\rangle|^2 = |\langle f, \mathcal{A}_j g\rangle|^2 \leq \langle f, f\rangle \langle \mathcal{A}_j g, \mathcal{A}_j g\rangle = 0,$$

whence $\langle A_j f, g\rangle = 0$. Since $\mathfrak{D}(A_j)$ is dense is $\mathcal{H}$, $\mathcal{A}_j f$ can be approximated in $\mathcal{H}$ by vectors $g_n \in \mathfrak{D}(\mathcal{A}_j)$. Then

$$\langle \mathcal{A}_j f, \mathcal{A}_j f\rangle = \lim_{n\to\infty} \langle \mathcal{A}_j f, g_n\rangle = 0,$$

by (6.43).

Accordingly, if $\hat{f}, \hat{g}, \ldots,$ denote the classes containing the elements $f, g, \ldots$ of $\mathcal{H}$ when the quotient by $\{h \in \mathcal{H} | \langle h, h\rangle = 0\}$ is taken, then $\hat{\mathcal{A}}_j \hat{f} = (\mathcal{A}_j f)\hat{}$ and $\mathfrak{D}(\hat{\mathcal{A}}_j) = (\mathfrak{D}(\mathcal{A}_j))\hat{}$. It is clear that the operators $\hat{\mathcal{A}}_1$ and $\hat{\mathcal{A}}_2$ are Hermitian and commute on $\mathfrak{D} \subseteq \mathfrak{D}(\hat{\mathcal{A}}_1) \cap \mathfrak{D}(\hat{\mathcal{A}}_2)$: $\hat{\mathcal{A}}_1\hat{\mathfrak{D}} \subseteq \mathfrak{D}(\hat{\mathcal{A}}_2), \mathcal{A}_2\hat{\mathfrak{D}} \subseteq \mathfrak{D}(\hat{\mathcal{A}}_1)$, and $\hat{\mathcal{A}}_1\hat{\mathcal{A}}_2 f = \hat{\mathcal{A}}_2\hat{\mathcal{A}}_1 f$ ($f \in \hat{\mathfrak{D}}$). The problem is to find conditions which imply that $\hat{\mathcal{A}}_1$ and $\hat{\mathcal{A}}_2$ are essentially selfadjoint and commute. Of course, Theorem 6.9 is applicable in certain cases, but nevertheless we shall establish two more results.

To formulate the first one we present an additional construction. Fix an $f_2 \in H_2$ and consider in H_1 the quasi-inner product $\langle f_1, g_1\rangle_{f_2} = \langle f_1 \otimes f_2, g_1 \otimes f_2\rangle$ ($f_1, g_1 \in H_1$). Identification and completion with respect to $\langle\cdot,\cdot\rangle_{f_2}$ yields the Hilbert space $\hat{H}_{1,f_2}$. The operator A_1 is Hermitian with respect to $\langle\cdot,\cdot\rangle_{f_2}$:

$$\begin{aligned}\langle A_1 f_1, g\rangle_{f_2} &= \langle (A_1 f_1) \otimes f_2, g_1 \otimes f_2\rangle = \langle \mathcal{A}_1(f_1 \otimes f_2), g_1 \otimes f_2\rangle \\ &= \langle f_1 \otimes f_2, \mathcal{A}_1(g_1 \otimes f_2)\rangle = \langle f_1, A_1 g_1\rangle_{f_2} \qquad (f_1, g_1 \in \mathfrak{D}(A_1)).\end{aligned}$$

It therefore gives rise, in a way analogous to the preceding, to a Hermitian operator $\hat{\mathcal{A}}_{1,f_2}$ acting in $\hat{\mathcal{H}}_{1,f_2}$. Similarly, fixing an $f_1 \in H_1$, we construct a space $\hat{H}_{2,f_1}$ and a Hermitian operator $\hat{A}_{2,f_1}$ acting in it.

THEOREM 6.10. *Suppose that the operators $\hat{A}_{1,f_2}$ and $\hat{A}_{2,f_1}$ are essentially selfadjoint for f_1 and f_2 in dense subsets M_1 and M_2 of H_1 and H_2, respectively. Then the operators $\hat{\mathcal{A}}_1$ and $\hat{\mathcal{A}}_2$ are also essentially selfadjoint.*

They commute if for some nonreal z each operator $\hat{\mathcal{A}}_{1,h_2}$ is essentially selfadjoint for $h_2 \in (A_2 - z1)\mathfrak{D}$ and

$$\operatorname{Ker}(A_j - z1) = \{g_j \in \mathfrak{D}(A_j) | (A_j - z1)g_j = 0\} = 0.$$

PROOF. We establish that $\hat{\mathcal{A}}_1$ is essentially selfadjoint. Let $f_1 \in H_1$ and $f_2 \in M_2$. It suffices to see that for a particular $\varepsilon > 0$ and a particular nonreal z there is a $g_1 \in \mathfrak{D}(A_1)$ such that

$$\|(f_1 \otimes f_2)\widehat{\ } - (\hat{\mathcal{A}}_1 - z1)(g_1 \otimes f_2)\widehat{\ }\|_{\hat{\mathcal{H}}} < \varepsilon.$$

But this norm is equal to

$$\begin{aligned}\langle\langle f_1 \otimes f_2 - (\mathcal{A}_1 - z1)(g_1 \otimes f_2)\rangle\rangle &= \langle\langle (f_1 - (A_1 - z1)g_1) \otimes f_2 \rangle\rangle \\ &= \|\hat{f}_1 - (\hat{A}_{1,f_2} - z1)\hat{g}_1\|_{\hat{H}_{1,f_2}}, \quad (6.44)\end{aligned}$$

where $\hat{f}_1$ and $\hat{g}_1$ are the classes containing f_1 and g_1 in the construction of the space $\hat{H}_{1,f_2}$. Since $\hat{A}_{1,f_2}$ is essentially selfadjoint, the right-hand side of (6.44) can be made arbitrarily small by properly choosing $g_1 \in \mathfrak{D}(A_1)$. The assertion is proved. It can be proved similarly that $\hat{\mathcal{A}}_2$ is essentially selfadjoint.

Let us prove that $\hat{\mathcal{A}}_1$ and $\hat{\mathcal{A}}_2$ commute. We first establish that the set of vectors

$$(((A_1 - z1)g_1) \otimes ((A_2 - z1)g_2))\widehat{\ }, \qquad g_1 \in \mathfrak{D}(A_1),\ g_2 \in \mathfrak{D}(A_2),$$

is dense in $\hat{\mathcal{H}}$. Take $f_1 \in M_1$ and $f_2 \in H_2$. With the obvious notation,

$$\begin{aligned}&\|(f_1 \otimes f_2)\widehat{\ } - (((A_1 - z1)g_1) \otimes ((A_2 - z1)g_2))\widehat{\ }\|_{\hat{\mathcal{H}}} \\ &\quad = \langle\langle f_1 \otimes f_2 - ((A_1 - z1)g_1) \otimes ((A_2 - z1)g_2)\rangle\rangle \\ &\quad \le \langle\langle f_1 \otimes (f_2 - (A_2 - z1)g_2)\rangle\rangle + \langle\langle (f_1 - (A_1 - z1)g_1) \\ &\qquad\qquad \otimes (A_2 - z1)g_2\rangle\rangle \\ &\quad = \|\hat{f}_2 - (\hat{A}_{2,f_1} - z1)\hat{g}_2\|_{\hat{H}_{2,f_1}} \\ &\qquad + \|\hat{f}_1 - (\hat{A}_{1,(A_z - z1)g_2} - z1)\hat{g}_1\|_{\hat{H}_{1,(A_2 - z1)g_2}}. \qquad (6.45)\end{aligned}$$

Since $\hat{\mathcal{A}}_{2,f_1}$ is essentially selfadjoint, the first term can be made arbitrarily small by properly choosing $g_2 \in \mathfrak{D}(A_2)$. The same is true for the second term, because $\hat{A}_{1,(A_2-z1)g_2}$ is essentially selfadjoint. The required density is established.

Since $\operatorname{Ker}(A_1 - z1) = 0$, the algebraic inverse operator $(A_1 - z1)^{-1}$ is defined in H_1. But then the algebraic inverse operator $(\mathcal{A}_1 - z1)^{-1} = (A_1 - z1)^{-1} \otimes 1$ is also defined in $\mathcal{H}$. The action of the resolvent $R_z(\tilde{\hat{\mathcal{A}}}_1)$ on the vector $(f_1 \otimes f_2)\widehat{\ }$ ($f_1 \in H_1$, $f_2 \in H_2$) is obviously equal to $((\mathcal{A}_1 - z1)^{-1}(f_1 \otimes f_2))\widehat{\ }$. The picture is analogous for A_2. Therefore, for $g_1 \in \mathfrak{D}(A_1)$ and $g_2 \in \mathfrak{D}(A_2)$

$$\begin{aligned} &R_z(\tilde{\hat{\mathcal{A}}}_2)R_z(\tilde{\hat{\mathcal{A}}}_1)(((A_1 - z1)g_1) \otimes ((A_2 - z1)g_2))\widehat{} \\ &= R_z(\tilde{\hat{\mathcal{A}}}_2)(((A_1 - z1)^{-1} \otimes 1)((A_1 - z1)g_1) \otimes ((A_2 - z1)g_2)))\widehat{} \\ &= R_z(\tilde{\hat{\mathcal{A}}}_2)(g_1 \otimes ((A_2 - z1)g_2))\widehat{} \\ &= ((1 \otimes (A_2 - z1)^{-1})(g_1 \otimes ((A_2 - z1)g_2))\widehat{} = (g_1 \otimes g_2)\widehat{} \\ &= R_z(\tilde{\hat{\mathcal{A}}}_1)R_z(\tilde{\hat{\mathcal{A}}}_2)(((A_1 - z1)g_1) \otimes ((A_2 - z1)g_2))\widehat{}. \end{aligned} \tag{6.46}$$

The set of vectors $(((A_1 - z1)g_1) \otimes (A_2 - z1)g_2))\widehat{}$ is dense in $\hat{\mathcal{H}}$, so (6.46) means that $R_z(\tilde{\hat{\mathcal{A}}}_1)$ and $R_z(\tilde{\hat{\mathcal{A}}}_2)$ commute. It remains to apply Lemma 6.4. ■

The next theorem is closed connected with Theorem 6.1.

THEOREM 6.11. *Assume that uniqueness holds for strong solutions of the equations*

$$(du/dt)(t) \pm (iA_j^*)u(t) = 0 \qquad (t \in [0, b_j)) \tag{6.47}$$

in H_j for some $b_j \in (0, \infty]$. Then the operator $\hat{\mathcal{A}}_j$ is essentially selfadjoint $(j = 1, 2)$. If in addition $\operatorname{Ker}(A_j - z_j 1) = 0$ $(j = 1, 2)$ *for some nonreal z_j, then $\hat{\mathcal{A}}_1$ and $\hat{\mathcal{A}}_2$ commute.*

PROOF. Let us establish that $\tilde{\hat{\mathcal{A}}}_1$ is selfadjoint; the proof that $\tilde{\hat{\mathcal{A}}}_2$ is selfadjoint is similar. Observe first that for each particular $f_2 \in H_2$ there exists a continuous operator $\hat{\mathcal{H}} \ni w \mapsto T_{f_2} w \in H_1$ such that

$$\langle w, (f_1 \otimes f_2)\widehat{}\rangle = (T_{f_2} w, f_1)_{H_1} \qquad (w \in \hat{\mathcal{H}},\ f_1 \in H_1). \tag{6.48}$$

Indeed, the expression on the left-hand side of (6.48) is a bilinear form with respect to $w \in \hat{\mathcal{H}}$ and $f_1 \in H_1$ and is continuous by virtue of the Cauchy-Schwarz-Bunyakovskiĭ inequality for $\langle \cdot, \cdot \rangle$ and (6.43). Therefore, it can be written as indicated in (6.48).

Consider a strong solution $[0, b_1) \ni t \mapsto v(t) \in \hat{\mathcal{H}}$ of equation

$$(dv/dt)(t) + (i\hat{\mathcal{A}}_1)^* v(t) = 0, \qquad v(0) = 0$$

in $\hat{\mathcal{H}}$. Writing this equation in the weak form of type (6.3), where the role of f is played by $(f_1 \otimes f_2)\widehat{}$ $(f_1 \in \mathfrak{D}(A_1),\ f \in H_2)$, we get

$$\langle (dv/dt)(t), (f_1 \otimes f_2)\widehat{}\rangle + \langle v(t), i\hat{\mathcal{A}}_1 (f_1 \otimes f_2)\widehat{}\rangle = 0 \qquad (t \in [0, b)). \tag{6.49}$$

The function $[0, b_1) \ni t \mapsto u(t) = T_{f_2} v(t) \in H_1$ is once strongly continuously differentiable. With the help of (6.48) and the relation $\hat{\mathcal{A}}_1 (f_1 \otimes f_2)\widehat{} = ((A_1 f_1) \otimes f_2)\widehat{}$ we can rewrite (6.49) in the form

$$((du/dt)(t), f_1)_{H_1} + (u(t), iA_1 f_1) = 0, \qquad u(0) = T_{f_2} v(0) = 0 \qquad (f_1 \in \mathfrak{D}(A_1);\ t \in [0, b_1)).$$

In other words, we obtain an expression for (6.47) in the weak form with $j = 1$ and the + sign. By the assumed uniqueness, $u(t) = 0$ ($t \in [0, b_1)$). Returning to $v(t)$ with the help of (6.48), we get that $\langle v(t), (f_1 \otimes f_2)^\frown \rangle = 0$ ($t \in [0, b_1)$) for all $f_1 \in \mathfrak{D}(A_1)$ and $f_2 \in H_2$. This gives us that $v(t) = 0$ ($t \in [0, b_1)$). Uniqueness is established similarly in the case of the $-$ sign. To prove that $\tilde{\hat{\mathcal{A}}}_1$ is selfadjoint it remains to use Theorem 6.1.

In the given proof that $\tilde{\hat{\mathcal{A}}}_1$ is selfadjoint it is essential that $f_1 \in \mathfrak{D}(A_1)$; the vector f_2 is not allowed to run through the whole of H_2 but only through a subset of it for which the vectors $(f_1 \otimes f_2)^\frown (f_1 \in \mathfrak{D}(A_1))$ form a total subset of $\hat{\mathcal{H}}$. In particular, we can take $f_2 \in (A_2 - z_2 1)\mathfrak{D}(A_2)$, because then the vectors $(f_1 \otimes f_2)^\frown$ run through the collection of vectors of the form $(\hat{\mathcal{A}}_2 - z_2 1)(f_1 \otimes g_2)^\frown$, where $g_2 \in \mathfrak{D}(A_2)$, and this is total because $\hat{\mathcal{A}}_2$ is essentially selfadjoint. This observation leads to the conclusion that the restriction

$$\hat{\mathcal{A}}_1 \restriction (\text{a.}(\mathfrak{D}(A_1) \otimes \mathfrak{R}(A_2 - z_2 1)))^\frown$$

is essentially selfadjoint. From this it follows that the vectors

$$(((A_1 - z_1 1)g_1) \otimes ((A_2 - z_2 1)g_2))^\frown \qquad (g_1 \in \mathfrak{D}(A_1),\ g_2 \in \mathfrak{D}(A_2)) \tag{6.49'}$$

form a dense subset of $\hat{\mathcal{H}}$.

Computing $R_{z_2}(\tilde{\hat{\mathcal{A}}}_2)R_{z_1}(\tilde{\hat{\mathcal{A}}}_1)$ and $R_{z_1}(\tilde{\hat{\mathcal{A}}}_1)R_{z_2}(\tilde{\hat{\mathcal{A}}}_2)$ on the vectors (6.49′), we arrive at the same result $(g_1 \otimes g_2)^\frown$. Thus, the operators $R_{z_1}(\tilde{\hat{\mathcal{A}}}_1)$ and $R_{z_2}(\tilde{\hat{\mathcal{A}}}_2)$ commute on a dense subset of $\hat{\mathcal{H}}$, and by Lemma 6.4, this implies that $\tilde{\hat{\mathcal{A}}}_1$ and $\tilde{\hat{\mathcal{A}}}_2$ commute. ■

We remark that uniqueness in the theorem can often be proved with the help of Lemma 6.2.

The case when the operators $\hat{\mathcal{A}}_1$ and $\hat{\mathcal{A}}_2$ are not themselves selfadjoint after closure but admit selfadjoint commuting extensions is considered (in a somewhat different formulation) in a book of the author (Berezanskiĭ [4], Chapter 3, §2.3).

6.5. ***Parabolic criteria for selfadjointness.*** Along with the "hyperbolic" and "Schrödinger" criteria for selfadjointness presented in subsection 1 it is very useful to have "parabolic" criteria, in which the evolution equation is parabolic. Here we present the corresponding facts, analogous in large part to the results in subsection 1. For simplicity we formulate them in the case of an interval $[0, b) = [0, \infty)$.

Let H be a Hilbert space, and A a Hermitian operator acting in it with domain $\mathfrak{D}(A)$. Consider the differential equation

$$(du/dt)(t) + A^* u(t) = 0 \qquad (t \in [0, \infty)). \tag{6.50}$$

THEOREM 6.12. *Let A be an Hermitian operator acting in H. A necessary condition for it to be essentially selfadjoint is that strong solutions of the*

Cauchy problem for equation (6.12) *be unique. This uniqueness condition is also sufficient in the case when A is semibounded below.*

PROOF. Suppose that the closure $\tilde{A}$ is selfadjoint, and let E be its resolution of the identity. We apply Lemma 6.2 with $r = 1$, $B = \tilde{A}$, and $\Phi = \bigcup_1^\infty E((-n,n))H$. A strong solution of the corresponding Cauchy problem, which now has the form

$$\varphi'(t) - \tilde{A}\varphi(t) = 0 \quad (t \in [0,T]), \qquad \varphi(T) = \varphi_0 \in \Phi,$$

exists and is equal to

$$\varphi(t) = \int_{\mathbf{R}^1} e^{\lambda(t-T)}\, dE(\lambda)\varphi_0 \qquad (t \in [0,T]). \tag{6.51}$$

(The integral in (6.51) is actually over a finite interval; thus the function $[0,T] \ni t \mapsto \varphi(t)$ is strongly continuously differentiable and is a solution of the required problem.) Lemma 6.2 implies that strong solutions of the Cauchy problem for (6.50) are unique.

We establish sufficiency. Assume that $\tilde{A}$ is not selfadjoint. Then A has two different selfadjoint extensions A_1 and A_2 acting in H and semibounded below by a number $c > -\infty$. Let E_1 and E_2 be the corresponding resolutions of the identity. For any $g \in \mathfrak{D}(A) \subseteq \mathfrak{D}(A_1)$ the integral $\int_c^\infty \lambda^2\, d(E_1(\lambda)g, g)_H$ converges; therefore, the vector-valued function

$$[0,\infty) \ni t \mapsto u_1(t) = \int_c^\infty e^{-\lambda t}\, dE_1(\lambda)g \tag{6.52}$$

is once strongly continuously differentiable, and $u_1(t) \in \mathfrak{D}(A_1) \subseteq \mathfrak{D}(A^*)$. The derivative $u_1'(t)$ can be expressed by the integral (6.52) with the factor $(-\lambda)$ in front of $e^{-\lambda t}$, and $A^*u_1(t) = A_1u_1(t)$ has a similar form. Thus, (6.52) is a strong solution of (6.50) with $u_1(0) = g$. A function $u_2(t)$ is constructed similarly, with E_1 replaced by E_2 in (6.52). The difference $u(t) = u_1(t) - u_2(t)$ is also a strong solution of (6.50), with $u(0) = 0$. By the uniqueness assumption, $u(t) = 0$ $(t \in [0,\infty))$. Taking the inner product of this equality with an $h \in H$, we get that

$$\int_c^\infty e^{-\lambda t}\, d((E_1(\lambda) - E_2(\lambda))g, h)_H = 0 \qquad (t \in [0,\infty)).$$

Since a measure is uniquely determined by its Laplace-Stieltjes transform, this implies that

$$((E_1(\lambda) - E_2(\lambda))g, h)_H = 0, \qquad (\lambda \in \mathbf{R}^1;\ g \in \mathfrak{D}(A),\ h \in H).$$

Thus, $E_1 = E_2$, which is absurd. ∎

We underscore that if A is a selfadjoint operator that is semibounded below, then a strong solution $u(t)$ of (6.50) satisfying the initial condition $u(0) = u_0 \in \mathfrak{D}(A)$ exists and can be expressed in terms of the semigroup generated by A:

$$u(t) = e^{-tA}u_0 = \int_c^\infty e^{-\lambda t}\, dE(\lambda)u_0 \qquad (t \in [0,\infty)).$$

(E is the resolution of the identity corresponding to A, and $c \in \mathbb{R}^1$ is such that $A \geq c1$.)

The uniqueness of strong solutions of the Cauchy problem for (6.50), and thereby the selfadjointness of A, can often be established with the help of the following consideration, which is analogous to Theorem 6.5.

THEOREM 6.13. *Let A be an Hermitian operator acting in H and semibounded below. Assume that there exist a sequence $(A_n)_1^\infty$ of operators acting in H with domains $\mathfrak{D}(A_n)$, a dense subset Φ of H, and a $b \in (0, \infty]$ with the following properties:*

1) *for any $T \in (0, b)$ and any $\varphi_0 \in \Phi$ there exists a sequence $(\varphi_{0,n})_1^\infty$ of vectors in H with $\varphi_{0,n} \to \varphi_0$ in H as $n \to \infty$ along with corresponding strong solutions of the Cauchy problems*

$$(d\varphi_n/dt)(t) - A_n\varphi_n(t) = 0 \qquad (t \in [0, T]), \tag{6.53}$$
$$\varphi_n(T) = \varphi_{0,n}, \qquad \varphi_n(t) \in \mathfrak{D}(A);$$

2) *for any $T \in (0, b)$ and any $\varphi_0 \in \Phi$*

$$\int_0^T (u(t), (A_n - A)\varphi_n(t))_H \, dt \underset{n\to\infty}{\to} 0, \tag{6.54}$$

where $u(t)$ is an arbitrary strong solution of (6.50) *with $u(0) = 0$.*

Then A is essentially selfadjoint.

PROOF. Let us show that $u(t) = 0$ ($t \in (0, b)$). Suppose that $T \in (0, b)$ and $\varphi_0 \in \Phi$. We construct an approximating sequence $(\varphi_{0,n})_1^\infty$ and corresponding solutions $\varphi_n(t)$. Integration by parts gives us

$$\int_0^T ((u'(t), \varphi_n(t))_H + (u(t), \varphi_n'(t))_H) \, dt = (u(T), \varphi_{0,n})_H, \qquad (n = 1, 2, \ldots). \tag{6.55}$$

Considering (6.50) and the inclusion $\varphi_n(t) \in \mathfrak{D}(A)$, we find that

$$(u'(t), \varphi_n(t))_H = -(A^*u(t), \varphi_n(t))_H = -(u(t), A\varphi_n(t))_H.$$

From this and (6.53) it follows that the left-hand side of (6.55) is equal to the integral (6.54). Passing in (6.55) to the limit as $n \to \infty$, we conclude that $(u(T), \varphi_0)_H = 0$ ($\varphi_0 \in \Phi$). Accordingly, $u(T) = 0$.

Let $c = b/2$. The function $[0, \infty) \ni t \mapsto u_1(t) = u(t+c)$ is a strong solution of (6.50) such that $u_1(0) = 0$. Repeating our argument, we get that $u_1(t) = 0$ ($t \in (0, b)$).

Consider next the function $[0, \infty) \ni t \mapsto u_2(t) = u_1(t + c) = u(t + 2c)$ and repeat the argument. The final result is that $u(t) = 0$ ($t \in [0, \infty)$). But then A is essentially selfadjoint by Theorem 6.12. ■

6.6. ***Selfadjointness of perturbations of an operator by a potential.*** Consider the special situation in the constructions of §6.5 when $H = L_2(R, d\mu(x))$, where R is some space of points x with a σ-finite measure μ defined on a σ-algebra $\mathfrak{R}$ of subsets of R. Below in this part, L_p denotes the space $L_p(R, d\mu(x))$ $(p \in [1, \infty])$.

THEOREM 6.14. *Let A_0 be a nonnegative selfadjoint operator acting in L_2 such that $D = \mathfrak{D}(A_0) \cap L_p$ is a core and the semigroup e^{-tA_0} is a contractive semigroup in L_p (i.e., $\|e^{-tA_0}\varphi\|_{L_p} \leq \|\varphi\|_{L_p}$ $(\varphi \in L_2 \cap L_p,\ t \in [0, \infty))$ for some fixed $p \in (2, \infty]$. Assume that the real-valued function $q \in L_r$, where $r = 2p(p-2)^{-1}$, is semibounded below, i.e., $\inf_{x \in R} q(x) > -\infty$. Then the operator*

$$D \ni f \mapsto (Af)(x) = (A_0 f)(x) + q(x) f(x) \tag{6.56}$$

acting in L_2 is essentially selfadjoint.

The proof of the theorem is broken up into four steps.

I. Assume that $p = \infty$ and the potential q is nonnegative. Then $r = 2$, and thus the operator (6.56) with $q \in L_2$ is densely defined on $D = \mathfrak{D}(A_0) \cap L_\infty$, and of course nonnegative. We verify the conditions of Theorem 6.13 for it, taking A_n to be the closure of the operator defined by (6.56) with $q(x)$ replaced by $q_n(x) = \min(q(x), n)$. Here $b = \infty$, $\Phi = D$, and the sequence $(\varphi_{0,n})_1^\infty$ approximating the vector $\varphi_0 \in D$ coincides with it: $\varphi_{0,n} = \varphi_0$. Each operator A_n is obviously selfadjoint and nonnegative; therefore, the solution of the Cauchy problem (6.53) exists and has the form

$$\varphi_n(t) = e^{(t-T)A_n}\varphi_0 \qquad (t \in [0, T],\ n = 1, 2, \ldots). \tag{6.57}$$

Since the operator of multiplication by the potential q_n is bounded,

$$\mathfrak{D}(A_0) = \mathfrak{D}(A_n) \qquad (n = 1, 2, \ldots). \tag{6.58}$$

It follows from (6.57) and (6.58) that

$$\varphi_n(t) \in \mathfrak{D}(A_0) \qquad (t \in [0, T],\ n = 1, 2, \ldots). \tag{6.59}$$

Now we use the estimate

$$\|\varphi_n(t)\|_{L_\infty} \leq \|\varphi_0\|_{L_\infty}, \qquad (t \in [0, T],\ n = 1, 2, \ldots), \tag{6.60}$$

which we shall prove in II. This estimate and (6.59) imply the inclusion $\varphi_n(t) \in \mathfrak{D}(A_0) \cap L_\infty = D$ $(t \in [0, T];\ n = 1, 2, \ldots)$ required for application of Theorem 6.13.

It remains to verify condition 2) in Theorem 6.13. Due to the continuity of the vector-valued function $[0, T] \ni t \mapsto u(t) \in L_2$, we have an estimate $\|u(t)\|_{L_2} \leq c$ $(t \in [0, T])$. From this and (6.60) we get

$$\begin{aligned} \left| \int_0^T (u(t), (A_n - A)\varphi_n(t))_{L_2}\, dt \right| &\leq \int_0^T \|u(t)\|_{L_2} \|q_n - q\|_{L_2} \|\varphi_n(t)\|_{L_\infty}\, dt \\ &\leq Tc \|\varphi_0\|_{L_\infty} \|q_n - q\|_{L_2} \underset{n \to \infty}{\longrightarrow} 0. \end{aligned} \tag{6.61}$$

Thus, all the conditions in Theorem 6.13 are verified, and the essential selfadjointness of A is thereby established in this case (after the proof of (6.60)).

II. Let us prove (6.60). By (6.57), it is a consequence of the more general inequality

$$\|e^{-sA_n}f\|_{L_\infty} \leq \|f\|_{L_\infty} \qquad (f \in L_2 \cap L_\infty;\ s \in [0,\infty);\ n = 1,2,\ldots), \quad (6.62)$$

which will now be established. Fix n. Since $A_n = A_0 + C_n$, where A_0, C_n, and A_n are nonnegative selfadjoint operators acting in L_2 and C_n is bounded, the multiplicative formula can be used (see Reed and Simon [2], Theorem VIII.31): in the sense of norm convergence in L_2,

$$e^{-sA_n}f + \lim_{m\to\infty}(e^{-(s/m)A_0}e^{-(s/m)C_n})^m f$$
$$(f \in L_2 \cap L_\infty;\ s \in [0,\infty)). \quad (6.63)$$

The operator $e^{-(s/m)C_n}$, as the operator of multiplication by the function $e^{-(s/m)q_n(\cdot)}$, has norms at most 1 on the spaces L_2 and L_∞. By the conditions of the theorem, the operator $e^{-(s/m)A_0}$ has the same property. Therefore,

$$e^{-(s/m)A_0}e^{-(s/m)C_n}f \in L_2 \cap L_\infty$$

and

$$\|e^{-(s/m)A_0}e^{-(s/m)C_n}f\|_{L_\infty} \leq \|f\|_{L_\infty}.$$

Integrating this estimate, we get

$$\|g_m\|_{L_\infty} \leq \|f\|_{L_\infty} \qquad (m = 1,2,\ldots), \quad (6.64)$$

where g_m denotes the vector after the lim sign in (6.63). We have that $g_m \to e^{-sA_n}f$ in L_2 as $m \to \infty$; therefore, there is a subsequence such that $g_{m_k}(x) \to (e^{-sA_n}f)(x)$ as $k \to \infty$ for almost all $x \in R$. From this and (6.64) we conclude that (6.62) holds.

III. Assume that $p \in (2,\infty)$ and $q(x) \geq 0$ $(x \in R)$. The operator (6.56) is now densely defined on $D = \mathfrak{D}(A_0) \cap L_p$, since $\|qf\|_{L_2} \leq \|q\|_{L_r}\|f\|_{L_p}$, $(f \in D)$ by Hölder's inequality, and it is clearly nonnegative. Subsequently we repeat the arguments in I, using instead of (6.60) the relation (it will be established below)

$$\|\varphi_n(t)\|_{L_p} \leq \|\varphi_0\|_{L_p} \qquad (t \in [0,T];\ n = 1,2,\ldots). \quad (6.65)$$

In the verification of condition 2) in Theorem 6.13 an estimate obtained with the help of Hölder's inequality and (6.65) is used instead of (6.61):

$$\begin{aligned}
&\left|\int_0^T (u(t),(A_n - A)\varphi_n(t))_{L_2}\,dt\right| \\
&\qquad \leq Tc\left(\int_R |(q_n(x) - q(x))\varphi_n(x,t)|^2\,d\mu(x)\right)^{1/2} \\
&\qquad \leq Tc\|q - q_n\|_{L_r}\|\varphi_n(t)\|_{L_p} \\
&\qquad \leq Tc\|\varphi_0\|_{L_p}\|q_n - q\|_{L_r} \underset{n\to\infty}{\to} 0. \qquad (6.66)
\end{aligned}$$

Let us prove (6.65). As before, this inequality follows from a more general estimate:

$$\|e^{-sA_n}f\|_{L_p} \leq \|f\|_{L_p} \qquad (f \in L_2 \cap L_p;\ s \in [0,\infty);\ n = 1,2,\ldots). \qquad (6.67)$$

The proof of (6.67) goes just as in II, except that L_∞ must be replaced by L_p. Instead of (6.64) we get that $\|g_m\|_{L_p} \leq \|f\|_{L_p}$ $(m = 1,2,\ldots)$. As in II, $g_{m_k}(x) \mapsto (e^{-sA_n}f)(x)$ as $k \to \infty$ for almost all $x \in R$. Consequently, $|g_{m_k}(x)|^p \to |(e^{-sA_n}f)(x)|^p$ as $k \to \infty$ for the same x. Applying Fatou's lemma to the last sequence, we get that

$$\|e^{-sA_n}f\|_{L_p}^p \leq \lim_{k\to\infty} \|g_{m_k}\|_{L_p}^p \leq \|f\|_{L_p}^p.$$

The estimate (6.67) is established.

IV. We show how to handle a potential q that is semibounded below. Let

$$q_1(x) = \min(q(x),0) \qquad q_2(x) = \max(q(x),0).$$

Then $q(x) = q_1(x) + q_2(x)$ $(x \in R)$, where $q_1(x)$ is a bounded function and $q_2(x)$ is nonnegative. The fact that q is in some L_α $(\alpha \in [1,\infty])$ implies that $q_1, q_2 \in L_\alpha$. Suppose, for example, that $p = \infty$. Then $q_2 \in L_2$, and the operator $D \ni f \mapsto A_2 f = (A_0 f)(x) + q_2(x)f(x)$ is essentially selfadjoint. But then the operator (6.56) is also essentially selfadjoint, since it differs from A_2 by a bounded operator. ■

REMARK 1. It is useful to keep the following circumstance in view when verifying that a semigroup is contractive. Let A_0 be a nonnegative selfadjoint operator acting in L_2. If the semigroup e^{-tA_0} is contractive in some space L_p $(p \in [1,\infty])$, i.e.,

$$\|e^{-tA_0}f\|_{L_p} \leq \|f\|_{L_p} \qquad (f \in L_2 \cap L_p;\ t \in [0,\infty)),$$

then it is also contractive in each L_α with $\alpha \in [p',p]$ and $p^{-1} + (p')^{-1} = 1$ (if $p \geq 2$; or $\alpha \in [p,p']$ if $p < 2$).

Indeed, let us first prove contractivity in $L_{p'}$. For definiteness suppose that $p > 2$ and $g \in L_2 \cap L_{p'}$. Then

$$|(e^{-tA_0}g, f)_{L_2}| = |(g, e^{-tA_0}f)_{L_2}| \leq \|g\|_{L_{p'}} \|e^{-tA_0}f\|_{L_p} \leq \|g\|_{L_{p'}} \|f\|_{L_p}.$$

It follows from this and the arbitrariness of $f \in L_2 \cap L_p$ that

$$\|e^{-tA_0}g\|_{L_{p'}} \leq \|g\|_{L_{p'}}T.$$

Thus,

$$\|e^{-tA_0}f\|_{L_\alpha} \leq \|f\|_{L_\alpha} \qquad (f \in L_2 \cap L_\alpha;\ t \in [0,\infty))$$

for $\alpha = p$ and $\alpha = p' < p$. By the Riesz-Thorin theorem (see Reed and Simon [2], Theorem IX.17), the same estimate holds for each $\alpha \in [p',p]$, i.e., the semigroup e^{-tA_0} is contractive in each L_α with $\alpha \in [p',p]$. ■

Consider the situation when μ is a probability measure: $\mu(R) = 1$. Now $L_\alpha \supseteq L_\beta$ and $\|\cdot\|_{L_\alpha} \leq \|\cdot\|_{L_\beta}$ for $1 \leq \alpha \leq \beta \leq \infty$. In this case Theorem 6.14 generalizes to potentials q such that instead of them the whole operator

A is required to be semibounded below (and a certain integrability condition is imposed on q). Thus, we have the following theorem.

THEOREM 6.15. *Suppose that A_0 is a nonnegative selfadjoint operator acting in the space L_2 constructed from a probability measure, and suppose that $D = \mathfrak{D}(A_0) \cap L_p$ is a core and the semigroup e^{-tA_0} is contractive in $L_{p+\varepsilon}$ for some fixed $p \in (2, \infty)$ and $\varepsilon > 0$. Assume that the real-valued function $q(\cdot)$ is in L_r with $r = 2p(p-2)^{-1}$, and $e^{-\delta q(\cdot)} \in L_1$ for some $\delta > 0$. If q is such that the operator* (6.56) *is semibounded below, then it is essentially selfadjoint.*

PROOF. The argument goes like the proof of Theorem 6.14, with the change that here $b > 0$ is finite, and q_n is defined by requiring that $q_n(x) = q(x)$ for $x \in R$ such that $|q(x)| \leq n$ and $q_n(x) = n$ otherwise. Let $h > 1$ be defined by the equality $h^{-1} + p_1^{-1} = p^{-1}$, where $p_1 = p + \varepsilon$. A little later we shall establish the following estimate for the solution (6.57) of equation (6.53):

$$\|\varphi_n(t)\|_{L_p} \leq c\|\varphi_0\|_{L_{p_1}}$$
$$(\varphi_0 \in D;\ t \in [0, T];\ T \in (0, \delta h^{-1}),\ n = 1, 2, \ldots). \quad (6.68)$$

This estimate is used to verify the conditions in Theorem 6.12 just as was done in proving Theorem 6.14. Here $b = \delta h^{-1}$.

Let us proceed to the derivation of (6.68). Note at once that, by Remark 1, the semigroup e^{-tA_0} is contractive in each L_α with $\alpha \in [p_1', p_1]$. The estimate (6.68) has the form (see (6.57))

$$\|e^{-sA_n}\varphi_0\|_{L_p} \leq c\|\varphi_0\|_{L_{p_1}} \qquad (\varphi_0 \in D;\ s \in [0, \delta h^{-1}];\ n = 1, 2, \ldots) \quad (6.69)$$

Let

$$A_{n,m} = (e^{-(s/m)q_n(\cdot)}e^{-(s/m)A_0})^m,$$

where $s \in [0, \delta h^{-1}]$ and $m = 1, 2, \ldots$; n is fixed. This operator is regarded as acting from L_{p_1} to L_p. It can be understood as the superposition of the following operators:

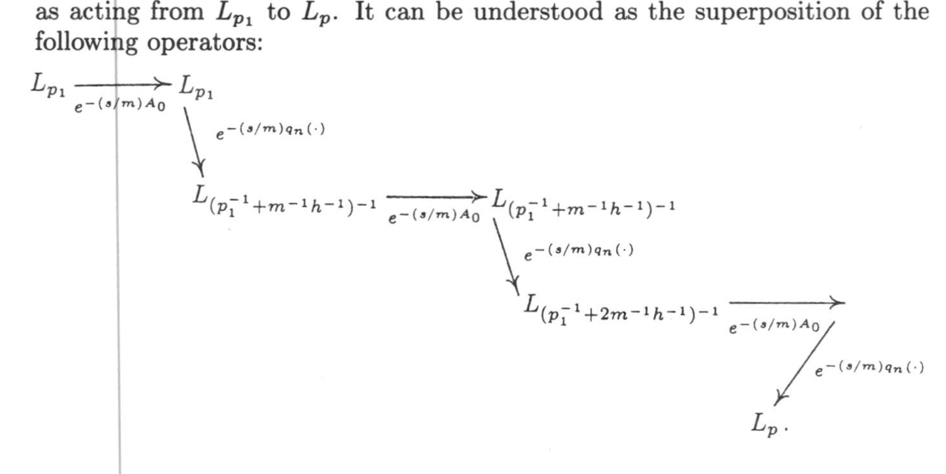

Since $p \in [p_1', p_1]$, the semigroup e^{-sA_0} is contractive in each space L_α with $\alpha \in [p, p_1]$. In particular, it is contractive for $\alpha = (p_1^{-1} + jm^{-1}h^{-1})^{-1}$ ($j = 0, 1, \dots, m$). Further, it follows from Hölder's inequality that each mapping

$$L_{(p_1^{-1}+jm^{-1}h^{-1})^{-1}} \to L_{(p_1^{-1}+(j+1)m^{-1}h^{-1})^{-1}} \qquad (j = 0, 1, \dots, m-1)$$

has norm at most $\|e^{-(s/m)q_n(\cdot)}\|_{L_{mh}}$. Therefore,

$$\begin{aligned}\|A_{n,m}\varphi_0\|_{L_p} &\le \|e^{-(s/m)q_n(\cdot)}\|_{L_{mh}}^m \|\varphi_0\|_{L_{p_1}} \\ &= \left(\int_R e^{-(s/m)q_n(x)hm}\, d\mu(x)\right)^{1/h} \|\varphi_0\|_{L_{p_1}} \\ &= \|e^{-sq_n(\cdot)}\|_{L_h} \|\varphi_0\|_{L_{p_1}}. \end{aligned} \tag{6.70}$$

We estimate $\|e^{-sq_n(x)}\|_{L_h}$ for $s \in [0, \delta h^{-1}]$. Let

$$q_{n,+}(x) = \max(q_n(x), 0), \qquad q_{n,-}(x) = -\min(q_n(x), 0),$$
$$q_-(x) = -\min(q(x), 0) \qquad (x \in R) \text{ and } sh = \sigma \in [0, \delta].$$

Then

$$\begin{aligned}\|e^{-sq_n(\cdot)}\|_{L_h}^h &= \int_R e^{-\sigma q_n(x)}\, d\mu(x) = \int_R e^{-\sigma q_{n,+}(x)} e^{\sigma q_{n,-}(x)}\, d\mu(x) \\ &\le \int_R e^{\delta q_{n,-}(x)}\, d\mu(x) \le \int_R e^{\delta q_-(x)}\, d\mu(x) \\ &\le \int_R (1 + e^{-\delta q(x)})\, d\mu(x) = 1 + \|e^{-\delta q(\cdot)}\|_{L_1}. \end{aligned}$$

From (6.70) and this estimate we get

$$\|A_{n,m}\varphi_0\|_{L_p} \le c\|\varphi_0\|_{L_p}, \qquad (c = (1 + \|e^{-\delta q(\cdot)}\|_{L_1})^{1/h}; m = 1, 2, \dots). \tag{6.71}$$

According to the multiplicative formula just used, $A_{n,m}\varphi_0 \to e^{-sA_n}\varphi_0$ in the sense of norm convergence in L_2 as $m \to \infty$. Therefore, there is a subsequence such that $(A_{n,m_k}\varphi_0)(x) \to (e^{-sA_n}\varphi_0)(x)$ as $k \to \infty$ for almost all $x \in R$. Consequently,

$$|(A_{n,m_k}\varphi_0)(x)|^p \underset{k\to\infty}{\longrightarrow} |(e^{-sA_n}\varphi_0)(x)|^p$$

for the same x. Applying Fatou's lemma to the last sequence and using (6.71), we conclude that (6.68) holds even for $s \in [0, \delta^{-1}h]$. ■

In connection with checking the conditions in the theorems of this subsection we mention the following general fact (see Reed and Simon [2], Theorem X.55) enabling us in a number of cases to establish the required contractivity of the semigroups in a simple way. Let A_0 be a nonnegative selfadjoint operator acting in a space L_2 constructed from a probability measure. Assume that the semigroup e^{-tA_0} preserves positivity (i.e., if $f \in L_2$ and $f(x) \ge 0$ almost everywhere, then $(e^{-tA_0}f)(x) \ge 0$ almost everywhere for $t \in [0, \infty)$), and that $e^{-tA_0}1 = 1$ ($t \in [0, \infty)$). Then this semigroup is contractive in each L_p ($p \in [1, \infty]$).

CHAPTER 3

Some Classes of Operators Acting in Spaces of Functions of Infinitely Many Variables

§0. Introduction

Some of the simplest and most important classes of commuting operators are formed by operators of differentiation with respect to different variables or, more generally, differential operators acting with respect to different variables. In essence, this chapter is devoted to the study of such operators in the case of infinitely many variables, functions of them, and perturbations of these functions by potentials. Here we do not try to consider operators which are important from a physical point of view: this would lead us too far afield, since the mathematical model constructions presented below are already sufficiently laborious. Examples of physical interest will be covered in part in the book of the author and Yu. G. Kondrat′ev mentioned earlier, which is being prepared for publication.

It is convenient to present this circle of questions on a more abstract level: instead of the space $L_2(\mathbb{R}^\infty)$ with respect to a product measure, we consider an infinite tensor product $\mathcal{H}_e = \bigotimes_{n=1;e}^{\infty} H_n$, and instead of differential operators acting with respect to one-dimensional variables we consider operators A_n acting in H_n. To obtain a commuting family of operators acting in $\mathcal{H}_e$ we must take the tensor products of the operators A_n by the identity operators in the spaces H_m with $m \neq n$. The countable family of commuting selfadjoint operators $\mathcal{A}_n$ constructed and studied in §1 (in particular, from the point of view of expansions in joint generalized eigenvectors) is of this type (if the A_n are selfadjoint). There we construct and study functions $\mathcal{L} = L(\mathcal{A}_1, \mathcal{A}_2, \ldots)$ of them defined from scalar functions $L(\lambda_1, \lambda_2, \ldots)$ by means of spectral integrals.

Of course, the definition of a function of a family of commuting selfadjoint operators by a spectral integral is simple and unambiguous, but is totally ineffective, as a rule, from the point of view of computing the action of the operator obtained, because it requires knowledge of the resolutions of the identity for the original operators. In essence, §2 can be perceived as a description of a technique for computing the action of a function $\mathcal{L}$ of operators in the simplest case of "separation of infinitely many variables", when $L(\lambda_1, \lambda_2, \dots) = \lambda_1 + \lambda_2 + \cdots$. Despite the simplicity of the question, there are subtleties and beautiful constructions involved here. Thus, on a formal level it is easy to understand that the resolution of the identity for $\mathcal{L}$ is equal to the infinite convolution of the resolutions of the identity for the operators A_n; we must determine when such a convolution exists, what its properties are, and so on. On the other hand, on the same formal level it is completely obvious that

$$\mathcal{L} = \mathcal{A}_1 + \mathcal{A}_2 + \cdots; \tag{0.1}$$

however, there are problems in how to describe the domain of this infinite sum, whether the operator obtained by closing the sum with this domain is selfadjoint, etc.

In §3 we first of all illustrate the constructions in §§1 and 2 in the case of differential operators A_n acting in spaces H_n equal to $L_2(\mathbb{R}^1)$ with respect to the probability measure generated by a smooth positive weight $p_n(x_n)$ and Lebesgue measure. (The measure has to be a probability measure in order to interpret $\mathcal{H}_e = \bigotimes_{n=1;e}^{\infty} H_n$ as $L_2(\mathbb{R}^\infty)$; the identity stabilization $e = (1, 1, \dots,)$ is used here.) It is most natural to start by considering the second-derivative operators $-d^2u/dx_n^2$ as A_n; however, the necessity of passing to $L_2(\mathbb{R}^1)$ with respect to a probability measure and still preserving the selfadjointness of the operator leads to the necessity of "sandwiching" with the weight, i.e., of considering the operator generated by the expression

$$-p_n^{-1/2}(x_n)\frac{d^2}{dx_n^2}(p_n^{1/2}(x_n)u(x_n)).$$

Then (0.1) gives us

$$\mathcal{L} = -p_1^{-1/2}(x_1)\frac{\partial^2}{\partial x_1^2}(p_1^{1/2}(x_1)\cdot) - p_2^{-1/2}(x_2)\frac{\partial^2}{\partial x_2^2}(p_2^{1/2}(x_2)\cdot) - \cdots. \tag{0.2}$$

If we use the "differential expression with infinitely many partial derivatives" (0.2) to try to define the operator, for example, on the cylindrical functions (i.e., on the functions that depend only on finitely many of the variables x_n and are thus constant with respect to the remaining variables), then the inclusion $\mathcal{L}1 \in L_2(\mathbb{R}^\infty)$ becomes necessary, and if we strengthen the last requirement somewhat, this becomes the requirement that the following series converge:

$$\sum_{n=1}^{\infty} \left\| \frac{\partial^2}{\partial x_n^2}\sqrt{p_n} \right\|_{L_2(\mathbb{R}^1, dx_n)} < \infty. \tag{0.3}$$

Condition (0.3) is a restriction on the weights p_n. For example, if $p_n(x_n) = \sqrt{\varepsilon_n/\pi}e^{-\varepsilon_n x_n^2}$ ($\varepsilon_n > 0$) is a Gaussian weight, then (0.3) is equivalent to the restriction $\sum_1^\infty \varepsilon_n < \infty$.

Restrictions of the type (0.3) can be removed by modifying the operators A_n by subtracting certain potentials from them; the remaining differential terms then will annihilate 1, and it is clear that $\mathcal{L}1 = 0 \in L_2(\mathsf{R}^\infty)$. This construction of the operators A_n is also used in §3 ("the case of annihilation"); however, it is then necessary to keep in mind that the spectral properties of the modified A_n are very different, as a rule, from those of the original operator. For instance, after a return to the space $L_2(\mathsf{R}^1)$ with respect to Lebesgue measure, the modified operator A_n in the case of a Gaussian weight becomes (to within coefficients) a harmonic oscillator with a discrete spectrum instead of a second derivative with the spectrum $[0, \infty)$. (Sums of type (0.2) of harmonic oscillators are of interest also from a physical point of view.)

The main part of §3 deals with the determination of selfadjointness for operators of the type (0.2) (with or without annihilation) that are perturbed by a potential $q(x_1, x_2, \ldots)$. This potential must be in $L_2(\mathsf{R}^\infty)$ in order that it be possible to define the operator (we emphasize that the domain of $\mathcal{L}$ contains the constants); but it is inappropriate to impose on it requirements like regularity, smoothness, etc., because R^∞ is, in essence, considered without a topology. We are thus led unwillingly to the case of differential operators with singular coefficients, whose spectral theory presents difficulties and surprises even for finitely many variables (for example, the invalidity of Povzner's theorem on essential selfadjointness of a semibounded Schrödinger operator with potential locally in L_2).

These difficulties are more pronounced in the case of infinitely many variables. The crux of the matter is that the abstract theorems giving selfadjointness of an operator ensure this selfadjointness on some domain that is difficult to describe, and then it is necessary to use a regularization procedure that is not always simple to drop down, for example, to smooth compactly supported functions on which the operator under investigation is defined in a natural way. In addition to the drop in the two directions of checking smoothness and compact support, there is also a drop in the third direction of checking the property of being cylindrical in the case of infinitely many variables.

Some of the results in §3 are presented without complete arguments when the proofs of selfadjointness are too laborious. In particular, only the scheme of the proof is given for the proper generalization of Povzner's theorem to the infinite-dimensional case and the case of a singular potential. (Both this theorem and the Titchmarsh-Sears tests are proved in §§3.4, 3.5 and 3.8 for finitely many variables with a singular potential.) Here it should be noted that the difficulties arising from this cycle of theorems in the infinite-dimensional case are connected (apparently in an essential way) with the fact that the wave equation with infinitely many spatial variables has been little studied.

The proof of selfadjointness in §3 is by means of the evolution approach in §6 of Chapter 2, with the use of both parabolic and hyperbolic equations.

A combination of these two classes of equations makes it possible to obtain sufficiently complete results; the "parabolic" approach can be used to weaken the local restrictions on the potential, and the "hyperbolic" approach can be used to weaken the restrictions on the behavior of the potential at infinity.

Of course, the results in §3 are in essence only illustrative: for example, we do not investigate selfadjointness of concrete Hamiltonians in field theory. However, the constructions presented show in any case how we can contend with another singular feature of such Hamiltonians: the appearance of infinitely many variables. As mentioned before, some results for such Hamiltonians will be presented in a book with Yu. G. Kondrat′ev. In the same place the results in §3 will be applied to the investigation of selfadjointness of differential operators with infinitely many variables defined in "invariant" form, i.e., defined on functions, for example, on a Hilbert space.

§1. Operators acting with respect to different variables and functions of them

In essence, the constructions in this section generalize the construction of differential operators with constant coefficients. The generalization touches on four points: an abstract Hilbert space is considered instead of the space L_2 for a single variable, the number of variables is infinite, an arbitrary selfadjoint operator is considered instead of differentiation with respect to the nth variable, and finally, more general functions of the differentiations are introduced in place of polynomials.

1.1. *Operators acting with respect to different variables.* Suppose that $(H_{0,n})_1^\infty$ is a sequence of Hilbert spaces, and A_n is a selfadjoint operator acting in $H_{0,n}$ with domain $\mathfrak{D}(A_n)$, with $\mathcal{B}(\mathsf{R}^1) \ni \delta \mapsto E_n(\delta)$ the corresponding resolution of the identity. We consider a stabilizing sequence $e = (e^{(n)})_1^\infty$ of unit vectors $e^{(n)} \in H_{0,n}$ and form the infinite tensor product $\mathcal{H}_{0,e,1} = \bigotimes_{n=1;e,1}^\infty H_{0,n}$. Each operator A_n induces an operator $\mathcal{A}_n'$ in $\mathcal{H}_{0,e,1}$ in the natural way. Namely, let

$$\mathcal{A}_n' = \underbrace{1 \otimes \cdots \otimes 1}_{n-1} \otimes A_n \otimes 1 \otimes 1 \otimes \cdots,$$

$$\mathfrak{D}(\mathcal{A}_n') = \text{a.}(H_{0,1} \otimes \cdots \otimes H_{0,n-1} \otimes \mathfrak{D}(A_n) \otimes H_{0,n+1} \otimes H_{0,n+2} \otimes \cdots),$$

$$\mathcal{A}_n'(f^{(1)} \otimes \cdots \otimes f^{(n-1)} \otimes f^{(n)} \otimes f^{(n+1)} \otimes f^{(n+2)} \otimes \cdots)$$
$$= f^{(1)} \otimes \cdots \otimes f^{(n-1)} \otimes (A_n f^{(n)}) \otimes f^{(n+1)} \otimes f^{(n+2)} \otimes \cdots$$
$$(f^{(j)} \in H_{0,j},\ j \neq n;\ f^{(n)} \in \mathfrak{D}(A_n);\ n = 1, 2, \ldots). \quad (1.1)$$

Since

$$\mathcal{A}_n' - z1 = \underbrace{1 \otimes \cdots \otimes 1}_{n-1} \otimes (A_n - z1) \otimes 1 \otimes 1 \otimes \cdots \qquad (z \in \mathbb{C}^1),$$

it follows that each of the operators $\mathcal{A}'_n$ is essentially selfadjoint. Denote by $\mathcal{A}_n$ the closure of $\mathcal{A}'_n$. It is easy to see that $\mathcal{A}_n$ has resolution of the identity

$$\mathcal{B}(\mathsf{R}^1) \ni \delta \mapsto \mathcal{E}_n(\delta) = \underbrace{1 \otimes \cdots \otimes 1}_{n-1} \otimes E_n(\delta) \otimes 1 \otimes 1 \otimes \cdots. \tag{1.2}$$

By (2.8) in Chapter 1, these resolutions of the identity commute.

Thus, we have constructed from $(A_n)_1^\infty$ (or from $(E_n)_1^\infty$) a countable family $(\mathcal{A}_n)_1^\infty$ of commuting selfadjoint operators acting in $\mathcal{H}_{0,e,1}$. We say that they act with respect to different variables. Let $\mathcal{C}_\sigma(\mathsf{R}^\infty) \ni \alpha \mapsto \mathcal{E}(\alpha)\colon \mathcal{H}_{0,e,1} \to \mathcal{H}_{0,e,1}$ be the joint resolution of the identity of this family constructed in §1 of Chapter 2. (Now $X = \{1, 2, \ldots\}$ and $\mathsf{R}^X = \mathsf{R}^\infty$; it is convenient to denote a point $\lambda(\cdot)$ of R^∞ by $\lambda = (\lambda_n)_1^\infty$ in what follows.)

This simple construction is analogous to passing from ordinary derivatives to partial derivatives. Indeed, suppose that we have finitely many spaces H_n ($n = 1, \ldots, p$) coinciding with $L_2(\mathsf{R}^1, dx_n)$, constructed with respect to Lebesgue measure. The operator A_n generated by the derivative $-i(d/dx_n)$ as the minimal operator acts in $L_2(\mathsf{R}^1, dx_n)$, i.e., A_n is equal to the closure of the operator $C_0^\infty(\mathsf{R}^1) \ni u(x_n) \mapsto -iu'(x_n)$ in $L_2(\mathsf{R}^1, dx_n)$. Then $\bigotimes_1^p H_n = L_2(\mathsf{R}^p, dx)$, and $\mathcal{A}_n$ is easily seen to be generated in a similar way by the derivative $-i(\partial/\partial x_n)$. Besides studying such a family $(\mathcal{A}_n)_1^\infty$, we shall construct operators from the "elementary" operators $\mathcal{A}_n$ in a way analogous to the way in which partial differential operators are constructed from $-i(\partial/\partial x_n)$. Expansion in joint generalized eigenvectors of the family $(\mathcal{A}_n)_1^\infty$ is analogous to such expansion for the family $(-i(\partial/\partial x_n))_1^p$, i.e., to the p-dimensional Fourier transformation.

Assume that we know the expansion in generalized eigenvectors for each operator A_n ($n = 1, 2, \ldots$): we have a chain of separable spaces

$$H_{-,n} \supseteq H_{0,n} \supseteq H_{+,n} \tag{1.3}$$

with quasinuclear imbedding $O_n\colon H_{+,n} \to H_{0,n}$, and we know the spectral measure $\mathcal{B}(\mathsf{R}^1) \ni \delta \mapsto \rho_n(\delta) = \operatorname{Tr}(O_n^+ E_n(\delta) O_n)$ for A_n along with the generalized projection operator $P_n(\lambda_n)$ ($\lambda_n \in \mathsf{R}^1$). From this data we construct an expansion for the whole family $(\mathcal{A}_n)_1^\infty$. It is assumed that $e^{(n)} \in H_{+,n}$ and $\|e^{(n)}\|_{H_{+,n}} = \|e^{(n)}\|_{H_{0,n}} = 1$ ($n = 1, 2, \ldots$). Without loss of generality we can suppose that

$$\sum_{n=1}^{\infty} (|O_n|^2 - 1) < \infty. \tag{1.4}$$

(To ensure (1.4), renorm each $H_{+,n}$ according to Lemma 2.1 in Chapter 1.)

Taking the tensor product of the chains (1.3), we get the chain (see Theorem 2.4 in Chapter 1)

$$\mathcal{H}_{-,e,1} = \bigotimes_{n=1;e,1}^{\infty} H_{-,n} \supseteq \underset{\substack{\| \\ \mathcal{H}_{0,e,1}}}{\bigotimes_{n=1;e,1}^{\infty} H_{0,n}} \supseteq \bigotimes_{n=1;e,1}^{\infty} H_{+,n} = \mathcal{H}_{+,e,1} \tag{1.5}$$

with quasinuclear (because of condition (1.4)) imbedding $\mathcal{H}_{+,e,1} \to \mathcal{H}_{0,e,1}$. As explained in §2.7 of Chapter 1, the equality $\mathcal{O} = \bigotimes_1^\infty O_n$ holds for the imbedding operator $\mathcal{O}\colon \mathcal{H}_{+,e,1} \to \mathcal{H}_{0,e,1}$, and the analogous equalities hold for $\mathcal{O}^+$ and $\mathbf{O} = \mathcal{O}^+\mathcal{O}$. The chain (1.5) plays the role of the chain (2.1) in Chapter 2 in the construction of an expansion for $(\mathcal{A}_n)_1^\infty$.

THEOREM 1.1. *The spectral measure $C_\sigma(\mathsf{R}^\infty) \ni \alpha \mapsto \rho(\alpha)$ and the generalized projection operator $\mathcal{P}(\lambda)$ of the family $(\mathcal{A}_n)_1^\infty$ are given by the formulas*

$$\rho = \bigotimes_{n=1}^{\infty} \rho_n, \quad \mathcal{P}(\lambda) = \bigotimes_{n=1}^{\infty} P_n(\lambda_n) \qquad (\lambda = (\lambda_n)_{n=1}^\infty). \tag{1.6}$$

The first product in (1.6) exists because $\rho_n(\mathsf{R}^1) = |O_n|^2$, and the product $\prod_1^\infty |O_n|^2$ converges due to (1.4). The existence of the second product in (1.6) follows from Theorem 2.6 in Chapter 1 and the remark after it.

PROOF. The spectral measure ρ is defined by the relation

$$\rho(\alpha) = \mathrm{Tr}\,(\mathcal{O}^+\mathcal{E}(\alpha)\mathcal{O}) \qquad (\alpha \in C_\sigma(\mathsf{R}^\infty)).$$

Let us compute its value in the case when $\alpha = \mathfrak{C}(1,\dots,p;\ \times_1^p \delta_n) = (\times_1^p \delta_n)\times \mathsf{R}^\infty$ $(\delta_n \in \mathcal{B}(\mathsf{R}^1))$. Introducing the unit vectors e_α for the space $\mathcal{H}_{+,e,1}$ (see (2.10) in Chapter 1), we get with the help of (1.2) and the computation (2.30) in Chapter 1 (see also the remark after Theorem 2.6 in Chapter 1) that

$$\begin{aligned}
\rho(\alpha) &= \rho\left(\left(\bigtimes_{n=1}^{p}\delta_n\right)\times\mathsf{R}^\infty\right) = \mathrm{Tr}\left(\mathcal{O}^+\mathcal{E}\left(\left(\bigtimes_{n=1}^{p}\delta_n\right)\times\mathsf{R}^\infty\right)\mathcal{O}\right)\\
&= \mathrm{Tr}\left(\mathcal{O}^+\mathcal{E}\left(\bigcap_{n=1}^{p}\mathfrak{C}(n;\delta_n)\right)\mathcal{O}\right) = \mathrm{Tr}\left(\mathcal{O}^+\left(\prod_{n=1}^{p}\mathcal{E}(\mathfrak{C}(n;\delta_n))\mathcal{O}\right)\right)\\
&= \mathrm{Tr}\left(\mathcal{O}^+\left(\prod_{n=1}^{p}\mathcal{E}_n(\delta_n)\right)\mathcal{O}\right)\\
&= \mathrm{Tr}\left(\mathcal{O}^+\left(\left(\bigotimes_{n=1}^{p}E_n(\delta_n)\right)\otimes 1\otimes 1\otimes\cdots\right)\mathcal{O}\right)\\
&= \mathrm{Tr}\left(\left(\bigotimes_{n=1}^{p}O_n^+E_n(\delta_n)O_n\right)\otimes\mathbf{O}_{p+1}\otimes\mathbf{O}_{p+2}\otimes\cdots\right)
\end{aligned}$$

$$= \left(\prod_{n=1}^{p} \operatorname{Tr}(O_n^+ E_n(\delta_n) O_n)\right) \left(\prod_{n=p+1}^{\infty} |O_n|^2\right)$$

$$= \left(\prod_{n=1}^{p} \rho_n(\delta_n)\right) \left(\prod_{n=p+1}^{\infty} \rho_n(\mathbf{R}^1)\right) = \left(\bigotimes_{k=1}^{\infty} \rho_k\right) \left(\left(\bigtimes_{n=1}^{p} \delta_n\right) \times \mathbf{R}^\infty\right)$$

$$= \left(\bigotimes_{k=1}^{\infty} \rho_k\right)(\alpha).$$

This equality gives us that the measures ρ and $\bigotimes_1^\infty \rho_k$ coincide on any $B \in \mathcal{C}_\sigma(\mathbf{R}^\infty)$, i.e., the first of the relations (1.6).

We prove the second relation in (1.6). By (2.13) in Chapter 2, the computations already carried out, and the first equality in (1.6),

$$\int_{(\bigtimes_1^p \delta_n) \times \mathbf{R}^\infty} \mathcal{P}(\lambda)\, d\rho(\lambda) = \mathcal{O}^+ \mathcal{E}\left(\left(\bigtimes_{n=1}^{p} \delta_n\right) \times \mathbf{R}^\infty\right) \mathcal{O}$$

$$= \left(\bigotimes_{n=1}^{p} O_n^+ E_n(\delta_n) O_n\right) \otimes \mathbf{O}_{p+1} \otimes \mathbf{O}_{p+2} \otimes \cdots$$

$$= \left(\bigotimes_{n=1}^{p} \int_{\delta_n} P_n(\lambda_n)\, d\rho_n(\lambda_n)\right) \otimes \left(\int_{\mathbf{R}^1} P_{p+1}(\lambda_{p+1})\, d\rho_{p+1}(\lambda_{p+1})\right)$$

$$\otimes \left(\int_{\mathbf{R}^1} P_{p+2}(\lambda_{p+2})\, d\rho_{p+2}(\lambda_{p+2})\right) \otimes \cdots$$

$$= \int_{(\bigtimes_1^p \delta_n) \times \mathbf{R}^\infty} \left(\bigotimes_{n=1}^{\infty} P_n(\lambda_n)\right) d\left(\bigotimes_{n=1}^{\infty} \rho_n\right)(\lambda)$$

$$= \int_{(\bigtimes_1^p \delta_n) \times \mathbf{R}^\infty} \left(\bigotimes_{n=1}^{\infty} P_n(\lambda_n)\right) d\rho(\lambda).$$

for any $\delta_n \in \mathcal{B}(\mathbf{R}^1)$. Since $(\bigtimes_1^p \delta_n) \times \mathbf{R}^\infty$ is arbitrary, this implies that $\mathcal{P}(\lambda)$ and $\bigtimes_1^\infty P_n(\lambda_n)$ coincide for ρ-almost all $\lambda \in \mathbf{R}^\infty$. ■

From the first equality in (1.6) and relation (1.15) in Chapter 2 we obtain for the joint spectrum $s((\mathcal{A}_n)_1^\infty) = \operatorname{supp} \mathcal{E} = \operatorname{supp} \rho$ of the family under consideration the formula

$$s((\mathcal{A}_n)_{n=1}^\infty) = \bigtimes_{n=1}^{\infty} s(A_n). \tag{1.7}$$

We shall not construct the expansion in individual generalized eigenvectors of the family $(\mathcal{A}_n)_1^\infty$ for the general case, though this is not hard to do. We mention only that these eigenvectors all lie in $\mathfrak{R}(\mathcal{P}(\lambda)) \subseteq \mathcal{H}_{-,e,1}$ and, since the c.l.s. of the vectors $u^{(1)} \otimes u^{(2)} \otimes \cdots$ ($u^{(n)} \in H_{+,n}$ and $u^{(n)} = e^{(u)}$ from some point on) coincides with $\mathcal{H}_{+,e,1}$, they are in the c.l.s. of the vectors

$$\mathcal{P}(\lambda)(u^{(1)} \otimes u^{(2)} \otimes \cdots) = (P_1(\lambda_1)u^{(1)}) \otimes (P_2(\lambda_2)u^{(2)}) \otimes \cdots. \tag{1.8}$$

The last vectors are infinite tensor products of the generalized eigenvectors of the operators A_n. (Note that, since $\bigotimes_1^\infty P_n(\lambda_n)$ is defined, the infinite tensor product of the vectors $P_n(\lambda_n)u^{(n)}$ exists as a vector in $\mathcal{H}_{-,e,1}$.)

We dwell briefly on the construction of an extension of the rigging (1.5) from such extensions for (1.3). Suppose that for each $n = 1, 2, \ldots$ we have a rigging extension

$$H_{-,n} \supseteq H_{0,n} \supseteq H_{+,n} \supseteq D_n, \tag{1.9}$$

that works for A_n. If the D_n are Hilbert spaces containing $e^{(n)}$, and $\|e^{(n)}\|_{D_n} = 1$, then

$$\mathcal{H}_{-,e,1} \supseteq \mathcal{H}_{0,e,1} \supseteq \mathcal{H}_{+,e,1} \supseteq \mathcal{D}, \tag{1.10}$$

where $\mathcal{D} = \bigotimes_{n=1;e,1}^\infty D_n$, can be taken as an extension of the rigging (1.5). It is easily seen to work for the family $(\mathcal{A}_n)_1^\infty$. If the D_n are nuclear spaces which are projective limits of Hilbert spaces (containing $e^{(n)}$ as a unit vector), then the role of $\mathcal{D}$ can be played by a weighted infinite tensor product of them (see Chapter 1, §4.3). The continuity of $\mathcal{A}_n \restriction \mathcal{D}$: $\mathcal{D} \to \mathcal{H}_{+,e,1}$ now follows easily from Theorem 4.4 in Chapter 1.

Let us dwell on the more special situation when each operator A_n has a strong cyclic vector coinciding with $e^{(n)}$. More precisely, assume that for each $n = 1, 2, \ldots$ we have a rigging extension (1.9) with nuclear D_n, $A_n^m e^{(n)} \in D_n$ $(m = 0, 1, \ldots)$, and the l.s. of these vectors is dense in $H_{+,n}$. (Such D_n can always be constructed; see Maurin [2], Chapter 17, §5.) Then *the family* $(\mathcal{A}_n)_1^\infty$ *of operators has the strong cyclic vector* $\Omega = e^{(1)} \otimes e^{(2)} \otimes \cdots$: $\mathcal{A}_1^{m_1} \cdots \mathcal{A}_p^{m_p}\Omega \in \mathcal{D}$ $(m_1, \ldots, m_p = 0, 1, \ldots;\ p = 1, 2, \ldots)$, *and the* l.s. *of these vectors is dense in* $\mathcal{H}_{+,e,1}$. (Here $\mathcal{D}$ is the weighted tensor product of the spaces D_n.) Indeed,

$$\begin{aligned}\mathcal{A}_1^{m_1} \cdots \mathcal{A}_p^{m_p}\Omega &= (A_1^{m_1}e^{(1)}) \otimes \cdots \otimes (A_p^{m_p}e^{(p)}) \otimes e^{(p+1)} \otimes e^{(p+2)} \otimes \cdots \\ &\in \mathrm{a.}(D_1 \otimes \cdots \otimes D_p \otimes e^{(p+1)} \otimes e^{(p+2)} \otimes \cdots) \subseteq \mathcal{D}.\end{aligned}$$

This implies at once that the l.s. of these vectors is dense in $\mathcal{H}_{+,e,1}$. ∎

According to Theorem 2.4 in Chapter 2,

$$\mathcal{C}_\sigma(\mathbb{R}^\infty) \ni \alpha \mapsto \sigma(\alpha) = (\mathcal{E}(\alpha)\Omega, \Omega)_{\mathcal{H}_{0,e,1}}$$

can now serve as the spectral measure of the family $(\mathcal{A}_n)_1^\infty$. It is easy to see that this measure coincides with $\bigotimes_1^\infty \sigma_n$, where $\mathcal{B}(\mathbb{R}^1) \ni \delta \mapsto \sigma_n(\delta) = (E_n(\delta)e^{(n)}, e^{(n)})_{H_{0,n}}$ is the corresponding spectral measure of A_n:

$$\sigma = \bigotimes_{n=1}^\infty \sigma_n. \tag{1.11}$$

(It suffices to verify (1.11) on the sets $(\times_1^p \delta_n) \times \mathbb{R}^\infty$ $(\delta_n \in \mathcal{B}(\mathbb{R}^1);\ p = 1, 2, \ldots)$, where it is obvious.)

By Theorem 5.8 in Chapter 2, the multiplicity of an eigenvalue is equal to 1 here: $N_\lambda = 1$ for ρ-almost all λ. Thus, the one-dimensional space $\mathfrak{R}(\mathcal{P}(\lambda))$ lies in the c.l.s. of the vectors (1.8), and each $\mathfrak{R}(P_n(\lambda_n))$ is also one-dimensional. From this it follows that

$$\varphi_1(\lambda) = \varphi_1^{(1)}(\lambda_1) \otimes \varphi_1^{(2)}(\lambda_2) \otimes \cdots \in \mathcal{H}_{-,e,1} \qquad (\lambda = (\lambda_n)_{n=1}^{\infty}) \tag{1.12}$$

for the corresponding individual generalized eigenvectors. ($\varphi_1(\lambda)$ is a joint eigenvector for the family $(\mathcal{A}_n)_1^\infty$, and $\varphi_1^{(n)}(\lambda_n)$ is the same for the operator A_n.)

The equality (1.12) implies a simple formula for the Fourier transformation associated with $(\mathcal{A}_n)_1^\infty$. Thus, formula (2.53) in Chapter 2, together with (1.7) and (1.12), gives us that

$$\begin{aligned} \mathsf{R}^\infty \supseteq s((\mathcal{A}_n)_{n=1}^\infty) = \bigtimes_{n=1}^{\infty} s(A_n) \ni \lambda = (\lambda_n)_{n=1}^\infty \mapsto \tilde{u}(\lambda) \\ = \left(u, \bigotimes_{n=1}^{\infty} \varphi_1^{(n)}(\lambda_n)\right)_{\mathcal{H}_{0,e,1}} \qquad (u \in \mathcal{H}_{+,e,1}). \end{aligned} \tag{1.13}$$

Parseval's equality (2.54) in Chapter 2 now has the form

$$(\mathcal{E}(\alpha)u, v)_{\mathcal{H}_{0,e,1}} = \int_\alpha \tilde{u}(\lambda)\overline{\tilde{v}(\lambda)}\, d\rho(\lambda) = \int_\alpha \tilde{u}(\lambda)\overline{\tilde{v}(\lambda)} \frac{d\rho(\lambda)}{d\sigma(\lambda)}\, d\sigma(\lambda)$$
$$(\alpha \in \mathcal{C}_\sigma(\mathsf{R}^\infty);\ u, v \in \mathcal{H}_{+,e,1}). \tag{1.14}$$

Of course, it can be extended also to vectors in $\mathcal{H}_{0,e,1}$ by continuity.

All the constructions in this subsection generalize easily to the case when each operator A_n is normal. As a result, we obtain a family $(\mathcal{A}_n)_1^\infty$ of commuting normal operators.

Observe that, according to Theorem 2.6 in Chapter 1, $\bigotimes_1^\infty P_n(\lambda_n)$ exists as the weak limit of the operators $P_1(\lambda_1) \otimes \cdots \otimes P_p(\lambda_p) \otimes \mathbf{O}_{p+1} \otimes \mathbf{O}_{p+2} \otimes \cdots$ as $p \to \infty$. Therefore, the vectors (1.8) and (1.12) can be approximated weakly by the corresponding vectors. However, it can be proved that the indicated limit exists also in the strong sense.

Indeed, it is necessary to verify the conditions formulated at the end of §2.10 in Chapter 1. The convergence of the product of traces is obvious: $\operatorname{Tr}(P_n(\lambda_n)) = 1$ for ρ_n-almost all $\lambda_n \in \mathsf{R}^1$. We show that for ρ-almost all $\lambda = (\lambda_n)_1^\infty \in \mathsf{R}^\infty$ there is a $q = 1, 2, \ldots$ such that the product $\prod_q^\infty (P_n(\lambda_n)e^{(n)}, e^{(n)})_{H_{0,n}}$ converges. Since

$$\operatorname{Tr}(\mathcal{P}(\lambda)) = \sum_{\alpha \in \mathrm{A}} (\mathcal{P}(\lambda)e_\alpha, e_\alpha)_{\mathcal{H}_{0,e,1}} = 1,$$

for ρ-almost all $\lambda \in \mathsf{R}^\infty$ there is a vector e_α (α depends on λ in general) such that

$$(\mathcal{P}(\lambda)e_\alpha, e_\alpha)_{\mathcal{H}_{0,e,1}} = \prod_{n=1}^{\infty} (P_n(\lambda_n)e_{\alpha_n}^{(n)}, e_{\alpha_n}^{(n)})_{H_{0,n}} \in (0, \infty).$$

It remains to let $q = \nu(\alpha) + 1$, because $e^{(n)}_{\alpha_n} = e^{(n)}$ for $n \geq q$. ■

1.2. *The Fourier-Wiener transformation.* We pass to consideration of a concrete situation which amounts to the transformation (1.13) turning out to be the infinite-dimensional Fourier-Wiener transformation. First of all, we establish the required facts in the case of one dimension.

Consider the space $H_0 = L_2(\mathsf{R}^1, \pi^{-1/2}e^{-t^2}\,dt)$, or $G^0(\mathsf{R}^1)$, in the notation of §4.2 in Chapter 1; as earlier, the Gaussian measure is denoted by $\pi^{-1/2}e^{-t^2}\,dt = \gamma(t)\,dt = dg(t)$. In this space let us define the operator A equal to the closure of the operator

$$C_0^\infty(\mathsf{R}^1) \ni u(t) \mapsto -i\gamma^{-1/2}(t)(\gamma^{1/2}(t)u)'(t) = -i(u'(t) - tu(t)). \tag{1.15}$$

The mapping

$$H_0 \ni u(t) \mapsto \gamma^{1/2}(t)u(t) = U(t) \in L_2(\mathsf{R}^1) \tag{1.16}$$

is an isometry between H_0 and $L_2(\mathsf{R}^1)$ under which $-iU'(t)$ is the image of the differential expression in (1.15). Since $C_0^\infty(\mathsf{R}^1)$ is preserved under this mapping, the closure B in $L_2(\mathsf{R}^1)$ of the operator $C_0^\infty(\mathsf{R}^1) \ni U(t) \to -iU'(t)$ is the image of A. It is well known that B is selfadjoint; therefore, so is A.

We remark that *the $C^\infty(\mathsf{R}^1)$-functions $u(t)$ that are bounded along with their derivatives are in $\mathfrak{D}(A)$, and the action of A on them is given by the same formula* (1.15). To prove this, consider a sequence $(\chi_n(t))_1^\infty$ of functions $\chi_n \in C_0^\infty(\mathsf{R}^1)$ that are equal to 1 for $|t| \leq n$ and to zero for $|t| \geq n+1$, while on the intervals $(-n-1, -n)$ and $(n, n+1)$ they take values in $[0,1]$ and are obtained by translating the appropriate pieces of the graph of $\chi_1(t)$. Then $u_n(t) = u(t)\chi_n(t) \in C_0^\infty(\mathsf{R}^1)$, and it is easy to compute that $u_n \to u$ as $n \to \infty$, and Au_n has a limit equal to the action of the differential expression (1.15) on u. This proves the assertion. ■

It is known (see, for example, Berezanskiĭ [4], Chapter 6, §5.3) that B is a Carleman operator, and its spectral measure ρ_B and spectral kernel $P_B(t, s; \lambda)$ have the representation

$$P_B(t, s; \lambda)\,d\rho_B(\lambda) = (1/2\pi)e^{i\lambda(t-s)}\,d\lambda \qquad (t, s \in \mathsf{R}^1;\ \lambda \in s(B) = \mathsf{R}^1). \tag{1.17}$$

Returning to A with the help of the mapping (1.16), we get from (1.17)

$$\begin{aligned} P(t, s; \lambda)\,d\rho(\lambda) &= (\gamma(t)\gamma(s))^{-1/2}P_B(t, s; \lambda)\,d\rho_B(\lambda) \\ &= (1/2)\pi^{-1/2}\exp(i\lambda(t-s) + (1/2)(t^2 + s^2))\,d\lambda \\ &\qquad (t, s \in \mathsf{R}^1;\ \lambda \in s(A) = S(B) = \mathsf{R}^1) \end{aligned} \tag{1.18}$$

for its spectral measure ρ and spectral kernel $P(t, s; \lambda)$. (The operator $P(\lambda)$ for A can be recovered from (1.18) according to (5.13) in Chapter 2, where the integration is taken with respect to the Gaussian measure $\gamma(s)ds$.)

The operator A has the cyclic vector $\Omega = 1 \in H_0$. Indeed, by (1.18), the resolution E of the identity for A satisfies

$$\begin{aligned}(E(\delta)1,1)_{H_0} &= \int_\delta \left(\int_{\mathbf{R}^1}\int_{\mathbf{R}^1} P(t,s;\lambda)\gamma(s)\gamma(t)\,ds\,dt\right) d\rho(\lambda) \\ &= \frac{1}{\sqrt{\pi}}\int_\delta e^{-\lambda^2}\,d\lambda \qquad (\delta\in\mathcal{B}(\mathbf{R}^1)). \end{aligned} \tag{1.19}$$

(We have used the fact that $\int_{\mathbf{R}^1} e^{i\lambda x - x^2/2}dx = \sqrt{2\pi}e^{-\lambda^2/2}$ $(\lambda\in\mathbf{R}^1)$.) Therefore, $\int_{\mathbf{R}^1}\lambda^{2m}d(E(\lambda)1,1)_{H_0} < \infty$ for any $m = 0,1,\ldots$, i.e., $\Omega\in\bigcap_0^\infty \mathfrak{D}(A^m)$. Further, by induction on $m = 0,1,\ldots$, it follows easily from (1.15) that

$$\text{l.s.}(\Omega, A\Omega,\ldots,A^m\Omega) = \text{l.s.}(1,t,\ldots,t^m).$$

But the collection of all polynomials is well known to be dense in H_0; therefore,

$$\text{c.l.s.}(\{A^m\Omega \mid m = 0,1,\ldots\}) = H_0,$$

which proves the assertion. According to Theorem 2.4 in Chapter 2 and (1.19), the spectral measure in our case is given by the formula

$$\sigma(\delta) = \frac{1}{\sqrt{\pi}}\int_\delta e^{-\lambda^2}\,d\lambda \qquad (\delta\in\mathcal{B}(\mathbf{R}^1)). \tag{1.20}$$

Of course, A is a Carleman operator, but here it is useful for us to construct a rigging of H_0 not of the type (5.10) in Chapter 2, but of the form (2.1) in that chapter, with quasinuclear imbedding $H_+\to H_0$. The space $G^2(\mathbf{R}^1)$ can be taken as H_+, and then Theorem 4.1 in Chapter 1 ensures that $H_+\to H_0$ is quasinuclear. Note that $\|\Omega\|_{H_0} = \|\Omega\|_{H_+} = 1$. We renorm H_+ with the help of Lemma 2.1 in Chapter 1, ensuring the inequality $|O^{\#_n}|^2 - 1 < n^{-2}$ ($n = 1,2,\ldots$ is fixed) for the imbedding operator $O^{\#_n}\colon H_+^{\#_n}\to H_0$. (It is convenient to mark the dependence of the renorming on n; see also the first part of §4.4 in Chapter 1.)

Accordingly, we have a rigging

$$H_-^{\#_n} \supseteq L_2\left(\mathbf{R}^1, \pi^{-1/2}e^{-t^2}\,dt\right) \supseteq (G^2(\mathbf{R}^1))^{\#_n}, \tag{1.21}$$

where $H_-^{\#_n}$ denotes the corresponding negative space. This rigging is not hard to extend in connection with A. For example, let

$$D = \operatorname*{pr\,lim}_{\tau\in\{0,1,\ldots\}} G^\tau(\mathbf{R}^1) = \mathcal{G}(\mathbf{R}^1).$$

It follows easily that Theorem 4.2 in Chapter 1 that $A\restriction D$ carries D into D continuously: use the mapping (1.16) and the well-known fact that the operator $\mathcal{S}(\mathbf{R}^1)\ni\Phi(t)\mapsto\Phi'(t)\in\mathcal{S}(\mathbf{R}^1)$ is continuous. Therefore, all the more so, $A\restriction D$ carries D continuously into $G^2(\mathbf{R}^1)$, i.e., we have actually constructed for A an extension of the rigging (1.21).

The cyclic vector Ω for A is strong. This is most conveniently checked with the help of the mapping (1.16) and Theorem 4.2 of Chapter 1. Thus, all the vectors $A^m\Omega$ $(m = 0, 1, \ldots)$ are in D, since this is equivalent to the assertion that the function $\pi^{-1/4}e^{-t^2/2}$ and all its derivatives are in $S(\mathbb{R}^1)$. Further, since $\mathrm{l.s.}(\Omega, A\Omega, \ldots, A^m\Omega) = \mathrm{l.s.}(1, t, \ldots, t^m)$, application of (1.16) and the observation that the collection of all polynomials in t, multiplied by $e^{-t^2/2}$, is dense in $S(\mathbb{R}^1)$ leads to the conclusion that $\mathrm{l.s.}(\{A^m\Omega \mid m = 0, 1, \ldots\})$ is dense in D and, all the more so, in $G^2(\mathbb{R}^1)$.

By considering (1.18) and (1.20), it is easy to write the transformation (2.53) of Chapter 2 and Parseval's equality (2.54) in the same chapter in our case. The corresponding transformation is called the one-dimensional Fourier-Wiener transformation and appears as follows:

$$\begin{aligned} \mathbb{R}^1 = s(A) \ni \lambda \mapsto \tilde{u}(\lambda) &= \frac{1}{\sqrt{2\pi}} \int_{\mathbb{R}^1} u(t) \exp(i\lambda t + \tfrac{1}{2}(\lambda^2 - t^2))\, dt \\ &= \int_{\mathbb{R}^1} u(t) \exp(i\lambda t + \tfrac{1}{2}\lambda^2) \frac{1}{\sqrt{2\pi}} e^{-t^2}\, dt \qquad (u \in G^2(\mathbb{R}^1)), \end{aligned} \tag{1.22}$$

while Parseval's equality (for $\alpha = \mathbb{R}^1$) has the form

$$\frac{1}{\sqrt{\pi}} \int_{\mathbb{R}^1} u(t)\overline{v(t)} e^{-t^2}\, dt = \int_{\mathbb{R}^1} \tilde{u}(\lambda)\overline{\tilde{v}(\lambda)}\, d\sigma(\lambda) = \frac{1}{\sqrt{\pi}} \int_{\mathbb{R}^1} \tilde{u}(\lambda)\overline{\tilde{v}(\lambda)} e^{-\lambda^2}\, d\lambda \qquad (u, v \in G^2(\mathbb{R}^1)). \tag{1.23}$$

The transformation (1.22) and the equality (1.23) can be extended by continuity to functions u, $v \in H_0$.

We pass to the infinite-dimensional case. Let $(H_{0,n})_1^\infty$ be a sequence of spaces all equal to the same space

$$H_{0,n} = L_2\left(\mathbb{R}^1, \pi^{-1/2}e^{-x_n^2}\, dx_n\right),$$

and let the operator A_n coinciding with the operator A constructed above act in each of these spaces. Take the stabilizing sequence $e = (e^{(n)}(x_n))_1^\infty$, where each $e^{(n)}(x_n) \equiv 1$, and form the infinite tensor product

$$\mathcal{H}_{0,e,1} = \bigotimes_{n=1;e,1}^{\infty} H_{0,n} = L_2(\mathbb{R}^\infty, d\mathfrak{g}(x)).$$

Here $d\mathfrak{g}(x)$, as in (4.13) of Chapter 1, is Gaussian measure in $\mathbb{R}^\infty$:

$$\begin{aligned} d\mathfrak{g}(x) &= (dg(x_1)) \otimes (dg(x_2)) \otimes \cdots \\ &= (\pi^{-1/2}e^{-x_1^2}\, dx_1) \otimes (\pi^{-1/2}e^{-x_2^2}\, dx_2) \otimes \cdots. \end{aligned} \tag{1.24}$$

From the operators A_n we construct a family $(\mathcal{A}_n)_1^\infty$ of commuting self-adjoint operators according to the general scheme in subsection 1 (roughly speaking, each $\mathcal{A}_n$ is the operator (1.15), acting with respect to the nth variable x_n), along with the corresponding joint resolution of the identity $\mathcal{C}_\sigma(\mathbf{R}^\infty) \ni \alpha \mapsto E(\alpha)$.

To construct a rigging of the space $L_2(\mathbf{R}^\infty, d\mathfrak{g}(x))$, it is necessary to take the product of the chain (1.21) for $n = 1, 2, \ldots$:

$$\mathcal{H}_{-,e,1} = \bigotimes_{n=1;e,1}^{\infty} H_-^{\#_n} \supseteq L_2(\mathbf{R}^\infty, d\mathfrak{g}(x))$$
$$\supseteq \bigotimes_{n=1;e,1}^{\infty} (G^2(\mathbf{R}^1))^{\#_n} = \mathcal{H}_{+,e,1}. \tag{1.25}$$

In this chain the imbedding of the positive space in the zero space is quasinuclear, because condition (1.4) holds due to the renorming. To extend the rigging (1.25) we must construct the weighted infinite tensor product of the spaces $\mathcal{G}(\mathbf{R}^1)$: here $\mathcal{D} = \text{w.} \bigotimes_{n=1;e}^{\infty} \mathcal{G}(\mathbf{R}^1) = \mathcal{G}(\mathbf{R}^\infty, \varepsilon)$ $(\varepsilon = (1, 1, \ldots))$, where the last space of cylindrical functions is described in §4.4 of Chapter 1.

As already noted, each operator A_n has the strong cyclic vector $e^{(n)}(x_n) \equiv 1$. But then the family $(\mathcal{A}_n)_1^\infty$ has the strong cyclic vector $\Omega(x) = 1 \otimes 1 \otimes \cdots \equiv 1$, according to the general construction in §1.1. Thus, in our case not only (1.6) and (1.7) are valid, but also (1.11)–(1.13).

We write out some of the relations obtained here. According to (1.7), $s((\mathcal{A}_n)_1^\infty) = \mathbf{R}^\infty$. The equalities (1.6) and (1.18), with the latter rewritten in the form

$$P(t, s; \lambda)\, d\rho(\lambda) = \exp(i\lambda(t - s) + \tfrac{1}{2}(t^2 + s^2) + \tfrac{1}{4}\lambda^2) \left(\frac{1}{2\sqrt{\pi}} e^{-\lambda^2/4}\, d\lambda \right),$$

give us that $\mathcal{P}(\lambda)$ is the "integral operator with kernel"

$$\mathcal{P}(x, y; \lambda)\, d\rho(\lambda) = \exp\left(i \sum_{n=1}^{\infty} \lambda_n (x_n - y_n) + \frac{1}{2} \sum_{n=1}^{\infty} (x_n^2 + y_n^2) + \frac{1}{4} \sum_{n=1}^{\infty} \lambda_n^2 \right)$$
$$\times \left(\frac{1}{2\sqrt{\pi}} e^{-\lambda_1^2/4}\, d\lambda_1 \right) \otimes \left(\frac{1}{2\sqrt{\pi}} e^{-\lambda_2^2/4}\, d\lambda_2 \right) \otimes \cdots$$
$$(x = (x_n)_1^\infty,\ y = (y_n)_1^\infty,\ \lambda = (\lambda_n)_1^\infty \in \mathbf{R}^\infty). \tag{1.26}$$

The series in (1.26) converge in the sense that the "cutoff" kernels corresponding to summations in (1.26) from 1 to p generate a sequence of operators weakly convergent to $\mathcal{P}(\lambda)\, d\rho(\lambda)$ as operators from $\mathcal{H}_{+,e,1}$ to $\mathcal{H}_{-,e,1}$ as $p \to \infty$.

It follows from (1.11) and (1.20) that the spectral measure σ is the Gaussian measure

$$d\sigma(\lambda) = (\pi^{-1/2} e^{-\lambda_1^2}\, d\lambda_1) \otimes (\pi^{-1/2} e^{-\lambda_2^2}\, d\lambda_2) \otimes \cdots = d\mathfrak{g}(\lambda). \tag{1.27}$$

The Fourier transformation (1.13), which is called the infinite-dimensional Fourier-Wiener transformation in this case, amounts to assigning to each

$u(x) \in \mathcal{H}_{+,e,1}$ the function

$$\begin{aligned} \mathbb{R}^\infty \ni \lambda = (\lambda_n)_1^\infty \mapsto \tilde{u}(\lambda) &= \int_{\mathbb{R}^\infty} u(x) \exp\left(i \sum_{n=1}^\infty \lambda_n x_n + \frac{1}{2} \sum_{n=1}^\infty \lambda_n^2 \right) \\ &\times \left(\frac{1}{\sqrt{2\pi}} e^{-x_1^2/2}\, dx_1 \right) \otimes \left(\frac{1}{\sqrt{2\pi}} e^{-x_2^2/2}\, dx_2 \right) \otimes \cdots . \end{aligned} \tag{1.28}$$

(The exponential in the integrand converges in the same way as (1.12).) For $u(x), v(x) \in \mathcal{H}_{+,e,1}$ Parseval's equality (with $\alpha = \mathbb{R}^\infty$) has the form

$$\begin{aligned} &\int_{\mathbb{R}^\infty} u(x)\overline{v(x)}\, d\mathfrak{g}(x) \\ &= \int_{\mathbb{R}^\infty} u(x)\overline{v(x)} \left(\frac{1}{\sqrt{\pi}} e^{-x_1^2}\, dx_1 \right) \otimes \left(\frac{1}{\sqrt{\pi}} e^{-x_2^2}\, dx_2 \right) \otimes \cdots \\ &= \int_{\mathbb{R}^\infty} \tilde{u}(\lambda)\overline{\tilde{v}(\lambda)}\, d\mathfrak{g}(\lambda) \\ &= \int_{\mathbb{R}^\infty} \tilde{u}(\lambda)\overline{\tilde{v}(\lambda)} \left(\frac{1}{\sqrt{\pi}} e^{-\lambda_1^2}\, d\lambda_1 \right) \otimes \left(\frac{1}{\sqrt{\pi}} e^{-\lambda_2^2}\, d\lambda_2 \right) \otimes \cdots . \end{aligned} \tag{1.29}$$

For functions in the class

$$\mathcal{H}_{+,e,1} = \bigotimes_{n=1;e,1}^\infty (G^2(\mathbb{R}^1))^{\#_n}$$

the integrals in (1.28) and (1.29) are absolutely convergent. We do not explain how to extend this class while preserving absolute convergence; we note only that for a more explicit description of such classes of functions it is useful to construct a rigging of $L_2(\mathbb{R}^\infty, d\mathfrak{g}(x))$ using the spaces $A_\tau(\mathbb{R}^\infty)$ and $H^\tau(\mathbb{R}^\infty)$ instead of the rigging (1.25) (see §§4.5 and 4.6 in Chapter 1). As already noted, if we pass to mean-square convergence of the integrals, then we can extend (1.28) and (1.29) to arbitrary u, $v \in L_2(\mathbb{R}^\infty, d\mathfrak{g}(x))$. It is also clear that the transformation (1.28) behaves just as the usual multidimensional Fourier transformation behaves with respect to differentiation; it is necessary only to replace the derivatives $-i(\partial/\partial x_n)$ by $-i((\partial/\partial x_n) - x_n)$.

1.3. ***Functions of operators acting with respect to different variables.*** Let $(\mathcal{A}_n)_1^\infty$ be the family of commuting selfadjoint operators in $\mathcal{H}_{0,e,1}$ constructed in §1.1, and let $\mathcal{C}_\sigma(\mathbb{R}^\infty) \ni \alpha \mapsto \mathcal{E}(\alpha)$ be its joint resolution of the identity. Consider a complex-valued function $\mathbb{R}^\infty \ni \lambda = (\lambda_n)_1^\infty \mapsto L(\lambda) = L(\lambda_1, \lambda_2, \ldots)$, which is measurable with respect to the σ-algebra $\mathcal{C}_\sigma(\mathbb{R}^\infty)$. Assume that it is defined and finite almost everywhere with respect to the measure $\mathcal{E}$, with its values at points of $\mathbb{R}^\infty$ outside the corresponding set taken to equal infinity. From this function we can construct a normal operator $\mathcal{L} = L(\mathcal{A}_1, \mathcal{A}_2, \ldots)$ acting in $\mathcal{H}_{0,e,1}$ by setting

$$\mathcal{L} = L(\mathcal{A}_1, \mathcal{A}_2, \ldots) = \int_{\mathbf{R}^\infty} L(\lambda)\, d\mathcal{E}(\lambda),$$
$$\mathfrak{D}(\mathcal{L}) = \left\{ f \in \mathcal{H}_{0,e,1} \mid \int_{\mathbf{R}^\infty} |L(\lambda)|^2\, d(\mathcal{E}(\lambda)f, f)_{\mathcal{H}_{0,e,1}} < \infty \right\}. \tag{1.30}$$

The fact that $\mathcal{L}$ is normal follows from general properties of the integral with respect to a resolution of the identity (see, for example, Plesner [1], Chapter 8). In connection with the check of normality we mention that $\mathfrak{D}(\mathcal{L})$ is dense in $\mathcal{H}_{0,e,1}$, because $\mathfrak{R}(\mathcal{E}(\alpha_n)) \subseteq \mathfrak{D}(\mathcal{L})$ for each $n = 1, 2, \ldots$, where $\alpha_n = \{\lambda \in \mathbf{R}^\infty \| L(\lambda)| \leq n\}$, and $\bigcup_1^\infty \alpha_n$ is a set of full $\mathcal{E}$-measure.

It is easy to describe an expression for the resolution of the identity of $\mathcal{L}$. To do this, consider the mapping

$$\mathbf{R}^\infty \setminus \{\lambda \in \mathbf{R}^\infty \mid L(\lambda) = \infty\} \ni \lambda \mapsto L(\lambda) \in \mathbf{C}^1$$

and the σ-algebra $\mathcal{B}(\mathbf{C}^1)$ of Borel subsets of $\mathbf{C}^1$. Since L is measurable, $L^{-1}(\delta) \in \mathcal{C}_\sigma(\mathbf{R}^\infty)$ for any $\delta \in \mathcal{B}(\mathbf{C}^1)$; therefore, it is possible to construct the L-image $\mathcal{E}_L$ of the measure $\mathcal{E}$, *which coincides with the resolution $\mathcal{E}_\mathcal{L}$ of the identity for $\mathcal{L}$*. Indeed, according to the formula in §7 of the Introduction, applied in the case of the function $\mathbf{C}^1 \ni z \mapsto F(z) \equiv z \in \mathbf{C}^1$, we get that

$$\mathcal{L} = \int_{\mathbf{R}^\infty} L(\lambda)\, d\mathcal{E}(\lambda) = \int_{\mathbf{C}^1} z\, d\mathcal{E}_L(z).$$

It is obvious that $\mathcal{E}_L$ is a resolution of the identity on $\mathbf{C}^1$; therefore, the last equality shows that $\mathcal{E}_\mathcal{L} = \mathcal{E}_L$. ■

Suppose now that we have the rigging (1.3). As before, we construct the chain (1.5). It certainly works for the construction of an expansion in the generalized eigenvectors of $\mathcal{L}$. (This chain will be used to construct all expansions in what follows.) It is almost obvious that *the spectral measure $\mathcal{B}(\mathbf{C}^1) \ni \delta \mapsto \rho_\mathcal{L}(\delta)$ of $\mathcal{L}$ is equal to the L-image of the spectral measure of the family $(\mathcal{A}_n)_1^\infty$*. Indeed,

$$\begin{aligned}\rho_\mathcal{L}(\delta) &= \operatorname{Tr}(\mathcal{O}^+ \mathcal{E}_\mathcal{L}(\delta)\mathcal{O}) = \operatorname{Tr}(\mathcal{O}^+ \mathcal{E}(L^{-1}(\delta))\mathcal{O}) \\ &= \rho(L^{-1}(\delta)) = \rho_L(\delta) \qquad (\delta \in \mathcal{B}(\mathbf{C}^1)). \quad ■\end{aligned}$$

From the definition (1.30) of the operator $\mathcal{L}$ it follows that for $z \in \mathbf{C}^1$

$$(\mathcal{L} - z1)^{-1} = \int_{\mathbf{R}^\infty} (L(\lambda) - z)^{-1}\, d\mathcal{E}(\lambda),$$
$$\|(\mathcal{L} - z1)^{-1}\| = \operatorname*{ess\,sup}_{\lambda \in \mathbf{R}^\infty} |(L(\lambda) - z)^{-1}|.$$

(The ess sup is taken with respect to the measure $\mathcal{E}$.) Thus, z is a regular point of $\mathcal{L}$ if and only if the last norm is finite. This implies that $z \in s(\mathcal{L})$

if and only if $z \in \bigcap(L(\operatorname{supp}\mathcal{E})\backslash\alpha))^{\sim}$, where the intersection is taken over all sets $\alpha \in \mathcal{C}_\sigma(\mathbb{R}^\infty)$ of zero $\mathcal{E}$-measure, and the tilde denotes the closure in $\mathbb{C}^1$. Considering formula (1.7) for $\operatorname{supp}\mathcal{E} = s((\mathcal{A}_n)_1^\infty)$, we conclude that the spectrum $s(\mathcal{L})$ of $\mathcal{L}$ coincides with the set

$$s(\mathcal{L}) = \bigcap_{\alpha\in\mathcal{C}_\sigma(\mathbf{R}^\infty),\ \mathcal{E}(\alpha)=0} \left(L\left(\left(\bigtimes_{n=1}^{\infty} s(A_n)\right)\backslash\alpha\right)\right)^{\sim}. \tag{1.31}$$

We show below how to express the generalized projection operator $\mathcal{P}_\mathcal{L}(z)$ corresponding to $\mathcal{L}$ in terms of the generalized projection operators $P_n(\lambda_n)$ $(n = 1, 2, \ldots)$. For this, some facts are needed about disintegration of measures (Bourbaki [1], §3 in Chapter 6). We recall them in the required form, using the notation in §7 of the Introduction for mappings of measures. Let $\mathbf{R}$ be a separable locally compact space, $\mathcal{B}(\mathbf{R}) \ni \boldsymbol{\alpha} \mapsto \boldsymbol{\sigma}(\boldsymbol{\alpha}) \geq 0$ a finite nonnegative measure on the σ-algebra of Borel subsets of $\mathbf{R}$, and $\mathbf{R} \ni \boldsymbol{\lambda} \mapsto \varphi(\boldsymbol{\lambda}) \in \mathbf{C}^1$ a particular mapping of $\mathbf{R}$ into the compactified complex plane $\mathbf{C}^1$ such that $\varphi^{-1}(\boldsymbol{\delta}) \in \mathcal{B}(\mathbf{R})$ for any $\boldsymbol{\delta} \in \mathcal{B}(\mathbf{C}^1)$. Consider the φ-image $\boldsymbol{\sigma}_\varphi$ of the measure $\boldsymbol{\sigma}$. We claim that there exists a family $(\boldsymbol{\sigma}_\mathbf{z})_{\mathbf{z}\in\mathbf{C}^1}$ of nonnegative finite measures, each defined on $\mathcal{B}(\mathbf{R})$ and concentrated on $\varphi^{-1}(\mathbf{z}) \in \mathcal{B}(\mathbf{R})$, such that

$$\int_{\mathbf{R}} F(\boldsymbol{\lambda})\, d\boldsymbol{\sigma}(\boldsymbol{\lambda}) = \int_{\mathbf{C}^1} \left(\int_{\varphi^{-1}(\mathbf{z})} F(\boldsymbol{\lambda})\, d\boldsymbol{\sigma}_\mathbf{z}(\boldsymbol{\lambda})\right) d\boldsymbol{\sigma}_\varphi(\mathbf{z}) \tag{1.32}$$

for any $F \in L_1(\mathbf{R}, d\boldsymbol{\sigma}(\boldsymbol{\lambda}))$.

THEOREM 1.2. *The following representation holds for the generalized projection operator $\mathcal{P}_\mathcal{L}(z)$ corresponding to $\mathcal{L}$ in terms of the generalized projection operators $P_n(\lambda_n)$:*

$$\mathcal{P}_\mathcal{L}(z) = \int_{L^{-1}(z)} \left(\bigotimes_{n=1}^{\infty} P_n(\lambda_n)\right) d\sigma_z(\lambda) \qquad (z \in \mathbb{C}^1;\ \lambda = (\lambda_n)_1^\infty). \tag{1.33}$$

Here $(\sigma_z)_{z\in\mathbb{C}^1}$ is a family of nonnegative finite measures, each defined on $\mathcal{C}_\sigma(\mathbb{R}^\infty)$ and concentrated on $L^{-1}(z) \in \mathcal{C}_\sigma(\mathbb{R}^\infty)$. The integral in (1.33) converges in the Hilbert norm of operators from $\mathcal{H}_{+,e,1}$ to $\mathcal{H}_{0,e,1}$.

PROOF. Consider the compactified joint resolution of the identity $\mathbf{E}$ of the family $(\mathcal{A}_n)_1^\infty$: $\mathcal{C}_\sigma(\mathbf{R}^\infty) \ni \boldsymbol{\alpha} \mapsto \mathbf{E}(\boldsymbol{\alpha})$ (see §1.8 in Chapter 2); $\mathbf{E}$ and $\mathcal{E}$ are connected by the usual formula (1.20) in Chapter 2. According to Theorem 2.2 in Chapter 2 and subsection 5 in the same section, the representation

$$\mathcal{O}^+\mathbf{E}(\boldsymbol{\alpha})\mathcal{O} = \int_{\boldsymbol{\alpha}} \mathbf{P}(\boldsymbol{\lambda})\, d\boldsymbol{\rho}(\boldsymbol{\lambda}) \qquad (\boldsymbol{\alpha} \in \mathcal{C}_\sigma(\mathbf{R}^\infty) = \mathcal{B}(\mathbf{R}^\infty)) \tag{1.34}$$

is valid, where $\boldsymbol{\rho}$ is the corresponding compactified spectral measure, and $\boldsymbol{\lambda} = (\boldsymbol{\lambda}_n)_1^\infty$ $(\boldsymbol{\lambda} \in \mathbf{R}^\infty)$. It is clear that $\mathbb{R}^\infty \in \mathcal{C}_\sigma(\mathbf{R}^\infty)$ and $\mathbf{E}(\mathbf{R}^\infty\backslash\mathbb{R}^\infty) = 0$.

Let us extend the function L from R^∞ to $\mathbf{R}^\infty$ by setting it equal to infinity on $\mathbf{R}^\infty \setminus \mathsf{R}^\infty$. Denote the extended function by $\mathbf{L}$. The mapping $\mathbf{R}^\infty \ni \boldsymbol{\lambda} \mapsto \mathbf{L}(\boldsymbol{\lambda}) \in \mathbf{C}^1$ has the property that $\mathbf{L}^{-1}(\boldsymbol{\delta}) \in \mathcal{C}_\sigma(\mathbf{R}^\infty) = \mathcal{B}(\mathbf{R}^\infty)$ for any $\boldsymbol{\delta} \in \mathcal{B}(\mathbf{C}^1)$; hence, what was noted about disintegration of measures applies to it.

Let $u,\ v \in \mathcal{H}_{+,e,1}$ and $\boldsymbol{\delta} \in \mathcal{B}(\mathbf{C}^1)$. Applying (1.32) to the case when $\mathbf{R} = \mathbf{R}^\infty$, $\varphi = \mathbf{L}$, $\boldsymbol{\sigma} = \boldsymbol{\rho}$, and $F(\boldsymbol{\lambda}) = \boldsymbol{\kappa}_{\boldsymbol{\delta}}(\mathbf{L}(\boldsymbol{\lambda}))(\mathbf{P}(\lambda)uv)_{\mathcal{H}_{0,e,1}}$, we get that

$$\int_{\mathbf{R}^\infty} \boldsymbol{\kappa}_{\boldsymbol{\delta}}(\mathbf{L}(\boldsymbol{\lambda}))(\mathbf{P}(\boldsymbol{\lambda})u,v)_{\mathcal{H}_{0,e,1}}\, d\boldsymbol{\rho}(\boldsymbol{\lambda})$$
$$= \int_{\boldsymbol{\delta}} \left(\int_{\mathbf{L}^{-1}(\mathbf{z})} (\mathbf{P}(\boldsymbol{\lambda})u,v)_{\mathcal{H}_{0,e,1}}\, d\boldsymbol{\rho}_{\mathbf{z}}(\boldsymbol{\lambda}) \right) d\boldsymbol{\rho}_{\mathbf{L}}(\mathbf{z}). \tag{1.35}$$

On the other hand, by (1.34), the left-hand side of this equality can be rewritten in the form

$$\int_{\mathbf{R}^\infty} \boldsymbol{\kappa}_{\boldsymbol{\delta}}(\mathbf{L}(\boldsymbol{\lambda}))\, d(\mathcal{O}^+\mathbf{E}(\boldsymbol{\lambda})\mathcal{O}u,v)_{\mathcal{H}_{0,e,1}}.$$

Applying to this integral the formula connecting the integrals under a mapping of measures, we get that it is equal to $\int_{\mathbf{C}^1} \boldsymbol{\kappa}_{\boldsymbol{\delta}}(\mathbf{z})\, d\omega_{\mathbf{L}}(\mathbf{z}) = \omega_{\mathbf{L}}(\boldsymbol{\delta})$, where $\omega_{\mathbf{L}}$ is the $\mathbf{L}$-image of the measure $\mathcal{B}(\mathbf{R}^\infty) \ni \boldsymbol{\alpha} \mapsto \omega(\boldsymbol{\alpha}) = (\mathcal{O}^+\mathbf{E}(\boldsymbol{\alpha})\mathcal{O}u,v)_{\mathcal{H}_{0,e,1}}$. Consider the $\mathbf{L}$-image $\mathbf{E}_{\mathbf{L}}$ of the resolution $\mathbf{E}$ of the identity. Obviously,

$$\omega_{\mathbf{L}}(\boldsymbol{\delta}) = (\mathcal{O}^+\mathbf{E}_{\mathbf{L}}(\boldsymbol{\delta})\mathcal{O}u,v)_{\mathcal{H}_{0,e,1}}.$$

Consequently, (1.35) can be rewritten in the form

$$(\mathcal{O}^+\mathbf{E}_{\mathbf{L}}(\boldsymbol{\delta})\mathcal{O}u,v)_{\mathcal{H}_{0,e,1}} = \int_{\boldsymbol{\delta}} \left(\int_{\mathbf{L}^{-1}(\mathbf{z})} (\mathbf{P}(\boldsymbol{\lambda})u,v)_{\mathcal{H}_{0,e,1}}\, d\boldsymbol{\rho}_{\mathbf{z}}(\boldsymbol{\lambda}) \right) d\boldsymbol{\rho}_{\mathbf{L}}(\mathbf{z}). \tag{1.36}$$

In this equality let $\boldsymbol{\delta} = \delta \in \mathcal{B}(\mathsf{C}^1)$. Since $\mathbf{L}$ coincides with L on R^∞ and $\mathbf{E}(\mathbf{R}^\infty \setminus \mathsf{R}^\infty) = 0$, the left-hand side of (1.36) becomes equal to

$$(\mathcal{O}^+\mathcal{E}_L(\delta)\mathcal{O}u,v)_{\mathcal{H}_{0,e,1}} = (\mathcal{O}^+\mathcal{E}_{\mathcal{L}}(\delta)\mathcal{O}u,v)_{\mathcal{H}_{0,e,1}}.$$

We transform the right-hand side of (1.36). As for $\mathbf{E}_{\mathbf{L}}$, we can write $\boldsymbol{\rho}_{\mathbf{L}}(\delta_1) = \boldsymbol{\rho}_{\mathbf{L}}(\delta_1) = \boldsymbol{\rho}_{\mathcal{L}}(\delta_1)$ for $\delta_1 \in \mathcal{B}(\mathsf{C}^1)$. Therefore, we can replace $d\boldsymbol{\rho}_{\mathbf{L}}(\mathbf{z})$ by $d\boldsymbol{\rho}_{\mathcal{L}}(z)$ in (1.36) when $\boldsymbol{\delta} = \delta$. Further, since $\mathbf{L}$ is equal to ∞ on $\mathbf{R}^\infty \setminus \mathsf{R}^\infty$ and coincides with L on R^∞, it follows that $\mathbf{L}^{-1}(z) = L^{-1}(z)$ $(z \in \mathsf{C}^1)$. Finally, $\mathbf{P} \restriction \mathsf{R}^\infty = \mathcal{P}$ (see (2.14)). Thus, by (1.36),

$$(\mathcal{O}^+\mathcal{E}_{\mathcal{L}}(\delta)\mathcal{O}u,v)_{\mathcal{H}_{0,e,1}} = \int_{\delta} \left(\int_{L^{-1}(z)} (\mathcal{P}(\lambda)u,v)_{\mathcal{H}_{0,e,1}}\, d\sigma_z(\lambda) \right) d\rho_{\mathcal{L}}(z)$$
$$(u, v \in \mathcal{H}_{+,e,1};\ \delta \in \mathcal{B}(\mathsf{C}^1)). \tag{1.37}$$

Here we have denoted by σ_z the restriction of the measure $\boldsymbol{\rho}_z$, defined on $\mathcal{B}(\mathbf{R}^\infty)$, to $\mathcal{B}(\mathsf{R}^\infty)$.

Writing the usual representation of $(\mathcal{O}^+\mathcal{E}_\mathcal{L}(\delta)\mathcal{O}u, v)_{\mathcal{H}_{0,e,1}}$ in terms of $\mathcal{P}_\mathcal{L}(z)$ and $d\rho_\mathcal{L}(z)$ and comparing it with (1.37), we conclude from the fact that δ, u, and v are arbitrary that the representation (1.33) is valid (taking into account the fact that $\mathcal{P}_\mathcal{L}(z)$ is defined for $\rho_\mathcal{L}$-almost all z, along with the second equality in (1.6)). The required convergence of the integral in (1.33) follows from the fact that

$$\operatorname{Tr}\left(\bigotimes_{n=1}^{\infty} P_n(\lambda_n)\right) = \operatorname{Tr}(\mathcal{P}(\lambda)) = 1. \quad ■$$

Of course, if L takes real values $\mathcal{E}$-almost everywhere, then $\mathcal{L}$ is selfadjoint.

Everything in this subsection carries over almost without change to the case of normal operators A_n.

1.4. *The case of functions which can be well approximated by cylindrical functions.* The definition (1.30) is usually ineffectual as far as computing the action of $\mathcal{L}$ goes. We modify it for a certain type of function L. An example of such a modification is given in the next section.

Denote by $L_0(\mathbb{R}^\infty, d\mathcal{E}(\lambda))$ the class of functions L for which we defined the operator $\mathcal{L}$, i.e., the collection of all complex-valued functions $\mathbb{R}^\infty \ni \lambda \mapsto L(\lambda)$ measurable with respect to $\mathcal{C}_\sigma(\mathbb{R}^\infty)$ and finite $\mathcal{E}$-almost everywhere.

Let $(E_p(z))_1^\infty$ be a sequence of resolutions $\mathcal{B}(\mathbb{C}^1) \ni \delta \mapsto E_p(\delta)$ of the identity. It will be said to converge weakly to a resolution $\mathcal{B}(\mathbb{C}^1) \ni \delta \mapsto E(\delta)$, of the identity if for any bounded continuous function $\mathbb{C}^1 \ni z \mapsto \varphi(z) \in \mathbb{C}^1$ the operators $\int_{\mathbb{C}^1} \varphi(z)\, dE_p(z)$ converge strongly to the operator $\int_{\mathbb{C}^1} \varphi(z)\, dE(z)$ as $p \to \infty$. Weak convergence of resolutions of the identity on $\mathbb{R}^1$ is defined similarly. (This definition is the operator variant of weak convergence of distribution functions $\mathbb{R}^1 \ni \lambda \mapsto \mu_p(\lambda)$ in probability theory or, in general, of bounded nondecreasing functions; for them replace $\mathbb{C}^1$, E_p, and E above by $\mathbb{R}^1$, μ_p, and μ. On the other hand, weak convergence of resolutions of the identity on $\mathbb{R}^1$ is equivalent to strong resolvent convergence of the corresponding selfadjoint operators (see Reed and Simon [1], Chapter 8, §7).)

THEOREM 1.3. *Let $L \in L_0(\mathbb{R}^\infty, d\mathcal{E}(\lambda))$ be such that there exists a sequence $(L_p)_1^\infty$ of cylindrical functions $L_p(\lambda) = L_{p,c}(\lambda_1, \ldots, \lambda_p)$, each in $L_0(\mathbb{R}^\infty, d\mathcal{E}(\lambda))$, for which the following two conditions hold:* a) *$L_p(\lambda) \to L(\lambda)$ as $p \to \infty$ for $\mathcal{E}$-almost all $\lambda \in \mathbb{R}^\infty$; and* b) *the resolutions $\mathcal{E}_{\mathcal{L}_p}$ of the identity for the normal operators $\mathcal{L}_p = L_{p,c}(\mathcal{A}_1, \ldots, \mathcal{A}_p)$ converge weakly to some resolution E of the identity as $p \to \infty$. Then $\mathcal{E}_\mathcal{L} = E$ $(\mathcal{L} = L(\mathcal{A}_1, \mathcal{A}_2, \ldots))$.*

If $f \in (\bigcap_1^\infty \mathfrak{D}(\mathcal{L}_p))$ is such that for each $\varepsilon > 0$ there is a $c > 0$ for which

$$\int_{|z|>c} |z|^2\, d(\mathcal{E}_{\mathcal{L}_p}(z)f, f)_{\mathcal{H}_{0,e,1}} < \varepsilon, \tag{1.38}$$

for any $p = 1, 2, \ldots$, then

$$\mathcal{L}f = \lim_{p\to\infty} \mathcal{L}_p f. \tag{1.39}$$

in the sense of convergence in $\mathcal{H}_{0,e,1}$.

We remark that the action of $\mathcal{L}_p$ often can be computed in a simple way, and then (1.39) leads to a procedure for computing $\mathcal{L}f$. It can be shown that for each $L \in L_0(\mathbf{R}^\infty, d\mathcal{E}(\lambda))$ there exists a sequence of cylindrical functions satisfying conditions a) and b).

PROOF. Since $\mathcal{E}_{\mathcal{L}_p} = \mathcal{E}_{L_p}$, we can write

$$\int_{\mathbf{C}^1} \varphi(z)\, d\mathcal{E}_{\mathcal{L}_p}(z) f = \int_{\mathbf{R}^\infty} \varphi(L_p(\lambda))\, d\mathcal{E}(\lambda) f \qquad (f \in \mathcal{H}_{0,e,1};\ p = 1, 2, \ldots) \quad (1.40)$$

for any function φ in the definition of weak convergence. Generally speaking, $\varphi(L_p(\lambda))$ is not defined for $\lambda \in \mathbf{R}^\infty$ such that $L_p(\lambda) = \infty$, but since the $\mathcal{E}$-measure of the set of these λ is equal to zero, the relation (1.40) is unambiguous.

We pass to the limit as $p \to \infty$ in (1.40). The left-hand side of this relation converges to $\int_{\mathbf{C}^1} \varphi(z)\, dE(z) f$ according to the definition of weak convergence. The right-hand side converges to $\int_{\mathbf{C}^1} \varphi(z)\, d\mathcal{E}_{\mathcal{L}}(z) f$. Indeed,

$$\left\| \int_{\mathbf{R}^\infty} \varphi(L_p(\lambda))\, d\mathcal{E}(\lambda) f - \int_{\mathbf{C}^1} \varphi(z)\, d\mathcal{E}_{\mathcal{L}}(z) f \right\|^2_{\mathcal{H}_{0,e,1}} = \int_{\mathbf{R}^\infty} |\varphi(L_p(\lambda)) - \varphi(L(\lambda))|^2\, d(\mathcal{E}(\lambda) f, f)_{\mathcal{H}_{0,e,1}}. \quad (1.41)$$

For each

$$\lambda \in \beta = \left(\bigcap_{p=1}^{\infty} \{\lambda \in \mathbf{R}^\infty \mid L_p(\lambda) \neq \infty\} \right) \cap \{\lambda \in \mathbf{R}^\infty \mid L(\lambda) \neq \infty\} \cap \{\lambda \in \mathbf{R}^\infty \mid \lim_{p \to \infty} L_p(\lambda) = L(\lambda)\} \quad (1.42)$$

we have that $\varphi(L_p(\lambda)) \to \varphi(L(\lambda))$ as $p \to \infty$ because φ is continuous, and $|\varphi(L_p(\lambda))| \leq \sup_{z \in \mathbf{C}^1} |\varphi(z)| < \infty$ ($\lambda \in \beta$; $p = 1, 2, \ldots$). Since β is a set of full $\mathcal{E}$-measure, we can pass to the limit under the integral sign on the right-hand side of (1.41), and we get zero as a result. Accordingly,

$$\int_{\mathbf{C}^1} \varphi(z)\, dE(z) f = \int_{\mathbf{C}^1} \varphi(z)\, d\mathcal{E}_{\mathcal{L}}(z) f \qquad (f \in \mathcal{H}_{0,e,1})$$

for any φ under consideration. This implies that $E = \mathcal{E}_{\mathcal{L}}$.

Let us prove the second part of the assertion. First of all, we see that $f \in \mathfrak{D}(\mathcal{L})$, i.e., that

$$\int_{\mathbf{R}^\infty} |L(\lambda)|^2\, d(\mathcal{E}(\lambda) f, f)_{\mathcal{H}_{0,e,1}} < \infty.$$

The integral in (1.38) is equal to

$$\int_{\{\lambda\in\mathbf{R}^\infty \mid |L_p(\lambda)|>c\}} |L_p(\lambda)|^2\, d(\mathcal{E}(\lambda)f,f)_{\mathcal{H}_{0,e,1}};$$

therefore, (1.38) ensures that

$$\int_{\mathbf{R}^\infty} |L_p(\lambda)|^2\, d(\mathcal{E}(\lambda)f,f)_{\mathcal{H}_{0,e,1}} \le c_1 < \infty \qquad (p=1,2,\ldots).$$

Since $L_p(\lambda)$ converges to $L(\lambda)$ almost everywhere with respect to the measure $\mathcal{C}_\sigma(\mathbf{R}^\infty) \ni \alpha \mapsto (\mathcal{E}(\alpha)f,f)_{\mathcal{H}_{0,e,1}}$ as $p\to\infty$, the required integral converges by Fatou's lemma.

Fix the indicated f and ε and choose $c>0$ large enough that (1.38) holds and, moreover, $\int_{|z|>c} |z|^2\, d(\mathcal{E}_{\mathcal{L}}(z)f,f)_{\mathcal{H}_{0,e,1}} < \varepsilon$. Let us construct a function $\varphi \in C(\mathbb{C}^1)$, $0\le\varphi(z)\le 1$ ($z\in\mathbb{C}^1$), equal to 0 for $|z|\ge c+1$ and equal to 1 for $|z|\le c$. Then

$$\begin{aligned}
\|\mathcal{L}_p f - \mathcal{L}f\|_{\mathcal{H}_{0,e,1}} &\le \left\|\mathcal{L}_p f - \int_{\mathbf{C}^1} z\varphi(z)\, d\mathcal{E}_{\mathcal{L}_p}(z)f\right\|_{\mathcal{H}_{0,e,1}} \\
&\quad + \left\|\int_{\mathbf{C}^1} z\varphi(z)\, d\mathcal{E}_{\mathcal{L}_p}(z)f - \int_{\mathbf{C}^1} z\varphi(z)\, d\mathcal{E}_{\mathcal{L}}(z)f\right\|_{\mathcal{H}_{0,e,1}} \\
&\quad + \left\|\int_{\mathbf{C}^1} z\varphi(z)\, d\mathcal{E}_{\mathcal{L}}(z)f - \mathcal{L}f\right\|_{\mathcal{H}_{0,e,1}} \\
&= \left(\int_{\mathbf{C}^1} |z|^2(1-\varphi^2(z))\, d(\mathcal{E}_{\mathcal{L}_p}(z)f,f)_{\mathcal{H}_{0,e,1}}\right)^{1/2} \\
&\quad + \left\|\int_{\mathbf{C}^1} z\varphi(z)\, d\mathcal{E}_{\mathcal{L}_p}(z)f - \int_{\mathbf{C}^1} z\varphi(z)\, d\mathcal{E}_{\mathcal{L}}(z)f\right\|_{\mathcal{H}_{0,e,1}} \\
&\quad + \left(\int_{\mathbf{C}^1} |z|^2(1-\varphi^2(z))\, d(\mathcal{E}_{\mathcal{L}}(z)f,f)_{\mathcal{H}_{0,e,1}}\right)^{1/2} \\
&\le 2\sqrt{\varepsilon} + \left\|\int_{\mathbf{C}^1} z\varphi(z)\, d\mathcal{E}_{\mathcal{L}_p}(z)f - \int_{\mathbf{C}^1} z\varphi(z)\, d\mathcal{E}_{\mathcal{L}}(z)f\right\|_{\mathcal{H}_{0,e,1}} \\
&\qquad (p=1,2,\ldots). \qquad (1.43)
\end{aligned}$$

Since $z\varphi(z)\in C(\mathbb{C}^1)$ and is bounded, the last norm in (1.43) can be made arbitrarily small for sufficiently large n, because $\mathcal{E}_{\mathcal{L}_p}$ converges weakly to $\mathcal{E}_{\mathcal{L}}$. ■

Assume now that we have the rigging (1.5) and that the corresponding spectral measure ρ of the family $(\mathcal{A}_n)_1^\infty$ is constructed. *If $L\in L_2(\mathbf{R}^\infty, d\rho(\lambda))$, then the conditions of the first part of Theorem* 1.3 *are satisfied for this function.*

Indeed, we approximate L in $L_2(\mathbf{R}^\infty, d\rho(\lambda))$ by a sequence of cylindrical functions and choose from it a subsequence $(L_p)_1^\infty$ converging to L ρ-almost everywhere. The functions L_p are the desired ones. Thus, requirement a) is certainly satisfied. To verify b), fix a bounded $\varphi \in C(\mathbb{C}^1)$. Considering that $\mathcal{E}_{\mathcal{L}_p} = \mathcal{E}_{L_p}$ and $\mathcal{E}_{\mathcal{L}} = \mathcal{E}_L$, we use the rule for transforming integrals under a transport of measures and the usual relation (2.13) in Chapter 2 to get that for $u \in \mathcal{H}_{+,e,1}$

$$\begin{aligned}
&\left\| \int_{\mathbb{C}^1} \varphi(z)\, d\mathcal{E}_{\mathcal{L}_p}(z)u - \int_{\mathbb{C}^1} \varphi(z)\, d\mathcal{E}_{\mathcal{L}}(z)u \right\|^2_{\mathcal{H}_{0,e,1}} \\
&= \left\| \int_{\mathbf{R}^\infty} (\varphi(L_p(\lambda)) - \varphi(L(\lambda)))\, d\mathcal{E}(\lambda)u \right\|^2_{\mathcal{H}_{0,e,1}} \\
&= \int_{\mathbf{R}^\infty} |\varphi(L_p(\lambda)) - \varphi(L(\lambda))|^2\, d(\mathcal{E}(\lambda)u, u)_{\mathcal{H}_{0,e,1}} \\
&= \int_{\mathbf{R}^\infty} |\varphi(L_p(\lambda)) - \varphi(L(\lambda))|^2 (\mathcal{P}(\lambda)u, u)_{\mathcal{H}_{0,e,1}}\, d\rho(\lambda) \\
&\le \|u\|^2_{\mathcal{H}_{+,e,1}} \int_{\mathbf{R}^\infty} |\varphi(L_p(\lambda)) - \varphi(L(\lambda))|^2\, d\rho(\lambda) \qquad (p = 1, 2, \ldots). \qquad (1.44)
\end{aligned}$$

Here, as in (1.40), the functions $\varphi(L_p(\lambda))$ and $\varphi(L(\lambda))$ are defined for ρ-almost all $\lambda \in \mathbf{R}^\infty$. Since the integrand in the last integral in (1.44) tends to zero on the set β (see (1.42)) of full ρ-measure and is bounded uniformly with respect to p, we can pass to the limit under the integral sign, and we get zero as the limit. Thus, in $\mathcal{H}_{0,e,1}$

$$\lim_{p\to\infty} \int_{\mathbb{C}^1} \varphi(z)\, d\mathcal{E}_{\mathcal{L}_p}(z)u = \int_{\mathbb{C}^1} \varphi(z)\, d\mathcal{E}_{\mathcal{L}}(z)u \qquad (u \in \mathcal{H}_{+,e,1}). \qquad (1.45)$$

The norms of all the operators represented by the integrals in (1.45) are bounded uniformly with respect to p by the number $\sup_{z\in\mathbb{C}^1} |\varphi(z)| < \infty$. Therefore, (1.45) and the fact that $\mathcal{H}_{+,e,1}$ is dense in $\mathcal{H}_{0,e,1}$ imply the same relation (1.45) with u replaced by an arbitrary vector $f \in \mathcal{H}_{0,e,1}$. ■

In conclusion we make a simple remark about the support of a limit of resolutions of the identity.

We introduce the following general definition. Let $(\alpha_p)_1^\infty$ be a sequence of sets $\alpha_p \subseteq \mathbb{C}^1$. Consider the set α of all points $z \in \mathbb{C}^1$ representable as a limit $z = \lim_{p\to\infty} z_p$, where z_p is a point in α_p. We call this set the limit of the sequence $(\alpha_p)_1^\infty$ and write $\alpha = \lim_{p\to\infty} \alpha_p$; α is clearly always closed.

It is easy to see that *if a sequence* $(E_p)_1^\infty$ *of resolutions of the identity converges weakly to the resolution* E *of the identity, then* $\operatorname{supp} E \subseteq \lim_{p\to\infty} \operatorname{supp} E_p$. (Simple examples show that equality need not hold here.) Indeed, let $z_0 \in \operatorname{supp} E$. We construct a nonnegative function $\varphi \in C(\mathbb{C}^1)$, $\varphi(z_0) = 1$, equal to zero outside some neighborhood of z_0. Then

$$0 \ne \int_{\mathbb{C}^1} \varphi(z)\, dE(z) = \lim_{p\to\infty} \int_{\mathbb{C}^1} \varphi(z)\, dE_p(z),$$

and so $\int_{\mathbf{C}^1} \varphi(z)\, dE_p(z) \neq 0$ for sufficiently large p. This means that the indicated neighborhood contains points of $\operatorname{supp} E_p$ for these p. ■

As earlier, the assertions of this subsection are preserved for selfadjoint operators $\mathcal{L}$, as well as for normal A_n.

§2. Operators admitting separation of infinitely many variables

We continue our study of functions of operators acting with respect to different variables introduced in §§1.3 and 1.4. The problem amounts to the following. Assume that $L(\lambda)$ is a function on $\mathbf{R}^\infty$ given by a formal expression. How do we find a family $(A_n)_1^\infty$ of selfadjoint operators (or a family $(E_n)_1^\infty$ of resolutions of the identity) such that L turns out to be a $\mathcal{C}_\sigma(\mathbf{R}^\infty)$-measurable function on $\mathbf{R}^\infty$ that is $\mathcal{E}$-almost everywhere finite? Here $\mathcal{E}$ is the joint resolution of the identity constructed from the corresponding family $(\mathcal{A}_n)_1^\infty$. We study this question for the function $L(\lambda) = \lambda_1 + \lambda_2 + \cdots$, and we even obtain conditions on $(E_n)_1^\infty$ that are necessary and sufficient in a certain respect. As is easy to see, the function $L(\lambda) = \lambda_1 + \lambda_2 + \cdots$ corresponds to the case when $\mathcal{L}$ admits separation of infinitely many variables.

2.1. *Convolution of finitely many resolutions of the identity on the line.* We show how to generalize the concept of a convolution of distribution functions on the line to the case of resolutions of the identity. Only resolutions of the identity on $\mathbf{R}^1$ will be considered. (In this section it is convenient to understand them not as measures $\mathcal{B}(\mathbf{R}^1) \ni \delta \mapsto E(\delta)$ but as functions $E(\lambda) = E((-\infty, \lambda))$ $(\lambda \in \mathbf{R}^1)$.)

Let H_1 and H_2 be two separable Hilbert spaces, and let $\mathcal{H} = H_1 \otimes H_2$. Consider two resolutions of the identity: $\mathbf{R}^1 \ni \lambda \mapsto E_1(\lambda) \in \mathcal{L}(H_1 \to H_1)$ and $\mathbf{R}^1 \ni \lambda \mapsto E_2(\lambda) \in \mathcal{L}(H_2 \to H_2)$. The functions $\mathbf{R}^1 \ni \lambda \mapsto \mathcal{E}_1(\lambda) = (E_1(\lambda)) \otimes 1$ and $\mathbf{R}^1 \ni \lambda \mapsto \mathcal{E}_2(\lambda) = 1 \otimes (E_2(\lambda))$ are commuting resolutions of the identity in $\mathcal{H}$. We define their convolution.

THEOREM 2.1. *The convolution $(\mathcal{E}_1 \star \mathcal{E}_2)(\lambda) = \mathcal{F}(\lambda)$ is defined as follows. Let $(e^{(1)}_{\alpha_1})_0^\infty$ and $(e^{(2)}_{\alpha_2})_0^\infty$ be orthonormal bases in the respective spaces H_1 and H_2. Fix $\lambda \in \mathbf{R}^1$ and form the matrix*

$$A_{\alpha\beta}(\lambda) = \int_{\mathbf{R}^1} (E_1(\lambda - \mu) e^{(1)}_{\beta_1}, e^{(1)}_{\alpha_1})_{H_1}\, d(E_2(\mu) e^{(2)}_{\beta_2}, e^{(2)}_{\alpha_2})_{H_2}$$

$$(\alpha = (\alpha_1, \alpha_2),\ \beta = (\beta_1, \beta_2) \in N^2). \quad (2.1)$$

In the orthogonal basis $e_\alpha = e^{(1)}_{\alpha_1} \otimes e^{(2)}_{\alpha_2}$ $(\alpha \in N^2)$ of the space $\mathcal{H} = H_1 \otimes H_2$ this matrix gives rise to an operator that will be denoted by $(\mathcal{E}_1 \star \mathcal{E}_2)(\lambda)$. The operator-valued function $\mathbf{R}^1 \ni \lambda \mapsto (\mathcal{E}_1 \star \mathcal{E}_2)(\lambda)$ defined in this way is a resolution of the identity in $\mathcal{H}$, and its definition does not depend on the choice of bases.

PROOF. Note that the integral in (2.1) exists as the convolution of two functions of bounded variation. A bilinear form a.$(H_1 \otimes H_2) \ni f, g \mapsto B_\lambda(f, g)$

is then defined in the usual way from (2.1) by setting

$$B_\lambda(f,g) = \sum_{\alpha,\beta\in N^2} A_{\alpha\beta}(\lambda) f_\beta \bar{g}_\alpha = \sum_{j,k=1}^{p} \int_{\mathbf{R}^1} (E_1(\lambda-\mu) f^{(1),k}, g^{(1),j})_{H_1}\, d(E_2(\mu) f^{(2),k}, g^{(2),j})_{H_2} \quad (2.2)$$

for

$$f = \sum_{k=1}^{p} f^{(1),k} \otimes f^{(2),k} = \sum_{k=1}^{p} \sum_{\beta\in N^2} f^{(1),k}_{\beta_1} f^{(2),k}_{\beta_2} e_\beta = \sum_{\beta\in N^2} f_\beta e_\beta,$$

$$g = \sum_{j=1}^{p} g^{(1),j} \otimes g^{(2),j} = \sum_{j=1}^{p} \sum_{\alpha\in N^2} g^{(1),j}_{\alpha_1} g^{(2),j}_{\alpha_2} e_\alpha = \sum_{\alpha\in N^2} g_\alpha e_\alpha.$$

(In establishing the last equality in (2.2) it is necessary to interchange the infinite sum over α and β with the integral; it is easy to justify this interchange.)

For $\lambda' < \lambda''$

$$B_{\lambda'}(f,f) \le B_{\lambda''}(f,f) \qquad (f \in \text{a.}\,(H_1 \otimes H_2)). \quad (2.3)$$

Indeed, if $[\lambda'-\mu, \lambda''-\mu) = \delta_\mu$, then

$$B_{\lambda''}(f,f) - B_{\lambda'}(f,f) = \sum_{j,k=1}^{p} \int_{\mathbf{R}^1} (E_1(\delta_\mu) f^{(1),k}, f^{(1),j})_{H_1}\, d(E_2(\mu) f^{(2),k}, f^{(2),j})_{H_2}.$$

To establish the nonnegativity of the last sum it suffices to see that the similar expression with the integral replaced by an integral sum is nonnegative, i.e. it suffices to verify that a sum of the form

$$\sum_{j,k=1}^{p} \sum_{l=1}^{q} (E_1(\delta_{\mu_l}) f^{(1),k}, f^{(1),j})_{H_1} (E_2(\beta_l) f^{(2),k}, f^{(2),j})_{H_2} = \sum_{l=1}^{q} \sum_{j,k=1}^{p} (E_1(\delta_{\mu_l}) f^{(1),k}, f^{(1),j})_{H_1} (E_2(\beta_l) f^{(2),k}, f^{(2),j})_{H_2}$$

is nonnegative. Here $\mathbf{R}^1 = \bigcup_1^q \beta_l$ is a partition of the line into Borel sets β_l, and μ_l is a point in β_l. For fixed l each of the matrices

$$(a_{jk})_{j,k=1}^{p} = ((E_1(\delta_{\mu_l}) f^{(1),k}, f^{(1),j})_{H_1})_{j,k=1}^{p},$$
$$(b_{jk})_{j,k=1}^{p} = ((E_2(\beta_l) f^{(2),k}, f^{(2),j})_{H_2})_{j,k=1}^{p}$$

is positive-definite. But then the matrix $(a_{jk}b_{jk})_{j,k=1}^{p}$ is also positive-definite; therefore, $\sum_{j,k=1}^{p} a_{jk}b_{jk} \ge 0$. Inequality (2.3) is established.

Next, it is easy to see that

$$\lim_{\lambda\to+\infty} B_\lambda(f,f) = (f,f)_{\mathcal{H}} \qquad (f \in \text{a.}\,(H_1 \otimes H_2)). \tag{2.4}$$

Indeed, we can pass to the limit as $\lambda \to +\infty$ under the integral sign in the integral in (2.2) with $g = f$, because the integrand is bounded. The result of passing to the limit is that

$$\sum_{j,k=1}^{p} (f^{(1),k}, f^{(1),j})_{H_1} (f^{(2),k}, f^{(2),j})_{H_2} \\ = \sum_{j,k=1}^{p} (f^{(1),k} \otimes f^{(2),k}, f^{(1),j} \otimes f^{(2),j})_{\mathcal{H}} = (f,f)_{\mathcal{H}}.$$

It follows from (2.3) and (2.4) that $B_\lambda(f,f) \le (f,f)_{\mathcal{H}}$ $(f \in \text{a.}(H_1 \otimes H_2))$ for each $\lambda \in \mathbb{R}^1$. It follows immediately from the definition (2.2) that the bilinear function B_λ is Hermitian; therefore, the last inequality means that it is bounded. Thus, there exists a bounded selfadjoint operator $\mathcal{F}(\lambda) \in \mathcal{L}(\mathcal{H} \to \mathcal{H})$ such that $(\mathcal{F}(\lambda)f, g)_{\mathcal{H}} = B_\lambda(f,g)$ $(f, g \in \text{a.}(H_1 \otimes H_2))$.

Let us see that the function $\mathbb{R}^1 \ni \lambda \mapsto \mathcal{F}(\lambda)$ is a resolution of the identity in $\mathcal{H}$. To do this the main thing is to establish the orthogonality relation $\mathcal{F}(\lambda')\mathcal{F}(\lambda'') = \mathcal{F}(\min(\lambda', \lambda''))$ or, what is the same, the equality

$$\sum_{\beta\in N^2} A_{\alpha\beta}(\lambda')A_{\beta\gamma}(\lambda'') = A_{\alpha\gamma}(\min(\lambda', \lambda'')) \\ (\lambda', \lambda'' \in \mathbb{R}^1;\ \alpha, \gamma \in N^2). \tag{2.5}$$

Fix $\alpha, \gamma \in N^2$. According to (2.1),

$$\sum_{\beta\in N^2} A_{\alpha\beta}(\lambda')A_{\beta\gamma}(\lambda'') \\ = \sum_{\beta_2=0}^{\infty}\sum_{\beta_1=0}^{\infty} \int_{\mathbb{R}^2} (E_1(\lambda'-\mu)e^{(1)}_{\beta_1}, e^{(1)}_{\alpha_1})_{H_1} (E_1(\lambda''-\nu)e^{(1)}_{\gamma_1}, e^{(1)}_{\beta_1})_{H_1} \\ \times d(E_2(\mu)e^{(2)}_{\beta_2}, e^{(2)}_{\alpha_2})_{H_2}\, d(E_2(\nu)e^{(2)}_{\gamma_2}, e^{(2)}_{\beta_2})_{H_2} \\ = \sum_{\beta_2=0}^{\infty} \int_{\mathbb{R}^2} (E_1(\lambda''-\nu)e^{(1)}_{\gamma_1}, E_1(\lambda'-\mu)e^{(1)}_{\alpha_1})_{H_1} \\ \times d(E_2(\mu)e^{(2)}_{\beta_2}, e^{(2)}_{\alpha_2})_{H_2}\, d(E_2(\nu)e^{(2)}_{\gamma_2}, e^{(2)}_{\beta_2})_{H_2} = \Sigma.$$

Here we interchanged the sum over β_1 and the double integral with respect to (μ, ν) (which is easy to justify) and used Parseval's equality. Using the fact that convolution of functions of bounded variation is commutative (or, if desired, integrating by parts) and the orthogonality relation $E_n(t')E_n(t'') = E_n(\min(t', t''))$ $(t', t'' \in \mathbf{R}^1;\ n = 1, 2)$, we transform the last expression as follows:

$$\Sigma = \sum_{\beta_2=0}^{\infty} \int_{\mathbf{R}^1} \left(\int_{\mathbf{R}^1} (E_2(\lambda'' - \nu)e^{(2)}_{\gamma_2}, e^{(2)}_{\beta_2})_{H_2} \right.$$

$$\left. \times\, d_\nu (E_1(\nu)e^{(1)}_{\gamma_1}, E_1(\lambda' - \mu)e^{(1)}_{\alpha_1})_{H_1} \right) d_\mu (E_2(\mu)e^{(2)}_{\beta_2}, e^{(2)}_{\alpha_2})_{H_2}$$

$$= \sum_{\beta_2=0}^{\infty} \int_{\mathbf{R}^1} \left(\int_{-\infty}^{\lambda'-\mu} (E_2(\lambda'' - \nu)e^{(2)}_{\gamma_2}, e^{(2)}_{\beta_2})_{H_2}\, d_\nu (E_1(\nu)e^{(1)}_{\gamma_1}, e^{(1)}_{\gamma_1})_{H_1} \right)$$

$$\times\, d_\mu (E_2(\mu)e^{(2)}_{\beta_2}, e^{(2)}_{\alpha_2})_{H_2}$$

$$= \sum_{\beta_2=0}^{\infty} \int_{\mathbf{R}^1} (E_2(\lambda' - \mu)e^{(2)}_{\beta_2}, e^{(2)}_{\alpha_2})_{H_2}\, d_\mu \left(\int_{-\infty}^{\mu} (E_2(\lambda'' - \nu)e^{(2)}_{\gamma_2}, e^{(2)}_{\beta_2})_{H_2} \right.$$

$$\left. \times\, d_\nu (E_1(\nu)e^{(1)}_{\gamma_1}, e^{(1)}_{\alpha_1})_{H_1} \right)$$

$$= \sum_{\beta_2=0}^{\infty} \int_{\mathbf{R}^1} (E_2(\lambda' - \mu)e^{(2)}_{\beta_2}, e^{(2)}_{\alpha_2})_{H_2} (E_2(\lambda'' - \mu)e^{(2)}_{\gamma_2}, e^{(2)}_{\beta_2})_{H_2}$$

$$\times\, d_\mu (E_1(\mu)e^{(1)}_{\gamma_1}, e^{(1)}_{\alpha_1})_{H_1}$$

$$= \int_{\mathbf{R}^1} (E_2(\lambda'' - \mu)e^{(2)}_{\gamma_2}, E_2(\lambda' - \mu)e^{(2)}_{\alpha_2})_{H_2}\, d(E_1(\mu)e^{(1)}_{\gamma_1}, e^{(1)}_{\alpha_1})_{H_1}$$

$$= \int_{\mathbf{R}^1} (E_2(\min(\lambda', \lambda'') - \mu))(e^{(2)}_{\gamma_2}, e^{(2)}_{\alpha_2})_{H_2}\, d(E_1(\mu)e^{(1)}_{\gamma_1}, e^{(1)}_{\alpha_1})_{H_1}$$

$$= \int_{\mathbf{R}^1} (E_1(\min(\lambda', \lambda'') - \mu)e^{(1)}_{\gamma_1}, e^{(1)}_{\alpha_1})_{H_1}\, d(E_2(\mu)e^{(2)}_{\gamma_2}, e^{(2)}_{\alpha_2})_{H_2}$$

$$= A_{\alpha\gamma}(\min(\lambda', \lambda'')).$$

The relation (2.5) is established, i.e., $\mathcal{F}(\lambda')\mathcal{F}(\lambda'') = \mathcal{F}(\min(\lambda', \lambda''))$. In particular, for $\lambda' = \lambda'' = \lambda$ this implies that $\mathcal{F}(\lambda)$ $(\lambda \in \mathbf{R}^1)$ is a projection.

To prove that $\mathcal{F}(\lambda)$ is a resolution of the identity it remains to observe that the normalization conditions are satisfied. The relation $\mathcal{F}(+\infty) = 1$ follows from (2.4). In a way similar to the derivation of (2.4) it is easy to see that also $\mathcal{F}(-\infty) = 0$ and $\mathcal{F}(\lambda - 0) = \mathcal{F}(\lambda)$ $(\lambda \in \mathbf{R}^1)$. Accordingly, $\mathcal{F}(\lambda)$ is a resolution of the identity.

The fact that the definition of the convolution is independent of the choice of bases follows from the representation (2.2) of the form B_λ in terms of an integral: this representation does not change when we choose other bases $(e^{(1)}_{\alpha_1})_0^\infty$ and $(e^{(2)}_{\alpha_2})_0^\infty$. ■

In essence, the definition of the convolution of $\mathcal{E}_1(\lambda)$ and $\mathcal{E}_2(\lambda)$ reduces to the following. We write the formal integral

$$(\mathcal{E}_1 \star \mathcal{E}_2)(\lambda) = \int_{\mathbf{R}^1} \mathcal{E}_1(\lambda - \mu)\, d\mathcal{E}_2(\mu). \tag{2.6}$$

This integral converges in the following sense. Apply (2.6) to $f^{(1)} \otimes f^{(2)}$ and take the inner product in $\mathcal{H}$ of the result with $g^{(1)} \otimes g^{(2)}$ ($f^{(1)}, g^{(1)} \in H_1$; $f^{(2)}, g^{(2)} \in H_2$). For the formal expression

$$((\mathcal{E}_1 \star \mathcal{E}_2)(\lambda)(f^{(1)} \otimes f^{(2)}), g^{(1)} \otimes g^{(2)})_{\mathcal{H}}$$

we get a formula (see (2.2)) that enables us to extend this expression by bilinearity and passage to the limit from $f^{(1)} \otimes f^{(2)}$ and $g^{(1)} \otimes g^{(2)}$ to arbitrary $f \in \mathcal{H}$ and $g \in \mathcal{H}$ and to prove that it is the bilinear form of some continuous operator, namely, $(\mathcal{E}_1 \star \mathcal{E}_2)(\lambda)$.

This procedure is also suitable for constructing the convolution $(\mathcal{E}_2 \star \mathcal{E}_1)(\lambda)$; it is only necessary to change the proof of Theorem 2.1 in the appropriate way. Clearly, the resulting convolution is commutative: $\mathcal{E}_1 \star \mathcal{E}_2 = \mathcal{E}_2 \star \mathcal{E}_1$. This follows from the fact that the convolution of the functions of bounded variation in (2.2) is commutative.

The definition of the convolution generalizes easily to the case of finitely many resolutions of the identity, acting in different spaces. Let $H_1, \ldots, H_p$ be separable Hilbert spaces, let $\mathcal{H} = \bigotimes_1^p H_n$, and suppose that the resolution $\mathbf{R}^1 \ni \lambda \mapsto E_n(\lambda)$ of the identity acts in H_n. The functions

$$\mathbf{R}^1 \ni \lambda \mapsto \mathcal{E}_n(\lambda) = \underbrace{1 \otimes \cdots \otimes 1}_{n-1} \otimes E_n(\lambda) \otimes \underbrace{1 \otimes \cdots \otimes 1}_{p-n} \qquad (n = 1, \ldots, p)$$

are commuting resolutions of the identity in $\mathcal{H}$. Each resolution $\mathcal{E}_n(\lambda)$ of the identity can be interpreted in two ways: 1) as a resolution of the identity of the form $(E'(\lambda)) \otimes 1$ in $\mathcal{H} = \mathcal{H}' + \mathcal{H}''$, where $\mathcal{H}' = \bigotimes_1^n H_j$ and $\mathcal{H}'' = \bigotimes_{n+1}^p H_j$; and 2) as a resolution of the form $1 \otimes (E''(\lambda))$, where $\mathcal{H}' = \bigotimes_1^{n-1} H_j$ and $\mathcal{H}'' = \bigotimes_n^p H_j$ (of course, in the first case $n = 1, \ldots, p-1$, and in the second $n = 2, \ldots, p$). Such an interpretation and the definition of convolution given above for $p = 2$ enables us to construct successively the p-fold convolution

$$\mathcal{F}(\lambda) = \left(\mathop{\bigstar}_{n=1}^{p} \mathcal{E}_n\right)(\lambda) = (\cdots((\mathcal{E}_1 \star \mathcal{E}_2) \star \mathcal{E}_3) \star \cdots \star \mathcal{E}_p)(\lambda), \tag{2.7}$$

which is a resolution of the identity.

It is easy to see that associativity holds: the parentheses in (2.7) can be arranged arbitrarily. In general, the convolution (2.7) can be constructed as follows. Form the formal convolution $(\bigstar_1^p E_n)(\lambda)$ of type (2.6); the convergence of the corresponding integrals is understood in the same sense as in (2.6), except with $g^{(1)} \otimes g^{(2)}$ and $f^{(1)} \otimes f^{(2)}$ replaced by $\bigotimes_1^p g^{(n)}$ and $\bigotimes_1^p f^{(n)}$ ($f^{(n)}, g^{(n)} \in H_n$).

Assume now that we have a rigging of each $H_n = H_{0,n}$:

$$H_{-,n} \supseteq H_{0,n} \supseteq H_{+,n} \qquad (n = 1, \ldots, p), \tag{2.8}$$

with the imbedding operator O_n: $H_{+,n} \to H_{0,n}$ quasinuclear. (Of course, this is not a restriction, since the chains (2.8) can always be constructed.) With the help of (2.8) the spectral measure $\mathcal{B}(\mathbf{R}^1) \ni \delta \mapsto \rho_n(\delta) = \operatorname{Tr}(O_n^+ E_n(\delta) O_n)$ ($n = 1, \ldots, p$) can be introduced in the usual way.

Taking the tensor product of the chain (2.8), we obtain a rigging of $\mathcal{H} = \mathcal{H}_0$:

$$\mathcal{H}_- \supseteq \mathcal{H}_0 \supseteq \mathcal{H}_+, \qquad \mathcal{H}_- = \bigotimes_{n=1}^{p} H_{-,n}, \ \mathcal{H}_+ = \bigotimes_{n=1}^{p} H_{+,n}. \tag{2.9}$$

As is known, the imbedding operator $\mathcal{O}$: $\mathcal{H}_+ \to \mathcal{H}_0$ here has the form $\mathcal{O} = \bigotimes_1^p O_n$ and is also quasinuclear (see §2.5 in Chapter 1). With the help of (2.9) we introduce the spectral measure ρ of the resolution of the identity

$$\begin{aligned} \mathcal{F}(\lambda) = \mathcal{E}_{\mathcal{L}_p}(\lambda) = \left(\mathop{\bigstar}_{n=1}^{p} \mathcal{E}_n \right)(\lambda) &: \mathcal{B}(\mathbf{R}^1) \ni \delta \mapsto \rho_{\mathcal{L}_p}(\delta) \\ &= \operatorname{Tr}(\mathcal{O}^+ \mathcal{F}(\delta) \mathcal{O}). \end{aligned}$$

(The meaning of the notation $\mathcal{E}_{\mathcal{L}_p}$ becomes clear just below.) It is not hard to see that, *for the corresponding nondecreasing functions*,

$$\rho_{\mathcal{L}_p}(\lambda) = \left(\mathop{\bigstar}_{n=1}^{p} \rho_n \right)(\lambda) \qquad (\lambda \in \mathbf{R}^1). \tag{2.10}$$

Indeed, according to the definition (2.7), it suffices to prove (2.10) for $p = 2$. Let $(e_{\alpha_1}^{(1)})_0^\infty$, $(e_{\alpha_2}^{(2)})_0^\infty$ and $(e_\alpha)_{\alpha \in N^2}$ ($e_\alpha = e_{\alpha_1}^{(1)} \otimes e_{\alpha_2}^{(2)}$) be orthonormal bases in the respective spaces $H_{+,1}, H_{+,2}$, and $\mathcal{H}_+$. Then, by (2.6) and easily justified interchanges,

$$\begin{aligned} \rho_{\mathcal{L}_2}(\lambda) &= \operatorname{Tr}(\mathcal{O}^+ \mathcal{E}_{\mathcal{L}_2}(\lambda) \mathcal{O}) = \sum_{\alpha \in N^2} (\mathcal{O}^+ \mathcal{E}_{\mathcal{L}_2}(\lambda) \mathcal{O} e_\alpha, e_\alpha)_{\mathcal{H}_0} \\ &= \sum_{\alpha \in N^2} (O_1^+ \otimes O_2^+) \int_{\mathbf{R}^1} ((E_1(\lambda - \mu)) \otimes 1)\, d(1 \otimes (E_2(\mu))) \\ &\qquad \times (O_1 \otimes O_2)(e_{\alpha_1}^{(1)} \otimes e_{\alpha_2}^{(2)}), (e_{\alpha_1}^{(1)} \otimes e_{\alpha_2}^{(2)})_{H_1 \otimes H_2} \\ &= \sum_{\alpha \in N^2} \int_{\mathbf{R}^1} (O_1^+ E_1(\lambda - \mu) O_1 e_{\alpha_1}^{(1)}, e_{\alpha_1}^{(1)})_{H_1}\, d(O_2^+ E_2(\mu) O_2 e_{\alpha_2}^{(2)}, e_{\alpha_2}^{(2)})_{H_2} \end{aligned}$$

(*continues*)

(*continued*)

$$= \sum_{\alpha_2=0}^{\infty} \int_{\mathbb{R}^1} \rho_1(\lambda - \mu)\, d(O_2^+ E_2(\mu) O_2 e_{\alpha_2}^{(2)}, e_{\alpha_2}^{(2)})_{H_2}$$

$$= \sum_{\alpha_2=0}^{\infty} \int_{\mathbb{R}^1} (O_2^+ E_2(\lambda - \mu) O_2 e_{\alpha_2}^{(2)}, e_{\alpha_2}^{(2)})_{H_2}\, d\rho_1(\mu)$$

$$= \int_{\mathbb{R}^1} \rho_2(\lambda - \mu)\, d\rho_1(\mu) = (\rho_1 \star \rho_2)(\lambda) \qquad (\lambda \in \mathbb{R}^1). \qquad \blacksquare$$

It is well known that if the measures $\rho_{\mathcal{L}_p}$ and ρ_n are related by (2.10), then

$$\operatorname{supp} \rho_{\mathcal{L}_p} = \left(\sum_{n=1}^{p} \operatorname{supp} \rho_n \right)^{\sim}, \tag{2.11}$$

where the sum is understood to be an algebraic sum of subsets of the line. Since the supports of a resolution of the identity and the corresponding spectral measure coincide, (2.11) gives us

$$\operatorname{supp} \mathcal{E}_{\mathcal{L}_p} = \left(\sum_{n=1}^{p} \operatorname{supp} E_n \right)^{\sim},$$

$$\mathcal{E}_{\mathcal{L}_p}(\lambda) = \left(\mathop{\bigstar}_{n=1}^{p} \mathcal{E}_n \right)(\lambda) \qquad (\lambda \in \mathbb{R}^1). \tag{2.12}$$

Consider the selfadjoint operators in $\mathcal{H} = \bigotimes_1^n H_n$ corresponding to the resolutions $\mathcal{E}_n$ and $\mathcal{E}_{\mathcal{L}_p}$. Denote them by $\mathcal{A}_n$ and $\mathcal{A}$, respectively ($n = 1, \ldots, p$). It is easy to see that $\mathcal{A} = L_p(\mathcal{A}_1, \ldots, \mathcal{A}_p)$ *where the function* $L_p(\mathcal{A}_1, \ldots, \mathcal{A}_p)$ *of the operators* $\mathcal{A}_n$ *is constructed from the scalar function* $\mathbb{R}^p \ni \lambda = (\lambda_n)_1^p \mapsto L_p(\lambda) = \lambda_1 + \cdots + \lambda_p \in \mathbb{R}^1$.

We use the following simply verified general change-of-variables formula. Let $\mathbb{R}^1 \ni \lambda \mapsto \tau_n(\lambda) \in \mathbb{C}^1$ be a function of bounded variation ($n = 1, \ldots, p$), and $\mathbb{R}^1 \ni \lambda \mapsto \varphi(\lambda) \in \mathbb{C}^1$ a $\mathcal{B}(\mathbb{R}^1)$-measurable function. Then

$$\int_{\mathbb{R}^p} \varphi(\lambda_1 + \cdots + \lambda_p)\, d\left(\bigotimes_{n=1}^{p} \tau_n \right)(\lambda_1, \ldots, \lambda_p) = \int_{\mathbb{R}^1} \varphi(\lambda)\, d\left(\mathop{\bigstar}_{n=1}^{p} \tau_n \right)(\lambda). \tag{2.13}$$

The same formula is valid in the case of nondecreasing functions $\tau_n(\lambda)$ for an arbitrary $\mathcal{B}(\mathbb{R}^1)$-measurable nonnegative function $\varphi(\lambda)$.

We proceed to the proof of the assertion. Let $\mathcal{B}(\mathbb{R}^p) \ni \delta \mapsto \mathcal{E}(\delta)$ be the joint resolution of the identity constructed from the family $(\mathcal{E}_n)_1^p$. Then the resolution of the identity for the operator $L_p(\mathcal{A}_1, \ldots, \mathcal{A}_p)$ is equal to the L_p-image $\mathcal{E}_{L_p}$ of the joint resolution of the identity $\mathcal{E}$ (cf. §1.3, where this is explained for the more complicated situation when there are countably many $\mathcal{E}_n$). Thus, it must be proved that $\mathcal{E}_{\mathcal{L}_p} = \mathcal{E}_{L_p}$. Let $f = \bigotimes_1^p f^{(n)}$ and

$g = \bigotimes_1^p g^{(n)}$, where $f^{(n)}, g^{(n)} \in H_n$. By (2.7),

$$(\mathcal{E}_{\mathcal{L}_p}(\lambda)f, g)_{\mathcal{H}} = \left(\left(\mathop{\bigstar}_{n=1}^{p} \mathcal{E}_n\right)(\lambda)f, g\right)_{\mathcal{H}}$$
$$= \left(\mathop{\bigstar}_{n=1}^{p}(E_n(\cdot)f^{(n)}, g^{(n)})_{H_n}\right)(\lambda) \qquad (\lambda \in \mathbf{R}^1).$$

With the help of (2.13) and the rule for transforming integrals under a transport of measures, this gives us that for $\delta \in \mathcal{B}(\mathbf{R}^1)$

$$(\mathcal{E}_{\mathcal{L}_p}(\delta)f, g)_{\mathcal{H}} = \int_{\mathbf{R}^1} \kappa_\delta(\lambda)\, d\left(\mathop{\bigstar}_{n=1}^{p}(E_n(\cdot)f^{(n)}, g^{(n)})_{H_n}\right)(\lambda)$$
$$= \int_{\mathbf{R}^p} \kappa_\delta(\lambda_1 + \cdots + \lambda_p)\, d\left(\bigotimes_{n=1}^{p}(E_n(\cdot)f^{(n)}, g^{(n)})_{H_n}\right)(\lambda_1, \dots, \lambda_p)$$
$$= \left(\int_{\mathbf{R}^p} \kappa_\delta(L_p(\lambda))\, d\mathcal{E}(\lambda)f, g\right)_{\mathcal{H}} = \left(\int_{\mathbf{R}^1} \kappa_\delta(\lambda)\, d\mathcal{E}_{L_p}(\lambda), fg\right)_{\mathcal{H}}$$
$$= (\mathcal{E}_{L_p}(\delta)f, g)_{\mathcal{H}}.$$

Since f and g are sufficiently arbitrary, this implies that $\mathcal{E}_{\mathcal{L}_p}(\delta) = \mathcal{E}_{L_p}(\delta)$. ■

According to §§1.3 and 1.4, we can write $\mathcal{A} = \mathcal{L}_p$, and $\mathcal{E}_{\mathcal{A}} = \mathcal{E}_{\mathcal{L}_p}$ is its resolution of the identity.

We now introduce the operator $\mathcal{A}$ from a different point of view. Suppose that A_n is an Hermitian operator acting in H_n with domain $\mathfrak{D}(A_n)$ ($n = 1, \dots, p$). We construct the operators in $\mathcal{H}$

$$\mathcal{A}'_{\min} = \sum_{n=1}^{p} \underbrace{1 \otimes \cdots \otimes 1}_{n-1} \otimes A_n \otimes \underbrace{1 \otimes \cdots \otimes 1}_{p-n} = \sum_{n=1}^{p} \mathcal{A}'_n,$$

$$\mathcal{A}'_n = \underbrace{1 \otimes \cdots \otimes 1}_{n-1} \otimes A_n \otimes \underbrace{1 \otimes \cdots \otimes 1}_{p-n}, \qquad \mathfrak{D}(\mathcal{A}'_{\min}) = \text{a.}\bigotimes_{n=1}^{p} \mathfrak{D}(A_n)$$

$$\mathfrak{D}(\mathcal{A}'_n) = \text{a.}(H_1 \otimes \cdots \otimes H_{n-1} \otimes \mathfrak{D}(A_n) \otimes H_{n+1} \otimes \cdots \otimes H_p). \quad (2.14)$$

It is clear that the operators $\mathcal{A}'_n$, and $\mathcal{A}'_{\min}$ are Hermitian; denote their closures by $\mathcal{A}_n$ and $\mathcal{A}_{\min}$.

THEOREM 2.2. *Let $\mathcal{A}$ be constructed from the resolutions $(E_n)_1^p$ of the identity, where E_n corresponds to some selfadjoint extension of A_n acting in H_n. Then $\mathcal{A}$ is a selfadjoint extension of $\mathcal{A}_{\min}$ acting in $\mathcal{H}$. If each operator A_n is essentially selfadjoint in H_n ($n = 1, \dots, p$), then $\mathcal{A}_{\min}$ is selfadjoint and coincides with $\mathcal{A}$.*

PROOF. For the inclusion $\mathcal{A} \supseteq \mathcal{A}_{\min}$ to be valid it suffices that $\mathcal{A} \supseteq \mathcal{A}'_{\min}$, or that $f = \bigotimes_1^p f^{(n)} \in \mathfrak{D}(\mathcal{A})$ and $\mathcal{A}f = \mathcal{A}'_{\min}f$ for any $f^{(n)} \in \mathfrak{D}(A_n)$. By

(2.7) and (2.13),

$$\begin{aligned}\int_{\mathbf{R}^1} \lambda^2 d(\mathcal{E}_{\mathcal{A}}(\lambda)f, f)_{\mathcal{H}} &= \int_{\mathbf{R}^1} \lambda^2 d\left(\mathop{\bigstar}_{n=1}^{p} (E_n(\cdot)f^{(n)}, f^{(n)})_{H_n}\right)(\lambda) \\ &= \int_{\mathbf{R}^p} (\lambda_1 + \cdots + \lambda_p)^2 d\left(\bigotimes_{n=1}^{p} (E_n(\cdot)f^{(n)}, f^{(n)})_{H_n}\right)(\lambda_1, \dots, \lambda_p) \\ &\le p\sum_{j=1}^{p} \int_{\mathbf{R}^1} \lambda_j^2 d\left(\bigotimes_{n=1}^{p} (E_n(\cdot)f^{(n)}, f^{(n)})_{H_n}\right)(\lambda_1, \dots, \lambda_p) \\ &= p\sum_{j=1}^{p} \|f^{(1)}\|_{H_1}^2 \cdots \|f^{(j-1)}\|_{H_{j-1}}^2 \left(\int_{\mathbf{R}^1} \lambda_j^2 d(E_j(\lambda_j)f^{(j)}, f^{(j)})_{H_j}\right) \\ &\qquad\qquad \times \|f^{(j+1)}\|_{H_{j+1}}^2 \cdots \|f^{(p)}\|_{H_p}^2 < \infty,\end{aligned}$$

i.e., $f \in \mathfrak{D}(\mathcal{A})$. Further, for the previous f and $g = \bigotimes_1^p g^{(n)}$ $(g^{(n)} \in H_n)$ we get from (2.7), (2.13), and (2.14) that

$$\begin{aligned}(\mathcal{A}f, g)_{\mathcal{H}} &= \left(\int_{\mathbf{R}^1} \lambda\, d\mathcal{E}_{\mathcal{A}}(\lambda)f, g\right)_{\mathcal{H}} = \int_{\mathbf{R}^1} \lambda\, d\left(\mathop{\bigstar}_{n=1}^{p} (E_n(\cdot)f^{(n)}, g^{(n)})_{H_n}\right)(\lambda) \\ &= \int_{\mathbf{R}^p} (\lambda_1 + \cdots + \lambda_p)\, d\left(\bigotimes_{n=1}^{p} (E_n(\cdot)f^{(n)}, g^{(n)})_{H_n}\right)(\lambda_1, \dots, \lambda_p) \\ &= \sum_{j=1}^{p} (f^{(1)}, g^{(1)})_{H_1} \cdots (f^{(j-1)}, g^{(j-1)})_{H_{j-1}} \left(\int_{\mathbf{R}^1} \lambda_j d(E_j(\lambda_j)f^{(j)}, g^{(j)})_{H_j}\right) \\ &\qquad\qquad \times (f^{(j+1)}, g^{(j+1)})_{H_{j+1}} \cdots (f^{(p)}, g^{(p)})_{H_p} \\ &= \sum_{j=1}^{p} (\mathcal{A}'f, g)_{\mathcal{H}} = (\mathcal{A}'_{\min} f, g)_{\mathcal{H}}.\end{aligned}$$

Since $g^{(n)} \in H_n$ is arbitrary, this implies that $\mathcal{A}f = \mathcal{A}'_{\min} f$.

We prove the last assertion of the theorem. Consider the case $p = 2$. Since $\mathcal{A}_{\min} \subseteq \mathcal{A}$, to prove that $\mathcal{A}_{\min}$ is selfadjoint and coincides with $\mathcal{A}$ it suffices to see that $\mathcal{R}_z(f^{(1)} \otimes f^{(2)}) \in \mathfrak{D}(\mathcal{A}_{\min})$ for some nonreal $z \in \mathbf{C}^1$, where $f^{(1)} \in H_1$ and $f^{(2)} \in H_2$. The equality $\mathcal{R}_z = \int_{\mathbf{R}^1} (\lambda - z)^{-1}\, d(\mathcal{E}_1 \star \mathcal{E}_2)(\lambda)$ and (2.6) give us that

$$\mathcal{R}_z = \int_{\mathbf{R}^1} \mathcal{R}_{z-\mu}^{(1)}\, d\mathcal{E}_2(\mu),$$

where $\mathcal{R}_z^{(1)}$ is the resolvent of $\mathcal{A}_1$. The convergence of this integral is understood in the same sense as the convergence of the integral in (2.6). Let $\mathbf{R}^1 = \bigcup_1^q \delta_l$ be a partition of the line into Borel sets, and let $\mu_l \in \delta_l$. Replacing the integral by the corresponding integral sum, we get the expressions

$$\sum_{l=1}^{q}(\mathcal{R}^{(1)}_{z-\mu_l}f^{(1)}) \otimes (\mathcal{E}_2(\delta_l)f^{(2)}) \in \mathfrak{D}(\mathcal{A}'_{\min}) \subseteq \mathfrak{D}(\mathcal{A}_{\min}).$$

These expressions are easily seen to converge weakly to $\mathcal{R}_z(f^{(1)} \otimes f^{(2)})$ as the partition is refined (use the indicated convergence of the integral and the easily established boundedness of their norms). Therefore, there are linear combinations of them (also in $\mathfrak{D}(\mathcal{A}_{\min})$, of course) that converge to $\mathcal{R}_z(f' \otimes f'')$ in the norm of $\mathcal{H}$. The fact that $\mathcal{A}_{\min}$ is closed implies that $\mathcal{R}_z(f' \otimes f'') \in \mathfrak{D}(\mathcal{A}_{\min})$, which is what was required.

The proof is analogous in the case $p > 2$, except with $\mathcal{E}_2(\lambda)$ replaced by $(\bigstar_2^p E_n)(\lambda)$. ■

By analogy with differential operators, it is natural to say that the operator $\mathcal{A}_{\min}$ (in particular, $\mathcal{A}$) admits separation of finitely many variables. The formula

$$s(\mathcal{A}) = \left(\sum_{n=1}^{p} s(A_n)\right)^{\sim} \tag{2.15}$$

for the spectrum of $\mathcal{A}$ follows from (2.12). Since $\mathcal{A} = \mathcal{L}_p = L_p(\mathcal{A}_1, \dots, \mathcal{A}_p)$, a formula of type (1.33) can be proved for the generalized projection operator corresponding to A in a way analogous to that for Theorem 1.2.

We make a last remark. The constructions in §§1.1 and 1.3, with $A_{p+1} = A_{p+2} = \cdots = 0$ and $L(\lambda) = \lambda_1 + \cdots + \lambda_p$, relate in essence to the situation considered in this subsection.

2.2. ***Convolution of infinitely many resolutions of the identity on the line and its existence.*** Consider a sequence $(H_n)_1^\infty$ of separable Hilbert spaces H_n, let $e = (e^{(n)})_1^\infty$ ($e^{(n)} \in H_n$, $\|e^{(n)}\|_{H_n} = 1$) be a fixed stabilizing sequence, and let $\mathcal{H}_{e,1} = \bigotimes_{n=1;e,1}^{\infty} H_n$ be the corresponding infinite tensor product. Assume that $\mathsf{R}^1 \ni \lambda \mapsto E_n(\lambda)$ is a resolution of the identity acting in H_n. Then

$$\mathsf{R}^1 \ni \lambda \mapsto \mathcal{E}_n(\lambda) = \underbrace{1 \otimes \cdots \otimes 1}_{n-1} \otimes E_n(\lambda) \otimes 1 \otimes 1 \otimes \cdots$$

is a resolution of the identity in $\mathcal{H}_{e,1}$ $(n = 1, 2, \dots)$. The last space can be interpreted as the tensor product $\mathcal{H}_{e,1} = H_1 \otimes \cdots \otimes H_p \otimes H'_{p+1}$, where $H'_{p+1} = \bigotimes_{n=p+1;e',1}^{\infty} H_n$ ($e' = (e^{(n)})_{p+1}^{\infty}$); therefore, according to §2.1, the convolution $(\bigstar_1^p \mathcal{E}_n)(\lambda)$ exists and is a resolution of the identity in $\mathcal{H}_{e,1}$.

We define the infinite convolution $(\bigstar_1^\infty \mathcal{E}_n)(\lambda)$ $(\lambda \in \mathsf{R}^1)$ as the resolution of the identity in $\mathcal{H}_{e,1}$ such that

$$\left(\mathop{\bigstar}_{n=1}^{\infty} \mathcal{E}_n\right)(\lambda) = \lim_{p\to\infty} \left(\mathop{\bigstar}_{n=1}^{p} \mathcal{E}_n\right)(\lambda), \tag{2.16}$$

where the limit is understood in the sense of weak convergence of resolutions of the identity (see the beginning of §1.4), and we determine conditions for it to exist. In this connection we need the following general lemma, which carries over to a resolution of the identity a well-known fact about weak convergence of distribution functions. (The formulation of the well-known fact is similar to that of our lemma.)

LEMMA 2.1. *Let $(E_n(\lambda))_1^\infty$ be a sequence of resolutions of the identity acting in some Hilbert space H. Consider the integrals*

$$U_n(t) = \int_{\mathbb{R}^1} e^{i\lambda t}\, dE_n(\lambda) \qquad (t \in \mathbb{R}^1;\ n = 1, 2, \ldots), \tag{2.17}$$

which are the corresponding groups of unitary operators. The sequence $(E_n(\lambda))_1^\infty$ converges weakly as $n \to \infty$ to some resolution of the identity if and only if the vector sequence $(U_n(t)f)_1^\infty$ converges in H for each $f \in H$ and $t \in \mathbb{R}^1$ and, moreover, the functions $\mathbb{R}^1 \ni t \mapsto (U_n(t)f, f)_H$ converge uniformly on each bounded interval of the line $\mathbb{R}^1$.

PROOF. Suppose that $E_n \to E$ as $n \to \infty$ in the sense under consideration, where E is a resolution of the identity. According to the definition of weak convergence, the vectors $U_n(t)f$ converge in H for each $f \in H$ and $t \in \mathbb{R}^1$. Further, it follows from the definition of the same convergence that the sequence $\mathbb{R}^1 \ni \lambda \mapsto (E_n(\lambda)f, f)_H$ ($\|f\|_H = 1$; $n = 1, 2, \ldots$) of distribution functions converges weakly as $n \to \infty$ to the distribution function $\mathbb{R}^1 \ni \lambda \mapsto (E(\lambda)f, f)_H$. As is well known, the sequence of their Fourier transforms $\mathbb{R}^1 \ni t \mapsto (U_n(t)f, f)_H$ converges uniformly on each bounded interval of $\mathbb{R}^1$ as $n \to \infty$. Of course, this holds also for $\|f\|_H \neq 1$.

We prove the converse assertion of the theorem. Let

$$U(t)f = \lim_{n\to\infty} U_n(t)f \qquad (f \in H;\ t \in \mathbb{R}^1).$$

It is clear that each operator $U(t)$ is isometric. The equality $U(-t)U(t) = 1$ ($t \in \mathbb{R}^1$) is valid. Indeed,

$$U(-t)U(t)f = \lim_{n\to\infty} U_n(-t)U_n(t)f$$

for each $f \in H$, since

$$\begin{aligned} &\|U(-t)U(t)f - U_n(-t)U_n(t)f\|_H \\ &\quad\leq \|U(-t)U(t)f - U_n(t)U(t)f\|_H \\ &\qquad + \|U_n(-t)U(t)f - U_n(-t)U_n(t)f\|_H \\ &\quad\leq \|(U(-t) - U_n(-t))U(t)f\|_H \\ &\qquad + \|U(t)f - U_n(t)f\|_H \to 0 \quad \text{as } n \to \infty. \end{aligned}$$

But $U_n(-t)U_n(t) = 1$, which is what was required.

The equality $U(-t)U(t) = 1$ shows that the operators $U(t)$ are unitary, and $U(-t) = U^{-1}(t) = U^*(t)$ ($t \in \mathbf{R}^1$). For $t = 0$ and $f \in H$ we have that $U(0)f = \lim_{n\to\infty} U_n(0)f = f$, i.e., $U(0) = 1$.

For $f, g \in H$ and $t, s \in \mathbf{R}^1$

$$(U_n(t+s)f, g)_H = (U_n(s)f, U_n(-t)g)_H \underset{n\to\infty}{\to} (U(s)f, U(-t)g)_H = (U(t)U(s)f, g)_H.$$

On the other hand, the left-hand side of this relation tends to $(U(t+s)f, g)_H$. Consequently, $U(t+s) = U(t)U(s)$. Thus, the operator-valued function $\mathbf{R}^1 \ni t \mapsto U(t)$ is a unitary representation of the group $\mathbf{R}^1$, and for it the function $\mathbf{R}^1 \ni t \mapsto (U(t)f, f)_H \in \mathbf{C}^1$ is continuous for each $f \in H$, due to the last condition of the lemma. Representing $(U(t)f, g)_H$ ($f, g \in H$) in terms of quadratic forms, we conclude that the function $\mathbf{R}^1 \ni t \mapsto (U(t)f, g)_H \in \mathbf{C}^1$ is also continuous, i.e., the representation $\mathbf{R}^1 \ni t \mapsto U(t)$ is weakly continuous. Consequently, it is continuous, and Stone's theorem $U(t) = \int_{\mathbf{R}^1} e^{i\lambda t}\, dE(\lambda)$ ($t \in \mathbf{R}^1$) holds for it, where $E(\lambda)$ is a resolution of the identity (see, for example, Naĭmark [1], §§29 and 31 in Chapter 6). We show that $E_n \to E$ as $n \to \infty$ in the sense of weak convergence.

Fix $f \in H$ ($\|f\|_H = 1$) and consider the sequence of distribution functions $\mathbf{R}^1 \ni \lambda \mapsto (E_n(\lambda)f, f)_H$ ($n = 1, 2, \ldots$). The corresponding Fourier transformations are equal to $\mathbf{R}^1 \ni t \mapsto (U_n(t)f, f)_H$. They converge uniformly on each bounded interval to $(U(t)f, f)_H$ as $n \to \infty$. In view of the usual connection between weak convergence of distribution functions and convergence of their Fourier transforms, we conclude that $(E_n(\lambda)f, f)_H \to \mu(\lambda)$ in the weak sense as $n \to \infty$, where $\mathbf{R}^1 \ni \lambda \mapsto \mu(\lambda)$ is some distribution function. Obviously, $\mu(\lambda) = (E(\lambda)f, f)_H$ ($\lambda \in \mathbf{R}^1$). Accordingly, for any bounded $\varphi(\lambda) \in C(\mathbf{R}^1)$

$$\int_{\mathbf{R}^1} \varphi(\lambda)\, d(E_n(\lambda)f, f)_H \to \int_{\mathbf{R}^1} \varphi(\lambda)\, d(E(\lambda)f, f)_H \quad \text{as } n \to \infty. \tag{2.18}$$

Taking $f \in H$ to be arbitrary here and passing from quadratic forms to bilinear forms, we conclude that, in the sense of weak convergence of operators,

$$A_n = \int_{\mathbf{R}^1} \varphi(\lambda)\, dE_n(\lambda) \to \int_{\mathbf{R}^1} \varphi(\lambda)\, dE(\lambda) = A \quad \text{as } n \to \infty. \tag{2.19}$$

To prove strong convergence in (2.19) it suffices to see that $\|A_n f\|_H \to \|Af\|_H$ as $n \to \infty$ for each $f \in H$. Replacing $\varphi(\lambda)$ by $|\varphi(\lambda)|^2$ in (2.18), we get the required relation. ■

THEOREM 2.3. *For the infinite convolution* (2.16) *to exist* (*and be a resolution of the identity*) *it is necessary that the following conditions hold for all* $c > 0$, *and sufficient that they hold for some* $c > 0$:

a) $\sum_{n=1}^{\infty}(1 - (E_n([-c, c))e^{(n)}, e^{(n)})_{H_n}) < \infty$, (2.20)

b) *the series* $\sum_{n=1}^{\infty} \int_{-c}^{c} \lambda\, d(E_n(\lambda)e^{(n)}, e^{(n)})_{H_n}$ *converges; and*

c) $\sum_{n=1}^{\infty} \int_{-c}^{c} [\lambda - \int_{-c}^{c} \lambda \, d(E_n(\lambda)e^{(n)}, e^{(n)})_{H_n}]^2 \, d(E_n(\lambda)e^{(n)}, e^{(n)})_{H_n} < \infty$.

PROOF. Assume that the conditions (2.20) hold. We prove that the infinite convolution exists. Consider the following unitary operators acting in $\mathcal{H}_{e,1}$:

$$\mathcal{U}_p(t) = \int_{\mathbb{R}^1} e^{i\lambda t} \, d\left(\mathop{\bigstar}_{n=1}^{p} \mathcal{E}_n \right)(\lambda) = \prod_{n=1}^{p} \int_{\mathbb{R}^1} e^{i\lambda t} \, d\mathcal{E}_n(\lambda) \qquad (t \in \mathbb{R}^1;\ p = 1, 2, \cdots), \quad (2.21)$$

where the operators

$$\mathcal{V}_n(t) = \int_{\mathbb{R}^1} e^{i\lambda t} \, d\mathcal{E}_n(\lambda) \quad (2.22)$$

are also unitary.

Let $f \in \mathcal{H}_{e,1}$. It follows from (2.21) that

$$\begin{aligned} \|\mathcal{U}_p(t)f - \mathcal{U}_q(t)f\|^2_{\mathcal{H}_{e,1}} &= \left\| \left(1 - \prod_{n=q+1}^{p} \mathcal{V}_n(t) \right) f \right\|^2_{\mathcal{H}_{e,1}} \\ &\leq 2 \left| \|f\|^2_{\mathcal{H}_{e,1}} - \left(\left(\prod_{n=q+1}^{p} \mathcal{V}_n(t) \right) f, f \right)_{\mathcal{H}_{e,1}} \right| \end{aligned} \qquad (t \in \mathbb{R}^1;\ p > q). \quad (2.23)$$

Here we have used the general inequality

$$\begin{aligned} \|(1 - U)f\|^2_H &= 2 \operatorname{Re}(\|f\|^2_H - (Uf, f)_H) \\ &\leq 2 \left| \|f\|^2_H - (Uf, f)_H \right| \qquad (f \in H) \end{aligned}$$

for a unitary operator $U \in \mathcal{L}(H \to H)$. For a vector f of the form

$$f = f^{(1)} \otimes \cdots \otimes f^{(r)} \otimes e^{(r+1)} \otimes e^{(r+2)} \otimes \cdots \qquad (f^{(n)} \in H_n;\ n = 1, \ldots, r) \quad (2.24)$$

the inequality (2.23) becomes, for $q \geq r$,

$$\begin{aligned} &\|\mathcal{U}_p(t)f - \mathcal{U}_q(t)f\|_{\mathcal{H}_{e,1}} \\ &\quad \leq \sqrt{2} \|f\|_{\mathcal{H}_{e,1}} \left| 1 - \prod_{n=q+1}^{p} \int_{\mathbb{R}^1} e^{i\lambda t} \, d(E_n(\lambda)e^{(n)}, e^{(n)})_{H_n} \right| \end{aligned} \qquad (t \in \mathbb{R}^1;\ p > q). \quad (2.25)$$

Consider the distribution functions $\mathbb{R}^1 \ni \lambda \mapsto \mu_n(\lambda) = (E_n(\lambda)e^{(n)}, e^{(n)})_{H_n}$ $(n = 1, 2, \ldots)$. By Kolmogorov's three series theorem (see, for example, Gikhman and Skorokhod [2], §3 in Chapter 2, or Wintner [1], pp. 26–33), conditions a)–c) ensure the existence of the infinite convolution $(\bigstar_1^{\infty} \mu_n)(\lambda)$, which is defined as the distribution function that is the weak limit of the distribution functions $(\bigstar_1^p \mu_n)(\lambda)$ as $p \to \infty$. Passing to Fourier transforms, we conclude that the limit

$$\lim_{p\to\infty} \int_{\mathbf{R}^1} e^{i\lambda t}\, d\left(\bigstar_{n=1}^{p} \mu_n\right)(\lambda) = \lim_{p\to\infty} \prod_{n=1}^{p} \int_{\mathbf{R}^1} e^{i\lambda t}\, d\mu_n(\lambda) \tag{2.26}$$

exists for each $t \in \mathbf{R}^1$, and this limit is uniform for t varying over a bounded interval.

It follows from (2.26) that the expression on the right-hand side of (2.25) tends to zero as $p, q \to \infty$, i.e., $\lim_{p\to\infty} \mathcal{U}_p(t)f$ exists for each $t \in \mathbf{R}^1$ and f of the form (2.24). Since these vectors form a total set in $\mathcal{H}_{e,1}$ and $\|\mathcal{U}_p(t)\| = 1$ $(t \in \mathbf{R}^1;\ p = 1, 2, \dots)$, this implies that the limit exists for any $f \in \mathcal{H}_{e,1}$.

Further, for f of the form (2.24) it follows from (2.21) that for $p > r$

$$\begin{aligned}(\mathcal{U}_p(t)f, f)_{\mathcal{H}_{e,1}} = \prod_{n=1}^{r} &\left(\int_{\mathbf{R}^1} e^{i\lambda t} d(E_n(\lambda)f^{(n)}, f^{(n)})_{H_n}\right) \\ &\times \prod_{n=r+1}^{p} \left(\int_{\mathbf{R}^1} e^{i\lambda t}\, d\mu_n(\lambda)\right).\end{aligned}$$

Comparing this with (2.26), we conclude that $\lim_{p\to\infty}(\mathcal{U}_p(t)f, f)_{\mathcal{H}_{e,1}}$ exists uniformly with respect to t varying over a bounded interval. The totality of the set of vectors of the form (2.24) and the fact that

$$\|\mathcal{U}_p(t)\| = 1 \qquad (t \in \mathbf{R}^1;\ p = 1, 2, \dots),$$

give us the analogous convergence also for any $g \in \mathcal{H}_{e,1}$: it is necessary to estimate $|(\mathcal{U}_p(t)g, g)_{\mathcal{H}_{e,1}} - (\mathcal{U}_q(t)g, g)_{\mathcal{H}_{e,1}}|$ by adding and subtracting analogous terms in which one or both of the vectors g are replaced by a vector h in the l.s. of the vectors (2.24) that is close to g.

Thus, the conditions of Lemma 2.1 hold, and consequently, the limit (2.16) exists.

Conversely, suppose that (2.16) exists. Since

$$\left(\left(\bigstar_1^p \mathcal{E}_n\right)(\lambda)\varepsilon, \varepsilon\right)_{\mathcal{H}_{e,1}} = \left(\bigstar_1^p \mu_n\right)(\lambda),$$

where $\varepsilon = e^{(1)} \otimes e^{(2)} \otimes \cdots$, the convolution $(\bigstar_1^\infty \mu_n)(\lambda)$ also exists. On the basis of the same theorem of Kolmogorov we conclude that condition (2.20) is satisfied. ■

It is not hard to write simple sufficient conditions ensuring that the convolution (2.16) exists. *Such a condition is the convergence in the norm of* $\mathcal{H}_{e,1}$ *of the series*

$$\sum_{n=1}^{\infty} \mathcal{A}_n(e^{(1)} \otimes e^{(2)} \otimes \cdots), \qquad \mathcal{A}_n = \int_{\mathbf{R}^1} \lambda\, d\mathcal{E}_n(\lambda), \tag{2.27}$$

or the (more stringent) condition

$$\sum_{n=1}^{\infty} \|A_n e^{(n)}\|_{H_n} < \infty, \qquad A_n = \int_{\mathbf{R}^1} \lambda \, dE_n(\lambda). \tag{2.28}$$

Indeed, observe first of all that (2.28) implies (2.27): if $\varepsilon = e^{(1)} \otimes e^{(2)} \otimes \cdots$, then

$$\begin{aligned}\sum_{n=1}^{\infty} \|\mathcal{A}_n \varepsilon\|_{\mathcal{H}_{e,1}} &= \sum_{n=1}^{\infty} \|e^{(n)} \otimes \cdots \otimes e^{(n-1)} \otimes A_n e^n \otimes e^{(n+1)} \otimes e^{(n+2)} \otimes \cdots \|_{\mathcal{H}_{e,1}} \\ &= \sum_{n=1}^{\infty} \|A_n e^{(n)}\|_{H_n} < \infty.\end{aligned}$$

Further, it is not hard to show that (2.27) implies the conditions (2.20) with $c = \infty$: compute $\| \sum_{n=q+1}^{\infty} \mathcal{A}_n \varepsilon\|^2_{\mathcal{H}_{e,1}}$. However, in this case it is useful to prove Theorem 2.3 without using Kolmogorov's theorem. Thus, from the family $(\mathcal{E}_n)_1^{\infty}$ of commuting resolutions of the identity we construct the joint resolution of the identity $\mathcal{C}_\sigma(\mathbf{R}^\infty) \ni \alpha \mapsto \mathcal{E}(\alpha)$ according to §1.1. Then the operators (2.22) and (2.27) can be written in the form

$$\mathcal{V}_n(t) = \int_{\mathbf{R}^\infty} e^{i\lambda_n t} \, d\mathcal{E}(\lambda), \quad \mathcal{A}_n = \int_{\mathbf{R}^\infty} \lambda_n \, d\mathcal{E}(\lambda) \qquad (t \in \mathbf{R}^1;\ n = 1, 2, \ldots);$$

therefore, instead of (2.25) we can write the estimate

$$\begin{aligned}\|\mathcal{U}_p(t)f - \mathcal{U}_q(t)f\|^2_{\mathcal{H}_{e,1}} &= \left\| \left(1 - \prod_{n=q+1}^{p} \mathcal{V}_n(t)\right) \right\|^2_{\mathcal{H}_{e,1}} \\ &= \int_{\mathbf{R}^\infty} \left| 1 - \exp\left(it \sum_{n=q+1}^{p} \lambda_n \right) \right|^2 d(\mathcal{E}(\lambda)f, f)_{\mathcal{H}_{e,1}} \\ &\leq t^2 \int_{\mathbf{R}^\infty} \left(\sum_{n=q+1}^{p} \lambda_n \right)^2 d(\mathcal{E}(\lambda)f, f)_{\mathcal{H}_{e,1}} \\ &= t^2 \|f\|^2_{\mathcal{H}_{e,1}} \int_{\mathbf{R}^\infty} \left(\sum_{n=q+1}^{p} \lambda_n \right)^2 d(\mathcal{E}(\lambda)\varepsilon, \varepsilon)_{\mathcal{H}_{e,1}} \\ &= t^2 \|f\|^2_{\mathcal{H}_{e,1}} \left\| \sum_{n=q+1}^{p} \mathcal{A}_n \varepsilon \right\|^2_{\mathcal{H}_{e,1}} \end{aligned} \tag{2.29}$$

on each vector (2.24) for $p > q \geq r$. This implies that $\lim_{p\to\infty} \mathcal{U}_p(t)f$ exists for each $t \in \mathbf{R}^1$ and f of the form (2.24). As before, we conclude that this limit exists also for any $f \in \mathcal{H}_{e,1}$. Further, by (2.29),

$$|(\mathcal{U}_p(t)f, f)_{\mathcal{H}_{e,1}} - (\mathcal{U}_q(t)f, f)_{\mathcal{H}_{e,1}}| \\ \leq \|f\|_{\mathcal{H}_{e,1}} \|\mathcal{U}_p(t)f - \mathcal{U}_q(t)f\|_{\mathcal{H}_{e,1}} \to 0 \quad \text{as } p, q \to \infty$$

uniformly with respect to t varying over a finite interval. This implies the same convergence for any $f \in \mathcal{H}_{e,1}$. Thus, the conditions of Lemma 2.1 hold, and the limit (2.16) therefore exists. ∎

The conditions for the existence of the infinite convolution (2.16) simplify in the case when $\operatorname{supp} E_n \subseteq [0, \infty)$ for sufficiently large n. For example, we have

THEOREM 2.4. *If* $\operatorname{supp} E_n \subseteq [0, \infty)$ $(n = 1, 2, \ldots)$, *then the sequence* $((\bigstar_{n=1}^{p} \mathcal{E}_n)(\lambda))_{p=1}^{\infty}$ *of resolutions of the identity is nonincreasing for each* $\lambda \in \mathbb{R}^1$; *therefore, the limit* (2.16), *understood as a strong limit of operators, exists for* $\lambda \in \mathbb{R}^1$.

If, moreover, conditions a) *and* b) *of* (2.20) *are satisfied for some* $c > 0$, *then this limit is a resolution of the identity, and its existence can be understood in the sense of weak convergence of resolutions of the identity.*

PROOF. To see the first part of the theorem it obviously suffices to verify the inequality $(\mathcal{E}_1 \star \mathcal{E}_2)(\lambda) \leq \mathcal{E}_1(\lambda)$ $(\lambda \in \mathbb{R}^1)$, where the convolution is constructed in $\mathcal{H} = H_1 \otimes H_2$ according to (2.6), and $\operatorname{supp} E_n \subseteq [0, \infty)$ $(n = 1, 2)$. We have that

$$\begin{aligned}(\mathcal{E}_1 \star \mathcal{E}_2)(\lambda) &= \int_0^\infty (E_1(\lambda - \mu) \otimes 1)\, d(1 \otimes E_2(\mu)) \\ &\leq \int_0^\infty (E_1(\lambda) \otimes 1) d(1 \otimes E_2(\mu)) \\ &= E_1(\lambda) \otimes 1 = \mathcal{E}_1(\lambda) \qquad (\lambda \in \mathbb{R}^1).\end{aligned}$$

This estimate is easily justified as follows. In place of the integral we write a corresponding integral sum, and then use the fact that the function $\mathbb{R}^1 \ni \lambda \mapsto E_1(\lambda)$ is nondecreasing and the fact that if A, B, and C are three unbounded operators acting in H, $A \leq B$, and C is nonnegative and commutes with A and B, then $AC \leq BC$.

Hence, for each $\lambda \in \mathbb{R}^1$ the sequence $((\bigstar_{n=1}^{p} \mathcal{E}_n)(\lambda))_{p=1}^{\infty}$ of projections defined on $\mathcal{H}_{e,1}$ is nonincreasing and thus has a strong limit $\mathcal{F}(\lambda)$ which is also a projection. Obviously, $\mathcal{F}(\lambda') \leq \mathcal{F}(\lambda'')$ $(\lambda', \lambda'' \in \mathbb{R}^1;\ \lambda' < \lambda'')$ and $\mathcal{F}(-\infty) = 0$. However, $\mathcal{F}(+\infty) \neq 1$ in general. Therefore, additional conditions are needed for $\mathcal{F}$ to be a resolution of the identity (a simple example when $\mathcal{F}(+\infty) = 0$: let $E_n(\lambda) = 0$ for $\lambda \leq 1$ and $E_n(\lambda) = 1$ for $\lambda > 1$, $n = 1, 2, \ldots$).

We prove the second part of the theorem. Let

$$a_n = \int_0^c \lambda\, d(E_n(\lambda)e^{(n)}, e^{(n)})_{H_n} \geq 0.$$

Condition b) in (2.20) means that $\sum_1^\infty a_n < \infty$. Let us prove that this implies condition c). Indeed,

$$\sum_{n=1}^{\infty} \int_0^c \left[\lambda - \int_0^c \lambda \, d(E_n(\lambda)e^{(n)}, e^{(n)})_{H_n}\right]^2 d(E_n(\lambda)e^{(n)}, e^{(n)})_{H_n}$$
$$= \sum_{n=1}^{\infty} \left[\int_0^c \lambda^2 \, d(E_n(\lambda)e^{(n)}, e^{(n)})_{H_n} - 2a_n^2 + a_n^2 (E_n([0,c))e^{(n)}, e^{(n)})_{H_n}\right]$$
$$\le \sum_{n=1}^{\infty} (ca_n - a_n^2) < \infty.$$

Accordingly, in our case conditions a) and b) imply a)–c); therefore, the convolution (2.16) exists.

It is easy to see that the function $\mathcal{F}(\lambda)$ introduced above is equal to $(\bigstar_1^\infty \mathcal{E}_n)(\lambda)$ $(\lambda \in \mathbf{R}^1)$. Indeed, by Helly's theorem and the definition of the weak convergence in (2.16),

$$\int_{\mathbf{R}^1} \varphi(\lambda) \, d(\mathcal{F}(\lambda)f, f)_{\mathcal{H}_{e,1}} = \lim_{p \to \infty} \int_{\mathbf{R}^1} \varphi(\lambda) \, d\left(\left(\bigstar_{n=1}^{p} \mathcal{E}_n\right)(\lambda)f, f\right)_{\mathcal{H}_{e,1}}$$
$$= \int_{\mathbf{R}^1} \varphi(\lambda) \, d\left(\left(\bigstar_{n=1}^{\infty} \mathcal{E}_n\right)(\lambda)f, f\right)_{\mathcal{H}_{e,1}}$$

for an $f \in \mathcal{H}_{e,1}$ and for a $\varphi \in C(\mathbf{R}^1)$ with compact support. Since φ and f are arbitrary, this gives us the required equality. ■

As in the case of a finite convolution, we introduce the spectral measure corresponding to the resolution $\mathcal{E}_{\mathcal{L}}(\lambda) = (\bigstar_1^\infty \mathcal{E}_n)(\lambda)$ of the identity. To do this, consider a sequence

$$H_{-,n} \supseteq H_{0,n} \supseteq H_{+,n} \qquad (n = 1, 2, \ldots) \tag{2.30}$$

of riggings of the spaces $H = H_{0,n}$ such that $e^{(n)} \in H_{+,n}$, $\|e^{(n)}\|_{H_{+,n}} = 1$, the imbedding operators $O_n \colon H_{+,n} \to H_{0,n}$ are quasinuclear, and

$$\sum_{n=1}^{\infty} (|O_n|^2 - 1) < \infty. \tag{2.31}$$

Of course, the riggings (2.30) can always be constructed. Multiplying the chains (2.30), we get a rigging (1.5) of the space $\mathcal{H}_{e,1} = \mathcal{H}_{0,e,1}$ with the required properties (see §1.1). The spectral measure $\rho_{\mathcal{L}}$ is defined as follows:

$$\mathcal{B}(\mathbf{R}^1) \ni \delta \mapsto \rho_{\mathcal{L}}(\delta) = \operatorname{Tr}(\mathcal{O}^+ \mathcal{E}_{\mathcal{L}}(\delta) \mathcal{O}).$$

We show that, in a way similar to that in (2.10), the function $\rho_{\mathcal{L}}(\lambda)$ can be expressed as an infinite convolution of the functions $\rho_n(\lambda)$. The proof of Theorem 2.3 already called to mind the concept of an infinite convolution of distribution functions. The functions $\rho_n(\lambda)$ are not distribution functions, because $\rho_n(+\infty) = \rho_n(\mathbf{R}^1) = |O_n|^2 > 1$ $(n = 1, 2, \ldots)$. But the product $\prod_1^\infty |O_n|^2 \in (1, \infty)$ converges, by (2.31); therefore, if the convolution $(\bigstar_1^\infty \rho_n)(\lambda)$ is defined analogously, then it exists simultaneously with the convolution $(\bigstar_1^\infty \rho_n^*)(\lambda)$ of distribution functions, where $\rho_n^*(\lambda) = |O_n|^{-2}\rho_n(\lambda)$. It is not hard to show that the conditions (2.20) imply the conditions in Kolmogorov's theorem for the sequence $(\rho_n^*(\lambda))_1^\infty$, and hence both the last convolutions exist. However, we proceed in a simpler fashion.

THEOREM 2.5. *The formula*

$$\rho_{\mathcal{L}}(\lambda) = \left(\mathop{\bigstar}_{n=1}^{\infty} \rho_n\right)(\lambda) = \lim_{p\to\infty} \left(\mathop{\bigstar}_{n=1}^{p} \rho_n\right)(\lambda) \tag{2.32}$$

is valid for the spectral measure $\rho_{\mathcal{L}}$ corresponding to the resolution of the identity $\mathcal{E}_{\mathcal{L}}(\lambda) = (\bigstar_1^\infty \mathcal{E}_n)(\lambda)$. (The limit is understood in the sense of weak convergence of nondecreasing functions.)

PROOF. We use the following simple fact here. let $H_- \supseteq H_0 \supseteq H_+$ be a chain of separable Hilbert spaces, and $(A_p)_1^\infty$ a sequence of nonnegative operators A_p: $H_+ \to H_-$ strongly convergent to an operator A: $H_+ \to H_-$. Then A is also nonnegative, and $\mathrm{Tr}(A) = \lim_{p\to\infty} \mathrm{Tr}(A_p)$.

Since each bounded function in $C(\mathbf{R}^1)$ can be represented as a linear combination of four bounded nonnegative functions φ in $C(\mathbf{R}^1)$, to establish (2.32) it suffices to see that for such φ

$$\int_{\mathbf{R}^1} \varphi(\lambda)\, d\rho_{\mathcal{L}}(\lambda) = \lim_{p\to\infty} \int_{\mathbf{R}^1} \varphi(\lambda)\, d\left(\mathop{\bigstar}_{n=1}^{p} \rho_n\right)(\lambda). \tag{2.33}$$

With the help of (2.10) we have that

$$\begin{aligned}
\int_{\mathbf{R}^1} \varphi(\lambda)\, d\rho_{\mathcal{L}}(\lambda) &= \int_{\mathbf{R}^1} \varphi(\lambda)\, d(\mathrm{Tr}\,(O^+\mathcal{E}_{\mathcal{L}}(\lambda)O)) \\
&= \mathrm{Tr}\left(\int_{\mathbf{R}^1} \varphi(\lambda) d(O^+\mathcal{E}_{\mathcal{L}}(\lambda)O)\right) \\
&= \mathrm{Tr}\left(O^+\left(\int_{\mathbf{R}^1} \varphi(\lambda)\, d\mathcal{E}_{\mathcal{L}}(\lambda)\right) O\right) \\
&= \lim_{p\to\infty} \mathrm{Tr}\left(O^+\left(\int_{\mathbf{R}^1} \varphi(\lambda) d\left(\mathop{\bigstar}_{n=1}^{p} \mathcal{E}_n\right)(\lambda)\right) O\right) \\
&= \lim_{p\to\infty} \int_{\mathbf{R}^1} \varphi(\lambda) d\left(\mathrm{Tr}\left(O^+\left(\mathop{\bigstar}_{n=1}^{p} \mathcal{E}_n\right)(\lambda) O\right)\right) \\
&= \lim_{p\to\infty} \int_{\mathbf{R}^1} \varphi(\lambda) d\left(\mathop{\bigstar}_{n=1}^{p} \rho_n\right)(\lambda).
\end{aligned}$$

Here we interchanged Tr with the integral and the lim on the basis of the general fact mentioned, and we used relations in §2.1. The equality (2.33) is established. ■

2.3. *Separation of infinitely many variables.* In a way similar to that in the case of a finite convolution, we associate with an infinite convolution a certain selfadjoint operator $\mathcal{A}$ acting in $\mathcal{H}_{0,e,1}$ such that $\mathcal{E}_{\mathcal{L}} = \mathcal{E}_{\mathcal{A}}$ is its resolution of the identity.

Consider a family $(\mathcal{A}_n)_1^\infty$ of commuting selfadjoint operators acting with respect to different variables, as in §1.1, and let $\mathcal{C}_\sigma(\mathsf{R}^\infty) \ni \alpha \mapsto \mathcal{E}(\alpha)$ be the joint resolution of the identity constructed from the resolutions $\mathcal{B}(\mathsf{R}^1) \ni \delta \mapsto E_n(\delta)$ of the identity corresponding to the operators A_n acting in H_n $(n = 1, 2, \ldots)$.

THEOREM 2.6. *Assume that for some $c > 0$ the conditions* (2.20) *hold for the sequence $(E_n)_1^\infty$ of resolutions of the identity. Then the function*

$$\mathsf{R}^\infty \ni \lambda = (\lambda_n)_{n=1}^\infty \mapsto L(\lambda) = \sum_{n=1}^\infty \lambda_n \tag{2.34}$$

is defined and finite for almost all $\lambda \in \mathsf{R}^\infty$ with respect to the corresponding joint resolution of the identity $\mathcal{E}$. Moreover, a) $L_p(\lambda) = L_{p,c}(\lambda_1, \ldots, \lambda_p) = \lambda_1 + \cdots + \lambda_p \to L(\lambda)$ *as $p \to \infty$ for $\mathcal{E}$-almost all $\lambda \in \mathsf{R}^\infty$; and* b) *the resolutions of the identity $\mathcal{E}_{\mathcal{L}_p}$ of the selfadjoint operators $\mathcal{L}_p = L_{p,c}(\mathcal{A}_1, \ldots, \mathcal{A}_p)$ converge weakly to $(\bigstar_1^\infty \mathcal{E}_n)(\lambda)$ as $p \to \infty$, and thus the last convolution coincides with the resolution of the identity constructed according to* (1.30) *from $\mathcal{E}$ and* (2.34), *by Theorem* 1.3.

Conversely, if condition b) *holds for some sequence $(E_n)_1^\infty$ of resolutions of the identity, then for any $c > 0$ the conditions* (2.20) *also hold for it.*

PROOF. In essence, it need only be proved that (2.20) implies a). Indeed, this gives us the required convergence of the series (2.34). Further, according to §2.1, $\mathcal{E}_{\mathcal{L}_p}(\lambda) = (\bigstar_1^p \mathcal{E}_n)(\lambda)$; therefore, b) follows from Theorem 2.3. Conversely, if b) holds, then the convolution (2.16) exists, and the conditions (2.20) hold by virtue of the same Theorem 2.3.

Accordingly, suppose that the conditions (2.20) hold for some $c > 0$. We prove that $L_p(\lambda) \to L(\lambda)$ $\mathcal{E}$-almost everywhere as $p \to \infty$. Without loss of generality it can be assumed that riggings (2.30) with the property (2.31) have been constructed. The spectral measure ρ corresponding to $\mathcal{E}$ is introduced with the help of (1.5). It can be expressed in terms of the ρ_n according to (1.6): $\rho = \bigotimes_1^\infty \rho_n$. It suffices to prove that $L_p(\lambda) \to L(\lambda)$ ρ-almost everywhere as $p \to \infty$. The argument below is well known, in essence (Wintner [1], pp. 39–40, or Gikhman and Skorokhod [2], Chapter 2, §3).

We prove first of all that $(L_p(\lambda))_1^\infty$ is a Cauchy sequence with respect to the measure ρ. The equality (2.13) gives us that for $p > q$ and $\varepsilon > 0$

$$\begin{aligned}
&\rho(\{\lambda \in \mathbf{R}^\infty \mid |L_p(\lambda) - L_q(\lambda)| > \varepsilon\}) \\
&= \int_{|\sum_{q+1}^p \lambda_n| > \varepsilon} \cdots \int d\left(\bigotimes_{n=q+1}^p \rho_n\right)(\lambda_{q+1}, \dots, \lambda_p) \\
&= \int_{|\lambda| > \varepsilon} d\left(\mathop{\bigstar}_{n=q+1}^p \rho_n\right)(\lambda).
\end{aligned} \tag{2.35}$$

Since the convolution (2.32) exists here, we conclude from the scalar variant of Lemma 2.1 that $\prod_1^\infty \hat{\rho}_n(t)$ converges uniformly for t in any bounded interval of $\mathbf{R}^1$, where $\hat{\rho}_n(t) = \int_{\mathbf{R}^1} e^{i\lambda t}\, d\rho_n(t)$. Therefore, $\prod_q^p \hat{\rho}_n(t) \to 1$ as $p, q \to \infty$ in the sense of the same convergence. Returning to convolutions, we conclude that $(\bigstar_q^p \rho_n)(\lambda) \to \delta_0(\lambda)$ as $p, q \to \infty$ in the sense of weak convergence of nondecreasing functions, where δ_0 is the unit measure concentrated at $\lambda = 0$. This and (2.35) lead us to conclude that for any $\varepsilon, \delta > 0$ there exists an $m = m(\varepsilon, \delta)$ such that

$$\rho(\{\lambda \in \mathbf{R}^\infty \mid |L_p(\lambda) - L_q(\lambda)| > \varepsilon\}) < \delta; \qquad p, q > m. \tag{2.36}$$

Let us show that

$$\rho(\{\lambda \in \mathbf{R}^\infty \mid \max_{m < r \le p} |L_r(\lambda) - L_m(\lambda)| > 2\varepsilon\}) \le \delta / (\rho(\mathbf{R}^\infty) - \delta). \tag{2.37}$$

We introduce the following sets in $\mathcal{C}_\sigma(\mathbf{R}^\infty)$:

$$\begin{aligned}
\alpha_r = \{\lambda \in \mathbf{R}^\infty \mid |L_{m+1}(\lambda) - L_m(\lambda)| \le 2\varepsilon, \dots, |L_{r-1}(\lambda) - L_m(\lambda)| \le 2\varepsilon, \\
|L_r(\lambda) - L_m(\lambda)| > 2\varepsilon\},
\end{aligned}$$

$$\beta_r = \{\lambda \in \mathbf{R}^\infty \mid |L_p(\lambda) - L_m(\lambda)| \le \varepsilon\} \qquad (r = m+1, \dots, p).$$

It is easy to see that

$$\{\lambda \in \mathbf{R}^\infty \mid |L_p(\lambda) - L_m(\lambda)| > \varepsilon\} \supseteq \bigcup_{r=m+1}^p (\alpha_r \cap \beta_r),$$

$$\alpha_j \cap \alpha_k = \varnothing \qquad (j \ne k;\ j, k = m+1, \dots, p),$$

and $\rho(\alpha_r \cap \beta_r) = \rho(\alpha_r)\rho(\beta_r)$ $(r = m+1, \dots, p)$. Therefore,

$$\begin{aligned}
\delta &> \rho(\{\lambda \in \mathbf{R}^\infty \mid |L_p(\lambda) - L_m(\lambda)| > \varepsilon\}) \\
&\ge \rho\left(\bigcup_{r=m+1}^p (\alpha_r \cap \beta_r)\right) = \sum_{r=m+1}^p \rho(\alpha_r \cap \beta_r) \\
&= \sum_{r=m+1}^p \rho(\alpha_r)\rho(\beta_r) \ge (\rho(\mathbf{R}^\infty) - \delta) \sum_{r=m+1}^p \rho(\alpha_r) \\
&= (\rho(\mathbf{R}^\infty) - \delta)\rho(\{\lambda \in \mathbf{R}^\infty \mid \max_{m < r \le p} |L_r(\lambda) - L_m(\lambda)| > 2\varepsilon\}).
\end{aligned}$$

This implies (2.37).

Let $\alpha_{n,N} = \{\lambda \in \mathsf{R}^\infty \mid \sup_{j,k>n} |L_j(\lambda) - L_k(\lambda)| > N^{-1}\} \in \mathcal{C}_\sigma(\mathsf{R}^\infty)$. The series $\sum_1^\infty \lambda_n$ diverges on the set $\beta = \bigcup_{N=1}^\infty \bigcap_{n=1}^\infty \alpha_{n,N}$. We estimate $\rho(\beta)$. From the inclusion $\alpha_{n,N} \subseteq \{\lambda \in \mathsf{R}^\infty \mid \sup_{m<r\leq p} |L_r(\lambda) - L_m(\lambda)| > 2\varepsilon\}$ for $n < m$ and $N^{-1} \geq 2\varepsilon$ and from (2.37) it follows that

$$\begin{aligned}\rho(\alpha_{n,N}) &\leq \rho(\{\lambda \in \mathsf{R}^\infty \mid \sup_{m<r\leq p} |L_r(\lambda) - L_m(\lambda)| > 2\varepsilon\}) \\ &\leq \delta/(\rho(\mathsf{R}^\infty) - \delta) \qquad (n < m,\ N^{-1} \geq 2\varepsilon),\end{aligned}$$

and so $\rho(\bigcap_{n=1}^{m-1} \alpha_{n,N}) \leq \delta(\rho(\mathsf{R}^\infty) - \delta)^{-1}$. Since in (2.36) m can be made arbitrarily large and δ arbitrarily small, the last inequality shows that

$$\rho\left(\bigcap_{n=1}^\infty \alpha_{n,N}\right) = 0 \qquad (N^{-1} \geq 2\varepsilon).$$

The $\varepsilon > 0$ in (2.36) is arbitrary, so this equality holds for any $N = 1, 2, \ldots$. Therefore, $\rho(\beta) = 0$. Thus, $\sum_1^\infty \lambda_n$ converges for ρ-almost all $\lambda \in \mathsf{R}^\infty$. ■

Accordingly, we can interpret the infinite convolution $(\bigstar_1^\infty \mathcal{E}_n)(\lambda)$, (if it exists) as the resolution of the identity $\mathcal{E}_{\mathcal{L}}(\lambda)$ of the operator $\mathcal{L}$ acting in $\mathcal{H}_{0,e,1}$ constructed from $\mathcal{E}$ and (2.34) according to (1.30); this operator will also be denoted by $\mathcal{A}$, and $\mathcal{E}_{\mathcal{A}} = \mathcal{E}_{\mathcal{L}}$. (We thereby justify the notation introduced in §2.2.) Here, of course, all the results in §§1.3 and 1.4 are applicable. In particular, *Theorem* 1.2 *is valid for* $\mathcal{A}$, *where* $\mathbb{C}^1$ *is replaced by* R^1 *and* L *has the form* (2.34).

As in §2.1, we now pass to a direct construction of $\mathcal{A}$. Suppose that A_n is an Hermitian operator with domain $\mathfrak{D}(A_n)$ acting in $H_{0,n}$ $(n = 1, 2, \ldots)$. Fix some stabilizing sequence $e = (e^{(n)})_1^\infty$, $e^{(n)} \in \mathfrak{D}(A_n)$, and construct the tensor product $\mathcal{H}_{0,e,1} = \bigotimes_{n=1;e,1}^\infty H_{0,n}$. We define the Hermitian operators $\mathcal{A}'_n$ in $\mathcal{H}_{0,e,1}$ according to (1.1) and let

$$\begin{aligned}\mathcal{A}'_{\min} &= \lim_{p\to\infty} \sum_{n=1}^p (\underbrace{1 \otimes \cdots \otimes 1}_{n-1} \otimes A_n \otimes 1 \otimes 1 \otimes \cdots) f \\ &= \lim_{p\to\infty} \sum_{n=1}^p \mathcal{A}'_n f, \qquad f \in \mathfrak{D}(\mathcal{A}'_{\min}) = \text{a.}\left(\bigotimes_{n=1;e,1}^\infty \mathfrak{D}(A_n)\right) \qquad (2.38)\end{aligned}$$

in the sense of convergence in $\mathcal{H}_{0,e,1}$.

For $\mathcal{A}'_{\min}$ to be well defined it suffices to require conditions of the type (2.27) and (2.28): the series

$$\sum_{n=1}^\infty \mathcal{A}'_n (e^{(1)} \otimes e^{(2)} \otimes \cdots) \qquad (2.39)$$

converges in $\mathcal{H}_{0,e,1}$, or (a more stringent condition)

$$\sum_{n=1}^\infty \|A_n e^{(n)}\|_{H_n} < \infty. \qquad (2.40)$$

We explain that, since each vector $f \in \mathfrak{D}(\mathcal{A}'_{\min})$ has the form

$$f = f_q \otimes e^{(q+1)} \otimes e^{(q+2)} \otimes \cdots, \quad \text{where } f_q \in \text{a.}\left(\bigotimes_{n=1}^{q} \mathfrak{D}(A_n)\right),$$

convergence of (2.39) implies convergence of $\sum_1^\infty \mathcal{A}'_n f$ in $\mathcal{H}_{0,e,1}$, i.e., $\mathcal{A}'_{\min}$ is well defined. It is also clear that (2.40) implies convergence of (2.39)—even the series

$$\sum_{n=1}^{\infty} \|\mathcal{A}'_n(e^{(1)} \otimes e^{(2)} \otimes \cdots)\|_{\mathcal{H}_{0,e,1}}.$$

converges.

The condition (2.39) (or (2.40)) will always be required in what follows. The operator $\mathcal{A}'_{\min}$ is obviously Hermitian. Let $\mathcal{A}_{\min}$ be its closure. The following theorem generalizes Theorem 2.2.

THEOREM 2.7. *Let the operator $\mathcal{A} = \mathcal{L}$ be constructed from a sequence $(E_n)_1^\infty$ of resolutions of the identity according to Theorem 2.6, where each E_n corresponds to some selfadjoint extension of A_n in H_n (condition (2.39) is assumed).*

Then $\mathcal{A}$ is a selfadjoint extension of $\mathcal{A}_{\min}$ in $H_{0,e,1}$. If each operator A_n is essentially selfadjoint in H_n $(n = 1, 2, \ldots)$, then $\mathcal{A}_{\min}$ is selfadjoint and coincides with $\mathcal{A}$.

PROOF. Since $\int_{\mathbb{R}^1} \lambda \, dE_n(\lambda) \supseteq A_n$, it follows that $\int_{\mathbb{R}^1} \lambda \, d\mathcal{E}_n(\lambda) \supseteq \mathcal{A}'_n$ $(n = 1, 2, \ldots)$; therefore, (2.39) implies (2.27). Thus, the requirements (2.20) hold, and the operator $\mathcal{A} = \mathcal{L}$ can be constructed.

Let

$$f = f^{(1)} \otimes \cdots \otimes f^{(q)} \otimes e^{(q+1)} \otimes e^{(q+2)} \otimes \cdots \qquad (f^{(n)} \in \mathfrak{D}(A_n)), \qquad (2.41)$$

with $q = 1, 2, \ldots$ fixed. It is clear that $f \in \mathfrak{D}(\mathcal{A}'_{\min})$. We show that $f \in \mathfrak{D}(\mathcal{A})$. Let $\varphi \in C(\mathbb{R}^1)$ be nonnegative, at most 1, and compactly supported. If we set

$$L_p(\lambda) = L_{p,c}(\lambda_1, \ldots, \lambda_p) = \sum_{n=1}^{p} \lambda_n, \qquad \mathcal{L}_p = L_{p,c}(\mathcal{A}_1, \ldots, \mathcal{A}_p)$$

and take into account that $\mathcal{E}_{\mathcal{L}_p}(\lambda) = (\bigstar_1^p \mathcal{E}_n)(\lambda)$ (see §2.1), then with the help of the definition (2.16) we get that

$$\begin{aligned}
\int_{\mathbb{R}^1} \varphi(\lambda)\lambda^2 d(\mathcal{E}_{\mathcal{A}}(\lambda)f, f)_{\mathcal{H}_{0,e,1}} &= \lim_{p\to\infty} \int_{\mathbb{R}^1} \varphi(\lambda)\lambda^2 \, d(\mathcal{E}_{\mathcal{L}_p}(\lambda)f, f)_{\mathcal{H}_{0,e,1}} \\
&\leq \lim_{p\to\infty} \int_{\mathbb{R}^1} \lambda^2 d(\mathcal{E}_{\mathcal{L}_p}(\lambda)f, f)_{\mathcal{H}_{0,e,1}} \\
&= \lim_{p\to\infty} \|\mathcal{L}_p f\|^2_{\mathcal{H}_{0,e,1}} = \lim_{p\to\infty} \left\| \sum_{n=1}^{p} \mathcal{A}'_n f \right\|^2_{\mathcal{H}_{0,e,1}} \\
&= \|\mathcal{A}'_{\min} f\|^2_{\mathcal{H}_{0,e,1}} < \infty.
\end{aligned}$$

Here we have used the fact that $\mathcal{L}_p$ is an extension of the Hermitian operator acting in $\mathcal{H}_{0,e,1}$ by the relation $\mathfrak{D}(\mathcal{A}'_{\min}) \ni f \mapsto \sum_1^p \mathcal{A}'_n f$; this is an obvious consequence of Theorem 2.2. Since φ is arbitrary, the inequality obtained implies that $\int_{\mathbf{R}^1} \lambda^2 \, d(\mathcal{E}_{\mathcal{A}} f, f)_{\mathcal{H}_{0,e,1}} < \infty$, i.e., $f \in \mathfrak{D}(\mathcal{A})$.

Since $\mathfrak{D}(\mathcal{A}'_{\min})$ consists of the l.s. of the vectors of the form (2.41), it follows that $\mathfrak{D}(\mathcal{A}'_{\min}) \subseteq \mathfrak{D}(\mathcal{A})$; hence $\mathfrak{D}(\mathcal{A}_{\min}) \subseteq \mathfrak{D}(\mathcal{A})$.

Let us prove that $\mathcal{A}_{\min} \subset \mathcal{A}$. To do this it suffices to see that $\mathcal{A}'_{\min} f = \mathcal{A} f$ in the case of the vector f being considered here. By Theorem 2.6 and (1.30), we can write

$$\mathcal{A} = \int_{\mathbf{R}^\infty} \left(\sum_{n=1}^{\infty} \lambda_n \right) d\mathcal{E}(\lambda), \qquad \mathcal{L}_p = \int_{\mathbf{R}^\infty} \left(\sum_{n=1}^{p} \lambda_n \right) d\mathcal{E}(\lambda) \qquad (p = 1, 2, \ldots). \tag{2.42}$$

Let $\beta_N = \{\lambda \in \mathbf{R}^\infty \mid |\lambda_1| \leq N, |\lambda_1 + \lambda_2| \leq N, \ldots\} \in \mathcal{C}_\sigma(\mathbf{R}^\infty)$. It is clear from (2.42) that for any $g \in \mathcal{H}_{0,e,1}$

$$\begin{gathered}(\mathcal{E}(\beta_N)\mathcal{A} f, g)_{\mathcal{H}_{0,e,1}} = \int_{\beta_N} \left(\sum_{n=1}^{\infty} \lambda_n \right) d(\mathcal{E}(\lambda) f, g)_{\mathcal{H}_{0,e,1}}, \\ (\mathcal{E}(\beta_N)\mathcal{L}_p f, g)_{\mathcal{H}_{0,e,1}} = \int_{\beta_N} \left(\sum_{n=1}^{p} \lambda_n \right) d(\mathcal{E}(\lambda) f, g)_{\mathcal{H}_{0,e,1}} \\ (N, p = 1, 2, \ldots). \end{gathered} \tag{2.43}$$

For each fixed N the second integrals in (2.43) converge as $p \to \infty$ to the first integral due to Lebesgue's theorem on passing to the limit under the integral sign. Thus,

$$(\mathcal{E}(\beta_N)\mathcal{A} f, g)_{\mathcal{H}_{0,e,1}} = \lim_{p \to \infty} (\mathcal{E}(\beta_N)\mathcal{L}_p f, g)_{\mathcal{H}_{0,e,1}}.$$

But $\mathcal{L}_p f = \sum_1^p \mathcal{A}'_n f \to \mathcal{A}'_{\min} f$ as $p \to \infty$ in the norm of $\mathcal{H}_{0,e,1}$; therefore,

$$(\mathcal{E}(\beta_N)\mathcal{A} f, g)_{\mathcal{H}_{0,e,1}} = \mathcal{E}(\beta_N)\mathcal{A}'_{\min} f, g)_{\mathcal{H}_{0,e,1}} \qquad (N = 1, 2, \ldots). \tag{2.44}$$

Since $\sum_1^\infty \lambda_n$ converges for $\mathcal{E}$-almost all $\lambda \in \mathbf{R}^\infty$ by Theorem 2.6, its partial sums are bounded, and $\lambda \in \beta_N$ for some N. Thus, $\bigcup_1^\infty \beta_N$ has full $\mathcal{E}$-measure; hence $\mathcal{E}(\beta_1) \leq \mathcal{E}(\beta_2) \leq \cdots$, and $\lim_{n \to \infty} \mathcal{E}(\beta_n) = 1$ strongly. Passing to the limit in (2.44) as $N \to \infty$ and considering that g is arbitrary, we conclude that $\mathcal{A} f = \mathcal{A}'_{\min} f$.

Let us proceed to the proof of the last assertion in the theorem. We first establish a general fact. *Fix $p = 1, 2, \ldots$ and consider the following Hermitian operator acting in $\mathcal{H}_{0,e,1}$:*

$$\mathcal{A}^{(p)'}_{\min} f = \sum_{n=1}^{p} \mathcal{A}'_n f,$$

$$f \in \mathfrak{D}(\mathcal{A}^{(p)'}_{\min}) = \text{a.}\ (\mathfrak{D}(A_1) \otimes \cdots \otimes \mathfrak{D}(A_p) \otimes H_{p+1} \otimes H_{p+2} \otimes \cdots). \tag{2.45}$$

Let $\mathcal{A}_{\min}^{(p)}$ be its closure. If the vector $f = f^{(p)} \otimes e^{(p+1)} \otimes e^{(p+2)} \otimes \cdots$ is in $\mathfrak{D}(\mathcal{A}_{\min}^{(p)})$, where $f^{(p)} \in \bigotimes_1^p H_n$, then it is also in $\mathfrak{D}(\mathcal{A}_{\min})$.

Indeed, let $f \in \mathfrak{D}(\mathcal{A}_{\min}^{(p)})$. Then there exists a sequence $(f_m)_1^\infty$, $f_m \in \mathfrak{D}(\mathcal{A}_{\min}^{(p)'})$, of vectors such that $f_m \to f$ and $\sum_{n=1}^p \mathcal{A}'_n f_m \to \mathcal{A}_{\min}^{(p)} f$ in $\mathcal{H}_{0,e,1}$ as $m \to \infty$. Denote by P the projection from $\bigotimes_{n=p+1;l,1}^\infty H_n$ onto the vector $e^{(p+1)} \otimes e^{(p+2)} \otimes \cdots$, where $l = (e^{(n)})_{p+1}^\infty$. Then the function

$$g_m = (\underbrace{1 \otimes \cdots \otimes 1}_{p} \otimes P) f_m$$

converges to f, and

$$\begin{aligned} \mathcal{A}_{\min}^{(p)'} g_m &= \left(\sum_{n=1}^p \mathcal{A}'_n\right) g_m \\ &= (\underbrace{1 \otimes \cdots \otimes 1}_{p} \otimes P) \left(\sum_{n=1}^p \mathcal{A}'_n\right) f_m \underset{m\to\infty}{\longrightarrow} \mathcal{A}_{\min}^{(p)} f. \end{aligned} \tag{2.46}$$

Since $g_m \in \mathfrak{D}(\mathcal{A}'_{\min})$ and $g_m \to f$ in $\mathcal{H}_{0,e,1}$ as $m \to \infty$, to prove the assertion it suffices to show that the limit $\lim_{m\to\infty} \mathcal{A}'_{\min} g_m$ exists in $\mathcal{H}_{0,e,1}$. Let $f_m = f_m^{(p)} \otimes f_m^\infty$, where $f_m^{(p)} \in$ a. $\bigotimes_1^p \mathfrak{D}(\mathcal{A}_n)$ and $f_m^\infty \in \bigotimes_{n=p+1;l,1}^\infty H_n$ ($m = 1, 2, \ldots$). Then

$$g_m = f_m^{(p)} \otimes P f_m^\infty = c_m f_m^{(p)} \otimes e^{(p+1)} \otimes e^{(p+2)} \otimes \cdots,$$

where $P f_m^\infty = c_m e^{(p+1)} \otimes e^{(p+2)} \otimes \cdots$ ($c_m \in \mathbf{C}^1$), and

$$\begin{aligned} \mathcal{A}'_{\min} g_m = &\left(\left(\sum_{n=1}^p \underbrace{1 \otimes \cdots \otimes 1}_{n-1} \otimes A_n \otimes \underbrace{1 \otimes \cdots \otimes 1}_{p-n}\right) c_m f_n^{(p)}\right) \\ &\otimes \left(\left(\sum_{n=p+1}^\infty \underbrace{1 \otimes \cdots \otimes 1}_{n-1} \otimes A_n \otimes 1 \otimes 1 \otimes \cdots\right) e^{(p+1)} \right. \\ &\left. \otimes e^{(p+2)} \otimes \cdots \right). \end{aligned} \tag{2.47}$$

The convergence (2.46) in $\mathcal{H}_{0,e,1}$ obviously implies the convergence of the first factor in (2.47) in $\bigotimes_1^p H_n$, and the second factor does not depend on m. Therefore, it follows from (2.47) that $\lim_{m\to\infty} \mathcal{A}'_{\min} g_m$ exists in $\mathcal{H}_{0,e,1}$. ■

Accordingly, suppose that each operator A_n is selfadjoint in H_n ($n = 1, 2, \ldots$). To prove that $\mathcal{A}_{\min} \subseteq \mathcal{A}$ is selfadjoint and coincides with $\mathcal{A}$ it suffices to see that $\mathcal{R}_z f \in \mathfrak{D}(\mathcal{A}_{\min})$ for some nonreal $z \in \mathbf{C}^1$, where $\mathcal{R}_z$ is the resolvent of $\mathcal{A}$ and f is an arbitrary vector of the form (2.41).

By Theorem 2.6, the resolvent $\mathcal{E}_{\mathcal{A}} = \mathcal{E}_{\mathcal{L}}$ of the identity is the weak limit of the resolutions $\mathcal{E}_{\mathcal{L}_p}$ of the identity, so $\mathcal{R}_z f = \lim_{p\to\infty} h_p$ in $\mathcal{H}_{0,e,1}$, where

$$h_p = \int_{\mathbf{R}^1} (\lambda - z)^{-1} d\mathcal{E}_{\mathcal{L}_p}(\lambda) f. \tag{2.48}$$

It follows from Theorem 2.2 that the operator $\mathcal{A}^{(p)}_{\min}$ defined above is selfadjoint in $\mathcal{H}_{0,e,1}$, and $(\bigstar_1^p \mathcal{E}_n)(\lambda) = \mathcal{E}_{\mathcal{L}_p}(\lambda)$ is its resolution of the identity. Therefore, $h_p \in \mathfrak{D}(\mathcal{A}^{(p)}_{\text{in}})$, and $h_p \in \mathfrak{D}(\mathcal{A}_{\min})$ for $p \geq q$, by the assertion just proved. Rewriting (2.48) in the form of an integral with respect to the measure $\mathcal{E}$ and then applying the operator $\mathcal{A}_{\min} \subseteq \mathcal{A}$, we get with the help of Theorem 2.6 that

$$\mathcal{A}_{\min} h_p = \int_{\mathbf{R}^\infty} L(\lambda) \left(\left(\sum_{n=1}^{p} \lambda_n \right) - z \right)^{-1} d\mathcal{E}(\lambda) f \qquad (p = q, q+1, \ldots).$$

Hence,

$$\|\mathcal{A}_{\min} h_p - \mathcal{A}_{\min} h_r\|^2_{\mathcal{H}_{0,e,1}}$$
$$= \int_{\mathbf{R}^\infty} |L(\lambda)|^2 \left| \left(\left(\sum_{n=1}^{p} \lambda_n \right) - z \right)^{-1} - \left(\left(\sum_{n=1}^{r} \lambda_n \right) - z \right)^{-1} \right|^2 \times d(\mathcal{E}(\lambda) f, f)_{\mathcal{H}_{0,e,1}}$$
$$(p, r = q, q+1, \ldots,). \tag{2.49}$$

The integrand in (2.49) can be estimated from above by the function

$$4|\operatorname{Im} z|^{-2} |L(\lambda)|^2,$$

which is integrable with respect to $d(\mathcal{E}(\lambda)f, f)_{\mathcal{H}_{0,e,1}}$, because $f \in \mathfrak{D}(\mathcal{A}'_{\min}) \subseteq \mathfrak{D}(\mathcal{A})$. Therefore, we can pass to the limit as $p, r \to \infty$ under the integral sign in (2.49), and this gives us that $(\mathcal{A}_{\min} h_p)_q^\infty$ is a Cauchy sequence. Thus, if $p \geq q$, then $h_p \in \mathfrak{D}(\mathcal{A}_{\min})$, $h_p \to \mathcal{R}_z f$ as $p \to \infty$, and $\lim_{p\to\infty} \mathcal{A}_{\min} h_p$ exists. The fact that $\mathcal{A}_{\min}$ is closed implies that $\mathcal{R}_z f \in \mathfrak{D}(\mathcal{A}_{\min})$, and this is what was required. ■

The operator $\mathcal{A}_{\min}$ is said to be an operator admitting separation of infinitely many variables. In particular, $\mathcal{A}$ is the same, and the operators defined in H_n corresponding to the E_n play the role of the A_n $(n = 1, 2, \ldots)$.

It is easy to give a sufficient condition on a sequence $(A_n)_1^\infty$ of essentially selfadjoint operators A_n acting in H_n for it to be possible to choose a stabilization $e = (e^{(n)})_1^\infty$ such that (2.40) holds. Let $\rho(0, s(A_n))$ *be the distance from* 0 *to the spectrum* $s(A_n)$ *of* A_n. *If*

$$\sum_{n=1}^{\infty} \rho(0, s(A_n)) < \infty, \tag{2.50}$$

then such a choice is possible. Indeed, let $\lambda_n \in s(A_n)$ be such that $|\lambda_n| < \rho(0, s(A_n)) + n^{-2}$. We choose an "almost eigenvector $\varphi^{(n)}$ of A_n with eigenvalue λ_n". More precisely, we find a vector $\varphi^{(n)} \in H_n$, $\|\varphi^{(n)}\|_{H_n} = 1$, such that $\|A_n\varphi^{(n)} - \lambda_n\varphi^{(n)}\|_{H_n} < n^{-2}$ $(n = 1, 2, \ldots)$. It is well known that such vectors always exist. The last two inequalities imply that $\|A_n\varphi^{(n)}\|_{H_n} \leq \rho(0, s(A_n)) + 2n^{-2}$, and $\sum_1^\infty \|A_n\varphi^{(n)}\|_{H_n} < \infty$ by (2.50). The sequence $e = (e^{(n)})_1^\infty$, where $e^{(n)} = \varphi^{(n)}$, is the required one. ■

Hence, if (2.50) holds, then the results of this subsection are true for the sequence $(A_n)_1^\infty$ of operators when the stabilization is appropriately chosen. If (2.50) does not hold, then it can always be made to hold by replacing the A_n by translated operators $A_n - c_n 1$ $(c_n \in \mathbf{R}^1)$. Thus, the results indicated are preserved after such a "regularization" of the operators.

THEOREM 2.8. *The spectrum $s(\mathcal{A})$ of $\mathcal{A}$ has the form of the following limit of sets*:

$$s(\mathcal{A}) = \lim_{p\to\infty} \sum_{n=1}^{p} s(\mathcal{A}_n), \qquad s(\mathcal{A}_n) = \operatorname{supp} E_n. \tag{2.51}$$

PROOF. By Theorem 2.6, the remark at the end of §1.4, and (2.11) we can write

$$s(\mathcal{A}) \subseteq \lim_{p\to\infty} \left(\sum_{n=1}^{p} s(\mathcal{A}_n) \right)^{\sim}.$$

However, by the definition of a limit of sets, the last limit is equal to the limit in (2.51). Thus, the inclusion is proved in one direction.

Let us prove the inclusion

$$s(\mathcal{A}) \supseteq \lim_{p\to\infty} \sum_{n=1}^{p} s(A_n). \tag{2.52}$$

We use the following inequality, which is valid for any bounded nondecreasing functions $\mathbf{R}^1 \ni \lambda \mapsto \tau_1(\lambda)$, $\tau_2(\lambda) \in \mathbf{R}^1$, $\lambda_0 \in \mathbf{R}^1$ and $\varepsilon > 0$:

$$\int_{\lambda_0-\varepsilon}^{\lambda_0+\varepsilon} d(\tau_1 \star \tau_2)(\lambda) \geq \int_{\lambda_0-\varepsilon/2}^{\lambda_0+\varepsilon/2} d\tau_1(\lambda_1) \int_{-\varepsilon/2}^{\varepsilon/2} d\tau_2(\lambda_2). \tag{2.53}$$

This follows directly from (2.13), if we set $p = 2$ and $\varphi(\lambda) = \kappa_{(\lambda_0-\varepsilon, \lambda_0+\varepsilon)}(\lambda)$ and properly estimate $\varphi(\lambda_1 + \lambda_2)$ from below.

Consider the spectral measure (2.32) corresponding to the operator $\mathcal{L} = \mathcal{A}$; $s(\mathcal{A}) = \operatorname{supp} \rho_{\mathcal{L}}$. Since

$$\rho_{\mathcal{L}} = \mathop{\bigstar}_{n=1}^{\infty} \rho_n = \left(\mathop{\bigstar}_{n=1}^{p} \rho_n \right) \star \left(\mathop{\bigstar}_{n=p+1}^{\infty} \rho_n \right),$$

it follows from (2.53) that

$$\int_{\lambda_0-\varepsilon}^{\lambda_0+\varepsilon} d\rho_{\mathcal{L}}(\lambda) \geq \int_{\lambda_0-\varepsilon/2}^{\lambda_0+\varepsilon/2} d\left(\bigstar_{n=1}^{p} \rho_n\right)(\lambda_1) \int_{-\varepsilon/2}^{\varepsilon/2} d\left(\bigstar_{n=p+1}^{\infty} \rho_n\right)(\lambda_2)$$
$$(p = 1, 2, \ldots). \quad (2.54)$$

In proving Theorem 2.6 we already explained that $(\bigstar_{p+1}^{\infty} \rho_n)(\lambda)$ converges to $\delta_0(\lambda)$ as $p \to \infty$. Therefore, for $p > p_0$ the last integral in (2.54) can be estimated from below by $1/2$ when p_0 is sufficiently large.

Suppose now that λ_0 belongs to the right-hand set in (2.52) and not to the left-hand one. Then there is an $\varepsilon > 0$ such that the left-hand integral in (2.54) vanishes. By the inequality already proved,

$$\left(\bigstar_{n=1}^{p} \rho_n\right)\left(\left[\lambda_0 - \frac{\varepsilon}{2}, \lambda_0 + \frac{\varepsilon}{2}\right)\right) = 0$$

for $p > p_0$, and thus the interval $(\lambda_0 - \varepsilon/2, \lambda_0 + \varepsilon/2)$ does not contain points in the set

$$\operatorname{supp} \bigstar_{n=1}^{p} \rho_n = \operatorname{supp} \rho_{\mathcal{L}_p} = \operatorname{supp} \mathcal{E}_{\mathcal{L}_p} = \left(\sum_{n=1}^{p} s(A_n)\right)^{\sim}$$

(see (2.10) and (2.12)). We have arrived at a contradiction. Accordingly, (2.52) is proved. ■

§3. Differential operators with infinitely many variables

To a sufficient degree this section is of an illustrative nature: here we indicate what the results of §2 on separation of infinitely many variables are in the important case when the A_n are certain second-order differential operators, as well as how to use the evolution equation techniques in §6 of Chapter 2 for the simplest situations to prove selfadjointness of a perturbation of such an operator with separating variables.

3.1. ***Some facts about general operators with separating variables.*** We recall some constructions from §2 of Chapter 1 and §2 of Chapter 3 in a convenient form. As earlier, let A be the countable set of all sequences of the form $\alpha = (\alpha_n)_1^\infty$, where $\alpha_n = 0, 1, \ldots$, and $\alpha_{n+1} = \alpha_{n+2} = \cdots = 0$ beginning with some index n depending on α (i.e., each α is a finitely nonzero sequence). We make A into a metric space by setting the distance $\rho(\alpha, \beta)$ between $\alpha, \beta \in$ A equal to the minimal $m = 0, 1, \ldots$ such that $\alpha_{m+1} = \beta_{m+1}$, $\alpha_{m+2} = \beta_{m+2}, \ldots$. Since the sequences in A are finitely nonzero, the distance is defined for all $\alpha, \beta \in$ A; obviously, $\rho(\alpha, \beta) = \rho(\beta, \alpha) \geq 0$, with equality if and only if $\alpha = \beta$. The triangle inequality follows from the estimate

$$\rho(\alpha, \beta) \leq \max(\rho(\alpha, \gamma), \rho(\gamma, \beta)) \leq \rho(\alpha, \gamma) + \rho(\gamma, \beta) \qquad (\alpha, \beta, \gamma \in \mathrm{A}).$$

We write $0 = (0, 0, \ldots) \in \mathrm{A}$, and let B_n be the closed ball about 0 with radius n: $\mathrm{B}_n = \{\alpha \in \mathrm{A} | \rho(\alpha, 0) \leq n\}$ $(n = 0, 1, \ldots)$. Then $\mathrm{A} = \bigcup_0^\infty \mathrm{A}_n = \bigcup_0^\infty \mathrm{B}_n$, where the A_n are the disjoint layers $\mathrm{A}_n = \mathrm{B}_n \backslash \mathrm{B}_{n-1}$ $(n = 1, 2, \ldots;\ \mathrm{A}_0 = \mathrm{B}_0 = \{0\})$ introduced in §2.3 of Chapter 1. The number $\nu(\alpha) = \rho(\alpha, 0)$ is the minimal $m = 0, 1, \ldots$ such that $\alpha_{m+1} = \alpha_{m+2} = \cdots = 0$ $(\alpha \in \mathrm{A})$. Therefore,

$$\mathrm{B}_n = \{\alpha \in \mathrm{A} | \alpha_{n+1} = \alpha_{n+2} = \cdots = 0\}.$$

Let $(H_n)_1^\infty$ be a sequence of separable Hilbert spaces, and let $e = (e^{(n)})_1^\infty$ $(e^{(n)} \in H_n)$ be a particular sequence of unit vectors in the H_n. Consider in each H_n an orthonormal basis $(e_j^{(n)})_{j=0}^\infty$ such that $e_0^{(n)} = e^{(n)}$, form the formal product

$$e_\alpha = e_{(\alpha_1, \alpha_2, \ldots)} = e_{\alpha_1}^{(1)} \otimes e_{\alpha_2}^{(2)} \otimes \cdots \qquad (\alpha = (\alpha_n)_1^\infty \in \mathrm{A}), \tag{3.1}$$

and let the vectors e_α span a Hilbert space by regarding them as mutually orthogonal unit vectors. The Hilbert space constructed in this way was called the tensor product of the spaces H_n with stabilization e; as earlier, we denote it by $\mathcal{H}_e = \bigotimes_{n=1;e}^\infty H_n$. The vectors in $\mathcal{H}_e$ thus have the form

$$f = \sum_{\alpha \in \mathrm{A}} f_\alpha e_\alpha \quad (f_\alpha \in \mathbb{C}^1), \qquad \sum_{\alpha \in \mathrm{A}} |f_\alpha|^2 < \infty;$$

$$(f, g)_{\mathcal{H}_e} = \sum_{\alpha \in \mathrm{A}} f_\alpha \bar{g}_\alpha. \tag{3.2}$$

For vectors

$$f^{(n)} = \sum_{j=0}^\infty f_j^{(n)} e_j^{(n)} \in H_n \qquad (f_j^{(n)} = (f^{(n)}, e_j^{(n)})_{H_n};\ n = 1, \ldots, p)$$

we defined

$$\begin{aligned} f &= f^{(1)} \otimes \cdots \otimes f^{(p)} \otimes e^{(p+1)} \otimes e^{(p+2)} \otimes \cdots \\ &= \sum_{\alpha_1, \ldots, \alpha_p = 0}^\infty f_{\alpha_1}^{(1)} \cdots f_{\alpha_p}^{(p)} e_{\alpha_1}^{(1)} \otimes \cdots \otimes e_{\alpha_p}^{(p)} \otimes e^{(p+1)} \\ &\qquad \otimes e^{(p+2)} \otimes \cdots \in \mathcal{H}_e. \end{aligned} \tag{3.3}$$

This definition is unambiguous, and the vectors f themselves form a total subset of $\mathcal{H}_e$ (for $p = 1, 2, \ldots$). For two vectors f and g of the form (3.3) we have

$$(f, g)_{\mathcal{H}_e} = \prod_{n=1}^p (f^{(n)}, g^{(n)})_{H_n}. \tag{3.4}$$

We introduce the Hilbert space $\hat{\mathcal{H}}_e$ of functions $\hat{f}$ of the form $\mathrm{A} \ni \alpha \mapsto \hat{f}_\alpha \in \mathbb{C}^1$, where $\sum_{\alpha \in \mathrm{A}} |\hat{f}_\alpha|^2 < \infty$ and $(\hat{f}, \hat{g})_{\hat{\mathcal{H}}_e} = \sum_{\alpha \in \mathrm{A}} \hat{f}_\alpha \overline{\hat{g}_\alpha}$; $\mathcal{H}_e$ and $\hat{\mathcal{H}}_e$

are, of course, isomorphic, namely, $\mathcal{H}_e \ni f \mapsto \hat{f} \in \hat{\mathcal{H}}_e$. A vector $f \in \mathcal{H}_e$ is said to be cylindrical if the corresponding function $\hat{f}$ has bounded support (with respect to the metric ρ, i.e., $f_\alpha = 0$ outside some ball B_n); (3.3) is an example of such a vector. The linear span of the cylindrical vectors is clearly dense in $\mathcal{H}_e$. The collection $\mathcal{H}_e^{(m)}$ of all cylindrical vectors f such that $\operatorname{supp} \hat{f} \subseteq \mathrm{B}_m$ ($m = 0, 1, \ldots$ is fixed) is a subspace of $\mathcal{H}_e$ isomorphic to the finite tensor product $\bigotimes_1^m H_n$. Denote by $\mathfrak{C}_m$ the projection of $\mathcal{H}_e$ onto $\mathcal{H}_e^{(m)}$; its image $\hat{\mathfrak{C}}_m$ in $\hat{\mathcal{H}}_e$ is equal to the operator of multiplication by the characteristic function κ_{B_m} of the ball B_m: $(\hat{\mathfrak{C}}_m \hat{f})_\alpha = \kappa_{\mathrm{B}_{m,\alpha}} \hat{f}_\alpha$ ($\alpha \in \mathrm{A}$; $\hat{f} \in \hat{\mathcal{H}}_e$). Obviously, $\mathfrak{C}_m f \to f \in \mathcal{H}_e$ in $\mathcal{H}_e$ as $m \to \infty$.

Assume that an Hermitian operator A_n acts with domain $\mathfrak{D}(A_n) \ni e^{(n)}$ ($n = 1, 2, \ldots$) in each space H_n. In §2 we constructed an operator $\mathcal{A}'_{\min}$ acting in $\mathcal{H}_e$ from these A_n; in the sense of norm convergence in $\mathcal{H}_e$,

$$\mathcal{A}'_{\min} f = \lim_{p \to \infty} \sum_{n=1}^{p} (\underbrace{1 \otimes \cdots \otimes 1}_{n-1} \otimes A_n \otimes 1 \otimes 1 \otimes \cdots) f. \tag{3.5}$$

Here $f \in \mathfrak{D}(\mathcal{A}'_{\min}) = \mathrm{a.}\,(\bigotimes_{n=1;e}^{\infty} \mathfrak{D}(A_n))$. (The last set is defined as the linear span of the vectors of the form (3.3) with $f^{(n)} \in \mathfrak{D}(A_n)$ ($n = 1, 2, \ldots$)). If, for example,

$$\sum_{n=1}^{\infty} \|A_n e^{(n)}\|_{H_n} < \infty, \tag{3.6}$$

then the limit in (3.5) exists and defines an Hermitian operator. The closure $(\mathcal{A}'_{\min})^{\sim} = \mathcal{A}_{\min}$ is an operator admitting separation of infinitely many variables. This operator is selfadjoint if each A_n is essentially selfadjoint in H_n (Theorem 2.7).

Let us dwell on a very convenient particular situation which ensures the possibility of constructing a simpler way to compute the action of $\mathcal{A}_{\min}$. Namely, assume that the following condition holds:

$$A_n e^{(n)} = 0 \qquad (n = 1, 2, \ldots). \tag{3.7}$$

The operator $\mathcal{A}'_{\min}$ is well defined here, and its action on a vector $f \in (\mathfrak{D}(\mathcal{A}'_{\min})) \cap \mathcal{H}_e^{(m)}$ can obviously be written in the form

$$\mathcal{A}'_{\min} f = \Big(\sum_{n=1}^{m} \left(\underbrace{1 \otimes \cdots \otimes 1}_{n-1} \otimes A_n \otimes \underbrace{1 \otimes \cdots \otimes 1}_{m-n} \right) f_{(m)} \Big) \otimes e^{(m+1)} \otimes e^{(m+2)} \otimes \cdots \qquad (m = 1, 2, \ldots). \tag{3.8}$$

We clarify this equality. In correspondence with the decomposition

$$\mathcal{H}_e = \left(\bigotimes_{n=1}^{m} H_n \right) \otimes \left(\bigotimes_{n=m+1}^{\infty} H_n \right)$$

into factors, the vector f can be represented in the form $f = f_{(m)} \otimes e^{(m+1)} \otimes e^{(m+2)} \otimes \cdots$, where $f_{(m)} \in \bigotimes_1^m H_n$; the first factor in (3.8) is also in $\bigotimes_1^m H_n$.

Thus, in the case of annihilation the action of $\mathcal{A}'_{\min}$ on each linear set $(\mathfrak{D}(\mathcal{A}'_{\min})) \cap \mathcal{H}_e^{(m)}$ $(m = 1, 2, \ldots)$ is isomorphic to the action in $\bigotimes_1^m H_n$ of a naturally defined operator which admits separation of finitely many variables. The following lemma plays an important role in the case under consideration.

LEMMA 3.1. *Assume the annihilation condition* (3.7). *Then the operators* $\mathcal{A}_{\min}$ *and* $\mathfrak{C}_m$ *commute: if* $f \in \mathfrak{D}(\mathcal{A}_{\min})$, *then* $\mathfrak{C}_m f \in \mathfrak{D}(\mathcal{A}_{\min})$ *and* $\mathcal{A}_{\min}\mathfrak{C}_m f = \mathfrak{C}_m \mathcal{A}_{\min} f$ $(m = 1, 2, \ldots)$.

PROOF. We first prove the lemma in the case when $\mathcal{A}_{\min}$ is replaced by $\mathcal{A}'_{\min}$. Here it suffices to establish it for a vector f of the form (3.3), where $f_n \in \mathfrak{D}(A_n)$. If $p \le m$, then $f, \mathcal{A}'_{\min} f \in \mathcal{H}_e^{(m)}$ (see (3.8)), and the assertion is obvious. Suppose now that $p > m$. According to the rules for computing $\mathfrak{C}_m \leftrightarrow \hat{\mathfrak{C}}_m$,

$$\mathfrak{C}_m f = f^{(1)} \otimes \cdots \otimes f^{(m)} \otimes (f_0^{(m+1)} e^{(m+1)}) \otimes \cdots \otimes (f_0^{(p)} e^{(p)}) \otimes e^{(p+1)} \otimes e^{(p+2)} \otimes \cdots.$$

Thus, $\mathfrak{C}_m f \in \mathfrak{D}(\mathcal{A}'_{\min})$, and (3.8) gives us

$$\begin{aligned}\mathcal{A}'_{\min}\mathfrak{C}_m f = \left(\prod_{n=m+1}^{p} f_0^{(n)}\right) ((A_1 f^{(1)}) \otimes f^{(2)} \otimes \cdots \otimes f^{(m)} + \cdots + f^{(1)} \otimes \cdots \\ \otimes f^{(m-1)} \otimes (A_m f^{(m)})) \otimes e^{(m+1)} \otimes e^{(m+2)} \otimes \qquad (3.9)\end{aligned}$$

Again by (3.8) (with $m = p$),

$$\begin{aligned}\mathcal{A}'_{\min} f = ((A_1 f^{(1)}) \otimes f^{(2)} \otimes \cdots \otimes f^{(p)} + \cdots + f^{(1)} \otimes \cdots \otimes f^{(p-1)} \\ \otimes (A_p f^{(p)})) \otimes e^{(p+1)} \otimes \cdots \qquad (3.10)\end{aligned}$$

Apply the projection $\mathfrak{C}_m$ to this expression. Passing to $\hat{\mathfrak{C}}_m$ as before, we get that for $n = 1, \ldots, m$

$$\begin{aligned}&\mathfrak{C}_m(f^{(1)} \otimes \cdots \otimes f^{(n-1)} \otimes (A_n f^{(n)}) \otimes f^{(n+1)} \otimes \cdots \otimes f^{(p)} \otimes e^{(p+1)} \otimes \cdots) \\ &= f^{(1)} \otimes \cdots \otimes f^{(n-1)} \otimes (A_n f^{(n)}) \otimes f^{(n+1)} \otimes \cdots \otimes f^{(m)} \\ &\otimes (f_0^{(m+1)} e^{(m+1)}) \otimes \cdots \otimes (f_0^{(p)} e^{(p)}) \otimes e^{(p+1)} \otimes e^{(p+2)} \otimes \cdots \qquad (3.11)\end{aligned}$$

For $n = m + 1, \ldots, p$

$$\begin{aligned}&\mathfrak{C}_m(f^{(1)} \otimes \cdots \otimes f^{(n-1)} \otimes (A_n f^{(n)}) \otimes f^{(n+1)} \otimes \cdots \\ &\qquad \otimes f^{(p)} \otimes e^{(p+1)} \otimes e^{(p+2)} \otimes \cdots) \\ &= f^{(1)} \otimes \cdots \otimes f^{(m)} \otimes (f_0^{(m+1)} e^{(m+1)}) \otimes \cdots \otimes \otimes (f_0^{(n-1)} e^{(n-1)}) \\ &\otimes ((A_n f^{(n)})_0 e^{(n)}) \otimes (f_0^{(n+1)} e^{(n+1)}) \otimes \cdots \otimes (f_0^{(p)} e^{(p)}) \otimes \cdots \qquad (3.12)\end{aligned}$$

Considering the Hermitian property of the A_n and (3.7), we get that

$$(A_n f^{(n)})_0 = (A_n f^{(n)}, e^{(n)})_{H_n} = (f^{(n)}, A_n e^{(n)})_{H_n} = 0.$$

Thus, all the expressions (3.12) are equal to zero.

By (3.10), the vector $\mathfrak{C}_m \mathcal{A}'_{\min} f$ is equal to the sum of (3.11) and (3.12), which coincides with (3.9) because the vectors in (3.12) are equal to zero. Accordingly, the lemma is proved for the operator $\mathcal{A}'_{\min}$.

We proceed to prove the lemma for $\mathcal{A}_{\min}$. Let $f \in \mathfrak{D}(\mathcal{A}_{\min})$ and let $(f_n)_1^\infty$ be a sequence of vectors $f_n \in \mathfrak{D}(\mathcal{A}'_{\min})$ such that $f_n \to f$ and $\mathcal{A}'_{\min} f_n \to \mathcal{A}_{\min} f$ in $\mathcal{H}_e$ as $n \to \infty$. Then, by what has been proved and by the continuity of $\mathfrak{C}_m$, $\mathfrak{C}_m f_n \in \mathfrak{D}(\mathcal{A}'_{\min})$, and $\mathcal{A}'_{\min} \mathfrak{C}_m f_n = \mathfrak{C}_m \mathcal{A}'_{\min} f_n \to \mathfrak{C}_m \mathcal{A}'_{\min} f$. This means that $\mathfrak{C}_m f \in \mathfrak{D}(\mathcal{A}_{\min})$ and $\mathcal{A}_{\min} \mathfrak{C}_m f = \mathfrak{C}_m \mathcal{A}_{\min} f$. ■

3.2. *Differential operators with infinitely many separating variables.* Consider the particular case of the construction in §3.1 when the role of H_n is played by the spaces of square-integrable functions of a single variable, and the A_n are differential operators. For this we introduce the product measure

$$\begin{aligned} d\rho(x) &= (p_1(x_1)\,dx_1) \otimes (p_2(x_2)dx_2) \otimes \cdots \\ &= (d\rho_1(x_1)) \otimes (d\rho_2(x_2)) \otimes \cdots \\ &\qquad (p_n(x_n)\,dx_n = d\rho(x_n);\ n = 1, 2, \ldots) \end{aligned} \tag{3.13}$$

on the space $\mathsf{R}^\infty = \mathsf{R}^1 \times \mathsf{R}^1 \times \cdots$ of points $x = (x_n)_1^\infty$ $(x_n \in \mathsf{R}^1)$ where $(p_n)_1^\infty$ is a fixed sequence of positive twice continuously differentiable probability weights $\mathsf{R}^1 \ni x_n \mapsto p_n(x_n) \in (0, \infty)$. The space $L_2(\mathsf{R}^\infty, d\rho(x))$ can be understood as the infinite tensor product

$$\bigotimes_{n=1;e}^{\infty} L_2(\mathsf{R}^1, d\rho_n(x_n))$$

with the "identity" stabilization $e = (e^{(n)})_1^\infty$, $e^{(n)} = 1$. To say that a function $f \in L_2(\mathsf{R}^\infty, d\rho(x))$ is cylindrical now means that there exists an $m = m_f = 1, 2, \ldots$ and an $f_c \in L_2(\mathsf{R}^m, d\rho_{(m)}(x^{(m)}))$ $(x^{(m)} = (x_1, \ldots, x_m)$, $d\rho_{(m)}(x^{(m)}) = \bigotimes_1^m (d\rho_n(x_n)))$ such that $f(x) = f_c(x^{(m)})$ $(x \in \mathsf{R}^\infty)$. (The notation $x^{(m)} \in \mathsf{R}^m$ for a point and $d\rho_{(m)}(x^{(m)})$ for a finite product of measures introduced here will be used below in subsections 3 and 6 without mention.)

The mapping

$$L_2(\mathsf{R}^1, dx_n) \ni U(x_n) \mapsto u(x_n) = p_n^{-1/2}(x_n) U(x_n) \in L_2(\mathsf{R}^1, d\rho_n(x_n)) \tag{3.14}$$

is an isometry between these spaces of square-integrable functions. It carries $(\partial U/\partial x_n)(x_n)$ into the sandwiched (by means of p_n) derivative

$$\begin{aligned} (D_n u)(x_n) &= p_n^{-1/2}(x_n)(\partial/\partial x_n)(p_n^{1/2}(x_n) u(x_n)) \\ &= (\partial u/\partial x_n)(x_n) + (D_n 1)(x_n) u(x_n), \end{aligned} \tag{3.15}$$

and it carries $(\partial^2 U/\partial x_n^2)(x_n)$ into the corresponding square:

$$\begin{aligned} (D_n^2 u)(x_n) &= p_n^{-1/2}(x_n) \frac{\partial^2}{\partial x_n^2}(p_n^{1/2}(x_n) u(x_n)) \\ &= \frac{\partial^2 u}{\partial x_n^2}(x_n) + 2(D_n 1)(\bar{x}_n) \frac{\partial u}{\partial x_n}(x_n) + (D_n^2 1)(x_n) u(x_n). \end{aligned} \tag{3.16}$$

The mapping (3.14) obviously does not change the operation of multiplication by a function: $q(x_n)U(x_n)$ goes into $q(x_n)u(x_n)$.

From $-D_n^2$ we construct the minimal operator A_n acting in $L_2(\mathbf{R}^1, d\rho(x_n))$, i.e., we consider the closure of the operator $C_0^\infty(\mathbf{R}^1) \ni u(x_n) \mapsto -(D_n^2 u)(x_n)$ in this space. Since $C_0^\infty(\mathbf{R}^1)$ is carried into $C_0^2(\mathbf{R}^1)$ under the mapping (3.14), it follows from well-known properties of the minimal operator generated by the second derivative in $L_2(\mathbf{R}^1, dx_n)$ that A_n is selfadjoint in $L_2(\mathbf{R}^1, d\rho_n(x_n))$, is nonnegative, and has spectrum $s(A_n) = [0, \infty)$. It is essential for us that this operator is defined on bounded smooth functions. Thus, denote by $C_{b,l}^\infty(\mathbf{R}^p)$ the collection of functions in $C^\infty(\mathbf{R}^p)$ $(p = 1, 2, \ldots)$ that are bounded along with their derivatives through order $l = 0, 1, \ldots$.

LEMMA 3.2. *Assume that the weight p_n satisfies the condition*

$$D_n^2 1 \in L_2(\mathbf{R}^1, d\rho_n(x_n)). \tag{3.17}$$

Then the operator $C_{b,2}^\infty(\mathbf{R}^1) \ni u \mapsto A_n'' u = -D_n^2 u$ is defined in $L_2(\mathbf{R}^1, d\rho_n(x_n))$ and admits a closure; its closure coincides with A_n.

PROOF. Let us first see that (3.17) implies that

$$D_n 1 \in L_2(\mathbf{R}^1, d\rho_n(x_n)). \tag{3.18}$$

In the course of the proof we use the notation $t = x_n, \varphi = p_n^{1/2}, (\partial/\partial x_n) = {}'$ and $d\rho_n(x_n) = d\sigma(t) = \varphi^2(t)\, dt$. According to (3.15) and (3.16),

$$(D_n 1)(t) = \varphi^{-1}(t)\varphi'(t) \qquad (D_n^2 1)(t) = \varphi^{-1}(t)\varphi''(t),$$

and thus (3.17) and (3.18) pass into the respective inclusions

$$\varphi'' \in L_2(\mathbf{R}^1, dt), \qquad \varphi' \in L_2(\mathbf{R}^1, dt). \tag{3.19}$$

It must be verified that the first of them implies the second.

By the Cauchy-Schwarz-Bunyakovskiĭ equality, the equality

$$\varphi'(t) - \varphi'(0) = \int_0^t \varphi''(\tau)\, d\tau \qquad (t \in \mathbf{R}^1)$$

and the inclusion $\varphi'' \in L_2(\mathbf{R}^1, dt)$ give us that $|\varphi'(t)| \leq c_1 + c_2|t|^{1/2}$ $(t \in \mathbf{R}^1)$ for some $c_1, c_2 > 0$. Since $\varphi \in L_2(\mathbf{R}^1, dt)$, we conclude that $\varphi'\varphi(1+|t|^{1/2})^{-1} \in L_2(\mathbf{R}^1, dt)$. But the even function $(1+|t|^{1/2})^{-1}$ is not in $L_2(\mathbf{R}^1, dt)$; therefore, there exist sequences $(-r_m)_1^\infty$ and $(s_m)_1^\infty$ such that $r_m, s_m \to +\infty$ as $m \to \infty$, and $\varphi'(-r_m)\varphi(-r_m), \varphi'(s_m)\varphi(s_m) \to 0$.

By integration by parts,

$$\int_{-r_m}^{s_m} \varphi''(t)\varphi(t)\, dt = \varphi'(s_m)\varphi(s_m) - \varphi'(-r_m)\varphi(-r_m) - \int_{-r_m}^{s_m} (\varphi'(t))^2\, dt.$$

Passing to the limit as $m \to \infty$, we conclude that $\varphi' \in L_2(\mathbf{R}^1, dt)$.

We proceed to the proof of the lemma. By(3.16), it follows from (3.17), (3.18), and the fact that u, u', and u'' are bounded that the operator A_n'' is well defined. It suffices for us to verify that for any $u \in C^\infty_{b,2}(\mathsf{R}^1)$ there is a sequence $(u_R)_1^\infty$, $u_R \in C_0^\infty(\mathsf{R}^1)$, such that $u_R \to u$ and $D_n^2 u_R \to D_n^2 u$ in $L_2(\mathsf{R}^1, d\sigma(t))$ as $R \to \infty$.

Consider an infinitely differentiable function $\mathsf{R}^1 \ni t \mapsto \chi_R(t) \in [0,1]$ equal to 1 for $|t| \leq R$ and 0 for $|t| \geq R+1$ whose graphs on $(-R-1, -R)$ and $(R, R+1)$ are obtained by translating the corresponding graphs for $R = 1$. Let $u_R(t) = \chi_R(t)u(t)$ ($t \in \mathsf{R}^1$; $R = 1, 2, \ldots$). Obviously, $u_R \in C_0^\infty(\mathsf{R}^1)$ and $u_R \to u$ in $L_2(\mathsf{R}^1, d\sigma(t))$ as $R \to \infty$. Differentiating termwise in (3.16), we then find that

$$\begin{aligned} &\|D_n^2 u - D_n^2 u_R\|_{L_2(\mathsf{R}^1, d\sigma(t))} \\ &\quad \leq \|(D_n^2 u)(\cdot) - \chi_R(\cdot)(D_n^2 u)(\cdot)\|_{L_2(\mathsf{R}^1, d\sigma(t))} + 2\|\chi_R'(\cdot)u(\cdot)\|_{L_2(\mathsf{R}^1, d\sigma(t))} \\ &\qquad + \|\chi_R''(\cdot)\|_{L_2(\mathsf{R}^1, d\sigma(t))} + 2\|\chi_R'(\cdot)(D_n 1)(\cdot)u(\cdot)\|_{L_2(\mathsf{R}^1, d\sigma(t))}. \end{aligned} \tag{3.20}$$

It follows from the inclusions $D_n^2 u, u', (D_n 1)(\cdot)u(\cdot) \in L_2(\mathsf{R}^1, d\sigma(t))$ and the form of χ_R that the right-hand side of (3.20) tends to zero as $R \to \infty$. Accordingly, $D_n^2 u_R \to D_n^2 u$. ■

Thus, if (3.17) holds, then $1 \in \mathfrak{D}(A_n)$, and hence it is possible to construct the corresponding operator $\mathcal{A}_{\min}$ in the case of the identity stabilization. Condition (3.6) now has the form

$$\sum_{n=1}^\infty \|D_n^2 1\|_{L_2(\mathsf{R}^1, d\rho_n(x_n))} = \sum_{n=1}^\infty \left\| \frac{\partial^2}{\partial x_n^2} \sqrt{p_n} \right\|_{L_2(\mathsf{R}^1, dt)} < \infty. \tag{3.21}$$

This condition is always assumed to be satisfied when $\mathcal{A}_{\min}$ is constructed from the A_n introduced here. It implies, in particular, that the function

$$p(x) = \sum_{n=1}^\infty (D_n^2 1)(x) \tag{3.22}$$

is defined as a series convergent in $L_2(\mathsf{R}^\infty, d\rho(x))$ and is in this space; of course, (3.17) follows from (3.21).

Accordingly, assume (3.21). The operators A_n ($n = 1, 2, \ldots$) and the identity stabilization are used to construct the operator $\mathcal{A}_0 = \mathcal{A}_{\min}$ according to (3.5). (We write the index 0 because this operator will be perturbed in what follows.) By §3.1 it is selfadjoint and (since all the A_n are nonnegative) nonnegative. The general formula for the spectrum of an operator admitting separation of infinitely many variables (Theorem 2.8) gives us that the spectrum $s(\mathcal{A}_0)$ of $\mathcal{A}_0$ is $[0, \infty)$.

This operator can also be defined in a way somewhat different from the general scheme (by taking a different initial domain). Namely, denote by $C^\infty_{c,b,l}(\mathsf{R}^\infty)$ the linear set (dense in $L_2(\mathsf{R}^\infty, d\rho(x))$) of all cylindrical functions u of the form

$$\mathsf{R}^\infty \ni x \mapsto u(x) = u_c(x_1, \ldots, x_p)$$

where $u_c \in C^\infty_{b,l}(\mathsf{R}^p)$ ($p = p(u)$; $l = 0, 1, \ldots$). *The operator*

$$C^\infty_{c,b,2}(\mathsf{R}^\infty) \ni u(x) \mapsto (\mathcal{L}_0 u)(x) = -\sum_{n=1}^{\infty}(D_n^2 u)(x) \tag{3.23}$$

is well defined in $L_2(\mathsf{R}^\infty, d\rho(x))$ *and admits a closure, and this closure coincides with* $\mathcal{A}_0$.

Indeed, if $u = u_c \in C^\infty_{b,2}(\mathsf{R}^p)$, then

$$(\mathcal{L}_0 u)(x) = -\sum_{n=1}^{p}(D_n^2 u_c)(x) - u_c(x)\sum_{n=p+1}^{\infty}(D_n^2 1)(x),$$

and $\mathcal{L}_0 u \in L_2(\mathsf{R}^\infty, d\rho(x))$ in view of (3.22). This shows that the operator (3.23) is well defined. To finish the proof it suffices to see that if $u \in C^\infty_{c,b,2}(\mathsf{R}^\infty)$, then there exists a sequence

$$(u_m)_{m=1}^\infty, \qquad u_m \in \text{a.}\left(\bigotimes_{n=1}^{\infty} \mathfrak{D}(A_n'')\right)$$

such that $u_m \to u$ and $\mathcal{L}u_m \to \mathcal{L}u$ in $L_2(\mathsf{R}^\infty, d\rho(x))$ as $m \to \infty$. The function u has the form $u = u_c \in C^\infty_{b,2}(\mathsf{R}^p)$. But in the space $C^2_{b,2}(\mathsf{R}^p)$ of twice continuously differentiable functions on R^p that are bounded along with their first and second derivatives, the functions of the form $u^{(1)}(x_1)\cdots u^{(p)}(x_p)$, where $u^{(n)} \in C^\infty_{b,2}(\mathsf{R}^1)$, form a total set in the sense of uniform convergence on R^p of the functions and their first and second derivatives. This implies the possibility of choosing the required sequence, since the convergence $u_m \to u$ in $C^2_{b,2}(\mathsf{R}^p)$ as $m \to \infty$ implies the convergence $u_m \to u$ and $D_n^2 u_m \to D_n^2 u$ ($n = 1, 2, \ldots$) in $L_2(\mathsf{R}^\infty, d\rho(x))$ as $m \to \infty$. (The latter in view of (3.17) and (3.18), which hold for each $n = 1, 2, \ldots$ because of (3.21).) ■

REMARK 1. Denote by $C^\infty_{0,c}(\mathsf{R}^\infty)$ the (nonlinear) collection of all cylindrical infinitely differentiable compactly supported functions, i.e., all the functions of the form $\mathsf{R}^\infty \ni x \mapsto \varphi(x) = \varphi_c(x_1, \ldots, x_p)$, where $\varphi_c \in C^\infty_0(\mathsf{R}^p)$, $p = p(\varphi)$. Consider the operator of the form (3.23), but defined not on $C^\infty_{c,b,2}(\mathsf{R}^\infty)$ but on l.s. $(C^\infty_{0,c}(\mathsf{R}^\infty)) \subset C^\infty_{c,b,2}(\mathsf{R}^\infty)$. It is easy to show that its closure coincides with the closure $\mathcal{A}_0$ of the operator (3.23); this follows from the abstract definition of $\mathcal{A}_0$ if the operators A_n are restricted to $C^\infty_0(\mathsf{R}^1)$.

3.3. ***Differential operators with infinitely many separating variables.*** Generally speaking, $D_n^2 1 \neq 0$ ($n = 1, 2, \ldots$) in the constructions of §3.2, and thus the annihilation condition (3.7) is not satisfied. At the same time, as pointed out, this condition makes it possible to study more fully the corresponding operator with separating variables, perturbations of it, and so on. Therefore, it is possible to proceed as follows. We alter each operator A_n by adding a potential such that the annihilation condition is satisfied, and we study the corresponding operator with separating variables, along with perturbations of it. In the absence of annihilation, operators appear which differ from the preceding ones by the additive term (3.22).

In place of condition (3.21) *the following weaker restriction will now be assumed*: *the inclusions* (3.17) *hold for each* $n = 1, 2, \ldots$. Consider the Hermitian operator

$$\begin{aligned} C_0^\infty(\mathsf{R}^1) \ni u \mapsto B'_n u &= (A'_n u)(x_n) + (D_n^2 1)(x_n) u(x_n) \\ &= -(D_n^2 u)(x) + (D_n^2 1)(x_n) u(x_n) \\ &= -(d^2u/dx_n^2)(x_n) - 2(D_n 1)(x_n)(du/dx_n)(x_n) \\ &\qquad\qquad (n = 1, 2, \ldots) \quad (3.24) \end{aligned}$$

acting in $L_2(\mathsf{R}^1, d\rho_n(x_n))$, and let $B_n = (B'_n)^\sim$.

As in the case of the operators A_n, it is possible to define an operator B''_n by the same formula (3.24), but on the class $C_{b,2}^\infty(\mathsf{R}^1)$. *This definition is unambiguous, the operator admits a closure, and*

$$(B''_n)^\sim = B_n. \tag{3.25}$$

Indeed, the inclusion $D_n 1 \in L_2(\mathsf{R}^1, d\rho_n(x_n))$ implies that the definition of B''_n is unambiguous; it is clearly an Hermitian operator, and thus admits a closure. We establish (3.25). Since $B_n \subseteq (B''_n)^\sim$, it suffices to see that for each $u \in C_{b,2}^\infty(\mathsf{R}^1)$ there exists a sequence $(u_R)_1^\infty$, $u_R \in C_0^\infty(\mathsf{R}^1)$, such that $u_R \to u$ and

$$-(D_n^2 u_R)(x_n) + (D_n^2 1)(x_n) u_R(x_n) \to -(D_n^2 u)(x_n) + (D_n^2 1)(x_n) u(x_n) \tag{3.26}$$

in $L_2(\mathsf{R}^1, d\rho_n(x_n))$ as $R \to \infty$. This sequence can be taken to be the corresponding sequence in Lemma 2.2, since the first terms in (3.26) converge for it (by this lemma), while the second terms converge because of the nature of the convergence of u_R and the fact that $D_n^2 1 \in L_2(\mathsf{R}^1, d\rho_n(x_n))$. ■

The spectrum of the operators B_n can differ essentially from the spectrum of A_n. For example, in the case of a Gaussian weight $p_1(x_1) = \pi^{-1/2} e^{-x_1^2}$ the differential expression in (3.24) has the form

$$-(d^2u/dx_1^2)(x_1) + 2x_1(du/dx_1)(x_1),$$

and hence (as is well known) $s(B_1) = (\lambda_m)_0^\infty$, $\lambda_m = 2m$ (Lemma 4.2 in Chapter 1). The operator is thus nonnegative (although now $(D_1^2 1(x_1) = x_1^2 - 1 \leq 0$ for $x_1 \in [-1, 1])$. Condition (3.6) is obviously satisfied for the operators B_n: it follows from (3.25) and (3.24) and $1 \in \mathfrak{D}(B_n)$ and

$$B_n 1 = 0 \qquad (n = 1, 2, \ldots). \tag{3.27}$$

It is now possible to define the Hermitian operator $\mathcal{B}_0 = \mathcal{B}_{\min}$ according to (3.5) with the operators B_n instead of A_n; condition (3.7) is satisfied because of (3.27). Just below we shall see that the operators B_n ($n = 1, 2, \ldots$), hence also $\mathcal{B}_0$, are selfadjoint. We make a few more remarks about $\mathcal{B}_0$.

As in the case of the operator $\mathcal{A}_0 = \mathcal{A}_{\min}$, we prove that the operator

$$C^\infty_{\mathrm{c},b,2}(\mathsf{R}^\infty) \ni u(x) \mapsto (\mathcal{M}_0 u)(x)$$
$$= \sum_{n=1}^\infty (-(D_n^2 u)(x) + (D_n^2 1)(x_n) u(x))$$
$$= \sum_{n=1}^\infty \left(- \left(\frac{\partial^2 u}{\partial x_n^2} \right)(x) - 2(D_n 1)(x_n) \left(\frac{\partial u}{\partial x_n} \right)(x) \right). \tag{3.28}$$

is unambiguously defined in $L_2(\mathsf{R}^\infty, d\rho(x))$. It admits a closure, and this closure coincides with $\mathcal{B}_0$. It is essential to observe that, contrary to the case for $\mathcal{A}_0$,

$$\mathcal{B}_0 1 = 0. \tag{3.29}$$

If condition (3.21) also holds, then the function (3.22) is defined, and

$$(\mathcal{M}_0 u)(x) = - \sum_{n=1}^\infty (D_n^2 u)(x) + p(x) u(x)$$
$$= (\mathcal{L}_0 u)(x) + p(x) u(x) \qquad (u \in C^\infty_{\mathrm{c},b,2}(\mathsf{R}^\infty);\ x \in \mathsf{R}^\infty). \tag{3.30}$$

We dwell on the question of selfadjointness for the B_n ($n = 1, 2, \ldots$). For each n the operator B_n is nonnegative. This follows from the following equality, which can be obtained with the help of integration by parts and the relation $(dp_n/dx_n)(x_n) = 2(D_n 1)(x_n) p_n(x)$ ($x_n \in \mathsf{R}^1$) (see (3.15)):

$$\int_{\mathsf{R}^1} \left(- \left(\frac{d^2 u}{dx_n^2} \right)(x_n) - 2(D_n 1)(x_n) \left(\frac{du}{dx_n} \right)(x_n) \right) \bar{v}(x_n) p(x_n)\, dx_n$$
$$= \int_{\mathsf{R}^1} \left(\frac{du}{dx_n} \right)(x_n) \left(\frac{dv}{dx_n} \right)(x_n) p(x_n)\, dx_n \qquad (u, v \in C_0^\infty(\mathsf{R}^1)).$$

The minimal operator in $L_2(\mathsf{R}^1, dx_n)$ equal to the closure of the operator

$$C_0^\infty(\mathsf{R}^1) \ni U(x_n) \mapsto -(d^2 U/dx_n^2)(x_n) + (D_n^2 1)(x_n) U(x_n) \tag{3.31}$$

is carried into B_n under the mapping (3.14). The fact that B_n is nonnegative implies that this closure is also nonnegative. But then (as is well known: see, for example, Berezanskiĭ [4], Theorem 1.8 in Chapter 6) the closure is selfadjoint. Therefore, B_n is also selfadjoint.

The operators B_n ($n = 1, 2, \ldots$) are thus all selfadjoint and nonnegative. Consequently, so is $\mathcal{B}_0$.

LEMMA 3.3. *The nonnegative selfadjoint operator $\mathcal{B}_0$ acting in the space $L_2(\mathsf{R}^\infty, d\rho(x))$ is such that the semigroup $e^{-t\mathcal{B}_0}$ is a contractive semigroup in each space $L_p(\mathsf{R}^\infty, d\rho(x))$ ($p \in [1, \infty]$) and preserves positivity.*

PROOF. We apply the result (Reed and Simon [2], Theorem X.55) formulated at the end of §6.6 in Chapter 2. By (3.29), the nonnegative selfadjoint operator $\mathcal{B}_0$ satisfies the condition $\mathcal{B}_0 1 = 0$, so $e^{-t\mathcal{B}_0}1 = 1$ ($t \in [0,\infty)$), and it is thus only necessary to see that the semigroup $e^{-t\mathcal{B}_0}$ preserves positivity.

Let $\mathcal{B}_m$ be the closure of the operator (3.28) in $L_2(\mathbb{R}^\infty, d\rho(x))$, where the upper limit ∞ is replaced by $m = 1, 2, \ldots$ in the sums. The operator $\mathcal{B}_m$ corresponds to the case when $A_n = B_n$ ($n = 1, \ldots, m$) and $A_n = 0$ ($n \geq m+1$) in (3.5); hence it is selfadjoint according to the general facts mentioned in subsection 1. It is easy to understand that $\mathcal{B}_m \to \mathcal{B}_0$ in the strong resolvent sense as $m \to \infty$. Indeed, by a known result (see Reed and Simon [1], Theorem VIII.25) it suffices to see that $\mathcal{B}_m u \to \mathcal{B}_0 u$ as $m \to \infty$ in $L_2(\mathbb{R}^\infty, d\rho(x))$ for each $u \in C^\infty_{\mathrm{c},b,2}(\mathbb{R}^\infty)$. But $u(x) = u_{\mathrm{c}}(x_1, \ldots, x_p)$ ($p = p(u)$), and thus $\mathcal{B}_m u = \mathcal{B}_0 u$ for $m \geq p$, by (3.27). The required convergence has been established.

By the properties of strong resolvent convergence,

$$F(\mathcal{B}_m)f \underset{m\to\infty}{\longrightarrow} F(\mathcal{B}_0)f \qquad (f \in L_2(\mathbb{R}^\infty, d\rho(x)))$$

in $L_2(\mathbb{R}^\infty, d\rho(x))$, where $F(\lambda) = e^{-\lambda t}$ for $\lambda \geq 0$, and $F(\lambda) = 1$ for $\lambda < 0$ ($t \in [0,\infty)$ is fixed). In other words, in $L_2(\mathbb{R}^\infty, d\rho(x))$

$$e^{-t\mathcal{B}_m} f \underset{m\to\infty}{\longrightarrow} e^{-t\mathcal{B}_0} f \qquad (f \in L_2(\mathbb{R}^\infty, d\rho(x));\ t \in [0,\infty)). \tag{3.32}$$

(The relation (3.32), of course, follows at once from Theorem 2.6.)

Suppose that $f \in L_2(\mathbb{R}^\infty, d\rho(x))$ and $f(x) \geq 0$ almost everywhere; it must be seen that $(e^{-t\mathcal{B}_0} f)(x) \geq 0$ almost everywhere ($t \in [0,\infty)$). Since the operator $e^{-t\mathcal{B}_0}$ is continuous, density arguments give us that it suffices to see this for $f \in C^\infty_{\mathrm{c},b,2}(\mathbb{R}^\infty)$, i.e., for a function of the form $f(x) = f_{\mathrm{c}}(x_1, \ldots, x_p)$, $f_{\mathrm{c}} \in C^\infty_{b,2}(\mathbb{R}^p)$ ($p = p(f)$). If $m > p$ is fixed, then $u(x,t) = (e^{-t\mathcal{B}_m} f_{\mathrm{c}})(x)$ is a solution of the Cauchy problem

$$\left(\frac{\partial u}{\partial t}\right)(x,t) = \sum_{n=1}^{m} \left(\left(\frac{\partial^2 u}{\partial x_n^2}\right)(x,t) + 2(D_n 1)(x_n)\left(\frac{\partial u}{\partial x_n}\right)(x,t)\right),$$
$$u(x,0) = f_{\mathrm{c}}(x_1, \ldots, x_p) \qquad (x \in \mathbb{R}^p;\ t \in [0,\infty)) \tag{3.33}$$

and thus does not depend on $x_{m+1}, x_{m+2}, \ldots$. Accordingly,

$$u(x,t) = u_{\mathrm{c}}(x_1, \ldots, x_m, t),$$

where u_{c} satisfies the parabolic equation (3.33), with $D_n 1 \in C^1(\mathbb{R}^1)$, x replaced by $(x_1, \ldots, x_m)$, and the initial condition

$$u_{\mathrm{c}}(x_1, \ldots, x_m, 0) = f_{\mathrm{c}}(x_1, \ldots, x_p) \geq 0 \qquad ((x-1, \ldots, x_m) \in \mathbb{R}^m).$$

By well-known properties of solutions of parabolic equations (see, for example, Dynkin [1], Chapter 5, §6), $u(x_1, \ldots, x_m, t) \geq 0$ ($(x_1, \ldots, x_m) \in \mathbb{R}^m$, $t \in [0,\infty)$). Thus, for the f under consideration we have $(e^{-t\mathcal{B}_m} f)(x) \geq 0$ ($x \in \mathbb{R}^\infty$, $t \in [0,\infty)$) when m is sufficiently large. It follows from (3.32) that the same inequality holds also for $e^{-t\mathcal{B}_0}$ for almost all $x \in \mathbb{R}^\infty$. ■

We mention an important case of weights satisfying the requirements of subsections 2 and 3. We refer to Gaussian weights (see §4.2 in Chapter 1)

$$p_n(x_n) = \sqrt{\frac{\varepsilon_n}{\pi}} e^{-\varepsilon_n x_n^2} \qquad (x_n \in \mathbf{R}^1;\ \varepsilon_n > 0;\ n = 1, 2, \ldots),$$

$$\sum_{n=1}^{\infty} \varepsilon_n < \infty. \tag{3.34}$$

It is easy to compute that the last condition in (3.34) is equivalent to (3.21).

Our goal in what follows is to get selfadjointness conditions for the operators (3.28) or (3.23), perturbed by a potential.

3.4. ***First digression. Selfadjointness of differential operators in the case of finitely many variables.*** It is most convenient to demonstrate the application of the evolution technique of §6 in Chapter 2 to the proof of selfadjointness for differential operators first for the simpler case of second-order elliptic differential operators with finitely many variables. We consider the nonelementary case of singular coefficients and observe that this technique leads to sufficiently difficult results here. To understand the selfadjointness results presented in §3.6 in the case of infinitely many variables it is only necessary to become familiar with the proof of Lemma 3.4 below.

We consider Lebesgue measure dx on $\mathbf{R}^N$ and the spaces $L_p = L_p(\mathbf{R}^N, dx)$ ($p \in [1, \infty]$). As usual, $C^{l+\alpha}(\mathbf{R}^N)$ ($l = 0, 1, \ldots;\ \alpha \in (0, 1)$) denotes the class of functions in $C^l(\mathbf{R}^N)$ whose lth derivatives satisfy a Hölder condition with exponent α and with a constant not depending on the point under consideration.

Let us take an elliptic differential expression in $\mathbf{R}^N$ of the form

$$(\mathcal{L}u(x) = -\sum_{j,k=1}^{N} \left(\frac{\partial}{\partial x_j}\left(a_{jk}\frac{\partial u}{\partial x_k}\right)\right)(x) + q(x)u(x) \tag{3.35}$$

with real coefficients satisfying the following conditions:

1) the a_{jk} are locally bounded in $\mathbf{R}^N$ (i.e., are bounded on each compact set) and have derivatives $\partial a_{jk}/\partial x_j \in L_{p,\mathrm{loc}}$ for some $p > N$ ($j, k = 1, \ldots, N;\ N \geq 2$);

2) the potential q is in $L_{2,\mathrm{loc}}$.

The following ellipticity condition is assumed: there is a $\sigma > 0$ such that for any $\xi \in \mathbf{C}^N$

$$\sigma|\xi|^2 \leq \sum_{j,k=1}^{N} a_{jk}(x)\xi_j\bar{\xi}_k \qquad (x \in \mathbf{R}^N). \tag{3.36}$$

Let

$$c(r) = \sup_{|x| \leq r} \|(a_{jk}(x))_{j,k=1}^{N}\|^{1/2} \qquad (r \in [0, \infty)),$$

where $\|\cdot\|$ denotes the norm of the nonnegative operator on $\mathbb{C}^N$ generated by the matrix under consideration (i.e., its maximal eigenvalue), and require also the condition

$$\int_0^\infty c^{-1}(r)\,dr = \infty. \tag{3.37}$$

(This condition can actually be weakened: the supremum in the definition of $c(r)$ can be taken only over the sphere $|x| = r$; see the Comments on the Literature for references to papers with precise formulations.)

From the expression for $\mathcal{L}$ we construct in the usual way the minimal operator L: $L = \tilde{L}'$, where $C_0^\infty(\mathbb{R}^N) \ni u \mapsto L'u = \mathcal{L}u$. If all the coefficients of $\mathcal{L}$ are sufficiently smooth, then L is selfadjoint (Berezanskiĭ [4], Chapter 6, Theorem 1.8, and [11]). Such a result fails in general for the case when the a_{jk} are smooth but the potential is not. For example, according to Simon [2], for $N \geq 5$ the operator L constructed from the expression

$$(\mathcal{L}u)(x) = -(\Delta u)(x) - \nu|x|^{-2} \cdot u(x) \qquad (\nu \in \mathbb{R}^1),$$

which has the form (3.35), is semibounded below if and only if $\nu \leq \frac{1}{4}(N-1)(N-3)+\frac{1}{4}$, and is selfadjoint if and only if $\nu < \frac{1}{4}(N-1)(N-3)-\frac{3}{4}$. Thus, there exist ν such that L is semibounded below but not selfadjoint.

On the other hand, in the case of sufficiently smooth a_{jk} it is known that if not L but q is semibounded below, then L is selfadjoint. We show in Theorem 3.1 that local semiboundedness of q and semiboundedness of L instead of q actually suffice for L to be selfadjoint.

THEOREM 3.1. *Assume that the potential q is locally semibounded below (i.e., $\inf_{|x|\leq r} q(x) > -\infty$ for each $r > 0$). If the operator L is semibounded below (i.e., $(\mathcal{L}u, u)_{L_2} \geq l\|u\|^2_{L_2}$ $(l \in \mathbb{R}^1;\ u \in C_0^\infty(\mathbb{R}^N))$), then it is selfadjoint.*

We proceed as follows in order not to encumber the main part of the proof with details connected with weak smoothness of the leading coefficients. We first give a proof in subsection 4 under the assumption that $a_{jk} \in C^{1+\alpha}(\mathbb{R}^N)$ for some $\alpha \in (0,1)$. We then show in subsection 5 how the proof goes without this smoothness assumption.

First of all, let us prove the following special case of the theorem by means of the parabolic approach.

LEMMA 3.4. *Suppose that $a_{jk} \in C^{1+\alpha}(\mathbb{R}^N)$ for some $\alpha \in (0,1)$ and these coefficients are bounded on $\mathbb{R}^N$ along with their first derivatives. If q is in L_2 and is semibounded below, then L is selfadjoint.*

We present the proof in steps.

I. Denote by A_0 the minimal operator in L_2 generated from the expression with $q = 0$. It is known that A_0 is a nonnegative selfadjoint operator. (If the a_{jk} are smooth, then the proof of this fact is simple; see Remark 1 below in §3.5; in the same subsection we give a proof that A_0 is selfadjoint even with smoothness restrictions weaker than $C^{1+\alpha}(\mathbf{R}^N)$ on the coefficients, namely, with the restrictions for which Theorem 3.1 is stated.)

The set $D = \mathfrak{D}(A_0) \cap L_\infty$ contains $C_0^\infty(\mathbf{R}^N)$ and is thus a core for A_0. Since $q = 0$, $A_0 1 = 0$; by the same arguments as in Lemma 3.3, the semigroup e^{-tA_0} is contractive in L_∞. Therefore, according to Theorem 6.14 in Chapter 2, the operator $D \ni f \mapsto Af = (A_0 f)(x) + q(x)f(x)$ is essentially selfadjoint. To prove that L is selfadjoint it suffices to show that

$$D \subseteq \mathfrak{D}((A_0 \restriction C_0^\infty(\mathbf{R}^N))^\sim). \tag{3.38}$$

The proof of (3.38) consists of establishing that any function $f \in D$ can be approximated by functions in $C_0^\infty(\mathbf{R}^N)$ in the sense of the graph norm of the operator A_0.

II. First of all, let us prove that

$$\mathfrak{D}(A_0) \subseteq W_2^1(\mathbf{R}^N) \cap W_{2,\mathrm{loc}}^2(\mathbf{R}^N). \tag{3.39}$$

To prove that $\mathfrak{D}(A_0) = \mathfrak{D}(A_0 + 1) \subseteq W_2^1(\mathbf{R}^N)$ we use the inequality

$$((A_0 + 1)g, g)_{L_2} \geq c_1 \|g\|^2_{W_2^1(\mathbf{R}^N)} \qquad (g \in C_0^\infty(\mathbf{R}^N)),$$

which follows from (3.36) by integration by parts. If $f \in \mathfrak{D}(A_0)$ and $(g_n)_1^\infty$ is a sequence of $g_n \in C_0^\infty(\mathbf{R}^N)$ such that $g_n \to f$ and $(A_0 + 1)g_n \to (A_0 + 1)f$ in L_2 as $n \to \infty$, then this energy inequality implies that $(g_n)_1^\infty$ is a Cauchy sequence in $W_2^1(\mathbf{R}^N)$, i.e., $f \in W_2^1(\mathbf{R}^N)$. Accordingly, $\mathfrak{D}(A_0) \subseteq W_2^1(\mathbf{R}^N)$. Further, if $f \in \mathfrak{D}(A_0)$ and G is a precompact domain in $\mathbf{R}^N$ with infinitely differentiable boundary, then $\varphi = f \restriction G$ is a weak solution in $W_2^1(G)$ of the equation $\mathcal{L}u = \psi$ interior to G, where

$$\psi = (A_0 f) \restriction G \in L_2(G, dx);$$

that is,

$$(\varphi, \mathcal{L}u)_{L_2(G,dx)} = (\psi, u)_{L_2(G,dx)} \qquad (u \in C_0^\infty(G)).$$

According to results (automatically valid for the indicated smoothness; see Ladyzhenskaya and Ural′tseva [1], Chapter 3, Theorem 10.1) on interior regularity of weak solutions of elliptic equations, we conclude that $\varphi \in W_2^2(G')$, where G' is an arbitrary domain having compact closure in G. It follows from the foregoing that $f \in W_{2,\mathrm{loc}}^2(\mathbf{R}^N)$. The relation (2.5) is established.

III. We construct a sequence of infinitely differentiable functions $\mathbf{R}^1 \ni t \mapsto \theta_R(t) \in [0, 1]$ $(R = 1, 2, \dots)$ equal to 1 for $t \in [-R, R]$ and equal to 0 for $t \notin [-(R+1), R+1]$, and with graphs on $[-(R+1), -R]$ and $[R, R+1]$ obtained by translations of the corresponding graphs for $R = 1$. Define

$$\chi_R(x) = \chi_R(x_1, \dots, x_N) = \theta_R(x_1) \cdots \theta_R(x_n),$$

and let $f \in \mathfrak{D}(A_0) \cap L_\infty$ and $f_R(x) = \chi_R(x) f(x)$. Obviously, $f_R \to f$ in L_2 as $R \to \infty$. Further, $f \in W_{2,\mathrm{loc}}^2(\mathbf{R}^N)$ and $f_R \in W_2^2(\mathbf{R}^N)$ by (3.39). Termwise differentiation gives us an easy proof of the estimate

$$\|\mathcal{L}f - \mathcal{L}f_R\|_{L_2}^2 \leq c_2 \left[\int_{\mathbf{R}^N} |(A_0 f)(x) - \chi_R(x)(A_0 f)(x)|^2\, dx + \int_{\mathbf{R}^N} |q(x)f(x) - \chi_R(x)q(x)f(x)|^2\, dx + \int_{\mathbf{R}^N} \left|\sum_{j,k=1}^{N} a_{jk}(x)\left(\frac{\partial \chi_R}{\partial x_j}\right)(x)\left(\frac{\partial f}{\partial x_k}\right)(x)\right|^2 dx + \int_{\mathbf{R}^N} |f(x)|^2\, |(A_0\chi_R)(x)|^2\, dx\right] \quad (3.40)$$

with some $c_2 > 0$. Since $f, qf, A_0 f, \partial f/\partial x_k \in L_2$ (see (3.39)) and the a_{jk} are bounded along with their first derivatives, we get from the form of χ_R that the right-hand side of (3.40) tends to zero as $R \to \infty$. Thus, $\mathcal{L}f \in L_2$, and the sequence $(f_R)_1^\infty$ of compactly supported functions $f_R \in W_2^2(\mathbf{R}^N)$ is such that

$$f_R \to f, \qquad \mathcal{L}f_R \to \mathcal{L}f = Af \quad (3.41)$$

in L_2 as $R \to \infty$.

For a $\varphi \in L_{1,\text{loc}}$ we construct in the usual way the averaged function

$$\varphi_\varepsilon(x) = (S_\varepsilon\varphi)(x) = \int_{\mathbf{R}^N} \omega_\varepsilon(|x-y|)\varphi(y)\, dy \qquad (x \in \mathbf{R}^N,\ \varepsilon > 0), \quad (3.42)$$

where $\omega_\varepsilon(|x-y|)$ is an infinitely differentiable nonnegative averaging kernel. Recall (see, for example, §3.2 in Chapter 3 of Mikhaĭlov [1], or Chapter 1 of Mikhlin [1]) that if φ is in L_2 and has compact support, then $\varphi_\varepsilon \in C_0^\infty(\mathbf{R}^N)$ and $\varphi_\varepsilon \to \varphi$ in L_2 as $\varepsilon \to 0$; the same holds when L_2 is replaced by $W_2^2(\mathbf{R}^N)$. (If $\varphi \in W_2^2(\mathbf{R}^N)$, then $\varphi_\varepsilon \to \varphi$ in this space as $\varepsilon \to 0$.)

Let us now consider the function $f_{R,\varepsilon}$ for a particular R. Considering that the coefficients a_{jk} and their first derivatives are bounded, we get for some $c_3 > 0$ that

$$\|\mathcal{L}f_R - \mathcal{L}f_{R,\varepsilon}\|_{L_2}^2 \leq c_3 \left[\|f_R - f_{R,\varepsilon}\|_{W_2^2(\mathbf{R}^N)}^2 + \int_{\mathbf{R}^N} |q(x)(f_R(x) - f_{R,\varepsilon}(x))|^2\, dx\right] \quad (3.43)$$

According to the foregoing, the first term on the right-hand side of (3.43) tends to zero as $\varepsilon \to 0$. The second term also tends to zero. Indeed, $f_{R,\varepsilon} \to f_R$ in L_2 as $\varepsilon \to 0$; therefore, $qf_{R,\varepsilon} \to qf_R$ in L_1. This implies that there exists a sequence $(\varepsilon_n)_1^\infty$ convergent to zero such that

$$h_n(x) = |q(x)f_{R,\varepsilon_n}(x) - q(x)f_R(x)|^2 \to 0$$

for almost all x. But f_R is bounded, so f_{R,ε_n} is bounded uniformly with respect to n by the same constant c_4, and hence $h_n(x)$ is majorized by the function $4c_4^2 q^2 \in L_1$. This allows us to pass to the limit under the integral sign. Thus, $f_{R,\varepsilon} \to f_R$ and $\mathcal{L} f_{R,\varepsilon} \to \mathcal{L} f_R$ in L_2 as $\varepsilon \to 0$.

From this, considering (3.41), we get that for $f \in D = \mathfrak{D}(A_0) \cap L_\infty$ there exists a sequence $g_m = f_{R_m,\varepsilon_m} \in C_0^\infty(\mathbf{R}^N)$ $(m = 1, 2, \ldots)$ such that $g_m \to f$ and $Ag_m \to Af$ in L_2 as $m \to \infty$. In other words, (3.38) holds. ∎

To prove the theorem in the case of unbounded coefficients a_{jk} we must resort to hyperbolic equations. In this connection consider the following Cauchy problem with the selfadjoint lower semibounded operator L in Lemma 3.4:

$$(d^2u/dt^2)(t) + Lu(t) = 0 \qquad (t \in [0, \infty)),$$
$$u(0) = \varphi_0, \quad u'(0) = \varphi_1 \quad (\varphi_0, \varphi_1 \in L_2). \tag{3.44}$$

We shall use the strong solutions of this problem, i.e., the twice strongly continuously differentiable vector-valued functions $[0, \infty) \ni t \mapsto u(t) \in L_2$, $u(t) \in \mathfrak{D}(L)$, satisfying the equation in (3.44) and the indicated initial conditions. If $\varphi_0 \in \mathfrak{D}(L)$ and $\varphi_1 \in \mathfrak{D}(\sqrt{L})$, then a strong solution of problem (3.44) exists and has the form

$$u(t) = \int_l^\infty \cos\sqrt{\lambda} t dE(\lambda)\varphi_0 + \int_l^\infty \frac{1}{\sqrt{\lambda}} \sin\sqrt{\lambda} t\, dE(\lambda)\varphi_1 \qquad (t \in [0, \infty)), \tag{3.45}$$

where $l \le \inf_{x \in \mathbf{R}^N} q(x)$ is the infimum of L, and E is its resolution of the identity. In particular, it exists for $\varphi_0, \varphi_1 \in C_0^\infty(\mathbf{R}^N) \subset \mathfrak{D}(L)$. Denote by $K_r(a)$ the closed ball of radius $r > 0$ about a point $a \in \mathbf{R}^N$.

LEMMA 3.5. *Assume the conditions of Lemma* 3.4. *Let* $\varphi_0, \varphi_1 \in C_0^\infty(\mathbf{R}^N)$ *be such that* $\operatorname{supp}\varphi_0, \operatorname{supp}\varphi_1 \subseteq K_\rho(a)$ *for some* $a \in \mathbf{R}^N$ *and* $\rho > 0$. *Then the inclusion* $\operatorname{supp} u(t) \subseteq K_{\rho + r_{a,\rho}}(a)$ *holds for a solution* $u(t)$ *of* (3.44) *for each* $t \in [0, \infty)$, *where the function* $r_{a,\rho}(t)$ *is a solution of the following Cauchy problem*:

$$(dr/dt)(t) = c(r(t) + |a| + \rho) \quad (t \in [0, \infty)), \qquad r(0) = 0. \tag{3.46}$$

We remark that if the coefficients of $\mathcal{L}$ are sufficiently smooth, then (3.44) becomes the Cauchy problem for a hyperbolic equation, and Lemma 3.5 becomes a well-known fact about the rate of propagation of a perturbation for such equations (see, for example, Ladyzhenskaya [1], Chapter 4, Theorem 2.2). Our purpose is to generalize this fact to the case of nonsmooth coefficients, when (3.44) can only be understood as the Cauchy problem for an equation with operator coefficients.

PROOF. In the expression for $\mathcal{L}$ replace the coefficients a_{jk} and q by the coefficients averaged by means of the operator (3.42):

$$a_{jk}^{(n)} = S_{1/n} a_{jk} \in C^\infty(\mathbf{R}^N), \qquad q^{(n)} = S_{1/n} q \in C^\infty(\mathbf{R}^N).$$

The resulting differential expression is denoted by $\mathcal{L}^{(n)}$. Applying the averaging operator to (3.36) and using the fact that it carries nonnegative functions into nonnegative functions and preserves constants, we conclude that $\mathcal{L}^{(n)}$ has the same properties as $\mathcal{L}$, and the potentials $q^{(n)}$ are lower semibounded uniformly with respect to n by the number l. The properties of the averaging operator give us that, as $n \to \infty$, $a_{jk}^{(n)} \to a_{jk}$ uniformly in $\mathbb{R}^N$ along with their first derivatives, $q^{(n)} \to q$ in L_2, and the corresponding elements $c^{(n)}$ converge to c uniformly in $\mathbb{R}^N$.

Denote by $L^{(n)}$ the minimal operator in L_2 constructed from $\mathcal{L}^{(n)}$. According to Lemma 3.4, it is selfadjoint. It is easy to see that $L^{(n)} \to L$ in the strong resolvent sense as $n \to \infty$. Indeed, $C_0^\infty(\mathbb{R}^N)$ is in the domain of each of the operators $L^{(n)}$ and L, and these operators can be constructed as the closures from $C_0^\infty(\mathbb{R}^N)$. Therefore (as is well known; Reed and Simon [1], Theorem VIII.25), the indicated convergence takes place if $L^{(n)} f \to Lf$ $(f \in C_0^\infty(\mathbb{R}^N))$ in L_2 as $n \to \infty$. But $L^{(n)} f = \mathcal{L}^{(n)} f$ and $Lf = \mathcal{L} f$, and this convergence follows from the approximation mentioned above for the coefficients of $\mathcal{L}$.

The strong resolvent convergence of $L^{(n)}$ to L gives us that $F(L^{(n)})g \to F(L)g$ $(g \in L_2)$ in L_2 as $n \to \infty$ for an arbitrary bounded continuous function $\mathbb{R}^1 \ni \lambda \mapsto F(\lambda) \in \mathbb{C}^1$. In particular, this is valid for the function $F(\lambda) = \cos\sqrt{\lambda}t$ when $\lambda \geq l$ and $F(\lambda) = \cos\sqrt{l}t$ when $\lambda < l$ ($t \in [0, \infty)$ is fixed) and for the analogous function with $\cos\sqrt{\lambda}t$ replaced by $(1/\sqrt{\lambda})\sin\sqrt{\lambda}t$. Consider now the problem (3.44), with L is replaced by $L^{(n)}$ and $\varphi_0, \varphi_1 \in C_0^\infty(\mathbb{R}^N)$; let $[0, \infty) \ni t \mapsto u^{(n)}(t) \in L_2$ be a solution of it. Because of (3.45) and what was just noted, we can assert that

$$u^{(n)}(t) \to u(t) \tag{3.47}$$

in L_2 as $n \to \infty$ for each $t \in [0, \infty)$.

Since the expression for $\mathcal{L}^{(n)}$ has infinitely differentiable coefficients, $u^{(n)}(t)$ can be understood as a solution $u^{(n)}(x, t)$ of the Cauchy problem for the hyperbolic equation

$$\begin{gathered}\left(\frac{\partial^2 u^{(n)}}{\partial t^2}\right)(x,t) - (\mathcal{L}^{(n)} u^{(n)})(x,t) = 0, \qquad (x \in \mathbb{R}^N;\ t \in [0,\infty)),\\ u^{(n)}(x,t)|_{t=0} = \varphi_0(x), \qquad \left(\frac{\partial u^{(n)}}{\partial t}\right)(x,t)|_{t=0} = \varphi_1(x).\end{gathered} \tag{3.48}$$

According to the fact just mentioned about the rate of propagation of a perturbation, we can assert on the basis of (3.48) that

$$\operatorname{supp} u^{(n)}(t) \subseteq K_{\rho + r_{a,\rho}^{(n)}}(a), \tag{3.49}$$

where $r_{a,\rho}^{(n)}(t)$ is a solution of the Cauchy problem (3.46) with $c(\cdot)$ replaced by $c^{(n)}(\cdot)$. As $n \to \infty$ we have that $c^{(n)} \to c$ uniformly on $\mathbb{R}^1$; therefore, $r_{a,\rho}^{(n)}(t) \to r_{a,\rho}(t)$ as $n \to \infty$ for each $t \in [0, \infty)$. Considering this circumstance and (3.47) and passing to the limit as $n \to \infty$ in (3.49), we prove the lemma. ■

PROOF OF THE THEOREM. The idea of it is to approximate the operator L generated by $\mathcal{L}$ by the operators $L^{(n)}$ generated by $\mathcal{L}^{(n)}$, which satisfy the conditions of Lemma 3.4 and have coefficients coinciding with those of $\mathcal{L}$ in the ball of radius n about zero. By Lemma 3.4, the operators $L^{(n)}$ are selfadjoint. On the other hand, this approximation satisfies the requirements implying that L is selfadjoint in the hyperbolic approach. It is convenient to prove the theorem in steps.

I. Let $n = 1, 2, \ldots$, and replace the coefficients a_{jk} in (3.35) by coefficients $a_{jk}^{(n)}$ as follows: 1) $a_{jk}^{(n)}(x) = a_{jk}(x)$ for $|x| \le n$; 2) the $a_{jk}^{(n)}$ are in $C^{1+\alpha}(\mathbf{R}^N)$ and are bounded along with their first derivatives in $\mathbf{R}^N$; 3) the ellipticity condition (3.36) with a_{jk} replaced by $a_{jk}^{(n)}$ is satisfied; and 4) $c^{(n)}(r) \le c(r+1)$ $(r \in [0, \infty))$, where $c^{(n)}$ is determined from $(a_{jk}^{(n)}(x))_{j,k=1}^N$ in the same way that c is determined from $(a_{jk}(x))_{j,k=1}^N$.

For example, the $a_{jk}^{(n)}$ are easy to realize as follows. Let $b_{jk}^{(n)}(x)$ coincide with $a_{jk}(x)$ when $|x| \le n+1$, and with $a_{jk}(0)$ when $|x| > n+1$. Consider a function $[0, \infty) \ni r \mapsto \varepsilon(r) \in [0, 1]$ in $C^\infty([0, \infty))$ equal to 0 for $r \in [0, n]$ and to 1 for $r \in [n+1, \infty)$.

It will be assumed that $a_{jk}^{(n)}(x)$ coincides with $b_{jk}^{(n)}(x) = a_{jk}(x)$ for $|x| \le n$, and with $(S_{\varepsilon(|x|)} b_{jk}^{(n)})(x)$ for $|x| > n$ (thus, $a_{jk}^{(n)}(x) = a_{jk}(0)$ for $|x| > n+2$). Requirements 1)–3) are obviously satisfied. (A direct computation verifies 2).)

Let us check the inclusion 4). For $r \in [0, n]$ we have that $c^{(n)}(r) = c(r)$. Since c is a nondecreasing function,

$$(B^{(n)}(x)\xi, \xi)_{\mathbf{R}^N} \le c(|x|)|\xi|^2 \le c(n+1)|\xi|^2$$

for any vector $\xi \in \mathbf{R}^N$ when $n < |x| \le n+1$ ($B^{(n)}(x)$ is the operator corresponding to the matrix $(b_{jk}^{(n)}(x))_{j,k=1}^N$), and

$$(B^{(n)}(x)\xi, \xi)_{\mathbf{R}^N} \le c(0)|\xi|^2$$

when $|x| > n+1$. Applying the averaging operators $S_{\varepsilon(|x|)}$ to these two inequalities, we obtain in the natural notation that

$$(A^{(n)}(x)\xi, \xi)_{\mathbf{R}^N} \le c(n+1)|\xi|^2 \qquad (\xi \in \mathbf{R}^N,\ n < |x| \le n+2),$$
$$(A^{(n)}(x)\xi, \xi)_{\mathbf{R}^N} \le c(0)|\xi|^2 \qquad (\xi \in \mathbf{R}^N,\ |x| > n+2).$$

This implies that $c^{(n)}(r) \le c(n+1)$ $(r \in (n, \infty))$. Considering that $c^{(n)}(r)$ coincides with $c(r)$ for $r \in [0, n]$, along with the inequality and the fact that c is nondecreasing, we get that $c^{(n)}(r) \le c(r+1)$ $(r \in [0, \infty))$.

The coefficient $q(x)$ in $\mathcal{L}$ is replaced by the lower semibounded coefficient $q^{(n)}(x)$ equal to $q(x)$ for $|x| \le n$ and to 0 for $|x| > n$. Let $\mathcal{L}^{(n)}$ denote the expression (3.35) with the coefficients $a_{jk}^{(n)}$ and $q^{(n)}$ constructed in this way, and let $L^{(n)}$ be the corresponding minimal operator acting in L_2. The results of Lemmas 3.4 and 3.5 are valid for $\mathcal{L}^{(n)}$ and $L^{(n)}$.

II. To prove the theorem it suffices to establish the following result (Chapter 2, Theorem 6.5). Let $[0,1] \ni t \mapsto u(t) \in L_2$ be a strong solution of the Cauchy problem

$$\left(\frac{d^2u}{dt^2}\right)(t) + L^* u(t) = 0 \quad (t \in [0,1)), \qquad u(0) = u'(0) = 0, \tag{3.50}$$

and $[0,T] \ni t \mapsto u^{(n)}(t) \in L_2$ a strong solution of the Cauchy problem

$$\begin{gathered}\left(\frac{d^2u^{(n)}}{dt^2}\right)(t) + L^{(n)}u^{(n)}(t) = 0 \qquad (t \in [0,T]),\\ u^{(n)}(T) = \varphi_0, \qquad (u^{(n)})'(T) = \varphi_1,\end{gathered} \tag{3.51}$$

(which exists by Lemma 3.4) where $T \in (0,1)$ and $\varphi_0, \varphi_1 \in C_0^\infty(\mathbb{R}^N)$ are fixed. It must be proved that a) $u^{(n)}(t) \in \mathfrak{D}(L)$ $(t \in [0,T])$ for large indices n and b)

$$\int_0^T (u(t), (L^{(n)} - L)u^{(n)}(t))_{L_2}\, dt \underset{n\to\infty}{\longrightarrow} 0. \tag{3.52}$$

III. We prove first that, for each $t \in [0,T]$ and $n = 1,2,\ldots$, $\operatorname{supp} u^{(n)}(t)$ lies in the ball $K_R(0)$ of fixed radius R determined only by the initial data φ_0 and φ_1 (and $\mathcal{L}$, of course). Indeed, suppose that $\operatorname{supp}\varphi_0, \operatorname{supp}\varphi_1 \subseteq K_\rho(0)$ for some $\rho > 0$. Then $\operatorname{supp} u^{(n)}(t) \leq K_{\rho + r_{0,\rho}^{(n)}(t)}(0)$ $(t \in [0,T])$ by Lemma 3.5, where the function $r_{0,\rho}^{(n)}(t)$ is a solution of the Cauchy problem

$$(dr/dt)(t) = c^{(n)}(r(t) + \rho) = F^{(n)}(r(t)) \quad (t \in [0,\infty)), \qquad r(0) = 0. \tag{3.53}$$

Consider along with (3.53) the Cauchy problem

$$(ds/dt)(t) = c(s(t) + \rho + 1) = F(s(t)) \quad (t \in [0,\infty)), \qquad s(0) = 0. \tag{3.54}$$

By (3.37), $\int_0^\infty F^{-1}(s)\, ds = \infty$, which implies that a solution $s(t)$ of (3.54) is defined for all $t \in [0,\infty)$ and increases monotonically to $+\infty$ as $t \to +\infty$. According to 4), $F^{(n)}(r) \leq F(r)$ $(r \in [0,\infty))$; hence the solution $r_{0,\rho}(t)$ of (3.53) has the same properties, and $r_{0,\rho}^{(n)}(t) \leq s(t)$ $(t \in [0,\infty))$. As a result we obtain the required inclusion:

$$\begin{aligned}\operatorname{supp} u^{(n)}(t) &\subseteq K_{\rho + r_{0,\rho}^{(n)}(t)}(0) \subseteq K_{\rho+s(t)}(0)\\ &\subseteq K_{\rho+s(T)}(0) = K_R(0) \qquad (t \in [0,T];\ n = 1,2,\ldots).\end{aligned} \tag{3.55}$$

IV. We prove a). More precisely, we establish the inclusion $u^{(n)}(t) \in \mathfrak{D}(L)$ $(t \in [0,T])$ for $n > R$. According to (3.55), it suffices to prove that if $f \in \mathfrak{D}(L^{(n)})$ for some particular $n = 1,2,\ldots$ and $\operatorname{supp} f$ lies in the interior of a ball $K_r(0)$ with $r < n$, then $f \in \mathfrak{D}(L)$ and $Lf = L^{(n)}f$. For this it suffices to see that there is a sequence $(f_m)_1^\infty$ with $f_m \in C_0^\infty(\mathbb{R}^N)$ and

$\operatorname{supp} f_m \subseteq K_n(0)$ such that $f_m \to f$ and $\mathcal{L}^{(n)} f_m = L^{(n)} f_m \to L^{(n)} f$ in L_2 as $m \to \infty$ (since $\mathcal{L}^{(n)} f_m = \mathcal{L} f_m = L f_m$).

Let $(g_m)_1^\infty$ be a sequence of functions $g_m \in C_0^\infty(\mathbf{R}^N)$ such that $g_m \to f$ and $\mathcal{L}^{(n)} g_m \to L^{(n)} f$ in L_2 as $m \to \infty$. (It exists by virtue of the definition of $L^{(n)}$.) Consider an infinitely differentiable function $\mathbf{R}^N \ni x \mapsto \chi(x) \in [0,1]$ equal to 1 in the ball $K_r(0)$ and to 0 outside $K_n(0)$. We claim that $f_m(x) = \chi(x) g_m(x)$ ($x \in \mathbf{R}^N$; $m = 1, 2, \ldots$) is the desired sequence. Indeed, it is obvious that $f_m \to f$ in L_2 as $m \to \infty$. Further, in a way similar to that for (3.40) we get

$$\begin{aligned}
&\|\mathcal{L}^{(n)} g_m - \mathcal{L}^{(n)} f_m\|_{L_2}^2 \\
&\quad \le c_1 \Bigg[\|(\mathcal{L}^{(n)} g_m)(\cdot) - \chi(\cdot)(\mathcal{L}^{(n)} g_m)(\cdot)\|_{L_2}^2 \\
&\qquad + \int_{\mathbf{R}^N} \left| \sum_{j,k=1}^N a_{jk}^{(n)}(x) \left(\frac{\partial \chi}{\partial x_j}\right)(x) \left(\frac{\partial g_m}{\partial x_k}\right)(x) \right|^2 dx \\
&\qquad + \int_{\mathbf{R}^N} |g_m(x)|^2 \left| \sum_{j,k=1}^N \left(\frac{\partial}{\partial x_j}\left(a_{jk}^{(n)} \frac{\partial \chi}{\partial x_k}\right)\right)(x) \right|^2 dx \Bigg]
\end{aligned} \tag{3.56}$$

The first term on the right-hand side of (3.56) tends to 0 as $m \to \infty$, because $\mathcal{L}^{(n)} g_m \to L^{(m)} f$ in L_2 and $\operatorname{supp} L^{(n)} f \subseteq K_r(0)$. The second and third terms can be estimated from above by the respective expressions

$$c_2 \sum_{k=1}^N \int_{K_n(0)\setminus K_r(0)} \left| \left(\frac{\partial g_m}{\partial x_k}\right)(x) \right|^2 dx, \qquad c_3 \int_{K_n(0)\setminus K_r(0)} |g_m(x)|^2 \, dx. \tag{3.57}$$

From the last estimate and the facts that $g_m \to f$ in L_2 and $\operatorname{supp} f \subseteq K_r(0)$ it follows that the third term tends to 0. To estimate the first expression in (3.57) we make use of the following energy inequality: for $a \ge -\inf_{x \in \mathbf{R}^N} q^{(n)}(x) + 1$ we have that

$$(\mathcal{L}^{(n)} \varphi + a\varphi, \varphi)_{L_2} \ge c_4 \|\varphi\|_{W_2^1(\mathbf{R}^N)}^2 \qquad (c_4 > 0,\ \varphi \in C_0^\infty(\mathbf{R}^N))$$

(which is obtained by integration by parts). Substituting $g_m - g_l$ here in place of φ and letting $m, l \to \infty$, we find that $\|g_m - g_l\|_{W_2^1(\mathbf{R}^N)} \to 0$ as $l, m \to \infty$. Along with the convergence $g_m \to f$ in L_2 as $m \to \infty$, this shows that $f \in W_2^1(\mathbf{R}^N)$ and $g_m \to f$ in $W_2^1(\mathbf{R}^N)$ as $m \to \infty$. Since $\operatorname{supp} f \subseteq K_r(0)$, the first expression in (3.57), hence also the second term on the right-hand side of (3.56), tends to 0 as $m \to \infty$. We now conclude from (3.56) that

$$\lim_{m\to\infty} \mathcal{L}^{(n)} f_m = \lim_{m\to\infty} \mathcal{L}^{(n)} g_m = Lf$$

in L_2. Assertion a) is proved.

V. Assertion b) is now obvious: according to what has been proved, (3.55) with $n > R$ allows us to write $L^{(n)}u^{(n)}(t) = Lu^{(n)}(t)$ $(t \in [0,T])$. ■

3.5. *Continuation of the digression: Getting rid of the additional smoothness restrictions on the leading coefficients.* Let us show how to prove Theorem 3.1 with the smoothness restrictions on the a_{jk} indicated in its formulation. We have

LEMMA 3.6. *Consider a differential expression $\mathcal{L}$ of the form* (3.35) *with $q = 0$, and assume that the coefficients a_{jk} are bounded on $\mathbb{R}^N$, while $\partial a_{jk}/\partial x_j \in L_{p,\mathrm{loc}}(\mathbb{R}^N)$ for some $p > N$ $(j, k = 1, \ldots, N;\ N \geq 2)$, and they are bounded outside some ball. Then the corresponding minimal operator A_0 is selfadjoint in L_2.*

PROOF. Let $K_\rho(0)$ be a ball outside which the derivatives $\partial a_{jk}/\partial x_j$ $(j, k = 1, \ldots, N)$ are bounded. We first show that the definition of A_0 is unambiguous, i.e., $\mathcal{L}f \in L_2(\mathbb{R}^N)$ for any $f \in C_0^\infty(\mathbb{R}^N)$. This follows from the estimate

$$\|\mathcal{L}f\|_{L_2}^2 \leq c_1 \left(\|f\|_{W_2^2(\mathbb{R}^N)}^2 + \sum_{j,k=1}^{N} \left\| \frac{\partial a_{jk}}{\partial x_j} \right\|_{L_p(K_\rho(0),dx)}^2 \left\| \frac{\partial f}{\partial x_k} \right\|_{L_2(K_\rho(0),dx)}^2 \right)$$
$$(p^{-1} + r^{-1} = 2^{-1},\ c_1 > 0), \qquad (3.58)$$

which is obtained by termwise differentiation and Hölder's inequality. The operator A_0 is nonnegative, of course. Suppose that $F \supseteq A_0$ is the selfadjoint Friedrich's extension of it. It suffices to prove that $F \subseteq A_0$. We first establish an inclusion similar to (3.39):

$$\mathfrak{D}(F) \subseteq W_2^1(\mathbb{R}^N) \cap W_{2,\mathrm{loc}}^2(\mathbb{R}^N). \qquad (3.59)$$

Thus, $\mathfrak{D}(F)$ is in the completion of $C_0^\infty(\mathbb{R}^N)$ with respect to the inner product $(f,g) = (\mathcal{L}f, g)_{L_2} + (f,g)_{L_2}$, which is equivalent to $(f,g)_{W_2^1(\mathbb{R}^N)}$ because of the energy inequality and the fact that the a_{jk} are bounded. Thus, $\mathfrak{D}(F) \subseteq W_2^1(\mathbb{R}^N)$. Further, $\mathfrak{D}(F) \subseteq \mathfrak{D}(A_0^*)$, and so for each precompact domain $G \subset \mathbb{R}^N$ with infinitely differentiable boundary and each $f \in \mathfrak{D}(F)$ the restriction $f \restriction G$ is a weak solution in $W_2^1(G)$ of the equation $\mathcal{L}u = (A_0 f) \restriction G \in L_2(G, dx)$ interior to G. From this and Theorem 10.1 in the book [1] of Ladyzhenskaya and Ural′tseva (Chapter 3) we conclude that $\mathfrak{D}(F) \subseteq W_{2,\mathrm{loc}}^2(\mathbb{R}^N)$, i.e., (2.25) is proved. (The nature of the smoothness described in subsection 4 for the a_{jk} determined the possibility of applying this theorem.)

Let $f \in \mathfrak{D}(F)$. By (3.59), $\mathcal{L}f$ makes sense; it is easy to see that

$$\mathcal{L}f = Ff \in L_2. \qquad (3.60)$$

Indeed, since $F \subseteq A_0^*$, it follows that $Ff = A_0^* f = g$, and hence

$$(f, \mathcal{L}u)_{L_2} = (g, u)_{L_2} \qquad (u \in C_0^\infty(\mathbb{R}^N)).$$

Integration by parts gives us that $(f, \mathcal{L}u)_{L_2} = (\mathcal{L}f, u)_{L_2}$ because of (3.58) and the fact that u is compactly supported. Since u is arbitrary, the last two equalities give us that $\mathcal{L}f = g = Ff$.

To prove the inclusion $F \subseteq A_0$ it is now enough to show that for each $f \in \mathfrak{D}(F)$ there exists a sequence $(g_m)_1^\infty$ of functions $g_m \in C_0^\infty(\mathbf{R}^N)$ such that $g_m \to f$ and $\mathcal{L}g_m \to \mathcal{L}f$ in L_2 as $m \to \infty$. This can be done by repeating almost word-for-word the arguments proving (3.38). Indeed, let $f_R(x) = \chi_R(x)f(x)$ ($x \in \mathbf{R}^N$, $R = 1, 2, \ldots$), where the χ_R were constructed earlier. Obviously, $f_R \to f$ in L_2 as $R \to \infty$. The following estimate is obtained in a way similar to that for (3.40):

$$\|\mathcal{L}f - \mathcal{L}f_R\|_{L_2}^2 \leq c_1 \left[\int_{\mathbf{R}^N} |(\mathcal{L}f)(x) - \chi_R(x)(\mathcal{L}f)(x)|^2 \, dx \right.$$
$$+ \int_{\mathbf{R}^N} \left| \sum_{j,k=1}^N a_{jk}(x) \left(\frac{\partial \chi_R}{\partial x_j}\right)(x) \left(\frac{\partial f}{\partial x_k}\right)(x) \right|^2 dx$$
$$\left. + \int_{\mathbf{R}^N} |f(x)|^2 \, |(\mathcal{L}\chi_R)(x)|^2 \, dx \right] \qquad (c_1 > 0). \quad (3.61)$$

Since $f, \mathcal{L}f$, and $\partial f/\partial x_k$ are in L_2 (see (3.60) and (3.59)) and the a_{jk} and their first derivatives are bounded outside $K_\rho(0)$, it follows from the form of χ_R that the right-hand side of (3.61) tends to 0 as $R \to \infty$, i.e., $\mathcal{L}f_R \to \mathcal{L}f$ in L_2. Next, let us pass from f_R to their averages $f_{R,\varepsilon} \in C_0^\infty(\mathbf{R}^N)$ by means of (3.42). For some $c_2 > 0$

$$\|\mathcal{L}f_R - \mathcal{L}f_{R,\varepsilon}\|_{L_2}^2 \leq c_2(\|f_R - f_{R,\varepsilon}\|_{W_2^2(\mathbf{R}^N)}^2 + \|f_R - f_{R,\varepsilon}\|_{W_r^1(\mathbf{R}^N)}^2) \underset{\varepsilon \to 0}{\longrightarrow} 0,$$

because f_R is in $W_{2,\mathrm{loc}}^2(\mathbf{R}^N) \subset W_{r,\mathrm{loc}}^1(\mathbf{R}^N)$ and has compact support. (The proof is similar to that for (3.58).) Therefore, $\mathcal{L}f_R \to \mathcal{L}f$ and $\mathcal{L}f_{R,\varepsilon} \to \mathcal{L}f_R$ in L_2, which implies the existence of the required sequence $(g_m)_1^\infty$. ■

LEMMA 3.7. *The operator A_0 constructed in Lemma* 3.6 *from $\mathcal{L}$ has the property that the semigroup e^{-tA_0} is contractive in L_∞.*

PROOF. We use (3.42) to replace the coefficients a_{jk} in the expression for $\mathcal{L}$ by averaged coefficients $a_{jk}^{(n)} = S_{1/n}a_{jk} \in C^\infty(\mathbf{R}^N)$ ($n = 1, 2, \ldots$). The resulting expression $\mathcal{L}^{(n)}$ has the same properties as $\mathcal{L}$, of course; let $A_0^{(n)}$ be the minimal operator corresponding to $\mathcal{L}^{(n)}$. (It is nonnegative and selfadjoint.) In a way analogous to that for (3.58) we get the existence of a

$c_1 > 0$ such that for each $f \in C_0^\infty(\mathbf{R}^N)$

$$\|\mathcal{L}^{(n)} f - \mathcal{L} f\|_{L_2}^2 \leq c_1 \left(\sum_{j,k=1}^{N} \|a_{jk} - a_{jk}^{(n)}\|_{L_4(\operatorname{supp} f, dx)}^2 \|f\|_{W_4^2(\mathbf{R}^N)}^2 + \sum_{j,k=1}^{N} \left\| \frac{\partial}{\partial x_j}(a_{jk} - a_{jk}^{(n)}) \right\|_{L_p(\operatorname{supp} f, dx)}^2 \|f\|_{W_2^1(\mathbf{R}^N)}^2 \right).$$

This implies that

$$A_0^{(n)} f = \mathcal{L}^{(n)} f \underset{n\to\infty}{\longrightarrow} \mathcal{L} f = A_0 f$$

in L_2, and A_0 is nonnegative and selfadjoint by Lemma 3.6. As in the proof of Lemma 3.5, this gives us that $A_0^{(n)} \to A_0$ as $n \to \infty$ in the strong resolvent sense. Therefore, if for fixed $t \in [0, \infty)$ we set $F(\lambda) = e^{-\lambda t}$ for $\lambda \in [0, \infty)$ and $F(\lambda) = 1$ for $\lambda \in (-\infty, 0)$, then $F(A_0^{(n)})g \to F(A_0)g$ ($g \in L_2$) in L_2 as $n \to \infty$. In other words, $e^{-tA_0^{(n)}} g \to e^{-tA_0} g$ in L_2 as $n \to \infty$ for each $t \in [0, \infty)$ and each $g \in L_2$.

We fix a $t \in [0, \infty)$ and a $g \in L_2$ and choose a subsequence $(e^{-tA_0^{(n_k)}} g)(x)$ converging to $(e^{-tA_0} g)(x)$ as $k \to \infty$ for almost all x. For each $k = 1, 2, \ldots$ the estimate

$$\|e^{-tA_0^{(n_k)}} g\|_{L_\infty} \leq \|g\|_{L_\infty}$$

is valid, because the smoothness of the coefficients of $\mathcal{L}^{(n_k)}$ gives us that the semigroup $e^{-tA_0^{(n_k)}}$ is contractive in L_∞. Passing to the limit as $k \to \infty$ in the last inequality, we get that $\|e^{-tA_0} g\|_{L_\infty} \leq \|g\|_{L_\infty}$. ∎

PROOF OF THEOREM 3.1 IN THE GENERAL CASE. The proof consists in repeating the arguments in subsection 4, which is possible because of Lemmas 3.6 and 3.7. Thus, it is possible to prove Lemma 3.4 for a_{jk} satisfying Lemma 3.6, since these lemmas now give us that A_0 is selfadjoint, and the semigroup e^{-tA_0} is contractive in L_∞. The difference from the previous proof of Lemma 3.4 is only that $R = \rho + 1, \rho + 2, \ldots$, and the estimate (3.43) is replaced by the inequality (for some $c_1 > 0$)

$$\|\mathcal{L} f_R - \mathcal{L} f_{R,\varepsilon}\|_{L_2(\mathbf{R}^N)}^2 \leq c_1 (\|f_R - f_{R,\varepsilon}\|_{W_2^2(\mathbf{R}^N)}^2 + \|f_R - f_{R,\varepsilon}\|_{W_r^1(\mathbf{R}^N)}^2 + \|q f_R - q f_{R,\varepsilon}\|_{L_2}^2),$$

which is obtained by termwise differentiation and Hölder's inequality. The proof that the right-hand side tends to 0 as $\varepsilon \to 0$ is similar to the corresponding proof in Lemma 3.6.

The proof of Lemma 3.5 for a_{jk} satisfying the condition in Lemma 3.6 differs from the proof given above in that $a_{jk}^{(n)} \to a_{jk}$ pointwise as $n \to \infty$; therefore, the corresponding $c^{(n)}(r)$ converge pointwise to some nondecreasing function $f(r) \leq c(r + 1)$, from which, as in Lemma 3.5, we get a finite rate of propagation for the solutions of (3.44), with $r_{a,\rho}(t)$ bounded above by a solution of the following Cauchy problem:

$$(dr/dt)(t) = c(r(t) + |a| + \rho + 1) \quad (t \in [0, \infty)), \qquad r(0) = 0.$$

The construction with a_{jk} replaced by $a_{jk}^{(n)}$ (step I in the proof of Theorem 3.1 in §3.4) is preserved with the difference that $b_{jk}^{(n)}(x)$ is required to coincide with $a_{jk}(x)$ not in the ball of radius $n+1$ but in the cube with edges $2(n+1)$, and the averaging is carried out over each variable separately. (The last point is connected with the existence for each a_{jk} of derivatives only with respect to x_j.) The conditions on the derivatives of a_{jk} play the role of assertion a) in the proof. We show that it is true also in the given situation. For this it suffices to see that the third term on the right-hand side of (3.56) tends to 0 as $m \to \infty$; but this is really so because of the following arguments. Since $g_m \to f$ in $W_2^1(\mathbf{R}^N)$, imbedding theorems give us that $g_m \to f$ in $L_{r,\mathrm{loc}}$. Estimating the third term in the natural notation by

$$c_3 \sum_{j,k=1}^{N} \left\| \frac{\partial}{\partial x_j} a_{jk}^{(n)} \right\|_{L_p(K_n(0),dx)}^2 \|g_m\|_{L_2(K_n(0)\setminus K_r(0),dx)}^2$$
$$(p^{-1} + r_1^{-1} = 2^{-1})$$

with the help of Hölder's inequality, we get the required assertion. ■

REMARK 1. There is, of course, a certain awkwardness in the proof of Theorem 3.1 due to the consideration of a singular potential q and insufficiently smooth a_{jk}. It is useful to demonstrate how this theorem can be obtained at once from Theorem 6.3 in Chapter 2 in the case of smooth coefficients, namely, when $a_{jk}, q \in C^{2+[N/2]}(\mathbf{R}^N)$ (see Berezanskiĭ [4], Chapter 6, Theorem 1.8). Thus, in this case the following Cauchy problem has a classical solution (see, for example, Petrovskiĭ [1], §16):

$$\frac{\partial^2 u}{\partial t^2}(x,t) - (\mathcal{L}u)(x,t) = 0, \qquad u(x,0) = \varphi_0(x),$$
$$\frac{\partial u}{\partial t}(x,0) = \varphi_1(x) \qquad (x \in \mathbf{R}^N;\ t \in [0,\infty);\ \varphi_0, \varphi_1 \in C_0^\infty(\mathbf{R}^N)). \quad (3.62)$$

By (3.37), $u(\cdot,t) \in C_0^\infty(\mathbf{R}^N)$ for each $t \in [0,\infty)$, and hence $u(\cdot,t)$ can be interpreted as a vector-valued function $[0,\infty) \ni t \mapsto u(t) \in L_2$ that is twice strongly continuously differentiable with respect to t. The solvability of problem (3.62) implies the existence of a strong solution of the Cauchy problem (3.44). But then L is selfadjoint if it is semibounded below, according to Theorem 6.3 of Chapter 2. Theorem 3.1 is proved in the case under consideration.

3.6. ***Selfadjointness of differential operators with infinitely many variables. The parabolic approach.*** Let us show how to use the evolution technique to prove selfadjointness of a differential operator with infinitely many variables that is perturbed by a potential. We give a proof for a result analogous to Lemma 3.4 relating to the perturbation of an operator of the form (3.28) (the annihilation case), along with some corollaries of it. The parabolic approach will be used. More complicated situations, with the use of the hyperbolic approach, will be described briefly in the next subsection.

We use the notation from subsections 1–3 and from §6.6 of Chapter 2. It is assumed that the conditions (3.17) hold and that the functions $D_n^2 1$ are semibounded below (each by its own constant) $(n = 1, 2, \ldots)$.

THEOREM 3.2. *If a is a real-valued potential in $L_2(\mathsf{R}^\infty, d\rho(x))$ that is semibounded below, then the operator*

$$\begin{aligned} C^\infty_{\mathrm{c},b,2}(\mathsf{R}^\infty) \ni u(x) &\mapsto (\mathcal{M}u)(x) \\ &= \sum_{n=1}^{\infty} \left(-\left(\frac{\partial^2 u}{\partial x_n^2}\right)(x) - 2(D_n 1)(x_n)\left(\frac{\partial u}{\partial x_n}\right)(x)\right) + a(x)u(x) \\ &= (\mathcal{M}_0 u)(x) + a(x)u(x) \end{aligned} \tag{3.63}$$

acting in $L_2(\mathsf{R}^\infty, d\rho(x))$ is essentially selfadjoint.

PROOF. Because of (3.27) and the inclusion $a \in L_2(\mathsf{R}^\infty, d\rho(x)) = L_2$ the operator under consideration is well defined; it is clearly Hermitian. Denote by $\mathcal{B}_0$ the operator acting in L_2 constructed according to (3.35) from $\mathcal{M}_0$ (i.e., the closure of the operator (3.28)). The set $D = \mathfrak{D}(\mathcal{B}_0) \cap L_\infty$ is a core for this operator, since $C^\infty_{\mathrm{c},b,2}(\mathsf{R}^\infty) \subset L_\infty$. According to Lemma 3.3, the semigroup $e^{-t\mathcal{B}_0}$ is contractive in L_∞. By Theorem 6.14 in Chapter 2 ($r = 2$ here) the operator

$$D \ni f \to Af = (\mathcal{B}_0 f)(x) + a(x)f(x) \tag{3.64}$$

acting in $L_2(\mathsf{R}^\infty, d\rho(x))$ is essentially selfadjoint. To prove the theorem it is necessary to see that essential selfadjointness of (3.36) implies the same for (3.35). This follows from the inclusion

$$D \subseteq \mathfrak{D}((A \upharpoonright C^\infty_{\mathrm{c},b,2}(\mathsf{R}^\infty))^\sim), \tag{3.65}$$

which must be proved. The proof of (3.65), like that of (3.38), consists in establishing that any $f \in D$ can be approximated (granted, in a more complicated way) by functions in $C^\infty_{\mathrm{c},b,2}(\mathsf{R}^\infty)$ in the sense of the graph norm of the operator A.

LEMMA 3.8. *In the space $L_2(\mathsf{R}^\infty, d\rho(x))$ (understood as the infinite tensor product described in §3.2) the operator $\mathfrak{C}_m$ has the form*

$$\begin{aligned} L_2(\mathsf{R}^\infty, d\rho(x)) \ni f(x) &\mapsto (\mathfrak{C}_m f)(x^{(m)}) \\ &= \int_{\mathsf{R}^\infty} f(x_1, \ldots, x_m, x_{m+1}, \ldots)(d\rho_{m+1}(x_{m+1})) \otimes (d\rho_{m+2}(x_{m+2})) \otimes \cdots \\ &\in L_2(\mathsf{R}^m, d\rho_{(m)}(x^{(m)})). \end{aligned} \tag{3.66}$$

(Here and below, $L_2(\mathbf{R}^m, d\rho_{(m)}(x^{(m)}))$ will be understood as $L_2(\mathbf{R}^\infty, d\rho(x))$.)

PROOF. The vectors $(e_\alpha)_{\alpha\in \mathrm{B}_m}$ form an orthonormal basis for the subspace $\mathcal{H}_e^{(m)}$ onto which $\mathfrak{C}_m$ projects. Therefore, for $\mathcal{H}_e = \bigotimes_{n=1;e}^\infty H_n \ni f = \sum_{\alpha\in \mathrm{A}} f_\alpha e_\alpha$ we have that $(\mathfrak{C}_m f)_\alpha = f_\alpha$ if $\alpha \in \mathrm{B}_m$, and $(\mathfrak{C}_m f)_\alpha = 0$ if $\alpha \in \mathrm{A}\backslash \mathrm{B}_m$. On the other hand, consider the function

$$\begin{aligned} g(x) = g_c(x^{(m)}) = \int_{\mathbf{R}^\infty} & f(x_1, \dots, x_m, y_{m+1}, y_{m+2}, \dots) \\ & \times (d\rho_{m+1}(y_{m+1})) \otimes (d\rho_{m+2}(y_{m+2})) \otimes \cdots \\ \in L_2(&\mathbf{R}^\infty, d\rho(x)). \end{aligned}$$

If $\alpha \in \mathrm{B}_m$, i.e., has the form $\alpha = (\alpha_1, \dots, \alpha_m, 0, \dots)$ $(\alpha_1, \dots, \alpha_m = 0, 1, \dots)$, then, considering that the measures ρ_n are probability measures, we get that

$$\begin{aligned} g_\alpha &= (g, e_\alpha)_{L_2(\mathbf{R}^\infty, d\rho(x))} \\ &= \int_{\mathbf{R}^\infty} \bigg(\int_{\mathbf{R}^\infty} f(x_1, \dots, x_m, y_{m+1}, y_{m+2}, \dots)(d\rho_{m+1}(y_{m+1})) \\ &\qquad \otimes (d\rho_{m+2}(y_{m+2})) \otimes \cdots \bigg) \bar{e}_{\alpha_1}^{(1)}(x_1) \cdots \bar{e}_{\alpha_m}^{(m)}(x_m)\, d\rho(x) \\ &= \int_{\mathbf{R}^\infty} f(x_1, \dots, x_m, y_{m+1}, y_{m+2}, \dots) \bar{e}_{\alpha_1}^{(1)}(x_1) \cdots \bar{e}_{\alpha_m}^{(m)}(x_m)\, d\rho(x) \\ &= f_\alpha. \end{aligned} \tag{3.67}$$

If $\alpha = (\alpha_n)_1^\infty \in \mathrm{A}\backslash\mathrm{B}_m$, then at least one α_{n_0} is ≥ 1 for $n_0 \geq m+1$. In the computation of g_α analogous to (3.67), $\bar{e}_{\alpha_1}^{(1)}(x_1)\bar{e}_{\alpha_2}^{(2)}(x_2) \cdots d\rho(x)$ appears instead of $\bar{e}_{\alpha_1}^{(1)}(x_1) \cdots \bar{e}_{\alpha_m}^{(m)}(x_m)\, d\rho(x)$. This expression contains the differential $\bar{e}_{\alpha_{n_0}}^{(n_0)}(x_{n_0})\, d\rho_{n_0}(x_{n_0})$, which gives zero when integrated over $\mathbf{R}^1$, because the basis $(e_{\alpha_{n_0}}^{(n_0)})_{n_0=0}^\infty$ $(e_0^{(n_0)} = 1)$ is orthonormal. Accordingly, $g_\alpha = 0$.

Thus, $g_\alpha = (\mathfrak{C}_m f)_\alpha$ $(\alpha \in \mathrm{A})$, i.e., $g = \mathfrak{C}_m f$. ■

We return to the proof of the theorem, i.e., of the relation (3.65). It will consist of several steps.

I. Let $f \in D$ and let $f_m = \mathfrak{C}_m f$ $(m = 1, 2, \dots)$. Then $f_m \to f$ in $L_2(\mathbf{R}^\infty, d\rho(x))$ as $m \to \infty$. Further, by Lemma 3.1 (which is applicable because of (3.27)), we conclude that

$$f_m \in \mathfrak{D}(\mathcal{B}_0), \qquad \mathcal{B}_0 f_m = \mathcal{B}_0 \mathfrak{C}_m f = \mathfrak{C}_m \mathcal{B}_0 f. \tag{3.68}$$

If $f \in L_\infty$ in (3.66), then $\mathfrak{C}_m f \in L_\infty$. Therefore, $f_m \in D$. Considering (3.64) and (3.68), we get that

$$\begin{aligned}\|Af - Af_m\|_{L_2(\mathbb{R}^\infty, d\rho(x))} &\le \|\mathcal{B}_0 f - \mathcal{B}_0 \mathfrak{C}_m f\|_{L_2(\mathbb{R}^\infty, d\rho(x))} \\ &\quad + \|a(\cdot)f(\cdot) - a(\cdot)(\mathfrak{C}_m f)(\cdot)\|_{L_2(\mathbb{R}^\infty, d\rho(x))} \\ &= \|\mathcal{B}_0 f - \mathfrak{C}_m \mathcal{B}_0 f\|_{L_2(\mathbb{R}^\infty, d\rho(x))} \\ &\quad + \left(\int_{\mathbb{R}^\infty} |a(x)(f(x) - (\mathfrak{C}_m f)(x))|^2 \, d\rho(x)\right)^{1/2}.\end{aligned}$$

The first term on the right-hand side of this inequality obviously tends to 0 as $m \to \infty$. The second term also tends to 0, by Lebesgue's theorem: it follows from (3.66) that

$$\|\mathfrak{C}_m f\|_{L_\infty} \le \|f\|_{L_\infty} \qquad (f \in L_\infty) \tag{3.69}$$

uniformly with respect to $m = 1, 2, \dots$, and $a \in L_2(\mathbb{R}^\infty, d\rho(x))$. Accordingly,

$$f_m \to f, \qquad Af_m \to Af \tag{3.70}$$

in $L_2(\mathbb{R}^\infty, d\rho(x))$ as $m \to \infty$.

II. Fix m and consider the differential expression

$$(\mathcal{M}_m u)(x^{(m)}) = \sum_{n=1}^{m} \left(- \left(\frac{\partial^2 u}{\partial x_n^2}\right)(x^{(m)}) - 2(D_n 1)(x_n) \left(\frac{\partial u}{\partial x_n}\right)(x^{(m)}) \right). \tag{3.71}$$

It generates in $L_2(\mathbb{R}^m, d\rho_{(m)}(x^{(m)}))$ the minimal operator $\mathcal{B}_m$ that is the closure of the operator $C_0^\infty(\mathbb{R}^m) \ni u(x^{(m)}) \mapsto (\mathcal{M}_m u)(x^{(m)})$. As in the construction of the operator (3.23), it is easy to show that $C_0^\infty(\mathbb{R}^m)$ can be replaced here by $C_{b,2}^\infty(\mathbb{R}^m)$.

We prove that $f_m \in \mathfrak{D}(\mathcal{B}_m)$. Since $f_m \in \mathfrak{D}(\mathcal{B}_0)$, it is easy to see in a way similar to that for Remark 1 in §3.2 that there exists a sequence $(f_{m,k})_{k=1}^\infty$ of functions $f_{m,k} \in \text{l.s.}(C_{0,c}^\infty(\mathbb{R}^\infty))$ such that $f_{m,k} \to f_m$ and $\mathcal{M}_0 f_{m,k} \to \mathcal{B}_0 f_m$ in $L_2(\mathbb{R}^\infty, d\rho(x))$ as $k \to \infty$. By (3.66), $\mathfrak{C}_m f_{m,k} \in C_{b,2}^\infty(\mathbb{R}^m)$. Because of the continuity of $\mathfrak{C}_m$ and (3.68),

$$\mathfrak{C}_m f_{m,k} \to \mathfrak{C}_m f_m = f_m$$

and

$$\begin{aligned}\mathcal{M}_m \mathfrak{C}_m f_{m,k} &= \mathcal{M}_0 \mathfrak{C}_m f_{m,k} = \mathcal{B}_0 \mathfrak{C}_m f_{m,k} \\ &= \mathfrak{C}_m \mathcal{B}_0 f_{m,k} = \mathfrak{C}_m \mathcal{M}_0 f_{m,k} \to \mathfrak{C}_m \mathcal{B}_0 f_m.\end{aligned}$$

as $k \to \infty$ in $L_2(\mathbb{R}^\infty, d\rho(x))$, or, what is the same in $L_2(\mathbb{R}^m, d\rho_{(m)}(x^{(m)}))$.

Since $\mathfrak{C}_m f_{m,k} \in \mathfrak{D}(\mathcal{B}_m)$, this implies that $f_m \in \mathfrak{D}(\mathcal{B}_m)$ and $\mathcal{B}_m f_m = \mathfrak{C}_m \mathcal{B}_0 f_m$.

III. Consider the mapping

$$\begin{aligned}L_2(\mathbb{R}^m, d\rho_{(m)}(x^{(m)})) \ni u(x^{(m)}) \mapsto v(x^{(m)}) \\ = \left(\prod_{n=1}^{m} p_n^{1/2}(x_n)\right) u(x^{(m)}) \\ \in L_2(\mathbb{R}^m, dx^{(m)}) = L_2(\mathbb{R}^m).\end{aligned} \tag{3.72}$$

It carries (3.71) into the differential expression

$$(\mathcal{N}v)(x^{(m)}) = -\sum_{n=1}^{m}\left(\frac{\partial^2 v}{\partial x_n^2}\right)(x^{(m)}) + \left(\sum_{n=1}^{m}(D_n^2 1)(x_n)\right) v(x^{(m)}), \tag{3.73}$$

and $\mathcal{B}_m$ into the minimal operator N generated by (3.73) in $L_2(\mathbb{R}^m)$. By the assumptions in this subsection, the potential $q(x^{(m)}) = \sum_1^m (D_n^2 1)(x_n)$ in (3.73) is semibounded below. Let $f \in \mathfrak{D}(N)$ be the image of f_m under the mapping (3.72). In a way similar to that in the proof of (3.39) we conclude that

$$f \in \mathfrak{D}(N) \subseteq W_2^1(\mathbb{R}^m) \cap W_{2,\mathrm{loc}}^2(\mathbb{R}^m).$$

We then construct f_R $(R = 1, 2, \ldots)$ as in Lemma 3.4; $f_R \to f$ in $L_2(\mathbb{R}^m)$ as $R \to \infty$. According to (3.40),

$$\begin{aligned}&\|\mathcal{N}f - \mathcal{N}f_R\|^2_{L_2(\mathbb{R}^m)} \\ &\le c_1 \left[\int_{\mathbb{R}^m} |(\mathcal{N}f)(x^{(m)}) - \chi_R(x^{(m)})(\mathcal{N}f)(x^{(m)})|^2\, dx^{(m)} \right.\\ &\quad + \int_{\mathbb{R}^m} \left|\sum_{n=1}^{m}\left(\frac{\partial \chi_R}{\partial x_n}\right)(x^{(m)})\left(\frac{\partial f}{\partial x_n}\right)(x^{(m)})\right|^2 dx^{(m)} \\ &\quad \left. + \int_{\mathbb{R}^m} |f(x^{(m)})|^2 \left|\sum_{n=1}^{m}\left(\frac{\partial^2 \gamma_R}{\partial x_n^2}\right)(x^{(m)})\right|^2 dx^{(m)}\right].\end{aligned} \tag{3.74}$$

By virtue of the inclusions $\mathcal{N}f = Nf \in L_2(\mathbb{R}^m)$ and $f \in W_2^1(\mathbb{R}^m)$, the right-hand side of (3.74) tends to 0 as $R \to \infty$. Thus,

$$f_R \to f, \qquad \mathcal{N}f_R \to \mathcal{N}f \tag{3.75}$$

in $L_2(\mathbb{R}^m)$ as $R \to \infty$.

Let $f_{m,R}$ be the inverse image of f_R under the mapping (3.72); it is obvious that $f_{m,R} \in W_2^2(\mathbb{R}^m)$ and $f_{m,R}(x^{(m)}) = \chi_R(x^{(m)}) f_m(x^{(m)})$. The inclusion $f_m \in L_\infty$ (see (3.69)) implies that $f_{n,R}$ is uniformly bounded in $\mathbb{R}^m$ with respect to $R = 1, 2, \ldots$. The relations (3.75) give us that

$$f_{m,R} \to f, \qquad \mathcal{M}_0 f_{m,R} \to \mathcal{M}_0 f_m \tag{3.76}$$

in $L_2(\mathbb{R}^\infty, d\rho(x))$ as $R \to \infty$. It is not hard to see that (3.76) holds with $\mathcal{M}_0$ replaced by $\mathcal{M}$. Indeed, we must see that

$$\int_{\mathbb{R}^\infty} |a(x)f_m(x) - a(x)f_{m,R}(x)|^2 \, d\rho(x) \underset{R\to\infty}{\longrightarrow} 0.$$

The proof of this relation is exactly the same as the proof that the integral with q in (3.43) tends to zero. (Use the fact that $f_{m,R}$ is uniformly bounded in $\mathbb{R}^\infty$ with respect to R.) Thus,

$$f_{m,R} \to f_m, \qquad \mathcal{M}f_{m,R} \to \mathcal{M}f_m \tag{3.77}$$

in $L_2(\mathbb{R}^\infty, d\rho(x))$ as $R \to \infty$.

IV. Fix R and then apply the averaging operator (3.42) to the function $f_{m,R}(x^{(m)})$ (with N replaced by m); let $f_{m,R,\varepsilon}(x^{(m)}) = (S_\varepsilon f_{m,R})(x^{(m)})$ ($x^{(m)} \in \mathbb{R}^m$). It is easy to see that

$$f_{m,R,\varepsilon} \to f_{m,R}, \qquad \mathcal{M}f_{m,R,\varepsilon} \to \mathcal{M}f_{m,R} \tag{3.78}$$

in $L_2(\mathbb{R}^\infty, d\rho(x))$ as $\varepsilon \to 0$. Indeed, the first of the relations in (3.78) follows from the properties of the averaging operator, because $f_{m,R,\varepsilon}$ is a cylindrical function with support bounded with respect to ε. We can see similarly that

$$\mathcal{M}_0 f_{m,R,\varepsilon} \underset{\varepsilon\to 0}{\longrightarrow} \mathcal{M}_0 f_{m,R}.$$

The equality

$$\lim_{\varepsilon\to 0} \int_{\mathbb{R}^\infty} |a(x)f_{m,R}(x) - a(x)f_{m,R,\varepsilon}(x)|^2 \, d\rho(x) = 0$$

can be proved just like the corresponding equality for (3.43).

V. It follows from the limits in (3.70), (3.77), and (3.78) that there exist sequences $(R_m)_1^\infty$ and $(\varepsilon_m)_1^\infty$ of positive integers and positive numbers, respectively, such that

$$f_{m,R_m,\varepsilon_m} \to f \qquad \mathcal{M}f_{m,R_m,\varepsilon_m} = Af_{m,R_m,\varepsilon_m} \to Af$$

in $L_2(\mathbb{R}^\infty, d\rho(x))$ as $m \to \infty$. This proves (3.65), and hence the theorem. ∎

By applying Theorem 6.15 of Chapter 2 instead of Theorem 6.14 of the same chapter it is possible to get the next result.

THEOREM 3.3. *Suppose that the potential a is such that there exist $\varepsilon, \delta > 0$ for which the real-valued function $a(\cdot)$ is in $L_{2+\varepsilon}(\mathbb{R}^\infty, d\rho(x))$ and $e^{-\delta a(\cdot)} \in L_1(\mathbb{R}^\infty, d\rho(x))$, and it is known that the operator* (3.63) *is semibounded below. Then this operator is essentially selfadjoint.*

PROOF. Let $r = 2+\varepsilon$ and let p be such that $r^{-1}+p^{-1} = 2^{-1}$; obviously, $p \in (2,\infty)$. Let $\mathcal{B}_0$ be the operator equal to the closure of (3.28) acting in L_2. The set $D = \mathfrak{D}(\mathcal{B}_0) \cap L_p$ is a core for it, since $C^\infty_{c,b,2}(\mathbb{R}^\infty) \subset L_\infty \subset L_p$. According to Lemma 3.3, the semigroup $e^{-t\mathcal{B}_0}$ is contractive in L_{p+1}. Therefore, it is possible to apply Theorem 6.15 of Chapter 2 with $\varepsilon = 1$, which gives us that the operator (3.64) is essentially selfadjoint. To prove the theorem we must establish (3.65). The latter can be proved precisely like (3.65) in the case of Theorem 3.2; it is just necessary to take the following facts into consideration.

a) Along with (3.69) we have the estimate

$$\|\mathfrak{C}_m f\|_{L_p} \leq \|f\|_{L_p} \qquad (f \in L_p;\ p \in (2,\infty);\ m = 1,2,\ldots). \tag{3.79}$$

Indeed, by (3.66), Hölder's inequality, and the fact that $\rho_n(\mathbb{R}^1) = 1$ $(n = 1,2,\ldots)$,

$$\|(\mathfrak{C}_m f)(x^{(m)})|^p \leq \int_{\mathbb{R}^\infty} |f(x_1,\ldots,x_m,x_{m+1},x_{m+2},\ldots)|^p (d\rho_{m+1}(x_{m+1})) \otimes (d\rho_{m+2},(x_{m+2})) \otimes \cdots. \tag{3.80}$$

Integrating (3.80) over $\mathbb{R}^\infty$ with respect to $d\rho(x)$, we obtain (3.79).

b) The estimate (3.79) shows that $f_m = \mathfrak{C}_m f \in L_p(\mathbb{R}^\infty, d\rho(x))$ for $f \in D$, and the norms of these functions in this space are uniformly bounded with respect to $m = 1,2,\ldots$. Since $a \in L_r(\mathbb{R}^\infty, d\rho(x))$, it follows that

$$\int_{\mathbb{R}^\infty} |a(x)(f(x) - (\mathfrak{C}_m f)(x))|^2\, d\rho(x) \leq c \qquad (m = 1,2,\ldots)$$

and thus it is possible to pass to the limit as $m \to \infty$ under the integral sign in the last integral. This enables us to show that $Af_m \to Af$ in $L_2(\mathbb{R}^\infty, d\rho(x))$.

c) The functions $f_{m,R}$ $(R = 1,2,\ldots)$ and $f_{m,R,\varepsilon}$ $(\varepsilon > 0)$ also have norms in $L_p(\mathbb{R}^\infty, d\rho(x))$ that are uniformly bounded with respect to R and ε, respectively. This makes it possible to pass to the limit under the integral sign in the corresponding terms with q just as in b). ■

Let us proceed to the consideration of perturbations of the operator (3.23) when annihilation does not hold. We now show that in the case of Gaussian weights a simple device can be used to reduce the problem to a perturbation of the operator (3.28) and obtain theorems analogous to Theorems 3.2 and 3.3. (Such a path is laborious in the case of general weights: there is no inequality of the type (3.83).)

Below, $dg_\varepsilon(x) = d\rho(x)$ is the Gaussian measure on $\mathbb{R}^\infty$ corresponding to the weights (3.34) $(\varepsilon = (\varepsilon_n)_1^\infty)$; since we are considering the operator $\mathcal{A}_0$, the series $\sum_1^\infty \varepsilon_n$ must converge, of course.

LEMMA 3.9. *In the case of Gaussian weights the expression* $\mathcal{M}_0$ *(see* (3.28)*) satisfies the inequality*

$$\left\|\left(\sum_{n=1}^\infty \varepsilon_n^2 x_n^2\right) u(x)\right\|^2_{L_2(\mathbb{R}^\infty, dg_\varepsilon(x))} \leq \left\|(\mathcal{M}_0 + \sum_{n=1}^\infty \varepsilon_n)u(x)\right\|^2_{L_2(\mathbb{R}^\infty, dg_\varepsilon(x))} + 2\sum_{n=1}^\infty \varepsilon_n^2 \|u\|^2_{L_2(\mathbb{R}^\infty, dg_\varepsilon(x))} \qquad (u \in C^\infty_{c,b,2}(\mathbb{R}^\infty)). \tag{3.81}$$

PROOF. Consider Hermitian operators A, B, and C defined on $C^{\infty}_{c,b,2}(\mathsf{R}^{\infty})$ in $L_2(\mathsf{R}^{\infty}, dg_{\varepsilon}(x))$ and with action on u coinciding with the expressions under the sign of the first and second norms in (3.81) and with $(2\sum_1^{\infty}\varepsilon_n^2)^{1/2}u(x)$, respectively. From (3.28) and the equality $(D_n^2 1)(x_n) = \varepsilon_n^2 x_n^2 - \varepsilon_n$, we get by termwise differentiation that for $u \in C^{\infty}_{c,b,4}(\mathsf{R}^{\infty})$

$$
\begin{aligned}
B^2 u &= \Big(-\sum_{n=1}^{\infty} D_n^2 + \sum_{n=1}^{\infty} D_n^2 1 + \sum_{n=1}^{\infty}\varepsilon_n)^2 u(x) \\
&= \left(-\sum_{n=1}^{\infty} D_n^2 + \sum_{n=1}^{\infty}\varepsilon_n^2 x_n^2\right)^2 u(x) \\
&= \left(\sum_{n=1}^{\infty} D_n^2\right)^2 u(x) - \sum_{n=1}^{\infty} D_n^2\left(\left(\sum_{m=1}^{\infty}\varepsilon_m^2 x_m^2\right)u(x)\right) \\
&\quad - \sum_{m=1}^{\infty}\varepsilon_m^2 x_m^2\left(\left(\sum_{n=1}^{\infty} D_n^2\right)u(x)\right) + \left(\sum_{n=1}^{\infty}\varepsilon_n^2 x_n^2\right)^2 u(x) \\
&= \left(\sum_{n=1}^{\infty} D_n^2\right)^2 u(x) + \left(\sum_{n=1}^{\infty}\varepsilon_n^2 x_n^2\right)^2 u(x) \\
&\quad - 2\sum_{m=1}^{\infty}\left(\varepsilon_m x_m \sum_{n=1}^{\infty} D_n^2(\varepsilon_m x_m u(x))\right) \\
&\quad + \left(\sum_{m=1}^{\infty}\left[-\varepsilon_m x_m, \left[-\varepsilon_m x_m, -\sum_{n=1}^{\infty} D_n^2\right]\right]\right) u(x).
\end{aligned}
$$

It is easy to compute that the double commutator satisfies the equality

$$
\left[-\varepsilon_m, x_m, \left[-\varepsilon_m x_m, -\sum_{n=1}^{\infty} D_n^2\right]\right] u(x) = -2\varepsilon_m^2 u(x),
$$

and, therefore,

$$
B^2 u = A^2 u - C^2 u + \left(\sum_{n=1}^{\infty} D_n^2\right)^2 u(x) - 2\sum_{m=1}^{\infty}(\varepsilon_m x_m \sum_{n=1}^{\infty} D_n^2(\varepsilon_m x_m u(x)). \tag{3.82}
$$

Considering that

$$
(D_n u, v)_{L_2(\mathsf{R}^{\infty}, d\rho(x))} = -(u, D_n v)_{L_2(\mathsf{R}^{\infty}, d\rho(x))} \qquad (u, v \in C^{\infty}_{c,b,1}(\mathsf{R}^{\infty});\ n = 1, 2, \ldots),
$$

we get by (3.82) that for $u \in C^{\infty}_{c,b,4}(\mathbb{R}^{\infty})$

$$\begin{aligned}(B^2u, u)_{L_2(\mathbb{R}^\infty, dg_\varepsilon(x))} &= ((A^2 - C^2)u, u)_{L_2(\mathbb{R}^\infty, dg_\varepsilon(x))} \\ &\quad + \left\| \sum_{n=1}^{\infty} D_n^2 u \right\|^2_{L_2(\mathbb{R}^\infty, dg_\varepsilon(x))} + 2 \left\| \sum_{m,n=1}^{\infty} D_n(\varepsilon_m x_m u(x)) \right\|^2_{L_2(\mathbb{R}^\infty, dg_\varepsilon(x))} \\ &\geq ((A^2 - C^2)u, u)_{L_2(\mathbb{R}^\infty, dg_\varepsilon(x))}.\end{aligned}$$

This implies (3.81) on $u \in C^{\infty}_{c,b,4}(\mathbb{R}^{\infty})$ hence also on $u \in C^{\infty}_{c,b,4}(\mathbb{R}^{\infty})$ (pass to the limit). The series in the computations are easily seen to converge because of the condition that $\sum_1^\infty \varepsilon_n < \infty$. ■

COROLLARY.

$$\left\| \left(\sum_{n=1}^{\infty} \varepsilon_n^2 x_n^2 \right) u(x) \right\|_{L_2(\mathbb{R}^\infty, dg_\varepsilon(x))} \leq \|(\mathcal{M}_0 + \alpha)u(x)\|_{L_2(\mathbb{R}^\infty, dg_\varepsilon(x))}$$

$$\left(\alpha = \sum_{n=1}^{\infty} \varepsilon_n + \left(2 \sum_{n=1}^{\infty} \varepsilon_n^2 \right)^{1/2}, \quad u \in C^{\infty}_{4,b,2}(\mathbb{R}^{\infty}) \right). \quad (3.83)$$

Indeed, let $\beta = \sum_1^\infty \varepsilon_n$ and $\gamma = (2\sum_1^\infty)\varepsilon_n^2)^{1/2}$, and let E be the resolution of the identity of the nonnegative selfadjoint operator $\mathcal{B}_0$. Then the right-hand side of (3.81) is equal to

$$\int_0^\infty ((\lambda + \beta)^2 + \gamma^2) d(E(\lambda)u, u)_{L_2(\mathbb{R}^\infty, dg_\varepsilon(x))}$$

and can be estimated from above by the square of the right-hand side of (3.83), since $(\lambda + \beta)^2 + \gamma^2 \leq (\lambda + \alpha)^2$ $(\lambda \in [0, \infty))$. ■

The differential expression (3.23), perturbed by the potential q, now has the form

$$\begin{aligned}(\mathcal{L}u)(x) &= -\sum_{n=1}^{\infty} (D_n^2 u)(x) + q(x)u(x) \\ &= -\sum_{n=1}^{\infty} \left(\left(\frac{\partial^2 u}{\partial x_n^2} \right)(x) - 2\varepsilon_n x_n \left(\frac{\partial u}{\partial x_n} \right)(x) \right. \\ &\qquad \left. + (\varepsilon_n^2 x_n^2 - \varepsilon_n)u(x) \right) + q(x)u(x) \\ &= (\mathcal{L}_0 u)(x) + q(x)u(x).\end{aligned} \quad (3.84)$$

In correspondence with (3.22), let

$$p(x) = \sum_{n=1}^{\infty} (D_n^2 1)(x) = \sum_{n=1}^{\infty} (\varepsilon_n^2 x_n^2 - \varepsilon_n) \in L_2(\mathbb{R}^\infty, dg_\varepsilon(x)). \quad (3.85)$$

The expression $\mathcal{M}$ in (3.63) corresponding to Gaussian weights has the form

$$\begin{aligned}(\mathcal{M}u)(x) &= \sum_{n=1}^{\infty}(-(D_n^2 u)(x) + (D_n^2 1)(x)u(x)) + a(x)u(x) \\ &= \sum_{n=1}^{\infty}\left(-\left(\frac{\partial^2 u}{\partial x_n^2}\right)(x) + 2\varepsilon_n x_n\left(\frac{\partial u}{\partial x_n}\right)(x)\right) + a(x)u(x) \\ &= (\mathcal{M}_0 u)(x) + a(x)u(x). \end{aligned} \tag{3.86}$$

Thus, it follows from (3.84)–(3.86) that

$$(\mathcal{L}u)(x) = (\mathcal{M}_0 u)(x) + a(x)u(x), \qquad a(x) = q(x) - p(x) \quad (x \in \mathsf{R}^\infty) \tag{3.87}$$

This relation and the inclusion (3.85) make it easy for us to reformulate Theorems 3.2 and 3.3 in connection with the determination of essential selfadjointness of the operator

$$C^\infty_{\mathrm{c},b,2}(\mathsf{R}^\infty) \ni u(x) \mapsto (\mathcal{L}u)(x) \tag{3.88}$$

acting in $L_2(\mathsf{R}^\infty, dg_\varepsilon(x))$.

We proceed to establish some deeper facts in this direction.

THEOREM 3.4. *Suppose that the real-valued potential q in $L_2(\mathsf{R}^\infty, dg_\varepsilon(x))$ is semibounded below. Then the operator* (3.88) *is essentially selfadjoint.*

PROOF. Let $c = \inf_{x\in\mathsf{R}^\infty} q(x) > -\infty$. It suffices to prove the essential selfadjointness of the operator $C^\infty_{\mathrm{c},b,2}(\mathsf{R}^\infty) \ni u(x) \mapsto ((\mathcal{L} - c1)u)(x)$, i.e., to consider (3.84), where $q(x)$ is replaced by $q_1(x) = q(x) - c$ $(x \in \mathsf{R}^\infty)$.

The operator $\mathcal{B}_0$ equal to the closure of the operator $C^\infty_{\mathrm{c},b,2}(\mathsf{R}^\infty) \ni u(x) \mapsto (\mathcal{M}_0)(x)$ is nonnegative and selfadjoint, and the semigroup $e^{-t\mathcal{B}_0}$ preserves positivity (see Lemma 3.3). The potential $q_1 \in L_2(\mathsf{R}^\infty, dg_\varepsilon(x))$ is nonnegative, and the operator $C^\infty_{\mathrm{c},b,2}(\mathsf{R}^\infty) \ni u(x) \mapsto (\mathcal{M}_0 u)(x) + q_1(x)u(x)$ is essentially selfadjoint according to Theorem 3.2. Therefore, the theorem of Davis and Faris is now applicable (see Reed and Simon [2], Theorem X.31) and gives us that if the function $\mathsf{R}^\infty \ni x \to b(x)$ is such that

$$\|b(x)u(x)\|_{L_2(\mathsf{R}^\infty, dg_\varepsilon(x))} \le \|(\mathcal{M}_0 + \alpha)u(x)\|_{L_2(\mathsf{R}^\infty, dg_\varepsilon(x))} \qquad (v \in C^\infty_{\mathrm{c},b,2}(\mathsf{R}^\infty)) \tag{3.89}$$

for some $\alpha \in \mathsf{R}^1$, then the same inequality holds also with $\mathcal{M}_0$ replaced by $\mathcal{M}_0 + q_1(x)$.

According to the corollary to Lemma 3.9, inequality (3.89) holds in the case

$$b(x) = \sum_{n=1}^{\infty} \varepsilon_n^2 x_n^2 = p(x) + \beta \qquad \left(\beta = \sum_{n=1}^{\infty} \varepsilon_n\right).$$

The same inequality with $\mathcal{M}_0$ replaced by $\mathcal{M}_0+q_1(x)$ means that the function $p(x)+\beta$ is $(\mathcal{M}_0+q_1)$-bounded with bound equal to 1. According to a theorem of R. Wüst (see Reed and Simon [2], Theorem X.14), the operator

$$C^{\infty}_{c,b,2}(\mathbb{R}^{\infty}) \ni u(x) \mapsto (\mathcal{M}_0 u)(x) + q_1(x)u(x) + (p(x)+\beta)u(x)$$

is essentially selfadjoint; but then the operator (3.88) is selfadjoint. ■

THEOREM 3.5. *Suppose that the real-valued potential* q *is such that*

$$q(\cdot) \in L_{2+\varepsilon}(\mathbb{R}^{\infty}, dg_{\varepsilon}(x)) \quad \text{and} \quad e^{-\delta q(\cdot)} \in L_1(\mathbb{R}^{\infty}, dg_{\varepsilon}(x))$$

for some $\varepsilon, \delta > 0$ *and it is known that the operator* (3.88) *is semibounded below. Then this operator is essentially selfadjoint.*

PROOF. The theorem follows immediately from Theorem 3.3 because of (3.87) and the fact that $p(\cdot) \in L_r(\mathbb{R}^{\infty}, dg_{\varepsilon}(x))$ $(r \in [2,\infty))$ and $e^{\gamma p(\cdot)} \in L_1(\mathbb{R}^{\infty}, dg_{\varepsilon}(x))$ $(\gamma \in [0, \beta^{-1}])$.

Indeed, if $q(\cdot) \in L_{2+\varepsilon}(\mathbb{R}^{\infty}, dg_{\varepsilon}(x))$ and $e^{-\delta q(\cdot)} \in L_1(\mathbb{R}^{\infty}, dg_{\varepsilon}(x))$, then

$$a(\cdot) = q(\cdot) - p(\cdot) \in L_{2+\varepsilon}(\mathbb{R}^{\infty}, dg_{\varepsilon}(x)), \qquad e^{(-\delta_1/2)a(\cdot)} \in L_1(\mathbb{R}^{\infty}, dg_{\varepsilon}(x)) \quad (0 < \delta_1 \le \beta^{-1}, \delta).$$

(The last inclusion follows from Hölder's inequality.) ■

3.7. *Selfadjointness of differential operators with infinitely many variables. Combination of the parabolic and hyperbolic approaches.* As already mentioned, the main result of §3.6 (Theorem 3.2) is analogous to Lemma 3.4 in the finite-dimensional case. We now sketch a proof of an assertion of the same type as Theorem 3.1 for a perturbation of the operator (3.23) in the infinite-dimensional case (with annihilation lacking).

It is known (see Yu. S. Samoĭlenko [1], Lemma 3 in §1.1 of Chapter 1) that for any probability measure on $\mathbb{R}^{\infty}$ there exists a sequence of positive numbers $\delta = (\delta_n)_1^{\infty}$, $\sum_1^{\infty} \delta_n < \infty$, such that $G = \{x \in \mathbb{R}^{\infty} \mid \sum_1^{\infty} \delta_n x_n^2 < \infty\}$ is a set of full measure. To describe the class of potentials considered below we fix a sequence δ chosen in this way from the measure (3.13) and introduce on $\mathbb{R}^{\infty}$ the norm

$$\|x\|_{\delta} = \left(\sum_{n=1}^{\infty} \delta_n x_n^2 \right)^{1/2} \le \infty.$$

Naturally, *condition* (3.21) *is now assumed; what is more, we assume that the function* $p(x)$ *defined here* (*see* (3.22)) *is semibounded below and is locally bounded above, i.e.,* $\sup_{\|x\|_{\delta} \le r} p(x) < \infty$ *for each* $r > 0$. The case of Gaussian weights (3.34) is an example of this situation. Here $\delta_n = \mu_n \varepsilon_n$, where $\mu_n \ge \varepsilon_n$ $(n = 1, 2, \ldots)$ and $\sum_1^{\infty} \mu_n < \infty$; it is then clear that $p(x) = \sum_1^{\infty} (\varepsilon_n^2 x_n^2 - \varepsilon_n)$ is locally bounded above. The equality $\rho(G) = 1$ follows from the Khintchine-Kolmogorov theorem (see for example, Shilov and Fan Dyk Tin′ [1], §3).

THEOREM 3.6. *Suppose that the real-valued potential $q \in L_2(\mathsf{R}^\infty, d\rho(x))$ is locally semibounded below in the sense that $\inf_{\|x\|_\delta \le r} q(x) > -\infty$ for each $r > 0$. Then the operator*

$$C^\infty_{\mathrm{c},b,2}(\mathsf{R}^\infty) \ni u(x) \mapsto (\mathcal{L}u)(x) = -\sum_{k=1}^{\infty}(D_k^2 u)(x) + q(x)u(x)$$
$$= (\mathcal{L}_0 u)(x) + q(x)u(x) \qquad (3.90)$$

is essentially selfadjoint provided that it is semibounded below.

The idea of the proof is as follows. Let L be the closure of the operator (3.90) in $L_2 = L_2(\mathsf{R}^\infty, d\rho(x))$. Denote by $a_n(x)$ the function of $x \in \mathsf{R}^\infty$ equal to $q(x) - p(x)$ for $x \in \{x \in \mathsf{R}^\infty | \, \|x\|_\delta \le n\}$ and equal to 0 outside this set $(n = 1, 2, \ldots)$. Let $\mathcal{M}^{(n)}$ be the differential expression (3.63) with a replaced by a_n, and let $M^{(n)}$ be the corresponding operator equal to the closure of the operator $C^\infty_{\mathrm{c},b,2}(\mathsf{R}^\infty) \ni u(x) \mapsto (\mathcal{M}^{(n)}u)(x)$. According to Theorem 3.2, it is selfadjoint and semibounded below.

To prove the theorem it suffices to establish the following (Chapter 2, Theorem 6.5). Assume that there is a set Φ which a) is dense in $L_2(\mathsf{R}^\infty, d\rho(x))$, and b) satisfies the condition $\Phi \subseteq \mathfrak{D}(M^{(n)})$ $(n = 1, 2, \ldots)$. Consider a strong solution $[0, 1) \ni t \mapsto u(t) \in L_2(\mathsf{R}^\infty, d\rho(x))$ of the Cauchy problem

$$(d^2u/dt^2)(t) + L^* u(t) = 0 \quad (t \in [0, 1)), \qquad u(0) = 0, \quad u'(0) = 0,$$

along with a solution $[0, T] \ni t \mapsto u^{(n)}(t) \in L_2(\mathsf{R}^\infty, d\rho(x))$ (which exists because $M^{(n)}$ is selfadjoint and semibounded below and $\Phi \subseteq \mathfrak{D}(\mathcal{M}^{(n)})$) of the Cauchy problem

$$(d^2u^{(n)}/dt^2)(t) + M^{(n)}u^{(n)}(t) = 0 \qquad (t \in [0, T]),$$

$u^{(n)}(T) = \varphi_0$ and $(u^{(n)})'(T) = \varphi_1$ $(T \in (0, 1);\ \varphi_0, \varphi_1 \in \Phi$ fixed).

It is required to choose Φ so that, in addition, c) $u^{(n)}(t) \in \mathfrak{D}(L)$ $(t \in [0, T])$ for sufficiently large n, and d)

$$\int_0^T (u(t), (M^{(n)} - L)u^{(n)}(t))_{L_2(\mathsf{R}^\infty, d\rho(x))}\, dt \underset{n\to\infty}{\longrightarrow} 0.$$

The proof of the theorem amounts to a construction of a set Φ with properties a)-d). This choice is fairly complicated; it is realized in the paper [2] of Berezanskiĭ and V. G. Samoĭlenko (Theorem 3.3).

The theorem, of course, implies an analogous assertion with regard to a perturbation of the operator (3.28), i.e., the operator (3.63): if the conditions assumed here are satisfied (convergence of the series (3.21) and the properties of the function p) and the potential a satisfies the same requirements as the potential q in Theorem 3.6, then the operator (3.63) is essentially selfadjoint provided only that it is semibounded below. The assertion follows from the equality $\mathcal{M}u = \mathcal{L}u$ if we set $q = a + p$ in the definition of $\mathcal{L}$ and note that q has the properties required in Theorem 3.6.

Observe that for Gaussian weights (3.34) the last assertion is valid also when the condition $\sum_1^\infty \varepsilon_n < \infty$ (i.e., (3.21)) is replaced by the requirement that the sequence $(\varepsilon_n)_1^\infty$ be bounded. (This fact, of course, no longer follows from Theorem 3.6, but can be proved in a similar way; see Berezanskiĭ and V. G. Samoĭlenko [2], Theorem 3.8.)

3.8. ***Second digression. Operators that are not semibounded below.*** In conclusion we explain how the techniques in §6 of Chapter 2 (more precisely, Theorem 6.4) enable us to establish selfadjointness in the case of operators not semibounded below and to obtain theorems like the Titchmarsh-Sears theorem. We confine ourselves to the case of the operators in §3.4 with finitely many variables. The three lemmas given below will be needed; the first of them generalizes Lemma 3.5. For simplicity the proofs of these lemmas are given under the assumption that $a_{jk} \in C^{1+\alpha}(\mathbf{R}^N)$ ($\alpha \in (0,1)$). This additional restriction can be removed just as described in §3.5.

LEMMA 3.10. *Let $\mathcal{L}$ be a differential expression like that in Theorem* 3.1, *but with $a_{jk} \in C^{1+\alpha}(\mathbf{R}^N)$ and with the potential q semibounded below, and let L be the corresponding minimal operator, which is semibounded below and selfadjoint according to that theorem. Then the assertion of Lemma* 3.5 *is valid for the Cauchy problem* (3.44).

PROOF. As in the proof of Lemma 3.5, we consider the differential expression $\mathcal{L}^{(n)}$ with averaged coefficients $a_{jk}^{(n)}, q^{(n)} \in C^\infty(\mathbf{R}^N)$ $(n = 1, 2, \ldots)$. Obviously, $a_{jk}^{(n)} \restriction K_R(0) \to a_{jk} \restriction K_R(0)$ uniformly along with their first derivatives as $n \to \infty$, and $q^{(n)} \restriction K_R(0) \to q \restriction K_R(0)$ in $L_2(K_R(0), dx)$ for each ball $K_R(0)$; and $c^{(n)} \to c$ uniformly on each interval $[0, R]$. As in the proof of Theorem 3.1, we conclude that $c^{(n)}(r) \le c(r+1)$ $(r \in [0, \infty);\ n = 1, 2, \ldots)$.

By the last inequality, it follows from condition (3.37) for c that (3.37) holds also for $c^{(n)}$ $(n = 1, 2, \ldots)$. The potential $q^{(n)}$ is semibounded below; therefore, the corresponding minimal operator $L^{(n)}$ is semibounded below and selfadjoint. Let $u^{(n)}(t) = u^{(n)}(x, t)$ be a solution of the Cauchy problem (3.44) with $\varphi_0, \varphi_1 \in C_0^\infty(\mathbf{R}^N)$, or, what is the same, of the problem of form (3.48). As in Lemma 3.5, the inclusion (3.49) holds. On the other hand, we conclude just as in that lemma that $L^{(n)} \to L$ in the strong resolvent sense. Consequently, (3.48) also holds. Passing to the limit as $n \to \infty$ in (3.49), we get a proof of the lemma. (It is easy to see that the nature of the convergence of $c^{(n)}$ to c suffices for convergence of the solutions $r_{a,\rho}^{(n)}(t) \to r_{a,\rho}(t)$ $(t \in [0, \infty))$ as $n \to \infty$.) ∎

LEMMA 3.11. *Let $\mathcal{L}$ be a differential expression as in Lemma* 3.6, *but with q locally semibounded below. (It is not assumed that L is semibounded below.) Then for each $T \in (0, \infty)$ the Cauchy problem*

$$(d^2u/dt^2)(t) + Lu(t) = 0 \qquad (t \in [0, T]),$$
$$u(0) = \varphi_0 \in C_0^\infty(\mathbf{R}^N), \qquad u'(0) = \varphi_1 \in C_0^\infty(\mathbf{R}^N) \tag{3.91}$$

has a strong solution.

PROOF. Suppose that $\operatorname{supp}\varphi_0 \operatorname{supp}\varphi_1 \subseteq K_\rho(0)$ for some $\rho > 0$. Consider the Cauchy problem (3.46) with c constructed from $\mathcal{L}$. By (3.37), the solution $r(t)$ of this problem is defined for all $t \in [0, \infty)$ and increases monotonically to $+\infty$ as $t \to +\infty$. Fix an $R_1 > R = \rho + r(T)$ and denote by $\mathcal{L}_1$ the differential expression differing from $\mathcal{L}$ in that the potential $q(x)$ is replaced by the potential $q_1(x)$ equal to $q(x)$ for $|x| \leq R_1$ and to 0 for $|x| > R_1$. Because q_1 is semibounded below, Theorem 3.1 gives us that the corresponding minimal operator L_1 is semibounded and selfadjoint. Consequently, the Cauchy problem (3.91) with L replaced by L_1 has a solution $u_1(t)$ ($t \in [0, T]$). By Lemma 3.10, $\operatorname{supp} u_1(t) \subseteq K_R(0)$, for each $t \in [0, T]$. The lemma is proved if we establish the following assertion of the same type as fact a) in the proof of Theorem 3.1.

If $f \in \mathfrak{D}(L_1)$ *and* $\operatorname{supp} f \subseteq K_R(0)$, *then*

$$f \in \mathfrak{D}(L), \qquad Lf = L_1 f. \tag{3.92}$$

Indeed, suppose that $(g_m)_1^\infty$ is a sequence $g_m \in C_0^\infty(\mathsf{R}^N)$ such that $g_m \to f$ and $\mathcal{L}_1 g_m \to \mathcal{L} f$ in L_2 as $m \to \infty$. Consider the infinitely differentiable function $\mathsf{R}^N \ni x \mapsto \chi(x) \in [0, 1]$ equal to 1 in the ball $K_R(0)$ and to 0 outside $K_{R_1}(0)$. The sequence $f_m(x) = \chi(x) g_m(x)$ ($x \in \mathsf{R}^N$; $m = 1, 2, \ldots$) is such that

$$f_m \to f, \qquad \mathcal{L}_1 f_m \to L_f \tag{3.93}$$

in L_2 as $m \to \infty$. This is proved exactly like a). Note here that the integral in the second and third terms in the estimate of type (3.56) for $\|\mathcal{L}_1 g_m - \mathcal{L}_1 f_m\|_{L_2}^2$ is over $K_{R_1}(0) \backslash K_R(0)$, where the a_{jk} and their first derivatives are bounded.

Since $\mathcal{L}_1 f_m = \mathcal{L} f_m$, (3.93) shows that $f \in \mathfrak{D}(L)$ and $Lf = L_1 f$. ∎

LEMMA 3.12. *The assertion of Lemma* 3.11 *remains in force when* $[0, T]$ *is replaced by* $[0, \infty)$. *Moreover, the solution* $[0, \infty) \ni t \mapsto u(t) \in L_2$ *of this problem on each interval* $[0, T]$ *is equal to the solution of* (3.91), *with* L *replaced by the minimal operator* L_T *generated by the expression* $\mathcal{L}_T$ *coinciding with* $\mathcal{L}$ *with* q *replaced by the lower semibounded potential* q_T *equal to* $q(x)$ *for* $|x| \leq \rho + r(T)$ *and to* 0 *for* $|x| > \rho + r(T)$. *Here* ρ *is the radius of a ball* $K_\rho(0) \supseteq \operatorname{supp}\varphi_0, \operatorname{supp}\varphi_1$, *and* r *is a solution of problem* (3.46).

PROOF. The proof consists in establishing the possibility of gluing together solutions of problem (3.91) with $L = L_T$ for different T. This gluing together is possible if the following uniqueness theorem is valid.

For $0 < T_1 < T_2 < \infty$ let $[0, T_1] \ni t \mapsto u_1(t) \in L_2$ and $[0, T_2] \ni t \mapsto u_2(t) \in L_2$ be solutions of the problems (3.91) with $L = L_{T_1}$ and $L = L_{T_2}$, respectively, and suppose that $\operatorname{supp}\varphi_0, \operatorname{supp}\varphi_1 \subseteq K_\rho(0)$. It is required to prove that $u_1(t) = u_2(t)$ ($t \in [0, T_1]$).

Since u_1 and u_2 are continuous, it suffices to verify this equality for $t \in [0, b)$, where $b \in (0, T_1)$ is arbitrary and fixed. According to Lemma 3.11,

$$\operatorname{supp} u_1(t), \operatorname{supp} u_2(t) \subseteq K_{\rho+r(b)}(0) = K_R(0)$$

for $t \in [0, b)$. We have that $\rho + r(T_1), \rho + r(T_2) > R$, so $u_j(t) \in \mathfrak{D}(L)$ and $L_{T_j} u_j(t) = L u_j(t)$ ($j = 1, 2$; $t \in [0, b)$) by (3.92). These relations show that $[0, b) \ni t \mapsto u_j(t) \in L_2$ can be understood as two solutions of the same Cauchy problem (3.91) on $[0, b)$ with the same initial conditions.

On the other hand, it follows from Lemma 3.11 that the Cauchy problem

$$(d^2\varphi/dt^2)(t) + L\varphi(t) = 0 \quad (t \in [0, T]), \qquad \varphi(T) = \psi_0, \quad \varphi'(T) = \psi_1,$$

where $t \in (0, b)$ and $\psi_0, \psi_1 \in C_0^\infty(\mathbb{R}^n)$, has a strong soluton. But then an appropriate modification of the Holmgren principle (Lemma 6.2 in Chapter 2) leads us to the conclusion that uniqueness holds for strong solutions of the Cauchy problem (3.91) on $[0, b)$ even when L is replaced in the equation by $L^* \supseteq L$. This gives us that $u_1(t) = u_2(t)$ ($t \in [0, b)$). ■

THEOREM 3.7. *Suppose that the coefficients a_{jk} are as indicated in* §3.4, *and the potential q is locally semibounded below. Assume that the following condition of Titchmarsh type is satisfied: there exists a nonincreasing function $[0, \infty) \ni s \mapsto Q(s) \in (-\infty, 0]$ such that $q(x) \geq Q(|x|)$ ($x \in \mathbb{R}^N$) and*

$$|Q(r(t))| \underset{t\to\infty}{=} O(t^2), \tag{3.94}$$

where $[0, \infty) \ni t \mapsto r(t) \in [0, \infty)$ is the function inverse to the function

$$[0, \infty) \in r \mapsto t(r) = \int_0^r c^{-1}(s)\, ds \in [0, \infty). \tag{3.95}$$

Then L is selfadjoint.

We remark that if the coefficients a_{jk} are bounded in the whole of $\mathbb{R}^N$, then $\sigma^{1/2} \leq c(r) \leq c_1 < \infty$ ($r \in [0, \infty)$), where c_1 is a constant. A solution r of the Cauchy problem $r'(t) = c(r(t))$ ($t \in [0, \infty)$), $r(0) = 0$, satisfies the estimate $\sigma^{1/2} t \leq r(t) \leq c_1(t)$ ($t \in [0, \infty)$). Therefore, (3.94) is equivalent to the condition $|Q(t)| = O(t^2)$ for $t \to \infty$. In the case of unbounded a_{jk} the function $r(t)$ increases faster than t as $t \to \infty$, and condition (3.94) is weaker than the condition just written.

Note also that we can set $Q(s) = -(a + b(t(s))^2)$ in the theorem for s sufficiently large, where $a, b \in \mathbb{R}^1$ are fixed. Then the condition of the theorem takes the form

$$q(x) \geq -(a + b(t(|x|))^2) \qquad (x \in \mathbb{R}^N); \tag{3.96}$$

(3.96) is essentially equivalent to (3.94), but the expression in the form (3.94) is more convenient for passing to a theorem like that of Sears (see Remark 1).

PROOF. Consider the Cauchy problem

$$(d^2\varphi/dt^2)(t) + L\varphi(t) = 0 \quad (t \in [0, \infty)), \qquad \varphi(0) = \varphi_0, \quad \varphi'(0) = \varphi_1. \tag{3.97}$$

According to Lemma 3.12, it has a strong solution for any $\varphi_0, \varphi_1 \in C_0^\infty(\mathbf{R}^N)$. Therefore, by Theorem 6.4 in Chapter 2, L is selfadjoint if a solution $\varphi(t)$ of (3.97) with $\varphi_1 = 0$ is such that

$$(\varphi(t), \varphi_0)_{L_2} \underset{t\to\infty}{=} O(e^{ct^2}) \tag{3.98}$$

with some $c = c_{\varphi_0} \geq 0$. Let us establish (3.98).

According to the second part of Lemma 3.12, $\varphi(t) = u_T(t)$ for $t \in [0, T]$, where u_T is a solution of (3.91) with $\varphi_1 = 0$ and L replaced by the lower semibounded selfadjoint operator L_T constructed according to the rule indicated in the lemma. It is obvious that $L_T \geq l_T 1$, where

$$l_T = \inf_{x\in\mathbf{R}^N} q_T(x) = \inf_{|x|\leq\rho+r(T)} q(x) \geq \inf_{s\leq\rho+r(T)} Q(s) \geq Q(\rho + r(T)) \leq 0$$

($\rho = \rho(\varphi_0) > 0$ is fixed). Denoting by E_T the resolution of the identity for L_T and using (3.45), we get

$$\begin{aligned}(\varphi(T), \varphi_0)_{L_2} &= (u_T(T), \varphi_0)_{L_2} \\ &= \int_{Q(p+r(T))}^{\infty} \cos\sqrt{\lambda}T d(E_T(\lambda)\varphi_0, \varphi_0)_{L_2} \\ &\leq (1 + \cosh T\sqrt{|Q(\rho + r(T))|}\,)\|\varphi_0\|_{L_2}^2.\end{aligned} \tag{3.99}$$

Since $r(T)$ tends montonically to ∞ as $T \to \infty$, while $r(T) \geq \sigma^{1/2}T$ ($T \in [0, \infty)$), there is a particular $T(\rho) > 0$ such that $\rho + r(T) \leq r(T + T(\rho))$ ($T \in [0, \infty)$). The fact that Q is nonincreasing and (3.94) give us that

$$\begin{aligned}|Q(\rho + r(T))| &\leq |Q(r(T + T(\rho)))| \\ &\leq c_1(T + T(\rho))^2 \leq c_2 T^2\end{aligned}$$

for $T \geq T(\rho) > 0$. This shows that the right-hand side of (3.99) has the form $O(e^{cT^2})$ as $T \to \infty$, with $c = c_{\varphi_0} \geq 0$. The estimate (3.98) is established. ∎

REMARK 1. The estimate (3.98) appearing here in connection with Theorem 6.4 can be weakened (see §6.2 in Chapter 2). Namely, the c in it can be replaced by a so-called slowly increasing function $c(t)$. The use of this strengthened version of Theorem 6.4 in our situation leads to a strengthening of Theorem 3.7: the estimate (3.94) in it can be replaced by an estimate like that of Sears, imposed on the function $Q(r(t))$.

REMARK 2. Theorem 3.7 does not encompass Theorem 3.1, since it gives restrictions on the growth of the potential q, not on the quadratic form of L. However, it can be generalized in such a way that Theorem 3.1 is encompassed. Namely, instead of the estimate $q(x) \geq Q(|x|)$ ($x \in \mathbf{R}^N$) of the potential we can require an analogous estimate of the function

$$\gamma(r) = \inf(L\varphi, \varphi)_{L_2}(\varphi, \varphi)_{L_2}^{-1},$$

where the infimum is over the functions $\varphi \in C_0^\infty(\mathbb{R}^N)$ with $\operatorname{supp} \varphi \subseteq K_r(0)$ (i.e., an estimate of the function equal to the "infimum of L on $K_r(0)$").

We do not give results of the same type as Theorem 3.7 for the case of differential operators with infinitely many variables that are not semibounded below. We mention only that they can be obtained for the operators in subsections 6 and 7 by a combination of the devices used here and in subsection 7. See Berezanskiĭ and V. G. Samoĭlenko [2] (Theorems 3.4 and 3.8) for their formulations and proofs.

Comments on the Literature

We give references supplementing and making more precise those in the text.

Introduction

Facts about topological spaces are presented in Pontryagin's book [1, Chapter 2]; it is also useful to be acquainted with some of the sections in the book of Arkhangel'skiĭ and Ponomarev [1]. See the books of Halmos [1] and Gikham and Skorokhod [1] for a presentation of measure and integration theory, and the book of Schaefer [1] about linear topological spaces. The scope of the theory of operators in Hilbert space used is roughly that of the book of Akhiezer and Glazman [1]; some points of view and approaches useful for us are contained in the monographs of Plesner [1], K. Maurin [2], and Birman and Solomyak [1]. It is advantageous to the familiar with some sections of the books of Dunford and Schwartz [1], [2], which contain a large amount of information about functional analysis, and with the books of Reed and Simon [1]–[4], which were written from the point of view of the needs of quantum field theory and mathematical physics as a whole. The theory of generalized functions is given in the aspect we require in the books of Sobolev [2], L. Schwartz [1], Gel'fand and Shilov [1], [2], and Vladimirov [1].

Chapter 1

§1. The presentation of the theory of spaces with negative norm follows Chapter 1 of Berezanskiĭ's book [4], though we have simplified some of the proofs. In writing the introduction of the chapter the references Leray [1] and Lax [1] were kept in mind. The techniques of spaces with negative norm later entered a number of areas of functional analysis and its applications: the theory of boundary value problems for partial differential equations, the spectral theory of differential operators with operator coefficients, the theory of extension of operators, and so on. We do not cite corresponding publications, since they are too far removed from the topic at hand. Here we mention only applications to measure theory and the theory of differential operators with infinitely many variables (Yu. L. Daletskiĭ [4]), to quantum field theory (see

Simon's book [2]), and to the spectral theory of differential expressions with operator coefficients (V. I. and M. L. Gorbachuk [1]); the last monograph contains, in particular, constructions of broad classes consisting of inductive and projective limits of spaces with negative norm and important in the investigation of boundary properties of solutions of differential equations. The theory of semibounded bilinear forms is presented in a natural way in an article of Berezanskiĭ [15] on the basis of the theory of spaces with negative norm.

§2. This is based on work of von Neumann [1] on infinite tensor products of Hilbert spaces; however, the presentation of the results needed from that article is modified in a way convenient for us: first a separable subspace of the complete product is introduced by specifying a basis (§2.3), and later the complete von Neumann product is introduced (§2.10). Weighted infinite tensor products and infinite tensor products of chains (§§2.4–2.6) were introduced and studied for the purposes of spectral theory in articles of Berezanskiĭ, Gali, and Zhuk [1], Berezanskiĭ and Gali [1], and Berezanskiĭ and Us [1], [2]. The tensor product of an infinite number of operators (§2.7) is constructed according to articles of Berezanskiĭ and Us [1], [2]; see the papers of Nakagami [1]–[3] and Araki and Nakagami [1] for other results on tensor products of infinitely many operators, including unbounded operators. A variant of the Schwartz kernel theorem (L. Schwartz [1], [2]) with a rigging by Hilbert spaces is presented in §§2.8 and 2.9. This presentation is basically the same as that in Berezanskiĭ's book [4], except that essential improvements are made in both the formulations of the theorems and their proofs. Triplets of spaces (§2.11) were introduced by Berezanskiĭ [9]. Marchenko [1]–[3] gave a broad and useful generalization of the construction of infinite tensor products both of Hilbert spaces and of nuclear spaces. Instead of the imbeddings $\bigotimes_1^p H_n \to (\bigotimes_1^p H_n) \otimes e^{(p+1)} \subseteq \bigotimes_1^{p+1} H_n$ in the construction of the infinite tensor product, this generalization uses a sequence of isometries $i_p\colon G_p \to G_{p+1}$ $(p = 1, 2, \ldots)$, where $(G_p)_1^\infty$ is a given sequence of Hilbert spaces which play the role of the spaces $\bigotimes_1^p H_n$. The role of the infinite tensor product here is played by $\lim \operatorname{ind}_{p\to\infty} G_p$.

§3. See Sobolev's books [1] and [2] concerning the theory of positive Sobolev spaces. Negative Sobolev spaces and their properties both in bounded and in unbounded domains (subsections 3, 4 and 6–8) are presented according to Berezanskiĭ's articles [4] and [8]; however, the results are strengthened and refined (see Wloka's note [1] in connection with refinements of these results). The kernel theorem in spaces of square-integrable functions on a bounded or unbounded domain is presented in §§3.5 and 3.11; these results refine somewhat the corresponding facts in Berezanskiĭ's book [4]. The same can be said about §3.12, which concerns chains constructed from a positive-definite kernel. The facts in §§3.9 and 3.10 are well known. We remark that Berezanskiĭ and Kalyuzhnyĭ [1] have constructed analogues of Sobolev spaces and projective limits of them in the case when the role of differentiation is played by a corresponding operation in a hypercomplex system with locally compact base.

§**4.** See the books of Gel′fand and Vilenkin [1], Shilov and Fan Dyk Tin′ [1], Gikhman and Skorokhod [2], Skorokhod [1], and Kuo [1] about the Gaussian measures in this section, §3 in Chapter 2, and §§1 and 3 in Chapter 3. Theorem 4.2 is due to L. Schwartz [1]. The concept and the properties of weighted infinite tensor products of nuclear spaces, on which the introduction of the test functions and generalized functions of infinitely many variables in this section is based (§4.3), are due to Berezanskiĭ and Yu. S. Samoĭlenko [1]. The introduction and study of important classes of spaces of test functions and generalized functions of infinitely many variables—the spaces $\mathcal{A}(\mathbf{R}^\infty)$ and $\mathcal{H}(\mathbf{R}^\infty)$ and their dual spaces (§§3.5 and 3.6)—are due to Kondrat′ev and Yu. S. Samoĭlenko [1], [2]; in essence, Theorem 4.6 develops a result of Piech [1]. The construction of weighted tensor products was generalized by Korsunskiĭ [2] to infinite tensor products of locally convex spaces. The Segal isomorphism carries the riggings of $L_2(\mathbf{R}^\infty, dg(x))$ constructed in §4 into riggings (moreover, nuclear riggings) of Fock space. Nonnuclear riggings of Fock spaces were introduced by Kristensen, Mejlbo, and Poulsen [1]; some facts about riggings of Fock space are contained in Napiórkowski's paper [1]. The Khintchine-Kolmogorov theorem was first presented in their paper [1]. Sobolev spaces of functions of infinitely many variables with a fixed smoothness were introduced and studied by Frolov [1], [2], but facts like quasinuclearity of imbeddings are not valid for them due to the absence of a weighting.

Kondrat′ev [1], [2] developed a variant of the theory of generalized functions in the framework of our approach to the introduction of generalized functions for the case of a dual nuclear space of arguments that no longer has a coordinate representation. Nuclear spaces of entire functions (in particular, an analogue of $\mathcal{A}(\mathbf{R}^\infty)$) were employed as spaces of test functions. Integral transformations of Fourier-Wiener type on test functions and generalized functions were studied and used to carry over a number of facts from the finite-dimensional theory to the infinite-dimensional case, for example, a description of nonnegative generalized functions. It was also shown that typical interactions in constructive field theory can be described as generalized functions in the spaces introduced. Another type of space of test functions of infinitely many variables was presented by Kondrat′ev and Tsikalenko [1]. These spaces can be regarded as infinite-dimensional analogues of the Gel′fand-Shilov spaces. The theory of generalized functions of infinitely many variables constructed in §4 and in the references just cited is still convenient in the respect that it uses techniques of ordinary mathematical analysis.

Another way of constructing a theory of generalized functions of infinitely many variables was presented by Fomin [1], [2], and was developed in a number of articles (Averbukh, Smolyanov, and Fomin [1], [2], Yu. L. Daletskiĭ and Fomin [1], Uglanov [1], and Dudin [1]). The main idea in this approach is that of passing (because of the absence of a standard measure of Lebesgue type on an infinite-dimensional space) from duality between test functions and generalized functions to duality between test functions and measures or to duality between test measures (which are smooth in a specific sense) and generalized functions. This approach is presented in detail in a recent book of Yu. L. Daletskiĭ and Fomin [2]; see also the bibliography there.

Chapter 2

§1. The approach developed for the study of commuting normal operators by constructing the joint resolution of the identity has been encountered earlier for countable families of operators (Plesner and Rokhlin [1], Plesner [1]). A joint resolution of the identity for a family that is more than countable was introduced and used by Berezanskiĭ [5]–[7]. The proof of Theorem 1.2 follows the book of Birman and Solomyak (§1.2 in Chapter 5). As already mentioned, contrary to the case of scalar measures, not every two commuting abstract resolutions of the identity can be multiplied; see Birman, Vershik and Solomyak [1] for a corresponding example. The concept of the support of a joint resolution of the identity is essential for §2. At the same time, the support of a measure is usually considered in the case of measures on locally compact spaces; unusual situations are possible for measures on Tychonoff products (see §1.7). A considerable number of the results in subsections 4–6 and 8–11 are contained, in essence, in Berezanskiĭ's articles [6] and [13]. The techniques for getting them are fairly standard in the theory of random processes (Gikhman and Skorokhod [2], Venttsel′ [1]), though the important (for us) Tychonoff topologization is rarely used in this theory, and that presents its own peculiarities. The original presentation of Kolmogorov's theorem is contained in his book [3], and a proof of Tychonoff's theorem (Tychonoff [1]) is in Pontryagin's book [1]. In connection with the study of measures on Tychonoff products see Bourbaki's book [2] and the articles of Prokhorov [1] and Smolyanov and Fomin [1]; the noncommutative constructions of Belavkin [1] are connected with the constructions in this section. In reading §1 it is advantageous to be acquainted with some sections in the books of Birman and Solomyak [1], Gikhman and Skorokhod [2], and Ventsel′ [1].

The example in §1.7 is a modification, contained in an article of Kosyak and Yu. S. Samoĭlenko [1], of an example in the Russian edition (§1.6 in Chapter 2) of this book. We mention the following additional results in connection with the constructions in §1. Kolomytsev and Yu. S. Samoĭlenko [2] proved that for countable families $A = (A_n)_1^\infty$ of commuting selfadjoint operators A_n acting in H the intersection $\bigcap_1^\infty \mathfrak{D}(A_n)$ is always dense in H. What is more, if the joint spectrum of the family A is E-simple (i.e., there exists a vector $f \in H$ such that c.l.s.$(E(\alpha)f) = H$, where E is the joint resolution of the identity for A and $\alpha \in \mathcal{C}_\sigma(\mathsf{R}^\infty)$ is arbitrary), then A always has a cyclic vector Ω. But if the number of unbounded operators is more than countable, then the situation can change. For example, Kosyak and Samoĭlenko [1] gave examples of selfadjoint representations $\mathcal{S}(\mathbf{R}^1) \ni x \mapsto A_x$ of the Schwartz space such that the joint spectrum of the operators in the representation $(A_x)_{x\in\mathcal{S}(\mathbf{R}^1)}$ is E-simple but the family does not have a cyclic vector. They also gave examples of families $(A_x)_{x\in X}$ of commuting selfadjoint operators such that $\bigcap_{x\in X}\mathfrak{D}(A_x) = \{0\}$. The results described are also contained in Yu. S. Samoĭlenko's book [2] (§§1.3 and 1.6 in Chapter 1).

§**2**. The results in this section were obtained and applied to representations of functionals of Wightman type in Berezanskiĭ's papers [5]–[7] in a less complete form; they were announced in his notes [13] and [17] in a form close to that presented here. We emphasize that the improvements in the results involve only the situation when there are more than countably many operators; this case gives rise to essential difficulties (see §2.7). An article of K. Maurin [4] played a significant and stimulating role here for the author. Lemma 2.4, which generalizes the so-called separability lemma (Naĭmark [1], §40.2 in Chapter 8), is due to Kondrat′ev. §2 itself is an extended version of an article of Berezanskiĭ [18]. Contrary to the Russian edition of this book, we were able to prove the spectral projection theorem (Theorem 2.7) for an arbitrary family of operators. (The result in the Russian edition was only for a countable family: in the general case we had not been able to prove earlier that $\mathfrak{R}(P(\lambda(\cdot)))$ consists of generalized joint eigenvectors of A; see p. 197 of the Russian edition.) See the article of Naĭmark and Fomin [1] and the book of Naĭmark [1] (§41 in Chapter 8) about direct integrals of Hilbert spaces.

The first paper on expansion in generalized eigenvectors of a selfadjoint operator was that of Gel′fand and Kostyuchenko [1], and then came an article of Berezanskiĭ [1] which considered expansions with a rigging by Hilbert spaces. There is a thorough bibliography on work in this direction up to 1965 in Berezanskiĭ's book [4]. Here we indicate only that the construction in Chapter 2 was influenced by papers of Gårding [1], K. Maurin [1], and Kats [1]–[3]. Some subsequent articles up to 1978 dealing both with general questions in the theory of generalized eigenvector expansions and with certain special situations and applications closely related to this book are Gerlach [1], [2], Hirschfeld [1], Giertz [1], Gould [1], Itagaki [1], Derzko [1], Wloka [2], Nussbaum [3], [4], Napiórkowski [1]–[5], Birman and Èntina [1], Yu. S. Samoĭlenko [1], Babbitt [1], Korsunskiĭ [1], [3], Prugovečki [1], [2], Fredricks [1], Koshmanenko and Yu. S. Samoĭlenko [1], and Koshmanenko [1], [2]. A number of questions related to expansions in generalized eigenfunctions are contained in the books of K. Maurin [3] and Cotlar [1], and the recent survey of Simon [5].

§**3**. The results here amount to a certain interpretation of well-known facts (Naĭmark [1], K. Maurin [3]) and of what was presented in §§1 and 2.

§**4**. The approach used here for obtaining spectral representations of families of commuting operators connected by relations was announced in Berezanskiĭ's articles [13] and [14], and Theorems 4.2 and 4.4 were announced in Berezanskiĭ [13] and Berezanskiĭ, Gali, and Kondrat′ev [1]. In connection with the concepts in this section see the books of Schaefer [1], Pontryagin [1], Gel′fand, Raĭkov and Shilov [1], and Naĭmark [1]. A number of classical results on groups and semigroups of operators (the Stone theorem and the Sz.-Nagy-Hille theorem) are contained in the books of Riesz and Sz.-Nagy [1] and Hille and Phillips [1]. Korsunskiĭ [1], [3] used arguments on shrinking the set where a joint resolution of the identity is concentrated in the case of a family $(A_x)_{x\in X}$ of operators having algebraic relations. (These papers used weaker facts about expansions than those in §2.) The results of Theorems 4.1, 4.2, and §4.6 (formula (4.24)) are known in the cases of a locally compact abelian group and a commutative normed algebra (Naĭmark [1], §31.7 in Chapter 6 and §17.4 in Chapter 4). The results on semigroup repesentations sketched in §4.3 are contained in Tishchenko's papers [1] and [2], while those in §4.7 concerning the differential equations (2) are presented in somewhat greater detail in Berezanskiĭ [14]. As a whole, the presentation in this section follows Berezanskiĭ [18].

The concept mentioned in §4.7 of a commutative hypercomplex system with a continuous base is contained along with the theory of such systems in articles of Berezanskiĭ and S. G. Kreĭn [1], Berezanskiĭ and Kalyuzhnyĭ [3], and Kalyuzhnyĭ [2]. The idea of these constructions is as follows. We interpret a vector x in the space $\mathbb{C}^d$ with fixed base Q as a complex-valued function on Q. Then an ordinary d-dimensional hypercomplex system can be understood as a space of functions $Q \ni p \mapsto x(p) \in \mathbb{C}^1$ with the usual operations of addition of functions and multiplication by a scalar and with the multiplication

$$(x * y)(z) = \sum_{p,q\in Q} x(p)y(q)\gamma(p,q,r) \qquad (r \in Q),$$

where $\gamma(p,q,r)$ is a certain function ("the cubic matrix of structure constants") determining the multiplication and having definite properties ensuring multiplicative associativity (and commutativity if necessary). The generalization here of such a hypercomplex system amounts to passing from a finite base Q to a locally compact space Q; in connection with the examples at hand, γ should now be replaced by a "structure measure" $\gamma(\alpha,\beta,r)$ ($\alpha,\beta \subseteq Q$; $r \in Q$) and not a function on $Q \times Q \times Q$.

The most complete results have been obtained in the case of hypercomplex systems with a nonnegative γ whose properties call to mind the group algebra of a locally compact group G (the so-called normal hypercomplex systems with base identity 0). We remark that the group algebra of a group G is a hypercomplex system if we take $Q = G$ and $\gamma(\alpha,\beta,r) = \mu(\alpha^{-1}r \cap \beta)$, where μ is Haar measure on G, and 0 the identity of the group.

Representations of such hypercomplex systems by commuting operators A_p are mappings $Q \ni p \mapsto A_p$ preserving the algebraic structure. For example, the equality $A_pA_q = \sum_{r\in Q}\gamma(p,q,r)A_r$ $(p,q \in Q)$ must hold in the finite-dimensional case. Berezanskiĭ and Kalyuzhnyĭ [2] have considered the determination of spectral representations for families of commuting normal operators connected in such a way.

Besides Stone's theorem and that of Sz.-Nagy and Hille, §4 encompasses and generalizes spectral representations from a number of other references dealing with the further development of these two theorems: Phillips [1], Devinatz [1], Nussbaum [1], [2], Getoor [1], Ionescu Tulcea [1], [2], Ionescu Tulcea and Simon [1], Kurepa [1], Devinatz and Nussbaum [2], and Maltese [1], [2].

In connection with the contents of §4 we note that Yu. S. Samoĭlenko's book [2] deals with spectral questions for countable families of operators connected by commutation relations, anticommutation relations, or more general algebraic relations.

We make some additional remarks about Theorem 4.2. Another generalization of Stone's theorem to the case when X is a real Hilbert space is possible. We can use the Minlos-Sazonov theorem (Minlos [1], Sazonov [1], Gikhman and Skorokhod [2], §5 in Chapter 5, or Skorokhod [1], §4 in Chapter 1) that a positive-definite function $k(x)$ on X which is continuous in the so-called $\mathcal{J}$-topology can be represented in the form $k(x) = \int_X e^{i(\lambda,x)}x\, d\rho(x)$ $(x \in X)$. By applying this representation to the positive-definite function $k(x) = (A_xf,f)_{H_0}$ $(f \in H_0)$ it is not hard to obtain a formula of the type (4.5). (This approach is essentially contained in the book of Gel′fand and Vilenkin [1], §5.4 in Chapter 4, and in Korsunskiĭ's article [3].) To obtain such a result it is necessary to require weak continuity of the representation $X \ni x \mapsto U_x$ in the $\mathcal{J}$-topology instead of the conditions in Theorem 4.2. On the other hand, Theorem 4.2 makes it possible to look at theorems of Minlos-Sazanov type from other points of view. This circle of questions, to which articles of Berezanskiĭ, Gali, and Zhuk [1] and Berezanskiĭ and Gali [1] are related, will be discussed in the book mentioned in the Foreword. There we shall explain the connections between results of the type in Theorem 4.4 and the generalized power moment problem (Berezanskiĭ [5], [7]–[9], Berezanskiĭ and Shifrin [1]). Here we should mention that an essential role in these questions is played by the results presented in §2.2 and used for the case of the Hilbert space constructed by means of a positive-definite kernel (see §1.7 in Chapter 1). The idea in such considerations is based on M. G. Kreĭn's papers [1] and [2]. We mention also a recent article of Kalyuzhnyĭ [1], where a theorem of Minlos-Sazonov type is obtained for exponentially convex functions on a nuclear space.

§5. The results in §5.4 form a generalization of the corresponding facts in Berezanskiĭ's book [4]. Theorem 5.1 is due to Kats [1], [3], and its proof is a modernized argument from Berezanskiĭ [4]. The results in §§5.2 and 5.3 are due to Berezanskiĭ and Geĭdarov (Geĭdarov [1]) and are a development of corresponding facts on eigenfunction expansions for Carleman operators in Berezanskiĭ [4], which makes it possible to apply them to broader classes of operators, in particular, to operators of Schrödinger type with singular potential indicated in §5.3. Somewhat later an expansion in eigenfunctions of a Schrödinger operator with singular potential was obtained in an article of Semenov [1] (see also the earlier article of Gestrin [1]). The local properties and the behavior at infinity were subsequently studied for eigenfunctions of a Schrödinger operator with singular potential in papers of Orochko [7], [6], Belyĭ, Kovalenko and Semenov [1], Kovalenko and Semenov [1], Kovalenko, Perel′muter and Semenov [1], and Simon [5]; an article of Perel′muter [1] is also closely related to these questions. The main idea in §5.2 is due to Kats and Berezanskiĭ (see Berezanskiĭ [4], §4 in Chapter 5). The papers of Dunford and Pettis [1] and Korotkov [1] can be useful in checking the Carleman property for an operator; see also Bukhvalov's survey [1]. Multiplicative formulas of type (5.22) go back to Sophius Lie. Later they appeared in 1958 in papers of Yu. L. Daletskiĭ [1]–[3] and Trotter [1]; the former author first studied perturbations of a semigroup by bounded operators, though f could be a generalized function, while the latter studied perturbations by unbounded operators, which is what we used; the last result is explicitly stated in Chernoff's paper [1].

The growth estimates mentioned at the end of §5 have the following form: for a Schrödinger operator A in R^N with lower semibounded continuous potential q, almost every (with respect to the spectral measure) generalized eigenfunction $\varphi_\gamma(x;\lambda)$ grows no more rapidly than $c_\varepsilon|x|^{N/2+\varepsilon}$ as $|x|\to\infty$ ($\varepsilon>0$ arbitrary). Shnol′ [1] obtained this result for $N=1$ with the help of classical methods in the theory of the Sturm-Liouville equation, and Berezanskiĭ, Kats and Kostyuchenko obtained it (see Berezanskiĭ [2], [4]) for $N\geq 2$ on the basis of the theory of expansions in generalized eigenfunctions. The spectrum of the operator A is determined by such estimates: if for some λ the equation $-\Delta u+q(x)u=\lambda u$ ($x\in\mathsf{R}^N$) has a solution with such an estimate (and with an even freer estimate), then λ is in the spectrum of A (Schnol′ [1]; see also the book of Glazman [1]). From physical considerations it is desirable to replace $c_\varepsilon|x|^{N/2+\varepsilon}$ by a constant. The clarification of this possibility is at present an open question with a fairly dramatic history; it is partially described in the survey of Simon [5]. (Faris and Simon made their own contributions to the drama.)

Connected with the results presented are earlier estimates obtained by Berezanskiĭ for integrals of eigenfunctions of a Carleman (in particular, an elliptic) operator, along with subsequent estimates of averaged eigenfunctions due to Berezanskiĭ and Orochko and to Orochko (Berezanskiĭ [4], §5 in Chapter 5; Berezanskiĭ and Orochko [1]; Orochko [3]).

The expression $c_\varepsilon|x|^{N/2+\varepsilon}$ in the eigenfunction estimates above is multiplied by a certain factor that depends on the negative part of the potential and increases to ∞ in the case of a Schrödinger operator with continuous potential that is not semibounded below (Berezanskiĭ [4], §2 in Chapter 6; Orochko [2]). Furthermore, there is a cycle of papers in which the Shnol′-Berezanskiĭ-Kats-Kostyuchenko estimates above are proved (with preservation of their forms) for singular potentials (Kovalenko and Semenov [1], Kovalenko, Perel′muter and Semenov [1], and Simon [5]). Finally, a result of the Shnol′ type [1] (i.e., the estimate implies membership in the spectrum) is contained in a paper of Simon [4] for potentials that are not semibounded below.

In conclusion we mention the recent book of Berezin and Shubin [1], which is devoted to a comprehensive study of the Schrödinger operator and which takes up a number of questions related to §§5.1–5.4 and to the growth estimates just discussed.

As already mentioned, the results in §5.5 are an outgrowth of the arguments used to prove the necessity in Theorems 4.2 and 4.4. They are connected with a construction, due to K. Maurin [3], of a rigging of H_0 by linear topological spaces invariant under the action of a given operator acting in H_0, and with some results of Yu. S. Samoĭlenko [1] (Theorem 11 in §1 of Chapter 1). Kosyak and Yu. S. Samoĭlenko [2] gave a construction of a rigging for representations of inductive limits of locally compact abelian groups that follows the classical construction of Gårding [1] but uses a quasi-invariant measure depending on the representation instead of an invariant measure, which does not exist on such groups.

See K. Maurin [1]–[4] about the nuclear spectral theorem in §5.6 and see Richter [1], [2] about the construction of expansions with the help of the Choquet theorem.

§**6.** The idea for the method of proving selfadjointness developed in this section appeared first for the case of the Schrödinger operator and the wave equation in Povzner's article [1]. A general scheme for such an approach to the question of selfadjointness was presented in Berezanskiĭ [3], [4]. A modernization and refinement of this scheme is given in §6.1; see S. G. Kreĭn's book [1] for information about evolution equations. Theorem 6.4 is an abstract presentation of the scheme in Levitan [1] and Orochko [1]; a closely related result was obtained also by Orochko [9] without the assumption that the deficiency numbers of the operator are equal. Theorems 6.5 and 6.6 are due to Berezanskiĭ [10]–[12]. For applications of this method of proving selfadjointness see Berezanskiĭ [3], [4], [10] and [12], Chaus [1] and [2], and Chumak [1]; see Orochko [1], [4] and [5] about the method and its combination with other devices. (The other articles of Levitan and Orochko contain a number of applications of the method of hyperbolic equations to other questions in spectral theory.) Chernoff [3], [5] has employed a closely related device which uses hyperbolic systems for proving selfadjointness. This device was developed by Kato [2] and Rauch and Taylor [1].

The concept of an analytic vector and a criterion for selfadjointness in terms of such vectors are due to Nelson [1]. The concept of a quasianalytic vector was introduced by Ismagilov [1] in somewhat different terms; the form presented in §6 is due to Nussbaum [3]. (After the latter paper the concept of such a vector became very widespread.) The same authors are credited with applications of quasianalytic vectors to selfadjointness questions; Devinatz [2] applied this technique to the proof of selfadjointness of degenerate elliptic operators. Other proofs and a generalization to Banach spaces are contained in papers of Hasegawa [1] and Chernoff [2]. Stieltjes vectors and their applications are dealt with in articles of Nussbaum [2], Masson and McClary [1], and Simon [1]. The simple reduction of "quasianalyticity criteria" to "evolution criteria" given in §6.3 (Berezanskiĭ [8], [9]) was essentially already contained in Berezanskiĭ's book [4] (§5.4 in Chapter 8). We mention Chernoff's survey [4] of "quasianalyticity criteria". Theorem 6.11 is due to Berezanskiĭ [3], [4]; see also the references in the latter. In the book of V. I. and M. L. Gorbachuk [1], essential use is made of the concept of an analytic vector in constructing spaces of test and generalized functions. Joint analytic vectors for families of operators were introduced and studied in Kosyak's paper [1] and Yu. S. Samoĭlenko's book [1] (§1 in Chapter 1).

The results on parabolic criteria for selfadjointness (§§6.5 and 6.6) are due to Berezanskiĭ and V. G. Samoĭlenko [1], [2]; Theorem 6.14 is very close to results of Semenov [1], Faris [1], Perel′muter and Semenov [1], and Perel′muter [2], and Theorem 6.15 is connected with the concept of hypercontractive semigroups and corresponding facts (see Reed and Simon [2], Theorem X.58).

Chapter 3

§1. The results here are due to Berezanskiĭ and Us [1], [2] and Us [1]. A further development of these results, which is touched on in §1.4 and the assertions at the end of §1.1, is contained in a paper of Us [2]. The Fourier-Wiener transformation (§1.2) in the "classical" presentation is contained in papers of Cameron and Martin [1] and Guseĭnov [1], [2]. Theorem 1.2 is due to L. and K. Maurin [1] and K. Maurin [3] in the case $L(\lambda_1, \lambda_2) = \lambda_1 + \lambda_2$ and in a somewhat different form.

§2. The results in §2.1 amount to a modernized and refined exposition of corresponding results in Berezanskiĭ's book [4] (§4 in Chapter 6). The results in §§1.2 and 1.3 are due to Berezanskiĭ and Us [1], [2]. Constructions close to Theorem 2.3 are contained in articles of Streit [1] and Reed [2]. The part of Theorem 2.7 relating to selfadjointness of $\mathcal{A}_{\min}$ was first established in another way by Reed [2]. For the original presentation of Kolmogorov's three series theorem see his papers [1] and [2]. Theorem 2.8 is essentially due to Wintner [1].

§**3.** The results in this section are due to Berezanskiĭ and V. G. Samoĭlenko [1], [2]. A large number of papers have dealt with differential operators with infinitely many variables. Such operators will be studied in more detail in the book under preparation. Here we confine ourselves to recalling some of the first mathematicians who occupied themselves with various aspects of this theory: E. M. Polishchuk and M. N. Feller, Yu. L. Daletskiĭ, G. E. Shilov, S. V. Fomin and O. G. Smolyanov, M. I. Vishik, L. Gross, M. A. Piech, P. Krée, and many others working in quantum field theory (see Simon [2] and Simon and Høegh-Krohn [1]). There are bibliographies covering the area in part in the books of Yu. L. Daletskiĭ and Fomin [2] and Kuo [1]. (The latter contains also a presentation from one of the first papers in this direction: Gross [1].)

At the same time, it is appropriate to throw more light here on questions of selfadjointness of such operators. First of all we dwell on the case of finitely many variables (subsections 4, 5 and 8), giving in addition to the references cited in the main text only those most closely connected with the exposition in the book. (The literature on selfadjointness and the spectral properties of elliptic operators is immense.) The theorem of Povzner and the Titchmarsh-Sears test mentioned in the Introduction to Chapter 3 were published in Povzner [1], Titchmarsh [1], and Sears [1]. A paper of Levitan [1] contains a proof of the Titchmarsh-Sears test with the help of hyperbolic equation techniques. This test, which relates to the Schrödinger operator, was generalized by Ikebe and Kato [1] to the case of a second-order elliptic expression $\mathcal{L}$ with smooth variable leading coefficients and with singular potential (satisfying certain conditions of Stummel type). Theorem 3.7, which is qualitatively close to the result of Ikebe and Kato, differs from it by a simpler formulation, the possibility for refinement indicated in Remark 2 of §3.8, and less stringent local requirements on the leading coefficients and the positive part of the potential. Generalizations of the Titchmarsh-Sears test close to Theorem 3.7 were also obtained by Chernoff [5] and Orochko [9].

It was mentioned in §3.4 that condition (3.37) can be weakened. Namely, Lemma 3.5 remains valid if the supremum in the definition of $c(r)$ extends only over the sphere $\{x \in \mathbf{R}^N \mid |x| = r\}$, and not over the corresponding ball. (Of course, this change strengthens the results in subsections 4, 5 and 8.) This fact follows from some results in Orochko [10], Rofe-Beketov [1], and Perel′muter and Semenov [2]. However, such a strengthening of Lemma 3.5 actually goes back to an old paper of Chumak [1].

The results in subsections 4, 5 and 8 were obtained with the help of the evolution technique in §6 of Chapter 2. Bordering on this technique is a method of Devinatz [3], [5] for proving selfadjointness of the operators L under discussion, based on consideration of the stochastic equation corresponding to $\mathcal{L}$ for construction of the semigroup generated by L (i.e., in other words, for proof of the existence of a strong solution of the corresponding parabolic equation). The existence of a solution of the stochastic equation is not affected by the singularity of the matrix of leading coefficients of $\mathcal{L}$, and this made it possible for Devinatz to obtain results on selfadjointness also in the case of degeneracy. Devinatz (see [4] and [6]) has investigated selfadjointness of analogous operators, proving Kato's inequality for them. (His proof also uses the stochastic equation technique.) (See Kato [1] and Reed and Simon [2] (Theorem X.27) for a presentation of the classical Kato inequality; we mention also a paper of Brezis and Kato [1] which is closely connected with the proof of selfadjointness based on Kato's inequality.) Belichkova [1] gave an application similar to that in §§3.4 and 3.8 of the evolution technique in §6 of Chapter 2 to the proof of selfadjointness of L in the degenerate case.

Papers of Knowles [1], Orochko [4], and Semenov [3] also border on the results in subsections 4, 5, and 8.

We proceed to the case of infinitely many variables. The articles of Umemura [1] and Hida [1] are devoted to an introduction of differential operators of the type in §§3.2 and 3.3 and to a clarification of some of their properties. The selfadjointness of the operators in §§3.2 and 3.3 follows from general facts in §2 about selfadjointness of operators admitting separation of infinitely many variables.

Yu. S. Samoĭlenko and Us ([1], [2]; see also Yu. S. Samoĭlenko [2], §§3.3 and 3.5 in Chapter 1) generalized these constructions by considering on $\mathbb{R}^\infty$ a certain measure μ that is quasi-invariant with respect to translations by finitely nonzero sequences in $\mathbb{R}^\infty$, along with a family of infinitesimal operators $\mathcal{D}_n$ for translation with respect to the nth coordinate instead of the differentiation (3.15) acting in the corresponding space $L_2(\mathbb{R}^\infty, d\mu(x))$. (The families $(\mathcal{D}_n)_1^\infty$ of differentiations that arise can fail to be unitarily equivalent due to the existence of nonequivalent quasi-invariant measures and nontrivial cocycles.) A number of facts similar to those proved in §§1 and 2 were established in these articles for such differentiations generalizing (3.15). On the other hand, the articles contain examples of functions of countable families of commuting selfadjoint operators that are "poorly" approximable by cylindrical functions. Their existence and the nontriviality of the corresponding operators depend essentially on the ergodic properties of the spectral measure of the family. Connected with the papers described is an article of A. Yu. Daletskiĭ [1], which studies the spectral properties of "regular" representations of the group of translations of a nuclear space Φ in the space $L_2(\Phi', d\mu(x))$, where μ is a measure on Φ' that is quasi-invariant with respect to translations by vectors in Φ.

Berezanskiĭ [12] first established selfadjointness of perturbations of the differential operators in §§3.2 and 3.3 with separating variables by a potential with the help of the hyperbolic equation technique (more precisely, Theorem 6.5 in Chapter 2). This technique yielded selfadjointness under mild restrictions on the weights p_n but under stringent restrictions on the potential. Thus, let $(p_n)_1^\infty$ be an arbitrary sequence of the weights introduced in §3.2 and assume that they satisfy condition (3.21). The operator (3.63) is essentially selfadjoint if there exists a sequence of real-valued functions $a_m(x_1, \ldots, x_m) \in C^\infty(\mathbb{R}^m)$ $(m = 1, 2, \ldots)$ converging in $L_2(\mathbb{R}^\infty, d\rho(x))$ to an $a \in L_2(\mathbb{R}^\infty, d\rho x))$ as $m \to \infty$ and such that

$$\inf_{x_1,\ldots,x_m \in \mathbb{R}^1;\ m=1,2,\ldots} \left(\sum_{n=1}^{m} (D_n^2 1)(x_n) + a_m(x_1, \ldots, x_m) \right) > -\infty, \qquad (*)$$

$$2^m \left(\int_{\mathbb{R}^\infty} |a(x) - a_m(x_1, \ldots, x_m)|^{2m} p_1^{-1}(x_1) \cdots p_m^{-1}(x_m)\, d\rho(x) \right)^{1/m} \underset{m\to\infty}{\to} 0 \qquad (**)$$

This result does not follow from Theorem 3.2, since condition $(*)$ does not, in general, imply that the potential a is semibounded below. On the other hand, in the case of the Gaussian weights (3.34) condition $(*)$ ensures that the function $q = p + a \in L_2(\mathbb{R}^\infty, d\rho(x))$ is semibounded below; therefore, Theorem 3.4 gives us that the operator (3.88) is essentially selfadjoint. But this means (see (3.30)) that the operator (3.63) is essentially selfadjoint, and condition $(**)$ becomes superfluous.

We did not present the cumbersome proof of this fact in the text; it uses a technique of Aronszajn and Smith [1] and Lizorkin [1]. Results in a recent note of Stein [1] apparently enable one to get rid of the factor 2^m in $(**)$. The proof itself is presented in detail in a paper of Berezanskiĭ [12] and in the Russian edition of this book (§3.2 in Chapter 3); see also Berezanskiĭ and V. G. Samoĭlenko [2].

After the appearance of Berezanskiĭ's paper [12], Perel′muter and Semenov [1] observed that for differential operators of the type $\mathcal{B}_0 + a$ with infinitely many variables it is possible as a first step in the proof of selfadjointness to use general theorems to prove selfadjointness initially on the domains $D = \mathfrak{D}(\mathcal{B}_0) \cap L_\infty$, in a way similar to that for finitely many variables. It is then necessary to pass (this is nontrivial) to the required domain. (The passage is not gone through in the cited reference.) This stimulated investigations and led to the results presented in §3 on selfadjointness both in the case of finitely many variables and in the case of infinitely many (§3.4–3.8). The generalization of these results to the case of a sufficiently general (but independent of x) matrix of second derivatives is contained in a paper of Yu. S. Samoĭlenko [1]. The idea for Lemma 3.9 goes back to Simon [2]; for the original presentation of the Davis-Faris theorem see Davis [1] and Faris [2].

In this book we did not consider differential operators that are infinite-dimensional in the sense that they are defined on functions of points in some infinite-dimensional (Hilbert or linear topological) space; such operators play an important role, in particular, in quantum field theory. Selfadjointness questions for such operators in the case of regular or "almost regular" potentials have been considered by Yu. L. Daletskiĭ in [5] and [6] (using stochastic equation methods; see also the book of Yu. L. Daletskiĭ and Fomin [2]) and by Frolov in [3]–[5]. In the case of singular potentials of interest for field theory see the survey of Simon and Høegh-Krohn [1] and the subsequent cycle of papers by Kondrat′ev. For example, in [3] he studied a certain class of second-order elliptic differential operators on functions of an infinite-dimensional argument in connection with a treatment of the Schrödinger representation for second quantization operators. This class is closely connected with the operators associated with the Dirichlet forms of measures on linear spaces. Potential perturbations of such operators were considered (Kondrat′ev [4]), and a certain canonical procedure was used to obtain an operator realization for some model Hamiltonians in field theory and quantum statistical physics in which the potential perturbations are given by generalized functions (Kondrat′ev [3]). Selfadjointness of potential perturbations of second quantization operators is established, in particular, with the help of an infinite-dimensional analogue of Kato's inequality (Kondrat′ev [5]). The scattering problem was considered by Kondrat′ev and Koshmanenko [1] for operators generated by Dirichlet forms.

The results presented in §3 on selfadjointness of infinite-dimensional differential operators acting in tensor products of spaces also make it possible to obtain selfadjointness conditions for differential operators acting on functions of points in a Hilbert space (Berezanskiĭ [16] and Berezanskiĭ and V. G. Samoĭlenko [3]). These facts will be presented in the book under preparation along with the results of Kondrat′ev mentioned above.

We mention the survey of V. I. Gorbachuk, Yu. S. Samoĭlenko and Us [1], that of Berezanskiĭ, Yu. S. Samoĭlenko, and Us [1], and that of Berezanskiĭ, Kondrat′ev and Yu. S. Samoĭlenko [1], where a number of questions touching on those in this book are considered.

Bibliography

AKHIEZER, N. I. AND GLAZMAN, I. M.

[1] *The theory of linear operators in Hilbert space*, 2nd rev. ed., "Nauka", Moscow, 1966; English transl. of 1st ed., Vols. I, II, Ungar, New York, 1961.

ARAKI, HUZIHIRO AND NAKAGAMI, YOSHIOMI

[1] *A remark on an infinite tensor product of von Neumann algebras*, Publ. Res. Inst. Math. Sci. **8** (1972), 363–374.

ARKHANGEL'SKIĬ, A. V. AND PONOMAREV, V. I.

[1] *Foundations of general topology in problems and exercises,* "Nauka", Moscow, 1974. (Russian)

ARONSZAJN, N. AND SMITH, K. T.

[1] *Theory of Bessel potentials.* I, Ann. Inst. Fourier (Grenoble) **11** (1961), 385–475.

AVERBUKH, V. I., SMOLYANOV, O. G. AND FOMIN, S. V.

[1] *Generalized functions and differential equations in linear spaces.* I: *Differentiable measures*, Trudy Moskov. Mat. Obshch. **24** (1971), 133–174; English transl. in Trans. Moscow Math. Soc. **24** (1971).

[2] *Generalized functions and differential equations in linear spaces.* II: *Differential operators and their Fourier transforms*, Trudy Moskov. Mat. Obshch. **27** (1972), 247–262; English transl. in Trans. Moscow Math. Soc. **27** (1972).

BABBITT, DONALD

[1] *Rigged Hilbert spaces and one-particle Schrödinger operators*, Rep. Math. Phys. **3** (1972), 37–42.

BELAVKIN, V. P.

[1] *A reconstruction theorem for a random quantum field*, Uspekhi Mat. Nauk **39** (1984), no. 2 (236), 137–138; English transl. in Russian Math. Surveys **39** (1984).

BELICHKOVA, N. V.

[1] *Selfadjointness of degenerate second-order elliptic operators*, Dokl. Akad. Nauk Ukrain. SSR Ser. A **1984**, no. 1, 3–6. (Russian)

BELYĬ, A. G., KOVALENKO, V. F. AND SEMENOV, YU. A.

[1] *On the continuity of generalized eigenfunctions of the Schrödinger operator*, Rep. Math. Phys. **12** (1977), 307–310.

BEREZANSKIĬ, YU. M.

[1] *On expansion in eigenfunctions of general selfadjoint differential operators*, Dokl. Akad. Nauk SSSR **108** (1956), 379–382. (Russian)

[2] *On expansion in eigenfunctions of selfadjoint operators*, Ukrain. Mat. Zh. **11** (1959), 16–24; English transl. in Amer. Math. Soc. Transl. (2) **93** (1970).

[3] *A generalization of the multidimensional Bochner theorem*, Dokl. Akad. Nauk SSSR **136** (1961), 1011–1014; English transl. in Soviet Math. Dokl. **2** (1961).

[4] *Expansion in eigenfunctions of selfadjoint operators*, "Naukova Dumka", Kiev, 1965; English transl., Amer. Math. Soc., Providence, R. I., 1968.

[5] *Integral representation of positive-definite functionals of Wightman type*, Ukrain. Mat. Zh. **19** (1967), 89–95; English transl. in Ukrainian Math. J. **19** (1967).

[6] *Expansion in generalized eigenvectors and integral representation of positive-definite kernels in the form of a functional integral*, Sibirsk. Mat. Zh. **9** (1968), 998–1013; English transl. in Sibirian Math. J. **9** (1968).

[7] *Representation of functionals of Wightman type by functional integrals*, Funktsional. Anal. i Prilozhen. **3** (1969), no. 2, 1–18; English transl. in Functional Anal. Appl. **3** (1969).

[8] *The generalized power moment problem*, Trudy Moskov. Mat. Obshch. **21** (1970), 47–102; English transl. in Trans. Moscow Math. Soc. **21** (1970).

[9] *On the generalized power moment problem*, Ukrain. Mat. Zh. **22** (1970), 435–460; English transl. in Ukrainian Math. J. **22** (1970).

[10] *Selfadjointness of elliptic operators with a singular potential*, Ukrain. Mat. Zh. **26** (1974), 579–590; English transl. in Ukrainian Math. J. **26** (1974).

[11] *A remark about essential selfadjointness of powers of an operator*, Ukrain. Mat. Zh. **26** (1974), 790–793; English transl. in Ukrainian Math. J. **26** (1974).

[12] *Selfadjointness of elliptic operators with infinitely many variables*, Ukrain. Mat. Zh. **27** (1975), 729–742; English transl. in Ukrainian Math. J. **27** (1975).

[13] *On expansions in joint generalized eigenvectors of an arbitrary family of commuting normal operators*, Dokl. Akad. Nauk SSSR **229** (1976), 531–533; English transl. in Soviet Math. Dokl. **17** (1976).

[14] *Spectral representations of solutions of certain classes of functional and differential equations*, Dokl. Akad. Nauk Ukrain. SSR Ser. A **1978**, 579–583. (Russian)

[15] *Bilinear forms and Hilbert riggings*, Spectral Analysis of Differential Operators (Yu. M. Berezanskiĭ, editor), Izdanie Inst. Mat. Akad. Nauk Ukrain. SSR, Kiev, 1980, pp. 83–106. (Russian)

[16] *On selfadjointness of infinite-dimensional elliptic differential operators with singular potentials*, Funktsional. Anal. i Prilozhen. **16** (1982), no. 4, 55–56; English transl. in Functional Anal. Appl. **16** (1982).

[17] *The spectral theory of commuting normal operators and applications of it*, Uspekhi Mat. Nauk **38** (1983), no. 5 (233), 133. (Russian)

[18] *The spectral projection theorem*, Uspekhi Mat. Nauk **39** (1984), no. 4 (238), 3–52; English transl. in Russian Math. Surveys **39** (1984).

BEREZANSKIĬ, YU. M. AND GALI, I. M.

[1] *Positive-definite functions of infinitely many variables in a layer*, Ukrain. Mat. Zh. **24** (1972), 435–464; English transl. in Ukrainian Math. J. **24** (1972).

BEREZANSKIĬ, YU. M., GALI, I. M. AND KONDRAT′EV, YU. G.

[1] *Stone's theorem for the additive group of a Hilbert space*, Funktsional. Anal. i Prilozhen. **11** (1977), no. 4, 68–69; English transl. in Functional Anal. Appl. **11** (1977).

BEREZANSKIĬ, YU. M., GALI, I. M. AND ZHUK, V. A.

[1] *On positive-definite functions of infinitely many variables*, Dokl. Akad. Nauk SSSR **203** (1972), 13–15; English transl. in Soviet Math. Dokl. **13** (1972).

BEREZANSKIĬ, YU. M. AND KALYUZHNYĬ, A. A.

[1] *Nuclear function spaces on the base of a hypercomplex system*, Ukrain. Mat. Zh. **35** (1983), 9–17; English transl. in Ukrainian Math. J. **35** (1983).

[2] *Representations of hypercomplex systems with locally compact base*, Ukrain. Mat. Zh. **36** (1984), 417–421; English transl. in Ukrain. Math. J. **36** (1984).

[3] *Hypercomplex systems with locally compact bases*, Selecta Math. Sovietica **4** (1985) (to appear).

BEREZANSKIĬ, YU. M., KONDRAT'EV, YU. G. AND SAMOĬLENKO, YU. S.

[1] *Generalized functions of infinitely many variables and their applications in spectral theory*, Generalized Functions and Their Applications in Mathematical Physics (Proc. Internat. Conf., Moscow, 1980), Vychisl. Tsentr Akad. Nauk SSSR, Moscow, 1981, pp. 50–70. (Russian)

BEREZANSKIĬ, YU. M. AND KREĬN, S. G.

[1] *Hypercomplex systems with a continuous base*, Uspekhi Mat. Nauk **12** (1957), no. 1 (73), 147–152; English transl. in Amer. Math. Soc. Transl. (2) **16** (1960).

BEREZANSKIĬ, YU. M. AND MIKHAĬLYUK, T. A.

[1] *On conditions for selfadjointness of elliptic operators with infinitely many variables*, Ukrain. Mat. Zh. **29** (1977), 157–165; English transl. in Ukrainian Math. J. **29** (1977).

BEREZANSKIĬ, YU. M. AND OROCHKO, YU. B.

[1] *A remark about the growth of eigenfunctions of selfadjoint operators*, Ukrain. Mat. Zh. **14** (1962), 180–184. (Russian)

BEREZANSKIĬ, YU. M. AND SAMOĬLENKO, V. G.

[1] *On selfadjointness of differential operators with finitely or infinitely many variables*, Dokl. Akad. Nauk Ukrain. SSR Ser. A **1979**, 691–695. (Russian)

[2] *Selfadjointness of differential operators with finitely or infinitely many variables, and evolution equations*, Uspekhi Mat. Nauk **36** (1981), no. 5 (221), 3–56; English transl. in Russian Math. Surveys **36** (1981).

[3] *Selfadjointness of infinite-dimensional elliptic differential operators*, Preprint 83.71, Inst. Mat. Akad. Nauk Ukrain. SSR, Kiev, 1983. (Russian)

BEREZANSKIĬ, YU. M. AND SAMOĬLENKO, YU. S.

[1] *Nuclear spaces of functions of infinitely many variables*, Ukrain. Mat. Zh. **25** (1973), 723–737; English transl. in Ukrainian Math. J. **25** (1973).

BEREZANSKIĬ, YU. M., SAMOĬLENKO, YU. S. AND US, G. F.

[1] *Selfadjoint operators in spaces of functions of infinitely many variables*, Operator Theory in Function Spaces (Proc. School, Novosibirsk, 1975; G. P. Akilov, editor), "Nauka", Novosibirsk, 1977, pp. 20–41. (Russian)

BEREZANSKIĬ, YU. M. AND SHIFRIN, S. N.

[1] *The generalized symmetric power moment problem*, Ukrain. Mat. Zh. **23** (1971), 291–306; English transl. in Ukrainian Math. J. **23** (1971).

BEREZANSKIĬ, YU. M. AND US, G. F.
[1] *On expansions in eigenfunctions of selfadjoint operators admitting separation of infinitely many variables*, Dokl. Akad. Nauk SSSR **213** (1973), 1005–1008; English transl. in Soviet Math. Dokl. **14** (1973).
[2] *Eigenfunction expansions of operators admitting separation of an infinite number of variables*, Rep. Math. Phys. **7** (1975), 103–126.

BEREZIN, F. A. AND SHUBIN, M. A.
[1] *The Schrödinger equation*, Izdat. Moskov. Univ., Moscow, 1983. (Russian)

BILLINGSLEY, PATRICK
[1] *Convergence of probability measures*, Wiley, 1968.

BIRMAN, M. SH. AND ÈNTINA, S. B.
[1] *A stationary approach to abstract scattering theory*, Izv. Akad. Nauk SSSR Ser. Mat. **31** (1967), 401–430; English transl. in Math. USSR Izv. **1** (1967).

BIRMAN, M. SH. AND SOLOMYAK, M. Z.
[1] *The spectral theory of selfadjoint operators in Hilbert spaces*, Izdat. Leningrad. Univ., Leningrad, 1980. (Russian)

BIRMAN, M. SH., VERSHIK, A. M. AND SOLOMYAK, M. Z.
[1] *A product of commutative spectral measures can fail to be countably additive*, Funktsional. Anal. i Prilozhen. **13** (1979), no. 1, 61–62; English transl. in Functional Anal. Appl. **13** (1979).

BOURBAKI, N.
[1] *Intégration*, Chaps. 6–8, Actualités Sci. Indust., nos. 1281, 1306, Hermann, Paris, 1959, 1963.
[2] *Intégration*, Chaps. 3–5, 9, Actualités Sci. Indust., nos. 1175 (2nd ed.), 1244 (2nd ed.), 1343, Hermann, Paris, 1965, 1967, 1969.

BREZIS, HAIM AND KATO, TOSIO
[1] *Remarks on the Schrödinger operator with singular complex potentials*, J. Math. Pures Appl. (9) **58** (1979), 137–151.

BUKHVALOV, V. B.
[1] *Applications of methods in the theory of order-bounded operators to the theory of operators in L^p-spaces*, Uspekhi Mat. Nauk **38** (1983), no. 6 (234), 37–83; English transl. in Russian Math. Surveys **38** (1983).

CAMERON, R. H. AND MARTIN, W. T.
[1] *Fourier-Wiener transforms of analytic functionals*, Duke Math. J. **12** (1945), 489–507.

CHAUS, N. N.

[1] *On classes of uniqueness of solutions of the Cauchy problem and representations of positive-definite kernels*, Dokl. Akad. Nauk SSSR **163** (1965), 36–39; English transl. in Soviet Math. Dokl. **6** (1965).

[2] *Classes of uniqueness for the solution of the Cauchy problem and representation of positive-definite kernels*, Proc. Sem. Functional. Anal., no. 1, Izdanie Inst. Mat. Akad. Nauk Ukrain. SSR, Kiev, 1968, pp. 176–270. (Russian)

CHERNOFF, PAUL. R.

[1] *Note on product formulas for operator semigroups*, J. Functional Anal. **2** (1968), 238–242.

[2] *Some remarks on quasi-analytic vectors*, Trans. Amer. Math. Soc. **167** (1972), 105–113.

[3] *Essential self-adjointness of powers of generators of hyperbolic equations*, J. Functional Anal. **12** (1973), 401–414.

[4] *Quasi-analytic vectors and quasi-analytic functions*, Bull. Amer. Math. Soc. **81** (1975), 637–646.

[5] *Schrödinger and Dirac operators with singular potentials and hyperbolic equations*, Pacific J. Math. **72** (1977), 361–382.

CHUMAK, A. A. [O. O.]

[1] *Selfadjointness of the Beltrami-Laplace operator on a complete paracompact Riemann manifold without boundary*, Ukrain. Mat. Zh. **25** (1973), 784–791; English transl. in Ukrainian Math. J. **25** (1973).

COTLAR, MISCHA

[1] *Equipación con espacios de Hilbert*, Univ. Buenos Aires, Buenos Aires, 1968.

DALETSKIĬ, A. YU.

[1] *Representations of translation groups of a nuclear space*, Dokl. Akad. Nauk Ukrain. SSR Ser. A **1982**, no. 12, 9–11.

DALETSKIĬ, YU. L.

[1] *On representability of solutions of operator equations as functional integrals*, Dokl. Akad. Nauk SSSR **134** (1960), 1013–1016; English transl. in Soviet Math. Dokl. **1** (1960).

[2] *Fundamental solutions of an operator equation and functional integrals*, Izv. Vyssh. Uchebn. Zaved. Math. **1961**, no. 3(22), 27–48. (Russian)

[3] *Functional integrals connected with operator evolution equations*, Uspekhi Mat. Nauk **17** (1962), no. 5(107), 3–115; English transl. in Russian Math. Surveys **17** (1962).

[4] *Infinite-dimensional elliptic operators and the parabolic equations connected with them*, Uspekhi Mat. Nauk **22** (1967), no. 4(136), 3–54; English transl. in Russian Math. Surveys **22** (1967).

[5] *On selfadjointness and maximal dissipativeness of differential operators for functions of an infinite-dimensional argument*, Dokl. Akad. Nauk SSSR **227** (1976), 784–787; English transl. in Soviet Math. Dokl. **17** (1976).

[6] *Probability methods in some problems of infinite-dimensional analysis*, Limit Theorems for Random Processes (Random Operators and Distributions in Banach Spaces) (A. V. Skorokhod, editor), Izdanie Inst. Mat. Akad. Nauk Ukrain. SSR, Kiev, 1977, pp. 108–124. (Russian)

DALETSKIĬ, YU. L. AND FOMIN, S. V.

[1] *Generalized measures in Hilbert space and the forward Kolmogorov equation*, Dokl. Akad. Nauk SSSR **205** (1972), 759–762; English transl. in Soviet Math. Dokl. **13** (1972).

[2] *Measures and differential equations in infinite-dimensional spaces*, "Nauka", Moscow, 1983. (Russian)

DAVIES, E. B.

[1] *Properties of the Green's functions of some Schrödinger operators*, J. London Math. Soc. (2) **7** (1973), 483–491.

DERZKO, N. A.

[1] *Generalized eigenfunctions and real axis limits of the resolvent*, Trans. Amer. Math. Soc. **174** (1972), 489–506.

DEVINATZ, A.

[1] *A note on semi-groups of unbounded self-adjoint operators*, Proc. Amer. Math. Soc. **5** (1954), 101–102.

[2] *Essential selfadjointness of certain partial differential operators on R^n*, Proc. Amer. Math. Soc. **60** (1976), 235–242.

[3] *Selfadjointness of second-order elliptic and degenerate elliptic differential operators*, Differential Equations (Proc. Internat. Conf., Uppsala, 1977), Almqvist & Wiksell, Stockholm, 1977, pp. 37–51.

[4] *Essential self-adjointness of Schrödinger-type operators*, J. Functional Anal. **25** (1977), 58–69.

[5] *Selfadjointness of second order degenerate-elliptic operators*, Indiana Univ. Math. J. **27** (1978), 255–266.

[6] *On an inequality of Tosio Kato for degenerate-elliptic operators*, J. Functional Anal. **32** (1979), 312–335.

DEVINATZ, A. AND NUSSBAUM, A. E.

[1] *On the permutability of normal operators*, Ann. of Math. (2) **65** (1957), 144–152.

[2] *Real characters of certain semi-groups with applications*, Duke Math. J. **28** (1961), 221–237.

DUDIN, D. I.

[1] *Generalized measures or distributions on Hilbert space*, Trudy Moskov. Mat. Obshch. **28** (1973), 134–158; English transl. in Trans. Moscow Math. Soc. **28** (1973).

DUNFORD, NELSON AND PETTIS, B. J.

[1] *Linear operations on summable functions*, Trans. Amer. Math. Soc. **47** (1940), 323–392.

DUNFORD, NELSON AND SCHWARTZ, JACOB T.

[1] *Linear operators.* Vol. I, Interscience, 1958.

[2] *Linear operators.* Vol. II, Interscience, 1963.

DYNKIN, E. B.

[1] *Markov processes*, Fizmatgiz, Moscow, 1963; English transl., Vols. I, II, Academic Press, New York, and Springer-Verlag, Berlin, 1965.

ERDÉLYI, A. ET AL.,

[1] *Higher transcendental functions.* Vol. 2, McGraw-Hill, 1953.

FARIS, WILLIAM G.

[1] *Quadratic forms and essential self-adjointness*, Helv. Phys. Acta **45** (1972/73), 1074–1088.

[2] *Essential self-adjointness of operators in ordered Hilbert space*, Comm. Math. Phys. **30** (1973), 23–34.

FOMIN, S. V.

[1] *Differentiable measures on linear spaces*, Uspekhi Mat. Nauk **23** (1968), no. 1(139), 221–222. (Russian)

[2] *Generalized functions of infinitely many variables and their Fourier transforms*, Uspekhi Mat. Nauk **23** (1968), no. 2(140), 215–216. (Russian)

FREDRICKS, DAVID

[1] *Tight riggings for a complete set of commuting observables*, Rep. Math. Phys. **8** (1975), 277–293.

FROLOV, N. N.

[1] *Imbedding theorems for functions of countably many variables and their application to the Dirichlet problem*, Dokl. Akad. Nauk SSSR **203** (1972), 39–42; English transl. in Soviet Math. Dokl. **13** (1972).

[2] *On a coercivity inequality for an elliptic operator with infinitely many independent variables*, Mat. Sb. **90(132)** (1973), 403–414; English transl. in Math. USSR Sb. **19** (1973).

[3] *On essential selfadjointness of an infinite-dimensional differential operator*, Mat. Zametki **24** (1978), 241–248; English transl. in Math. Notes **24** (1978).

[4] *Selfadjointness of elliptic operators with infinitely many variables*, Funktsional. Anal. i Prilozhen. **14** (1980), no. 1, 85–86; English transl. in Functional Anal. Appl. **14** (1980).

[5] *On selfadjointness of a Schrödinger operator with infinitely many variables*, Sibirsk. Mat. Zh. **22** (1981), no. 1, 198–204; English transl. in Sibirian Math. J. **22** (1981).

GÅRDING, LARS

[1] *Note on continuous representations of Lie groups*, Proc. Nat. Acad. Sci. U.S.A. **33** (1947), 331–332.

[2] *Eigenfunction expansions*, Supplement I to L. Bers, F. John and M. Schechter, *Partial differential equations* (Summer Sem., Boulder, Colo., 1957), Interscience, 1964; reprint, Amer. Math. Soc., Providence, R. I., 1974, pp. 301–325.

GEĬDAROV, A. G.

[1] *On expansion in eigenfunctions of selfadjoint elliptic operators with singular potential*, Proc. All-Union Conf. Partial Differential Equations (75th Anniversary of the Birth of I. G. Petrovskiĭ, Moscow, 1976), Mekh.-Mat. Fak. Moskov. Gos. Univ., Moscow, 1978, pp. 287–288. (Russian)

GEL′FAND, I. M. AND KOSTYUCHENKO, A. G.

[1] *On expansion in eigenfunctions of differential and other operators*, Dokl. Akad. Nauk SSSR **103** (1955), 349–352. (Russian)

GEL′FAND, I. M., RAĬKOV, D. A. AND SHILOV, G. E.

[1] *Commutative normed rings*, Fizmatgiz, Moscow, 1960; English transl., Chelsea, New York, 1964.

GEL′FAND, I. M. AND SHILOV, G. E.

[1] *Generalized functions.* Vol. 1: *Generalized functions and operations on them*, 2nd ed., Fizmatgiz, Moscow, 1959; English transl. of 1st ed., Academic Press, 1964.

[2] *Generalized functions.* Vol. 2: *Spaces of test functions*, Fizmatgiz, Moscow, 1958; English transls., Academic Press; Gordon and Breach, New York, 1968.

GEL′FAND, I. M. AND VILENKIN, N. YA.

[1] *Generalized functions.* Vol. 4: *Some applications of harmonic analysis*, Fizmatgiz, Moscow, 1961; English transl., Academic Press, 1964.

GERLACH, EBERHARD

[1] *On spectral representation for selfadjoint operators. Expansion in generalized eigenelements*, Ann. Inst. Fourier (Grenoble) **15** (1965), fasc. 2, 537–574.

[2] *On the analyticity of generalized eigenfunctions (case of real variables)*, Ann. Inst. Fourier (Grenoble) **18** (1968), fasc. 2, 11–16.

GESTRIN, G. N.
[1] *On expansion in eigenfunctions of a Schrödinger operator with singular potential*, Mat. Zametki **15** (1974), 455–465; English transl. in Math. Notes **15** (1974).

GETOOR, R. K.
[1] *On semi-groups of unbounded normal operators*, Proc. Amer. Math. Soc. **7** (1956), 387–391.

GIERTZ, MAGNUS
[1] *On generalized elements with respect to linear operators*, Pacific J. Math. **23** (1967), 47–67.

GIKHMAN, I. I. AND SKOROKHOD, A. V.
[1] *Introduction to the theory of random processes*, "Nauka", Moscow, 1965; English transl., Saunders, Philadelphia, Pa., 1969.
[2] *The theory of stochastic processes*. I, "Nauka", Moscow, 1971; English transl., Springer-Verlag, 1974.

GLAZMAN, I. M.
[1] *Direct methods of qualitative spectral analysis of singular differential operators*, Fizmatgiz, Moscow, 1963; English transl., Israel Program Sci. Transls., Jerusalem, 1965, and Davey, New York, 1966.

GORBACHUK, V. I. AND GORBACHUK, M. L.
[1] *Boundary value problems for differential-operator equations*, "Naukova Dumka", Kiev, 1984. (Russian)

GORBACHUK, V. I., SAMOĬLENKO, YU. S. AND US, G. F.
[1] *Spectral theory of selfadjoint operators and infinite-dimensional analysis*, Uspekhi Mat. Nauk **31** (1976), no. 1(187), 203–216; English transl. in Russian Math. Surveys **31** (1976).

GOULD, G. G.
[1] *The spectral representation of normal operators on a rigged Hilbert space*, J. London Math. Soc. **43** (1968), 745–754.

GROSS, LEONARD
[1] *Potential theory on Hilbert space*, J. Functional Anal. **1** (1967), 123–181.

GUSEĬNOV, R. V.
[1] *On the theory of the Fourier-Wiener transformation*, Vestnik Moskov. Univ. Ser. I Mat. Mekh. **1969**, no. 4, 17–25; English transl. in Moscow Univ. Math. Bull. **24** (1969).

[2] *Some properties of the Fourier-Wiener transformation*, Vestnik Moskov. Univ. Ser. I Mat. Mekh. **1970**, no. 4, 39–49; English transl. in Moscow Univ. Math. Bull. **25** (1970).

HALMOS, PAUL R.

[1] *Measure theory*, Van Nostrand, Princeton, N. J., 1950.

HASEGAWA, MINORU

[1] *On quasi-analytic vectors for dissipative operators*, Proc. Amer. Math. Soc. **29** (1971), 81–84.

HIDA, TAKEYUKI

[1] *Note on the infinite dimensional Laplacian operator*, Nagoya Math. J. **38** (1970), 13–19.

HILLE, EINAR

[1] *On semi-groups of transformations in Hilbert space*, Proc. Nat. Acad. Sci. U.S.A. **24** (1938), 159–161.

HILLE, EINAR AND PHILLIPS, RALPH S.

[1] *Functional analysis and semi-groups*, rev. ed., Amer. Math. Soc., Providence, R. I., 1957.

HIRSCHFELD, R. A.

[1] *Expansions in eigenfunctionals*, Nederl. Akad. Wetensch. Proc. Ser. A **68**=Indag. Math. **27** (1965), 513–520.

IKEBE, TERUO AND KATO, TOSIO

[1] *Uniqueness of the self-adjoint extension of singular elliptic differential operators*, Arch. Rational Mech. Anal. **9** (1962), 77–92.

IONESCU TULCEA, CASSIUS

[1] *Spectral representation of semigroups of normal operators*, Proc. Nat. Acad. Sci. U.S.A. **44** (1958), 44–45.

[2] *Spectral representation of certain semi-groups of operators*, J. Math. and Mech. **8** (1958), 95–109.

IONESCU TULCEA, C. AND SIMON, ARTHUR B.

[1] *Spectral representations and unbounded convolution operators*, Proc. Nat. Acad. Sci. U.S.A. **45** (1959), 1765–1767.

ISMAGILOV, R. S.

[1] *Selfadjoint extensions of a system of commuting symmetric operators*, Dokl. Akad. Nauk SSSR **133** (1960), 511–514; English transl. in Soviet Math. Dokl. **1** (1960).

ITAGAKI, YOSHIO

[1] *The eigenfunction expansion of the symmetric operators associated with Gelfand triplet*, Tôhoku Math. J. (2) **23** (1971), 259–271.

KALYUZHNYĬ, A. A.

[1] *On integral representation of exponentially convex functions*, Ukrain. Mat. Zh. **34** (1982), 370–373; English transl. in Ukrainian Math. J. **34** (1982).

[2] *A theorem on the existence of a multiplicative measure*, Ukrain. Mat. Zh. **35** (1983), 369–371; English transl. in Ukrainian Math. J. **35** (1983).

KATO, TOSIO

[1] *Schrödinger operators with singular potentials*, Israel J. Math. **13** (1972), 135–148.

[2] *A remark to the preceding paper by Chernoff*, J. Functional Anal. **12** (1973), 415–417.

KATS, G. I.

[1] *On expansion in eigenfunctions of selfadjoint operators*, Dokl. Akad. Nauk SSSR **119** (1958), 19–22. (Russian)

[2] *Generalized elements in Hilbert space*, Ukrain. Mat. Zh. **12** (1960), no. 1, 13–24. (Russian)

[3] *Spectral expansions of selfadjoint operators in generalized elements of a Hilbert space*, Ukrain. Mat. Zh. **13** (1961), no. 4, 13–33. (Russian)

KNOWLES, IAN

[1] *On essential self-adjointness for singular elliptic differential operators*, Math. Ann. **227** (1977), 155–172.

KHINTCHINE, A. AND KOLMOGOROFF, A. [KHINCHIN, A. YA. AND KOLMOGOROV, A. N.]

[1] *Über Konvergenz von Reihen, deren Glieder durch den Zufall bestimmt werden*, Mat. Sb. **32** (1925), 668–677.

KOLMOGOROFF, A. [KOLMOGOROV, A. N.]

[1] *Über die Summen durch den Zufall bestimmter unabhängiger Grössen*, Math. Ann. **99** (1928), 309–319.

[2] *Bemerkungen zu meiner Arbeit "Über die Summen zufälliger Grössen,"* Math. Ann. **102** (1929/30), 484–488.

[3] *Grundbegriffe der Wahrscheinlichkeitsrechnung*, Springer-Verlag, 1933; English transl., Chelsea, New York, 1950

KOLOMYTSEV, V. I. AND SAMOĬLENKO, YU. S.

[1] *On a countable set of commuting selfadjoint operators and the canonical commutation relations*, Methods of Functional Analysis in Problems of Mathematical Physics (Yu. M. Berezanskiĭ, editor), Inst. Mat. Akad. Nauk Ukrain. SSR, Kiev, 1978, 115–128. (Russian)

[2] *On a countable set of commuting selfadjoint operators and the algebra of local observables*, Ukrain. Mat. Zh. **31** (1979), 365–371; English transl. in Ukrainian Math. J. **31** (1979).

KONDRAT'EV, YU. G.

[1] *Nuclear spaces of entire functions in problems of infinite-dimensional analysis*, Dokl. Akad. Nauk SSSR **254** (1980), 1325–1330; English transl. in Soviet Math. Dokl. **22** (1981).

[2] *Spaces of entire functions of infinitely many variables associated with a rigging of Fock space*, Spectral Analysis of Differential Operators (Yu. M. Berezanskiĭ, editor), Inst. Mat. Akad. Nauk Ukrain. SSR, Kiev, 1980, pp. 18–37. (Russian)

[3] *Second-quantization operators and perturbations of them*, Preprint No. 82.28, Inst. Mat. Akad. Nauk Ukrain. SSR, Kiev, 1982. (Russian)

[4] *Perturbations of second-quantization operators*, Funktsional. Anal. i Prilozhen. **16** (1982), no. 4, 76–77; English transl. in Functional. Anal. Appl. **16** (1982).

[5] *The Kato inequality for second-quantization operators*, Ukrain. Mat. Zh. **35** (1983), 753–756; English transl. in Ukrainian Math. J. **35** (1983).

KONDRAT'EV, YU. G. AND KOSHMANENKO, V. D.

[1] *The scattering problem for operators associated with Dirichlet forms*, Dokl. Akad. Nauk SSSR **267** (1982), 285–288; English transl. in Soviet Math. Dokl. **26** (1983).

KONDRAT'EV, YU. G. AND SAMOĬLENKO, YU. S.

[1] *Integral representation of generalized positive-definite kernels of infinitely many variables*, Dokl. Akad. Nauk SSSR **227** (1976), 800–803; English transl. in Soviet Math. Dokl. **17** (1976).

[2] *The spaces of trial and generalized functions of infinite number of variables*, Rep. Math. Phys. **14** (1978), 325–350.

KONDRAT'EV, YU. G. AND TSIKALENKO, T. V.

[1] *Test functions and generalized functions on dual nuclear spaces*, Problems in the Theory of Probability Distributions (A. V. Skorokhod, editor), Inst. Mat. Akad. Nauk Ukrain. SSR, Kiev, 1983, pp. 123–133. (Russian)

KOROTKOV, V. B.

[1] *On integral operators with Carleman kernels*, Dokl. Akad. Nauk SSSR **165** (1965), 748–751; English transl. in Soviet Math. Dokl. **6** (1965).

KORSUNSKIĬ, L. M.

[1] *Expansions in generalized eigenvectors and integral representation of invariant positive-definite kernels*, Ukrain. Mat. Zh. **24** (1972), 341–351; English transl. in Ukrainian Math. J. **24** (1972).

[2] *On infinite tensor products of locally convex spaces*, Ukrain. Mat. Zh. **27** (1975), 13–23; English transl. in Ukrainian Math. J. **27** (1975).

[3] *Integral representations of invariant positive-definite kernels*, Spectral Theory of Operators and Infinite-dimensional Analysis (Yu. M. Berezanskiĭ et al., editors), Izdanie Inst. Mat. Akad. Nauk Ukrain. SSR, Kiev, 1976, pp. 26–87. (Russian)

KOSHMANENKO, V. D.

[1] *Scattering theory in terms of bilinear functionals*, Dokl. Akad. Nauk SSSR **224** (1975), 277–280; English transl. in Soviet Math. Dokl. **16** (1975).

[2] *Scattering theory with different state spaces of perturbed and free system*, Rep. Math. Phys. **14** (1978), 185–206.

KOSHMANENKO, V. D. AND SAMOĬLENKO, YU. S.

[1] *On an isomorphism between Fock space and a space of functions of infinitely many variables*, Ukrain. Mat. Zh. **27** (1975), 669–674; English transl. in Ukrainian Math. J. **27** (1975).

KOSYAK, A. V.

[1] *Analytic and entire vectors for families of operators*, Spectral Analysis of Differential Operators (Yu. M. Berezanskiĭ, editor), Inst. Mat. Akad. Nauk Ukrain. SSR, Kiev, 1980, pp. 3–18. (Russian)

KOSYAK, A. V. AND SAMOĬLENKO, YU. S.

[1] *On families of commuting selfadjoint operators*, Ukrain. Mat. Zh. **31** (1979), 555–558; English transl. in Ukrainian Math. J. **31** (1979).

[2] *A Gårding domain and entire vectors for inductive limits of commuting locally compact groups*, Ukrain. Mat. Zh. **35** (1983), 427–434; English transl. in Ukrainian Math. J. **35** (1983).

KOVALENKO, V. F., PEREL'MUTER, M. A. AND SEMENOV, YU. A.

[1] *Schrödinger operators with $L_W^{1/2}(\mathbf{R}')$-potentials*, J. Mathematical Phys. **22** (1981), 1033–1044.

KOVALENKO, V. F. AND SEMENOV, YU. A.

[1] *Some questions about expansion in generalized eigenfunctions of a Schrödinger operator with strongly singular potential*, Uspekhi Mat. Nauk **33** (1978), no. 4 (202), 107–140; English transl. in Russian Math. Surveys **33** (1978).

KREĬN, M. G.

[1] *On a general method of decomposing Hermite-positive nuclei into elementary products*, C.R. (Dokl.) Acad. Sci. URSS **53** (1946), 3–6.

[2] *On Hermitian operators with directed functionals*, Zbirnik Prats' Inst. Mat. Akad. Nauk Ukrain. RSR **10** (1948), 83–106. (Ukrainian)

KREĬN, S. G.

[1] *Linear differential equations in Banach space*, "Nauka", Moscow, 1967; English transl., Amer. Math. Soc., Providence, R. I., 1971.

[2] (editor) *Functional analysis*, 2nd rev. ed., "Nauka", Moscow, 1972; English transl. of 1st ed., Noordhoff, 1972.

KRISTENSEN, P., MEJLBO, L. AND POULSEN, E. THUE

[1] *Tempered distributions in infinitely many dimensions. I: Canonical field operators*, Comm. Math. Phys. **1** (1965), 175–214.

KUO, HUI-HSIUNG

[1] *Gaussian measures in Banach spaces*, Lecture Notes in Math., vol. 463, Springer-Verlag, 1975.

KUREPA, SVETOZAR

[1] *A cosine functional equation in Hilbert space*, Canad. J. Math. **12** (1960), 45–50.

LADYZHENSKAYA, O. A.

[1] *Boundary value problems of mathematical physics*, "Nauka", Moscow, 1973. (Russian)

LADYZHENSKAYA, O. A. AND URAL'TSEVA, N. N.

[1] *Linear and quasilinear elliptic equations*, 2nd ed., "Nauka", Moscow, 1973; English transl. of 1st ed., Academic Press, 1968.

LAX, PETER D.

[1] *On Cauchy's problem for hyperbolic equations and the differentiability of solutions of elliptic equations*, Comm. Pure Appl. Math. **8** (1955), 615–633.

LERAY, JEAN

[1] *Lectures on hyperbolic equations with variable coefficients*, Inst. Advanced Study, Princeton, N. J., 1952.

LEVITAN, B. M.

[1] *On a theorem of Titchmarsh and Sears*, Uspekhi Mat. Nauk **16** (1961), no. 4 (100), 175–178. (Russian)

LIZORKIN, P. I.

[1] *Generalized Liouville differentiation and the function spaces $L_p^r(E_n)$. Imbedding theorems*, Mat. Sb. **60(102)** (1963), 325–353. (Russian)

MALTESE, GEORGE

[1] *Spectral representations for solutions of certain abstract functional equations*, Compositio Math. **15** (1962/64), 1–22.

[2] *Spectral representations for some unbounded normal operators*, Trans. Amer. Math. Soc. **110** (1964), 79–87.

MANDELBROJT, S.

[1] *Séries adhérentes, régularisation des suites, applications*, Gauthier-Villars, Paris, 1952.

MARCHENKO, A. V.

[1] *On inductive limits of linear spaces and operators, and their applications*, Vestnik Moskov. Univ. Ser. I Mat. Mekh. **1974**, no. 2, 26–33; English transl. in Moscow Univ. Math. Bull. **29** (1974).

[2] *Selfadjoint differential operators with infinitely many independent variables*, Mat. Sb. **96**(**138**) (1975), 276–293; English transl. in Math. USSR Sb. **25** (1975).

[3] *Spaces of functions of infinitely many variables as inductive limits of locally convex function spaces*, Teor. Funktsiĭ Funktsional. Anal. i Prilozhen. Vyp. 24 (1975), 86–98. (Russian)

MASSON, D. AND MCCLARY, W. K.

[1] *Classes of C^∞ vectors and essential self-adjointness*, J. Functional Anal. **10** (1972), 19–32.

MAURIN, KRZYZTOF

[1] *Eine Bemerkung zur allgemeinen Eigenfunktionsentwicklungen für vertauschbare Operatorensysteme beliebiger Mächtigkeit*, Bull. Acad. Polon. Sci. Sér. Sci. Math. Astr. Phys. **8** (1960), 381–384.

[2] *Methods of Hilbert spaces*, PWN, Warsaw, 1959; English transl., 1967.

[3] *General eigenfunction expansions and unitary representations of topological groups*, PWN, Warsaw, 1968.

[4] *A remark on Berezanskiĭ version of spectral theorem*, Studia Math. **34** (1970), 165–167.

MAURIN, LIDIA AND KRZYZTOF, MAURIN

[1] *Spektraltheorie separierbarer Operatoren*, Studia Math. **23** (1963), 1–29.

MIKHAĬLOV, V. P.

[1] *Partial differential equations*, "Nauka", Moscow, 1976; English transl., "Mir", Moscow, 1978.

MIKHLIN, S. G.

[1] *Mathematical physics, an advanced course*, "Nauka", Moscow, 1968; English transl., North-Holland, 1970.

MINLOS, R. A.

[1] *Generalized random processes and their extension to measures*, Trudy Moskov. Mat. Obshch. **8** (1959), 497–518; English transl. in Selected Transl. Math. Statist. and Probab., vol. 3, Amer. Math. Soc., Providence, R. I., 1963.

NAĬMARK, M. A.

[1] *Normed rings*, 2nd rev. ed., "Nauka", Moscow, 1968; English transl., *Normed algebras*, Noordhoff, 1972.

NAĬMARK, M. A. AND FOMIN, S. V.

[1] *Continuous direct sums of Hilbert spaces and some of their applications*, Uspekhi Mat. Nauk **10** (1955), no. 2(64), 111–142; English transl. in Amer. Math. Soc. Transl. (2) **5** (1957).

NAKAGAMI, YOSHIOMI

[1] *Infinite tensor products of von Neumann algebras*. I, Kōdai Math. Sem. Rep. **22** (1970), 341–354.

[2] *Infinite tensor products of von Neumann algebras*. II, Publ. Res. Inst. Math. Sci. **6** (1970), 257–292.

[3] *Infinite tensor products of operators*, Publ. Res. Inst. Math. Sci. **10** (1974/75), 111–145.

NAPIÓRKOWSKI, KAZIMIERZ

[1] *Continuous tensor products of Hilbert spaces and product operators*, Studia Math. **39** (1971), 307–327.

[2] *On generalized eigenfunctions of operators in a Hilbert space*, Studia Math. **46** (1973), 79–82.

[3] *Good and bad generalized eigenvectors*, Bull. Acad. Polon. Sci. Sér. Sci. Math. Astr. Phys. **22** (1974), 1215–1218.

[4] *Good and bad generalized eigenvectors*. II, Bull. Acad. Polon. Sci. Sér. Sci. Math. Astr. Phys. **23** (1975), 251–252.

[5] *A characterization of families of good generalized eigenvectors*, Bull. Acad. Polon. Sci. Sér. Sci. Math. Astr. Phys. **24** (1976), 957–960.

NELSON, EDWARD

[1] *Analytic vectors*, Ann. of Math. (2) **70** (1959), 572–615.

VON NEUMANN, J.

[1] *On infinite direct products*, Compositio Math. **6** (1939), 1–77.

NUSSBAUM, A. E.

[1] *The Hausdorff-Bernstein-Widder theorem for semi-groups in locally compact abelian groups*, Duke Math. J. **22** (1955), 573–582.

[2] *Integral representation of semi-groups of unbounded self-adjoint operators*, Ann. of Math. (2) **69** (1959), 133–141.

[3] *Quasi-analytic vectors*, Ark. Mat. **6** (1965), 179–191.

[4] *A note on quasi-analytic vectors*, Studia Math. **33** (1969), 305–309.

[5] *Integral representation of functions and distributions positive definite relative to the orthogonal group*, Trans. Amer. Math. Soc. **175** (1973), 355–387.

[6] *On functions positive definite relative to the orthogonal group and the representation of functions as Hankel-Stieltjes transforms*, Trans. Amer. Math. Soc. **175** (1973), 389–408.

OROCHKO, YU. B.

[1] *A uniqueness theorem for a selfadjoint extension of a Schrödinger operator with an operator potential*, Dopovīdī Akad. Nauk Ukrain. RSR **1966**, 1391–1394. (Ukrainian)

[2] *Some estimates at infinity for eigenfunctions of the Schrödinger operator*, Ukrain. Mat. Zh. **19** (1967), 39–52; English transl. in Ukrainian Math. J. **19** (1967).

[3] *An estimate of the growth of averaged eigenfunctions of selfadjoint operators*, Ukrain. Mat. Zh. **20** (1968), 400–405; English transl. in Ukrainian Math. J. **20** (1968).

[4] *A remark on the essential selfadjointness of the Schrödinger operator with a singular potential*, Mat. Zametki **20** (1976), 571–580; English transl. in Math. Notes **20** (1976).

[5] *A sufficient condition for the essential selfadjointness of polynomials in the Schrödinger operator*, Mat. Sb. **99(141)** (1976), 192–210; English transl. in Math. USSR Sb. **28** (1976).

[6] *The Carleman estimates for the Schrödinger operator with a locally semibounded strongly singular potential*, Mat. Sb. **104(146)** (1977), 162–174; English transl. in Math. USSR Sb. **33** (1977).

[7] *The hyperbolic equation method in the spectral theory of a Schrödinger operator with a real singular potential*, Proc. All-Union Conf. Partial Differential Equations (75th Anniversary of the Birth of I. G. Petrovskiĭ, Moscow, 1976), Mekh.-Mat. Fak. Moskov. Gos. Univ., Moscow, 1978, pp. 403–404. (Russian)

[8] *On the theory of eigenfunction expansions for selfadjoint operators*, Ukrain. Mat. Zh. **30** (1978), 63–69; English transl. in Ukrainian Math. J. **30** (1978).

[9] *The operator cosine method in the problem of essential selfadjointness of an unsemibounded symmetric operator*, Ukrain. Mat. Zh. **33** (1981), 348–355; English transl. in Ukrainian Math. J. **33** (1981).

[10] *On the property of a finite global rate of propagation for a second-order elliptic differential expression*, Differentsial'nye Uravneniya **18** (1982), 1764–1772; English transl. in Differential Equations **18** (1982).

PEREL'MUTER, M. A.

[1] *Perturbations of operators with integral resolvent*, Funktsional. Anal. i Prilozhen. **14** (1980), no. 2, 75–76; English transl. in Functional Anal. Appl. **14** (1980).

[2] *Positivity preserving operators and one criterion of essential selfadjointness*, J. Math. Anal. Appl. **82** (1981), 406–419.

PEREL'MUTER, M. A. AND SEMENOV, YU. A.

[1] *Selfadjointness of elliptic operators with finitely and infinitely many variables*, Funktsional. Anal. i Prilozhen. **14** (1980), no. 1, 81–82; English transl. in Functional Anal. Appl. **14** (1980).

[2] *On finiteness of the rate of propagation of perturbations for hyperbolic equations*, Ukrain. Mat. Zh. **36** (1984), 56–63; English transl. in Ukrainian Math. J. **36** (1984).

PETROVSKIĬ, I. G.

[1] *Lectures on partial differential equations*, 3rd ed., Fizmatgiz, Moscow, 1961; English transl., Saunders, Philadelphia, Pa., 1967.

PHILLIPS, R. S.

[1] *Spectral theory for semi-groups of linear operators*, Trans. Amer. Math. Soc. **71** (1951), 393–415.

PIECH, M. A.

[1] *The Ornstein-Uhlenbeck semigroup in an infinite dimensional L^2 setting*, J. Functional Anal. **18** (1975), 271–285.

PLESNER, A. I.

[1] *Spectral theory of linear operators*, "Nauka", Moscow, 1965; English transl., vols. I, II, Ungar, New York, 1969.

PLESNER, A. I. AND ROKHLIN, V. A.

[1] *Spectral theory of linear operators.* II, Uspekhi Mat. Nauk **1** (1946), no. 1(11), 71–191; English transl., Amer. Math. Soc. Transl. (2) **62** (1967).

PONTRYAGIN, L. S.

[1] *Topological groups*, 2nd ed., GITTL, Moscow, 1954; English transl., Gordon and Breach, New York, 1966.

POVZNER, A. YA.

[1] *On expansion of arbitrary functions in eigenfunctions of the operator* $-\Delta u + cu$, Mat. Sb. **32 (74)** (1953), 109–156; English transl., Amer. Math. Soc. Transl. (2) **60** (1967).

PROKHOROV, YU. V.

[1] *Convergence of random processes and limit theorems in probability theory*, Teor. Veroyatnost. i Primenen. **1** (1956), 177–238; English transl. in Theor. Probab. Appl. **1** (1956).

PRUGOVEČKI, EDUARD

[1] *The bra and ket formalism in extended Hilbert space*, J. Mathematical Phys. **14** (1973), 1410–1422.

[2] *Eigenfunction expansions for stationary scattering theory in spaces with negative norm*, Rep. Math. Phys. **7** (1975), 127–151.

RAUCH, JEFFREY AND TAYLOR, MICHAEL

[1] *Essential self-adjointness of powers of generators of hyperbolic mixed problems*, J. Functional Anal. **12** (1973), 491–493.

REED, MICHAEL C.

[1] *The damped self-interaction*, Comm. Math. Phys. **11** (1968/69), 346–357.

[2] *On self-adjointness in infinite tensor product spaces*, J. Functional Anal. **5** (1970), 94–124.

REED, MICHAEL AND SIMON, BARRY

[1] *Methods of modern mathematical physics*, Vol. 1, Academic Press, 1972.

[2] *Methods of modern mathematical physics*, Vol. 2, Academic Press, 1975.

[3] *Methods of modern mathematical physics*, Vol. 3, Academic Press, 1979.

[4] *Methods of modern mathematical physics*, Vol. 4, Academic Press, 1978.

RICHTER, PETER

[1] *Zerlegung einer Familie symmetrischer Operatoren nach gemeinsamen verallgemeinerten Eigenvektoren*, Preprint, Karl-Marx-Univ., Leipzig, 1979.

[2] *Zerlegung positiv definiter Kerne und Entwicklung nach gemeinsamen verallgemeinerten Eigenfunktionen für Familien streng kommutierender symmetrischer Operatoren*, Wiss. Z. Karl-Marx-Univ. Leipzig Math.-Nat. Reihe **31** (1982), 63–68.

RIESZ, FRÉDÉRIC [FRIGYES] AND SZ.-NAGY, BÉLA

[1] *Leçons d'analyse fonctionelle*, 2nd ed., Akad. Kiadó, Budapest, 1952; English transl., Ungar, New York, 1955.

ROFE-BEKETOV, F. S.

[1] *On positive differential operators*, Preprint 23–83, Inst. Nizkikh Temperatur Akad. Nauk Ukrain. SSR, Kharkov, 1983. (Russian)*

SAMOĬLENKO, YU. S.

[1] *Matrix-valued kernels of the Wightman functional type*, Methods of Functional Analysis in Problems of Mathematical Physics (Yu. M. Berezanskiĭ, editor), Inst. Mat. Akad. Nauk Ukrain. SSR, Kiev, 1971, pp. 201–254. (Russian)

[2] *The spectral theory of collections of selfadjoint operators*, "Naukova Dumka", Kiev, 1984. (Russian)

SAMOĬLENKO, YU. S. AND US, G. F.

[1] *On differential operators with constant coefficients in spaces of functions of countably many variables*, Spectral Analysis of Differential Operators (Yu. M. Berezanskiĭ, editor), Izdanie Inst. Mat. Akad. Nauk Ukrain. SSR, Kiev, 1980, pp. 49–67. (Russian)

Editor's note.* See the same author's paper, *On positive differential operators (deficiency indices, factorization, perturbations)*, Proc. Roy. Soc. Edinburgh Sect. A **97 (1984), 247–257.

[2] *Differential operators with constant coefficients on functions of countably many variables*, Uspekhi Mat. Nauk **39** (1984), no. 1 (235), 155–156; English transl. in Russian Math. Surveys **39** (1984).

SAZONOV, V. V.

[1] *On characteristic functionals*, Teor. Veroyatnost. i Primenen. **3** (1958), 201–205; English transl. in Theor. Probab. Appl. **3** (1958).

SCHAEFER, HELMUT H.

[1] *Topological vector spaces*, Macmillan, 1966.

SCHWARTZ, LAURENT

[1] *Théorie des distributions*, Vols. 1, 2, Hermann, Paris, 1950.

[2] *Théorie des noyaux*, Proc. Internat. Congr. Math. (Cambridge, Mass., 1950), Vol. 1, Amer. Math. Soc., Providence, R. I., 1952, pp. 220–230.

SEARS, D. B.

[1] *Note on the uniqueness of the Green's functions associated with certain differential equations*, Canad. J. Math. **2** (1950), 314–325.

SEMENOV, YU. A.

[1] *On the Lie-Trotter theorems in L_p-spaces*, Preprint ITP-72-65E, Inst. Teoret. Fiz. Akad. Nauk Ukrain. SSR, Kiev, 1972.**

[2] *On the theory of eigenfunctions expansion of the Schrödinger operator*, Lett. Math. Phys. **1** (1977), 463–465.

[3] *Smoothness of generalized solutions of the equation $(\lambda - \sum_{i,j} \nabla_i a_{ij} \nabla_j)u = f$ with continuous coefficients*, Mat. Sb. **118(160)** (1982), 399–410; English transl. in Math. USSR Sb. **46** (1983).

SHILOV, G. E. AND TIN′, FAN DYK

[1] *Integral, measure, and derivative on linear spaces*, "Nauka", Moscow, 1967. (Russian)

SHNOL′, È. È.

[1] *On the behavior of eigenfunctions of the Schrödinger equation*, Mat. Sb. **42(84)** (1957), 273–286. (Russian)

SIMON, BARRY

[1] *The theory of semi-analytic vectors: A new proof of a theorem of Masson and McClary*, Indiana Univ. Math. J. **20** (1970/71), 1145–1151.

[2] *Essential self-adjointness of Schrödinger operators with positive potentials*, Math. Ann. **201** (1973), 211–220.

[3] *The $P(\Phi)_2$ Euclidean (quantum) field theory*, Princeton Univ. Press, Princeton, N. J., 1974.

[4] *Spectrum and continuum eigenfunctions of Schrödinger operators*, J. Functional Anal. **42** (1981), 347–355.

***Editor's note*. See also Lett. Math. Phys. **1** (1975/77), 379–385.

[5] *Schrödinger semigroups*, Bull. Amer. Math. Soc. (N. S.) **7** (1982), 447–526.

SIMON, BARRY AND HØEGH-KROHN, RAPHAEL

[1] *Hypercontractive semigroups and two dimensional self-coupled Bose fields*, J. Functional Anal. **9** (1972), 121–180.

SKOROKHOD, A. V.

[1] *Integration in Hilbert space*, "Nauka", Moscow, 1975; English transl., Springer-Verlag, 1974.

SMOLYANOV, O. G. AND FOMIN, S. V.

[1] *Measures on linear topological spaces*, Uspekhi Mat. Nauk **31** (1976), no. 4(190), 3–56; English transl. in Russian Math. Surveys **31** (1976).

SOBOLEV, S. L.

[1] *Applications of functional analysis in mathematical physics*, Izdat. Leningrad. Univ., Leningrad, 1950; English transl., Amer. Math. Soc., Providence, R. I., 1963.

[2] *Introduction to the theory of cubature formulas*, "Nauka", Moscow, 1974. (Russian)

STEIN, E. M.

[1] *Some results in harmonic analysis in* $\mathbf{R}^n$ *for* $n \to \infty$, Bull. Amer. Math. Soc. **9** (1983), 71–73.

STREIT, L.

[1] *Test function spaces for direct product representations of the canonical commutation relations*, Comm. Math. Phys. **4** (1967), 22–31.

SZ.-NAGY, BÉLA

[1] *Über messbare Darstellungen Liescher Gruppen*, Math. Ann. **112** (1935/36), 286–296.

[2] *On semi-groups of selfadjoint transformations in Hilbert space*, Proc. Nat. Acad. Sci. U.S.A. **24** (1938), 559–560.

TISHCHENKO, S. V.

[1] *Spectral representation of an infinite-dimensional semigroup of contractive selfadjoint operators*, Operators of Mathematical Physics and Infinite-dimensional Analysis (Yu. M. Berezanskiĭ, editor), Inst. Mat. Akad. Nauk Ukrain. SSR, Kiev, 1979, 115–120. (Russian)

[2] *Spectral representations of countably parametric semigroups*, Ukrain Mat. Zh. **34** (1982), 532–535; English transl. in Ukrainian Math. J. **34** (1982).

TITCHMARSH, E. C.

[1] *On the uniqueness of the Green's functions associated with a second-order differential equation*, Canad. J. Math. **1** (1949), 191–198.

TROTTER, H. F.

[1] *On the product of semi-groups of operators*, Proc. Amer. Math. Soc. **10** (1959), 545–551.

TYCHONOFF, A.

[1] *Sur les espaces abstraits*, C. R. Acad. Sci. Paris **182** (1926), 1519–1520.

UGLANOV, A. V.

[1] *Differentiable measures on rigged Hilbert space*, Vestnik Moskov. Univ. Ser. I Mat. Mekh. **1972**, no. 5, 14–24; English transl. in Moscow Univ. Math. Bull. **27** (1972).

UMEMURA, YASUO

[1] *On the infinite dimensional Laplacian operator*, J. Math. Kyoto Univ. **4** (1964/65), 477–492.

US, G. F.

[1] *Spectral decomposition of selfadjoint operators which are functions of operators acting with respect to different variables*, Dokl. Akad. Nauk SSSR **229** (1976), 812–815; English transl. in Soviet Math. Dokl. **17** (1976).

[2] *On functions of operators acting with respect to different variables*, Operators of Mathematical Physics and Infinite-dimensional Analysis (Yu. M. Berezanskiĭ, editor), Inst. Mat. Akad. Nauk Ukrain. SSR, Kiev, 1979, pp. 121–138. (Russian)

VENTTSEL′, A. D.

[1] *A course in the theory of random processes*, "Nauka", Moscow, 1975; German transl., A. D. Wentzell, *Theorie zufällige Prozesse*, Akademie-Verlag, Berlin, and Birkhäuser, Basel and Boston, Mass., 1979.

VILENKIN, N. YA.

[1] *Special functions and the theory of group representations*, "Nauka", Moscow, 1965; English transl., Amer. Math. Soc., Providence, R. I., 1968.

VINOKUROV, V. G.

[1] *Compact measures and products of Lebesgue spaces*, Mat. Sb. **74(116)** (1967), 434–472; English transl. in Math. USSR Sb. **3** (1967).

VLADIMIROV, V. S.

[1] *Generalized functions in mathematical physics*, "Nauka", Moscow, 1976; English transl. of 2nd ed., "Mir", Moscow, 1979.

VUL, E. B.

[1] *On uniqueness theorems for a class of integral representations*, Dokl. Akad. Nauk SSSR **129** (1959), 722–725. (Russian)

WINTNER, AUREL

[1] *The Fourier transforms of probability distributions: Lectures*, Baltimore, Md., 1947.

WLOKA, JOSEPH

[1] *Über eine Abschätzung der Sobolevschen Konstanten*, Arch. Math. (Basel) **18** (1967), 411–413.

[2] *Gelfand triplets and spectral theory*, Summer School on Topological Vector Spaces (Brussels, 1972), Lecture Notes in Math., vol. 331, Springer-Verlag, 1973, pp. 163–182.

Subject Index

ABCDEFGHIJ-AMS/AP-89876